PROCESSING AND IMPACT ON ANTIOXIDANTS IN BEVERAGES

PROCESSING AND IMPACT ON ANTIOXIDANTS IN BEVERAGES

Editor

VICTOR PREEDY
King's College London, London, UK

ELSEVIER

AMSTERDAM • BOSTON • HEIDELBERG • LONDON
NEW YORK • OXFORD • PARIS • SAN DIEGO
SAN FRANCISCO • SINGAPORE • SYDNEY • TOKYO
Academic Press is an imprint of Elsevier

Academic Press is an imprint of Elsevier
The Boulevard, Langford Lane, Kidlington, Oxford, OX5 1GB, UK
225 Wyman Street, Waltham, MA 02451, USA

British Library Cataloguing-in-Publication Data
A catalogue record for this book is available from the British Library

Library of Congress Cataloging-in-Publication Data
A catalog record for this book is available from the Library of Congress

ISBN: 978-0-12-404738-9

For information on all Academic Press publications
visit our website at store.elsevier.com

Printed and bound in USA

14 15 16 17 10 9 8 7 6 5 4 3 2 1

Contents

8. Effects of Aging on the Antioxidant Capacity of Red Wines

EDUARDO LISSI, ANA MARIA CAMPOS, CRISTIAN CALDERÓN, SERGIO LOBATO, CAMILO LÓPEZ-ALARCÓN

9. Effects of Varieties and Growing Conditions on Antioxidant Capacity of Coffee

DRAŽENKA KOMES, ALEKSANDRA VOJVODIĆ

10. Effects of Preparation Techniques on the Antioxidant Capacity of Coffee Brews

DRAŽENKA KOMES, ANA BELŠČAK-CVITANOVIĆ

11. Applications of Enzymes in Processing Green Tea Beverages: Impact on Antioxidants

C.S. MURUGESH, R. SUBRAMANIAN

12. Antioxidant Capacity of Tea: Effect of Processing and Storage

AGNIESZKA KOSIŃSKA, WILFRIED ANDLAUER

13. Antioxidant Quality of Tea (*Camellia sinensis*) as Affected by Environmental Factors

LOVEDEEP KAUR, SHIROMANI JAYASEKERA, PAUL J. MOUGHAN

14. Antioxidants of Rooibos Beverages: Role of Plant Composition and Processing

ELIZABETH JOUBERT, DALENE DE BEER

15. Antioxidant Activity of Maté Tea and Effects of Processing

KLEBER BERTÉ, DÉLIA B. RODRIGUEZ-AMAYA, ROSEMARY HOFFMANN-RIBANI, AGENOR MACCARI JUNIOR

16. Antioxidants in Goji Berry Juice (*Lycium barbarum*) and Effects of Processing Steps

HARUNOBU AMAGASE

26. Methodology for the Measurement of Antioxidant Capacity of Coffee: A Validated Platform Composed of Three Complementary Antioxidant Assays

SEBASTIAN E.W. OPITZ, SAMO SMRKE, BERNARD A. GOODMAN, CHAHAN YERETZIAN

27. Off-Line HPLC Integrated to Total Antioxidant Capacity Measurement of Beverages

ESMA TÜTEM, REŞAT APAK, KEVSER SÖZGEN BAŞKAN

28. Antioxidant Screening of Beverages using the Online HPLC-DPPH• Assay Incorporating Active Flow Technology Chromatography Columns

M. CAMENZULI, G.R. DENNIS, H.J. RITCHIE, R.A. SHALLIKER

29. Analytical Methods for Determination of Polyphenols in Beer

M.L. FERNÁNDEZ DE CÓRDOVA, A. RUIZ MEDINA

30. Deriving a Global Antioxidant Score for Commercial Juices by Multivariate Graphical and Scoring Techniques: Applications to Blackcurrant Juice

JESSICA TABART, CLAIRE KEVERS, NADIA DARDENNE, VALÉRIE SCHINI-KERTH, ADELIN ALBERT, JACQUES DOMMES, JEAN-OLIVIER DEFRAIGNE, JOËL PINCEMAIL

List of Contributors

Adelin Albert Department of Medical Informatics and Biostatistics B23, University of Liège, CHU Sart Tilman, Liège, Belgium

Harunobu Amagase FreeLife International, Phoenix, AZ, USA

Miryam Amigo-Benavent Department of Nutrition and Metabolism, Institute of Food Science and Technology and Nutrition (ICTAN-CSIC), Madrid, Spain

Wilfried Andlauer Institute of Life Technologies, University of Applied Sciences Valais, Sion, Switzerland

Reşat Apak Istanbul University, Faculty of Engineering, Department of Chemistry, Avcilar, Istanbul, Turkey

M. Fátima Barroso REQUIMTE, Instituto Superior de Engenharia do Porto, Instituto Politécnico do Porto, Porto, Portugal

Burcu Bekdeşer Istanbul University, Istanbul, Turkey

Ana Belščak-Cvitanović Department of Food Engineering, Faculty of Food Technology and Biotechnology, University of Zagreb, Pierottijeva, Zagreb, Croatia

Mustafa Bener Istanbul University, Istanbul, Turkey

Kleber Berté Graduate Program of Food Engineering – PPGEAL, Chemical Engineering Department, Federal University of Paraná (UFPR), Curitiba, PR, Brazil

Joshua A. Bomser Department of Human Sciences, The Ohio State University, Columbus, OH, USA

Oreste V. Brenna Department of Food, Environmental and Nutritional Sciences, University of Milan, Milan, Italy

Richard S. Bruno Department of Human Sciences, The Ohio State University, Columbus, OH, USA

Arijana Bušić Department of Food Engineering, Faculty of Food Technology and Biotechnology, University of Zagreb, Zagreb, Croatia

María D. Busto Department of Biotechnology and Food Science, Area of Biochemistry and Molecular Biology, University of Burgos, Burgos, Spain

Cristian Calderón Facultad de Química y Biología, Universidad de Santiago de Chile, Santiago Chile

M. Camenzuli Australian Centre for Research on Separation Science (ACROSS), School of Science and Health, University of Western Sydney (Parramatta), Sydney, NSW, Australia

Ana Maria Campos Facultad de Química y Biología, Universidad de Santiago de Chile, Santiago Chile

Mª. Dolores del Castillo Food Bioscience Group, Department of Food Bioactivity and Analysis, Institute of Food Science Research (CSIC-UAM), Madrid, Spain

Mónica Cavia-Saiz Department of Biotechnology and Food Science, Area of Biochemistry and Molecular Biology, University of Burgos, Burgos, Spain

Luísa Correia-Sá REQUIMTE, Instituto Superior de Engenharia do Porto, Instituto Politécnico do Porto, Porto, Portugal

Chiara Dall'Asta Department of Food Science, University of Parma, Parma, Italy

Nadia Dardenne Department of Medical Informatics and Biostatistics B23, University of Liège, CHU Sart Tilman, Liège, Belgium

Dalene de Beer Post-Harvest & Wine Technology Division, Agricultural Research Council, Infruitec-Nietvoorbij Institute, Stellenbosch, South Africa

Jean-Olivier Defraigne CREDEC, University of Liège, CHU Sart Tilman, Liège, Belgium

Cristina Delerue-Matos REQUIMTE, Instituto Superior de Engenharia do Porto, Instituto Politécnico do Porto, Porto, Portugal

Gui-Fang Deng Guangdong Provincial Key Laboratory of Food, Nutrition and Health, Department of Nutrition, School of Public Health, Sun Yat-Sen University, Guangzhou, China

G.R. Dennis Australian Centre for Research on Separation Science (ACROSS), School of Science and Health, University of Western Sydney (Parramatta), Sydney, NSW, Australia

Valentina F. Domingues REQUIMTE, Instituto Superior de Engenharia do Porto, Instituto Politécnico do Porto, Porto, Portugal

Jacques Dommes Plant Molecular Biology and Biotechnology B22, University of Liège, Plant Biology Institute, Sart Tilman, Liège, Belgium

Sandra A.V. Eremia National Institute for Biological Sciences, Bucharest, Romania

M.L. Fernández de Córdova Department of Physical and Analytical Chemistry, University of Jaén, Jaén, Spain

Isabel M.P.L.V.O. Ferreira REQUIMTE, Laboratório de Bromatologia e Hidrologia, Departamento de Ciências Químicas, Faculdade de Farmácia, Universidade do Porto, Porto, Portugal,

Mario G. Ferruzzi Department of Food Science, Purdue University, West Lafayette, IN, USA

Gianni Galaverna Department of Food Science, University of Parma, Parma, Italy

Cristina García-Viguera Phytochemistry Laboratory, Department of Food Science and Technology, CEBAS-CSIC, Espinardo, Murcia, Spain

Ramón Gervilla Departamento de Ciencia Animal y de los Alimentos, Universitat Autònoma de Barcelona, Edificio V. Campus de la UAB, Cerdanyola del Vallès, Spain

Bernard A. Goodman State Key Laboratory for Conservation and Utilization of Subtropical Agro-Bioresources, Guangxi University, Nanning, Guangxi, China

Kubilay Güçlü Istanbul University, Istanbul, Turkey

Rosemary Hoffmann-Ribani Graduate Program of Food Engineering – PPGEAL, Chemical Engineering Department, Federal University of Paraná (UFPR), Curitiba, PR, Brazil

Shiromani Jayasekera Riddet Institute, Massey University, Palmerston North, New Zealand

Elizabeth Joubert Post-Harvest & Wine Technology Division, Agricultural Research Council, Infruitec-Nietvoorbij Institute, Stellenbosch, South Africa

Lovedeep Kaur Riddet Institute, Massey University, Palmerston North, New Zealand

Claire Kevers Plant Molecular Biology and Biotechnology B22, University of Liège, Plant Biology Institute, Sart Tilman, Liège, Belgium

Draženka Komes Department of Food Engineering, Faculty of Food Technology and Biotechnology, University of Zagreb, Pierottijeva, Zagreb, Croatia

Agnieszka Kosińska Division of Food Sciences, Institute of Animal Reproduction and Food Research of the Polish Academy of Sciences, Olsztyn, Poland; Institute of Life Technologies, University of Applied Sciences Valais, Sion, Switzerland

Hua-Bin Li Guangdong Provincial Key Laboratory of Food, Nutrition and Health, Department of Nutrition, School of Public Health, Sun Yat-Sen University, Guangzhou, China

Sha Li Guangdong Provincial Key Laboratory of Food, Nutrition and Health, Department of Nutrition, School of Public Health, Sun Yat-Sen University, Guangzhou, China

Shu-Ke Li Guangdong Provincial Key Laboratory of Food, Nutrition and Health, Department of Nutrition, School of Public Health, Sun Yat-Sen University, Guangzhou, China

Eduardo Lissi Facultad de Química y Biología, Universidad de Santiago de Chile, Santiago Chile

Simona Carmen Litescu National Institute for Biological Sciences, Bucharest, Romania

Sergio Lobato Facultad de Química y Biología, Universidad de Santiago de Chile, Santiago Chile

Camilo López-Alarcón Pontificia Universidad Católica de Chile, Santiago, Chile

Agenor Maccari, Junior Chemical Engineering Department, Federal University of Paraná (UFPR), Curitiba, PR, Brazil

Nuria Martí Instituto de Biología Molecular y Celular, Universidad Miguel Hernández, Elche, Alicante, Spain

Pedro Mena Phytochemistry Laboratory, Department of Food Science and Technology, CEBAS-CSIC, Espinardo, Murcia, Spain

Liuping Miao Shanghai Institute of Pharmaceutical Industry, State Key Laboratory of New Drug & Pharmaceutical Process, Shanghai, China

Paul J. Moughan Riddet Institute, Massey University, Palmerston North, New Zealand

Pilar Muñiz Department of Biotechnology and Food Science, Area of Biochemistry and Molecular Biology, University of Burgos, Burgos, Spain

C.S. Murugesh Academy of Scientific and Innovative Research, and Food Engineering Department, CSIR-Central Food Technological Research Institute, Mysore, India

Anita Oberholster Department of Viticulture and Enology, University of California, Davis, CA, USA

M.B.P.P. Oliveira REQUIMTE, Departamento de Ciências Químicas, Faculdade de Farmácia, Universidade do Porto, Porto, Portugal

Sebastian E.W. Opitz Zurich University of Applied Sciences, Institute of Chemistry and Biological Chemistry, Wädenswil, Switzerland

Natividad Ortega Department of Biotechnology and Food Science, Area of Biochemistry and Molecular Biology, University of Burgos, Burgos, Spain

Mustafa Özyürek Istanbul University, Istanbul, Turkey

M. Trinidad Pérez-Palacios Tecnología de los Alimentos, Facultad de Veterinaria, Universidad de Extremadura, Cáceres, Spain

Joël Pincemail CREDEC, University of Liège, CHU Sart Tilman, Liège, Belgium

Alexandra Plácido REQUIMTE, Instituto Superior de Engenharia do Porto, Instituto Politécnico do Porto, Porto, Portugal

Gabriel-Lucian Radu National Institute for Biological Sciences, Bucharest, Romania

H.J. Ritchie Thermo Fisher Scientific, Runcorn, UK

Délia B. Rodriguez–Amaya Faculty of Food Engineering, University of Campinas – UNICAMP, Department of Food Science, Campinas, SP, Brazil

A. Ruiz Medina Department of Physical and Analytical Chemistry, University of Jaén, Jaén, Spain

Jordi Saldo Departamento de Ciencia Animal y de los Alimentos, Universitat Autònoma de Barcelona, Edificio V. Campus de la UAB, Cerdanyola del Vallès, Spain

Valérie Schini-Kerth Faculty of Pharmacy, University of Strasbourg, Illkirch, France

R.A. Shalliker Australian Centre for Research on Separation Science (ACROSS), School of Science and Health, University of Western Sydney (Parramatta), Sydney, NSW, Australia

José Manuel Silván Department of Nutrition and Metabolism, Institute of Food Science and Technology and Nutrition (ICTAN-CSIC), Madrid, Spain

Samo Smrke National Institute of Chemistry, Laboratory for Food Chemistry, Ljubljana, Slovenia

Kevser Sözgen Başkan Istanbul University, Faculty of Engineering, Department of Chemistry, Avcilar, Istanbul, Turkey

Ángela Suárez Jacobo CIATEJ, Unidad Noreste, Apodaca, Nuevo León, México

R. Subramanian Academy of Scientific and Innovative Research, and Food Engineering Department, CSIR-Central Food Technological Research Institute, Mysore, India

Jessica Tabart Plant Molecular Biology and Biotechnology B22, University of Liège, Plant Biology Institute, Sart Tilman, Liège, Belgium

Andreia Tache National Institute for Biological Sciences, Bucharest, Romania

Wessel du Toit Department of Viticulture and Oenology, Stellenbosch University, South Africa

Esma Tütem Istanbul University, Faculty of Engineering, Department of Chemistry, Avcilar, Istanbul, Turkey

Ioana Vasilescu National Institute for Biological Sciences, Bucharest, Romania

Aleksandra Vojvodić Department of Food Engineering, Faculty of Food Technology and Biotechnology, University of Zagreb, Zagreb, Croatia

Tong Wu Shanghai Institute of Pharmaceutical Industry, State Key Laboratory of New Drug & Pharmaceutical Process, Shanghai, China

Dong-Ping Xu Guangdong Provincial Key Laboratory of Food, Nutrition and Health, Department of Nutrition, School of Public Health, Sun Yat-Sen University, Guangzhou, China

Xiang-Rong Xu Key Laboratory of Marine Bio-resources Sustainable Utilization, South China Sea Institute of Oceanology, Chinese Academy of Sciences, Guangzhou, China

Chahan Yeretzian Zurich University of Applied Sciences, Institute of Chemistry and Biological Chemistry, Wädenswil, Switzerland

Haifeng Zhao College of Light Industry and Food Sciences, South China University of Technology, Guangzhou, China

Preface

Damage by oxidative stress is an important pathogenic step in the initiation of disease. Targets of oxidative stress and different free radical species include components of the cell, such as proteins, carbohydrate, lipids, nucleic acids and conjugated molecules. Organelles, pathways and metabolic processes are also affected. It is also becoming increasingly clear that some food components may be protective in those diseases caused by oxidative stress. For example, some epidemiological studies show that the lower prevalence of some cancers is associated with higher intake of foods that are rich in antioxidants. This is also supported by *in vitro* studies as well. On the other hand, excessive intake of some dietary antioxidants may be damaging. This suggests it is important to maintain a healthy balance between dietary levels of antioxidants and prooxidants. In the food industry antioxidants are added to preserve the shelf life of foods and prevent off-flavors developing. For example, vitamin E is used to prevent lipid peroxidation occurring. These *production-added* components also contribute to the overall intake of essential nutrients. Moreover, some production processes reduce the amount of naturally occurring antioxidants. Thus, there is an important need to understand not only the physiological importance of antioxidants, but their amount in the different food types, how they are reduced or enhanced by processing, what new antioxidants are being characterized and how they are measured. This is addressed in *Processing and Impact on Antioxidants in Beverages*. We cover wine, beer, coffee, tea, herbal infusions and other tea types, soft drinks, flavored waters and a wide variety of fruit juices.

The book is divided into three sections:

[1] **Composition and Characterization of Antioxidants**
[2] **Effects of Production and Processing**
[3] **Selective Assays for Antioxidants**

In **Section [1] Composition and Characterization of Antioxidants** we have wine, beer, coffee, tea, herbal infusions and soft drinks. In **Section [2] Effects of Production and Processing** we cover fermentation, aging, varieties and growing conditions, preparation techniques, enzymes, storage, environmental factors, plant composition, homogenization and debittering. As well as the beverages mention in the previous Section we also cover rooibos and mate tea, goji, acai, pomegranate, grapefruit and juices in general. In **Section** [3] **Selective Assays for Antioxidants** we describe assays in a variety of beverages. Methods include the cuprac methods, the Oxygen Radical Absorbance Capacity (ORAC) and Trolox Equivalent Antioxidant Capacity (TEAC) assays, Off-line and Online HPLC assays, and methods for the derivation of a global antioxidant score.

Processing and Impact on Antioxidants in Beverages is designed for food scientists, technologist, food industry workers, as well as research scientists. Contributions are from leading national and international experts including those from world renowned institutions.

Professor Victor R. Preedy,
King's College London

Biography

Victor R. Preedy BSc, PhD, DSc, FSB, FRSH, FRIPH, FRSPH, FRCPath, FRSC is a senior member of King's College London (Professor of Nutritional Biochemistry) and King's College Hospital (Professor of Clinical Biochemistry: Hon). He is attached to both the Diabetes and Nutritional Sciences Division and the Department of Nutrition and Dietetics. He is also Director of the Genomics Centre and a member of the School of Medicine. Professor Preedy graduated in 1974 with an Honours Degree in Biology and Physiology with Pharmacology. He gained his University of London PhD in 1981. In 1992, he received his Membership of the Royal College of Pathologists and in 1993 he gained his second doctoral degree, for his outstanding contribution to protein metabolism in health and disease. Professor Preedy was elected as a Fellow to the Institute of Biology in 1995 and to the Royal College of Pathologists in 2000. Since then he has been elected as a Fellow to the Royal Society for the Promotion of Health (2004) and The Royal Institute of Public Health (2004). In 2009, Professor Preedy became a Fellow of the Royal Society for Public Health and in 2012 a Fellow of the Royal Society of Chemistry. In his career Professor Preedy has carried out research at Imperial College London (National Heart Hospital) and the MRC Centre at Northwick Park Hospital. He is a leading expert on the science of health. He has lectured nationally and internationally. To his credit, Professor Preedy has published over 570 articles, which includes 165 peer-reviewed manuscripts based on original research, 100 reviews and over 50 books and volumes.

SECTION 1

Composition and Characterization of Antioxidants

CHAPTER

1

Anthocyanic Compounds and Antioxidant Capacity in Fortified Wines

*Isabel M.P.L.V.O. Ferreira**, *M. Trinidad Pérez-Palacios*[†]

*REQUIMTE, Laboratório de Bromatologia e Hidrologia, Departamento de Ciências Químicas, Faculdade de Farmácia, Universidade do Porto, Porto, Portugal, [†]Tecnología de los Alimentos, Facultad de Veterinaria, Universidad de Extremadura, Cáceres, Spain

CHAPTER POINTS

- Definition of fortified wines.
- Classification of Port, Sherry, Madeira, Moscatel, Marsala wines.
- Polyphenolic content of fortified wines.
- Anthocyanins in fortified wines and factors affecting degradation of these pigments.
- Antioxidant capacity of fortified wines.
- Aging process is a major factor influencing the antioxidant activity of fortified wines.

INTRODUCTION

Fortified wines contain additional alcohol that has been added to the base wine during fermentation, when part of the original sugar content has been converted to alcohol. Thus in these wines, must fermentation is stopped by the addition of a neutral grape spirit, reaching a final alcohol content around 17–22%. Many fortified wines are blends of various grapes and various vintages. Fortified wines are known for their long-standing contribution to the world of wine as both an aperitif and a dessert wine option (Jackson, 2008).

Fortified wines can be made in either dry or sweet styles (with the middle-ground of medium-sweet or medium-dry covered in virtually all types of fortified wine categories). The determining factor of the sweetness/dryness of fortified wine is the point at which the addition of alcohol occurs during fermentation (Jackson, 2008). A sweeter fortified wine is obtained by adding the alcohol within the first day and a half of fermentation; thus, the yeast stops converting sugar to alcohol and all of the remaining grape sugar is left in the wine as residual sugar. Conversely, to obtain a dry fortified wine, the full fermentation process must occur, consuming the remaining sugar and then the alcohol is added to the wine. Table 1.1 summarizes classification of the most common types of fortified wines (Port, Sherry, Madeira, Moscatel, and Marsala) according to their sugar content.

Biochemical reactions catalyzed by enzymes of yeasts and bacteria and chemical reactions between molecules present in the must, which were extracted from the grape solids during fermentation, derived from metabolism, or released by the wood, start as soon as the beginning of winemaking (crushing) and continue through fermentation and aging (Perestrelo *et al.*, 2011). Aging is an important feature for fortified wines and includes bulk storage maturation in barrels or tanks and in-bottle aging (Pinho *et al.*, 2012). The aging time depends on the fortified wine, but in general the cheaper the fortified wine, the less time it has spent aging in oak. As a result of deep wood aging, many fortified wines will benefit from decanting and aeration. Apart from location, grape varieties, soil, etc., there are also differences in the way the most common types of fortified wines (Port, Sherry, Madeira, Moscatel, and Marsala) are fortified and aged (Table 1.2).

Port Wine

Port wine is produced primarily from red grapes grown and fermented in the upper Douro Valley in northern Portugal. Although originating in the upper Douro, the wine is transported downriver to Porto for

Processing and Impact on Antioxidants in Beverages
http://dx.doi.org/10.1016/B978-0-12-404738-9.00001-5

maturation and aging. These processes occur in buildings called lodges in Vila Nova de Gaia, located at the mouth of the Douro River, opposite the city of Porto. The major red varieties are Touriga Nacional, Mourisco, Mourisco de Semente, Tinta Roriza, Tinta Cão, and Tinta Francisco, Tinta Barroca (Mateus *et al.*, 2002). A small amount of white Port is also produced. Codega, Malvasia, and Rabigato are the preferred white varieties. Most of the present-day wine is vinified by regional cooperatives using modern crushing, pressing, and fermenting equipment. When part of the original sugar content has been converted to alcohol, must fermentation is stopped by the addition of wine spirit obtaining around 20% of the final alcohol content (Esteves *et al.*, 2004). Aging includes bulk storage maturation in barrels or tanks and in-bottle aging.

Different types of Port wines are produced. Ruby Port wine is the most extensively produced type. After fermentation, it is stored (in general, for 2 years) in tanks made of stainless steel to prevent oxidative aging and to preserve its rich claret color. It is fined and cold filtered before bottling and does not generally improve with age. Ruby Port wine is usually blended to match the style of the brand to which it is to be sold. Tawny Ports are wines made from red grapes that are aged in wooden barrels and blended in such a way that the finished product is a mixture of ages. A Tawny without an indication of age (usually named as Reserve) has spent at least 2 years in wooden barrels, exposing it to gradual oxidation and evaporation until a golden-color is obtained, and it is blended in such a way that the finished product is a mixture of ages. Tawny wines with age categories (10, 20, 30, and over 40 years) indicate a target age profile. Vintage Port is made entirely from the grapes of a declared vintage year, aged in barrels for a maximum of 2½ years, and bottled unfined. Generally, it requires another 10 to 40 years of aging in the bottle. Late Bottle Vintage (LBV) is the product of a single year's harvest that was left in the barrel for 4–6 years. LBV can be fined and filtered or not before being bottled (Pinho *et al.*, 2012). White Port is made from white grapes and can be made in a wide variety of styles, from dry to very sweet (Jackson, 2008).

TABLE 1.1 Classification of Fortified Wines According to their Sugar Content

Wine	Category	Residual Sugar (g/l)
Port	Very sweet	>130
	Sweet	90–130
	Semi-dry	65–90
	Dry	45–65
	Extra-dry	<40
Sherry	Dulce/Sweet	>160
	Pedro Ximénez	>212
	Cream	115–140
	Medium	5–115
	Pale Cream	45–115
	Dry	5–45
	Fino/ Manzanilla/Oloroso/ Palo Cortado/Amontillado	0–5
Madeira	Sweet (Malvasia)	110
	Medium dry (Verdelho)	90
	Medium sweet (Boal)	65
	Dry (Sercial)	25
Moscatel de Setúbal		90–100
Marsala	Dolce	>100
	Semi-secco	40–100
	Secco	<40

Sherry Wine

Sherry is a fortified wine made from white grapes that are grown near the town of Jerez de la Frontera in Andalusia, Spain. Sherry is produced in a variety of dry styles made primarily from the Palomino grape. After fermentation is complete, the base wines are fortified with grape spirit in order to increase their final alcohol content. Wines classified as Fino and Amontillado are fortified until they reach a total alcohol content of 15–17%. Those wines classified as Oloroso are fortified to reach an alcohol content of at least 17–22% (BOJA, 2012; Stevenson, 2005). Sherry wines are aged and blended using the *solera* system: a series of 3–9 barrels are used, and the method involves moving the wine down from one barrel into the next one. At the end of the series only a portion of the final barrel is bottled and sold. Depending on the type of wine, the portion moved may be between 5 and 30% of each barrel. The amount added is equivalent to that removed for transfer to older wine or bottling. The age of the youngest wine going into the bottle is determined by the number of barrels in the series, and every bottle also contains some much older wine. Sherry is aged in the *solera* for a minimum of 3 years. Fino Sherries require many and frequent transfers and stages whereas Oloroso Sherries develop best with few and infrequent transfers. Amontillado Sherries begin similarly to Fino Sherry, but subsequently, the frequency of transfer is slowed. Sweet Sherries (Pedro Ximénez) are made either by fermenting sweet grape varieties (Pedro Ximénez or Moscatel) or by blending sweeter wines (Jackson, 2008).

Madeira Wine

Madeira wine evolved on the island of the same name on the coast of Portugal. It presents different characteristics to all other types of fortified wine due to its specific winemaking process, since it is obtained by intentional heating, and characterized by a distinct baked bouquet (Perestrelo *et al.*, 2011).

During the Age of Exploration, Madeira was a standard port of call for ships heading to the New World or East Indies. Neutral grape spirit was added to the wine to prevent it from spoiling, and on long sea voyages, the wine was were exposed to excessive heat and movement which transformed its properties, including the color and flavor (Perestrelo *et al.*, 2011). It was observed that when an unsold wine returned to the islands after a round trip the taste of the wine had improved significantly; consequently, merchants started shipping barrels of Madeira to the Indies with the sole objective of enriching it to be sold to Europe with added value. Later in the mid-eighteenth century, wineries invested in *estufagem* (baking process) chambers, and this technique is still used today (Perestrelo *et al.*, 2011).

Madeira wine is produced almost exclusively from white grapes. The preferred varieties are Malvasia, Sercial, Verdelho, and Bual de Madeira. Listrão, Tinta Negra Mole, and Negra are commonly used for inexpensive Madeira wines (Stevenson, 2005). The initial winemaking steps of Madeira are similar to other wines: grapes are harvested, crushed, pressed, and then fermented. The fermentation process is stopped by the addition of natural grape spirit in order to obtain an ethanol content of 18–19%. After fortification, the wines may be subjected to one of the two different heating processes: *estufagem* (baking process) or *canteiro* (wood casks). During the baking process, the wine is placed in large coated vats and the temperature is slowly increased by about 5°C per day and maintained at 45–50°C for 3 months. After this treatment, the wine is allowed to undergo a maturation process in oak casks for a minimum of 3 years. Finally, some Madeira wines undergo an aging process, from 3 to 20 years or even longer (Câmara *et al.*, 2006). The highest-quality Madeira wines, which are aged without the application of the baking process, undergo heating in wooden casks placed on wooden support beams called *canteiros*. In this winemaking procedure, Madeira wine aging usually occurs in the top floors of cellars, where the temperatures (30–35 ºC in the summer) and humidity levels (70–75%) are high, for a minimum of 2 years, developing complex aromas and intense flavors (Perestelo *et al.*, 2011). Concerning old Madeira wines, the wine bouquet is dictated by the particular aging process rather than by the grape variety used. Wines from different vintages and varieties usually are kept separated for at least the first 2 years of maturation. Subsequently, producers begin the process of blending. Madeira wines are usually labeled based on the length of time for which they were aged: Finest (3 years), Reserve (5 years), Special Reserve (10 years), Extra Reserve (over 15 years), *Colheita* or Harvest (wines from a single vintage and aged for a shorter period than Vintage), Fine Vintage (at least 20 years), Vintage or *Frasqueira* (Fine Vintage followed by 2 years in bottle) (Stevenson, 2005) (Table 1.2).

Moscatel de Setúbal Wine

Moscatel de Setúbal is a Portuguese wine produced around the Setúbal Municipality. These wines have to be composed primarily of Muscat of Alexandria or Moscatel Roxo grapes. The blend can include up to 30% of other varieties such as Arinto, Boais, Diagalves, or Vital. This fortified wine has a very short fermentation time that ends with the addition of spirit or vinous alcohol to the grape must. The resulting product then stays in contact with the grape skins (maceration period) and is later separated from the solid materials. The wine is stored for at least 2 years in tanks and eventually in oak wood barrels (maturation and aging periods). Moscatel de Setúbal wines can be made from grapes of a single vintage or in a 'non-vintage' style as a blend of several vintages. Before bottling, approval by the 'Comissão Vitivinícola Regional da Península de Setúbal' is required (Feliciano *et al.*, 2009).

Marsala Wines

Marsala is a Sicilian fortified wine with ancient tradition. It is exclusively produced in the province of Trapani. The vines grow in the typical red Sicilian earth, in conditions that are particularly dry and sunny (La Torre *et al.*, 2008). Marsala is produced from local, indigenous white grapes—Catarratto, Grillo, or the highly aromatic Inzolia grape. The ruby-colored Marsalas hail from a combination of red grape varietals, Perricone, Calabrese, Nero d'Avola, and Nerello Mascalese (Saunders, 2004). The fermentation of Marsala is halted by the addition of a grape brandy when the residual sugar content reaches the pre-determined levels according to the sweet/dry style. Similar to the *solera* system of blending various vintages of Sherry, Marsala often goes through process called in *perpetuum*. Marsalas are characterized by an average alcoholic content around 18° also and can be classified according to their contents of reducing sugars, color, and age. As summarized in Table 1.2 the sweetest Marsalas are called 'Dolce' (total sugars > 100 g/l), followed by 'Semi-secco' (total sugars around 40 g/l). Marsala wine can be classified according to their color: 'Oro' (golden) and 'Ambra' (amber) produced from the Grillo, Cataratto, Inzolia, and Damaschino grapevine varieties,

TABLE 1.2 Summary of the Fortification and Maturation Steps within the Winemaking of Fortified Wines

Wine	Sub-Type	Fortification	Maturation
Port	Ruby	Interrupt the fermentation	Two years in tanks of stainless steel
	Tawny		At least 2 years in wooden barrels
	Vintage		Maximum of 2½ years in the barrels. Aged in the bottle
	Late Bottled Vintage		4–6 years in the barrels. Aged in the bottle
Sherry	Fino	At the end of the fermentation	*Solera* system: many and frequent transfers
	Oloroso		*Solera* system: few and infrequent transfers
	Amontillado		*Solera* system: many and frequent transfers, which go slowing
Madeira	Finest		3 years
	Reserve	Interrupt the fermentation	3 years
	Special Reserve	Fortification is followed by a baking stage (40–45°C for 3 months)	10 years
	Extra Reserve		15 years
	Colheita		Lower than 20 years
	Fine Vintage		At least 20 years
	Vintage		At least 20 years plus years in bottle
Moscatel de Setúbal		At the beginning of the fermentation	At least 2 years
Marsala	Fine Superiore Superiore-Riserva Vergine and/or Soleras Stravecchio and/or Soleras	Fortified with must, alcohol and wine (13% v.v. ethanol content) Fortified only with alcohol and wine	>1 year >2 years >4 years >5 years >10 years

and 'Rubino' (ruby) from Pingatello, Nerello Mascalese, and Calabrese varieties. The age grades are 'Fine' (>1 year), 'Superiore' (>2 years), 'Superiore-Riserva' (>4 years), 'Vergine' (>5 years), and 'Stravecchio' (>10 years). During vinification, 'Fine,' 'Superiore,' and 'Superiore-Riserva' Marsalas, are fortified with must, alcohol and wine (13% v.v. ethanol content), while 'Vergine Soleras' Marsala is fortified only with alcohol and wine (La Torre *et al.*, 2008).

POLYPHENOLIC CONTENT OF FORTIFIED WINES

Polyphenols are the main compounds related to benefits of wine consumption due to antioxidant and free radical scavenging properties. These effects are related to flavonoids and stilbenes, namely quercetin, (+)-catechin, gallic acid, and trans-resveratrol (Paixão *et al.*, 2008). Polyphenols are affected by several factors including grape variety, sun exposure, vinification techniques, and aging. Significant changes in phenolic composition occur during aging, since these compounds can suffer diverse reactions, namely oxidation, condensation and polymerization, and extraction from wood, that are usually associated to the changes in color and colloidal stability, flavor, bitterness, and astringency (Perez-Magarino and González-San José, 2006).

Total polyphenolic content or polyphenolic compound index has been determined in Madeira and Sherry wines by using the Folin–Ciocalteu method using gallic acid as standard (Paixão *et al.*, 2008; Fernández-Pachón *et al.*, 2004, 2006). This method is based on the reduction of a phosphowolframate–phophomolybdate complex by phenolics to blue reaction products. The absorbance of analytes is determined at 700 or 750nm. The results are expressed as milligrams of gallic acid equivalents (GAE) per liter. Comparing the results of these studies, higher total polyphenolic content was observed in red Madeira wines (1724–1936GAE/l) than in white Madeira wines (282–770GAE/l) and Sherry wines (207–446GAE/l). Mean total polyphenolic content described for Moscatel wines was 1243GAE/l (Feliciano *et al.*, 2009) close to that obtained for Madeira red wines. Moreover, individual phenolic compounds in fortified wines have been identified and quantified.

The phenolic compounds determined in Port, Sherry, Madeira, and Marsala wines and the methodologies used are summarized in Table 1.3. García-Viguera *et al.* (1997) identified 10 phenolic compounds (gallic acid, 3,4-dihydroxybenzoic acid, tyrosol, *cis*-caffeoyl tartaric acid, vallinic acid, *trans*-coumaroyl tartaric acid, syringic acid, epicatechin, rutin, and myricetin) in Port wine by high-performance liquid chromatography with diode array detection (HPLC-DAD) using an ODS Hypersil column (100 × 2.1 mm, particle size 5 μm), acidified water (with 0.6% perchloric acid), and methanol as solvents A and B, respectively, and reporting the chromatograms at 280 nm. Andrade *et al.* (1998) also determined individual phenolic compounds in Port wine (tyrosol, epicatchin, catechin, syringic acid, p-coumaric acid, caffeic acid, gallic acid, 3,4-dihydroxybenzoic acid, *cis*-coumaroyl tartaric acid, and *trans*-coumaroyl tartaric acid) by capillary zone electrophoresis with DAD at 280 nm. A fused-silica capillary (57 cm total length × 75 μm ID), was used with a running buffer of 0.1 M sodium borate (pH 9.5). Ribeiro de Lima *et al.* (1999) determined four stilbenes (*trans*-astringin, *cis*- and *trans*-piceid, and *cis*- and *trans*-resveratrol) in Port wine by using HPLC-DAD with a Nucleosil 100 C18 column (4.0 × 250 mm) and the following solvents: A, acetic acid in H_2O, pH 2.4; B, 20% phase A with 80% of acetonitrile.

In Sherry wines, gallic acid, protocatechuic acid, syringic acid, caftaric acid, glucosidecutaric acid, cutaric acid, caffeic acid, p-coumaric acid, and ethylcaffecte were detected by using HPLC-DAD with a Superspher 100 RP-18 column (250 × 4 mm; 5 μm) and using the following solvents: A (glacial acetic acid/water, pH 2.65) and B (20% A + 80% acetonitrile) (Fernández-Pachón *et al.*, 2006). Schwarz *et al.* (2012), using UPLC-DAD with a C18 column (100 × 2.1 mm I.D., with 1.7 μm particle size) and the binary system of solvents: A, 3% acetonitrile, 2% acetic acid, 95% water; and B, 85% acetonitrile, 2% acetic acid, 13% water, detected protocatechuic acid, protocatechualdehyde, p-hydroxybenzoic acid, tyrosol, vallinic acid, p-hydroxybenzaldehyde, syringic acid, vanillin, p-coumaric, and syring aldehyde in Sherry wines.

Generally, the gallic acid is one of the highest phenolic compounds in all fortified wines. In Port wine, syringic acid, *cis*-caffeoyl tartaric acid, rutin, tyrosol, epicatchin, and catechin are also important quantitatively (García-Viguera *et al.*, 1997; Andrade *et al.*, 1998). Caftaric and cutaric acid are also found in high quantities in Sherry wines (Fernández-Pachón *et al.*, 2006) whereas quercetin, vallinic acid, and siringic acid are abundant in Madeira wines (Paixão *et al.*, 2008). Grape varieties influence the quantity of phenolic compounds. Port from the Touriga (Nacional and Francesa) varietal contained twice the total phenolic concentration (over 220 mg/l) of the wines prepared from the Tinta Barroca and Tinta Roriz grapes (around 100 mg/l). The major differences were due to tyrosol (Andrade *et al.*, 1998). Variations in the processing also lead to differences in the phenolic compound concentration. In Port wine the greatest total phenolic content was found when traditional foot treading had been applied. It was also reported that the addition of fortifying grape spirit prior to fermentation has a neutral or negative effect on the yield of phenolic compounds in Port wine (García-Viguera *et al.*, 1997).

The aging also affects the content of phenolic compounds. In Port wine, it was noted that concentrations of the tyrosol and gallic acid increased with aging of wine (Andrade *et al.*, 1998). In Sherry Vintage the concentration of each phenolic compound tended to be higher in the wines aged for longer periods (Schwarz *et al.*, 2012), although there were evident differences in the behavior of the individual phenolic compounds between the Fino, Oloroso, and Amontillado Sherry wines, i.e., levels of gallic acid increased with the maturation process in the Fino and decreased in the others; p-hydroxybenzoic acid was not detected in the samples of Fino wine but was present, albeit at a low level, in the samples of Oloroso and Amontillado wine.

ANTHOCYANIC COMPOUNDS IN FORTIFIED WINES

Grape skins are amongst the best natural sources of anthocyanins that occur as 3-*O*-monoglucosides and 3-*O*-acylated monoglucosides of five main anthocyanidins (the aglycone forms)—delphinidin, cyanidin, petunidin, peonidin, and malvidin (Mateus *et al.*, 2002; Welch *et al.*, 2008; Heredia *et al.*, 1998; He *et al.*, 2006). Anthocyanins are the major color components in red wine. During vinification of fortified wines anthocyanins react with other molecules, such as pyruvic acid, vinylphenol, vinylcatechol, α-ketoglutaric acid, acetone, 4-vinylguaiacol, and glyoxylic acid, leading to the formation of more stable pigments that stabilize wine color (Heredia *et al.*, 1998; Oliveira *et al.*, 2006; Mateus *et al.*, 2006; Castañeda-Ovando *et al.*, 2009; Kähkonen and Heinonen, 2003; Mateus and Freitas, 2001; Vivar-Quintana *et al.*, 2002; Rentzsch *et al.*, 2007).

Many factors affect the stability of anthocyanins, including temperature, pH, sugars and sugar degradation products, the presence of oxygen, enzymes, co-pigments, metal ions, ascorbic acid, and sulfur dioxide (Romero and Bakker, 2000; Queiroz *et al.*, 2009; Oliveira *et al.*, 2010; Pinho *et al.*, 2011). Ethanol content affects the mechanism of Maillard reaction and produces new browning products. Degradation of anthocyanins in ethanolic solutions is faster than in aqueous solution (Tseng *et al.*, 2006), consequently, it is expected that fortified wines in general present low content of

TABLE 1.3 Individual Phenolic Compounds in Fortified Wines Analyzed using Different Methodologies

	Fortified Wines				Methodology				
	Port	*Sherry*	*Madeira*	*Marsala*	*HPLC-DAD*	*RP-HPLC*	*UPLC-DAD*	*Capillary Electrophoresis*	*HPLC/MS*
3,4-Dihydroxybenzoic acid	✓				✓			✓	
Caffeic acid	✓		✓	✓		✓		✓	✓
Caftaric acid		✓			✓				
Catechin	✓			✓				✓	✓
cis-Caffeoyl tartaric acid	✓				✓				
cis-Coumaroyl tartaric acid	✓				✓			✓	
cis-Piceid	✓			✓	✓				✓
cis-Resveratrol	✓			✓	✓				✓
Cutaric acid		✓			✓				
Epicatechin	✓	✓	✓	✓	✓	✓		✓	✓
Ethyl caffeate		✓			✓				
Ferrulic acid				✓					✓
Gallic acid	✓	✓	✓	✓	✓	✓		✓	✓
Glucoside cutaric acid		✓			✓				
Myricetin	✓	✓	✓	✓	✓	✓			✓
p-Coumaric acid	✓	✓	✓	✓	✓	✓	✓	✓	✓
p-Hydroxybenzaldehyde		✓					✓		
p-Hydroxybenzoic acid									
Procyanindins B1 and B2				✓					✓
Protocatechuic acid		✓		✓	✓		✓		✓
Protocateuchualdehyde		✓					✓		
Quercetin			✓	✓		✓			✓
Rutin	✓			✓	✓				✓
Syringaldehyde		✓					✓		
Syringic acid	✓	✓		✓	✓	✓	✓	✓	✓
trans-Astring	✓				✓				
trans-Coumaroyl tartaric acid	✓				✓			✓	
trans-Piceid	✓			✓	✓				✓
trans-Resveratrol	✓	✓	✓	✓	✓	✓			✓
Tyrosol	✓	✓		✓	✓		✓	✓	✓
Vanillin		✓					✓		
Vallinic acid	✓	✓	✓	✓	✓	✓	✓		✓

Summarized from Paixão et al. *(2008); Alonso* et al. *(2004); García-Viguera* et al. *(1997); Andrade* et al. *(1998); Ribeiro de Lima* et al. *(1999); Fernández-Pachón* et al. *(2006); Schwarz* et al. *(2012).*
HPLC-DAD, high-performance liquid chromatography with diode array detection; RP-HPLC, Reversed phase high-performance liquid chromatography; UPLC-DAD, ultra-performance liquid chromatography method with diode array detection; HPLC/MS, high-performance liquid chromatography–mass spectrometry.

those compounds. Research pertaining to the degradation of anthocyanins in fortified wines is scarce (Tseng *et al.*, 2006) although some studies were performed in Port wine (Mateus and Freitas, 2001; Mateus *et al.*, 2002; Pinho *et al.*, 2012).

Anthocyanin–pyruvic acid adducts are more abundant in fortified red wines than in red table wines as reported by Romero and Bakker (2000), probably because when wine spirit is added in order to stop fermentation, the pyruvic acid concentration is higher than when the fermentation is allowed to go to dryness. Anthocyanin–pyruvic acid adduct formation is influenced by the pyruvic acid excreted by the yeast at the beginning of the fermentation and the slightly higher pH values of fortified wines. Additionally, the higher content of ethanol, which is known to be a good solvent for polyphenols, increases the pigment solubility and can favor the formation of new pigments (Mateus and Freitas, 2001).

HPLC has been a method of choice for the analysis of anthocyanins (Mateus and Freitas, 2001; Mateus *et al.*, 2006; Queiroz *et al.*, 2009; Oliveira *et al.*, 2010; Pinho *et al.*, 2011, 2012). The major challenge for HPLC quantification of individual anthocyanins is often the difficulty in obtaining the anthocyanin reference compounds, since a large number of peaks appear on the chromatogram and it is difficult to identify all the individual anthocyanins. The chromatographic profile of Port wines presented both anthocyanins from must and newly formed anthocyanins corresponding to pyruvic acid adducts (Figure 1.1) (Mateus and Freitas 2001, Vivar-Quintana *et al.*, 2002, Pinho *et al.*, 2012)

The aging conditions, including temperature, pH, sugar content, presence or absence of oxygen, influence anthocyanin degradation as summarized in Figure 1.2.

For 3 years, the evolution of the three major anthocyanidin monoglucosides (malvidin 3-glucoside, malvidin 3-acetylglucoside, and malvidin 3-coumaroylglucoside) and their anthocyanin–pyruvic acid adducts was monitored in varietal Port wines stored in oak barrels (Mateus and Freitas, 2001). The anthocyanin–pyruvic acid adducts were found to be much more stable than the original anthocyanidin monoglucosides. The levels of malvidin 3-glucoside–pyruvic acid adduct and its acylated forms increased right after wine fortification with wine spirit until 100 days and start to decrease after this period. The initial formation of anthocyanin–pyruvic acid adducts was concurrent with the degradation of anthocyanidin monoglucosides. Additionally, several commercial Port wines were analyzed after 6 months of processing (when bottled) and after 2 years of aging regarding their anthocyanic content. The results obtained point to the much higher stability of anthocyanin-derived pigments when well protected from any air contact comparatively to the stability of the original anthocyanidin monoglucosides. During bottle aging, wines develop in a reducing environment and the oxidation–reduction potential decreases regularly until it reaches a minimum value that prevents oxidation reactions (Mateus and Freitas, 2001).

The anthocyanin content of Ruby, LBV, and Tawny Port wines was assayed by Pinho *et al.* (2012). Higher amounts of anthocyanins were present in Ruby samples followed by LBV samples. These Port wines mature in stainless steel tanks (at least, for 2 and 4 years, respectively) and in sealed glass bottles (for longer periods in the case of LBV), with no exposure to air, undergoing 'reductive' aging. This process leads to the wine losing its anthocyanins very slowly. No anthocyanins were quantified in Tawny Port wines, except Tawny reserve. Tawny Port wines are matured in wooden barrels, whose permeability allows a small exposure to oxygen, undergoing 'oxidative' aging. This type of wine loses anthocyanins at a faster pace. They also lose volume to evaporation, leading to a wine that is slightly more viscous and containing only traces or not detectable amounts of anthocyanins (Pinho *et al.*, 2012).

ANTIOXIDANT CAPACITY IN FORTIFIED WINES

The potential beneficial effects of wine consumption have been related to its antioxidant and free radical scavenging properties (Pascual-Teresa *et al.*, 2010). For assessing the antioxidant activity in fortified wines, several different methods have been used: 1,1-diphenyl-2-picrylhydrazyl (DPPH) radical scavenging assay; 2,2′-azino-bis-3-ethylbenzothiazoline-6-sulfonic acid (ABTS) radical cation decolorization; ferric reducing/antioxidant power (FRAP); oxygen radical absorbance capacity (ORAC); and by means of an electrochemical method (Larrauri *et al.*, 1999; Pellegrini *et al.*, 2000; Paião etal., 2008; Fernández-Pachón *et al.*, 2006; Alonso *et al.*, 2004).

The DPPH method utilizes the stable free radical $DPPH^-$ by the addition of scavenging compounds. Wine samples are mixed with DPPH solutions and absorbance of the remaining $DPPH^-$ at 515 nm is measured at different time intervals until the reaction reaches the equilibrium. The ABTS method is based on the generation of the highly stable chromophoric cation-radical of $ABTS^+$. A wine sample is mixed with $ABTS^+$ solution and the absorbance is read at 734 nm over 20 min. In the FRAP method, the reducing power of wines is measured by the determination of a colored product formed during a redox reaction between the ferrous (Fe^{2+}) ion and the added 2,20-dipyridyl (Paixão *et al.*, 2008). In the ORAC procedure, the fluorescence response of the reaction between β-Phycoerithrin, 2,2-azobis (2-amidino-propane) dihydrochloride (AAPH) and wine (previously dealcoholized) is measured (Fernández-Pachón *et al.*, 2006). For determining the antioxidant activity

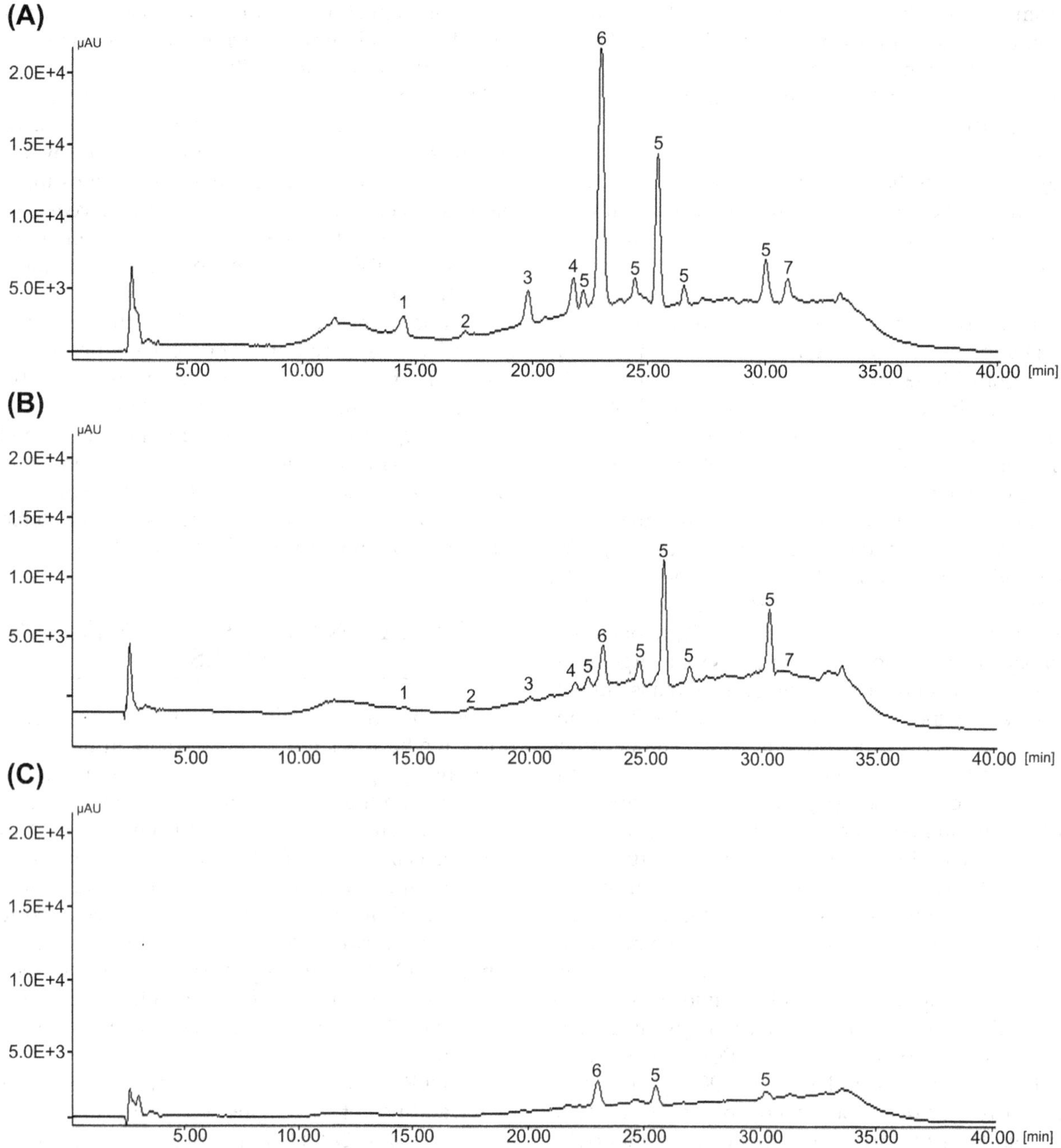

FIGURE 1.1 Chromatographic profiles of anthocyanins in Port wines at 520 nm: (A) Ruby; (B) Late Bottle Vintage (LBV); (C) Tawny. Peak identification: (1) delphinidin 3-*O*-glucoside (Dp-3g); (2) cyanidin 3-*O*-glucoside (Cy-3g); (3) petunidin 3-*O*-glucoside (Pt-3g); (4) peonidin 3-*O*-glucoside (Pn-3g); (5) malvidin 3-*O*-glucoside (Mv-3g); (6) malvidin 3-*O*-acetylglucoside (Mv-3ac); (7) newly formed anthocyanins corresponding to pyruvic acid adducts.

electrochemically, an electrolytic device is used which measures the coulombs consumed in the oxidation of the sample with ABTS (Alonso *et al.*, 2004). The antioxidant activity calculated by the DPPH, ABTS, ORAC, and electrochemical methods is usually expressed as equivalent concentration of 6-hydroxy-2,5,7,8-tetramethylchroman-2-carboxylic acid (Trolox) in mM. The antiradical activity can also be expressed as percentage inhibition of DPPH radical caused by a wine sample or the amount of wine sample necessary to decrease the initial DPPH concentration by 50% (EC50). Results from the FRAP procedure are presented by some authors as quercetin equivalents (QE) in mM.

Table 1.4 summarizes data from literature concerning total polyphenols, total anthocyanins and *in vitro* antiradical activity of red wines and fortified wines. Considerable variations in phenolic, anthocyanins, DPPH, FRAP, ABTS, and ORAC values were observed in the

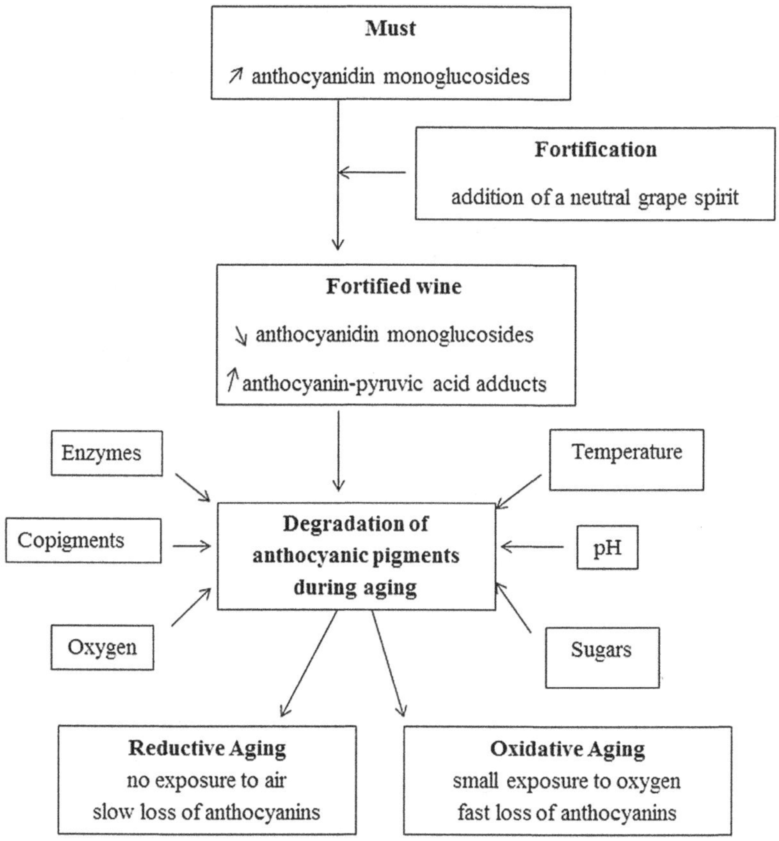

FIGURE 1.2 Influence of aging conditions in degradation of anthocyanin compounds of fortified wines.

literature. In general antiradical activity of red wines is higher than that of fortified wines. The magnitude of the difference depends of the method employed.

Differences were found when comparing the composition and antioxidant activity of different fortified wines. For example, in Madeira wines, higher antioxidant activity, measured by both DPPH and FRAP methods, was obtained in red (0.56–0.71 Trolox mM and 3.45–3.86 QE mM, respectively) than in white wines (0.04–0.08 Trolox mM and 0.44–0.67 QE mM, respectively) (Paixão *et al.*, 2008). Results of the electrochemical procedure applied to Sherry wines with different aging showed an increase of antioxidant activity with the increase of maturing in oak casks (Schwarz *et al.*, 2012). According to these authors, aging in wood of spirits is characterized by the diffusion of compounds, such as aromatic benzoic and cinnamic aldehydes, from within the wood. It has been generally recognized that these compounds are the result of the degradation of the lignin (Barroso *et al.*, 1996). Comparable values of antioxidant activity were obtained in Sherry and Spanish white table wines with DPPH (0.49–2.21 Trolox mM) and ABTS (0.08–1.45 Trolox mM) methods (Fernández-Pachón *et al.*, 2006).

The correlation between phenolic compounds and the antioxidant activity has also been investigated in Sherry wines. Results of these studies confirmed the strong influence of polyphenols on the antioxidant activity (Schwarz *et al.*, 2012). However, not all polyphenols have the same influence on the antioxidant activity and the method applied also influences the results. Ethyl caffeate is the phenolic compound that presents larger correlation coefficient with ORAC, ABTS, and DPPH tests. The ORAC assay presents the highest significant correlation coefficients and gallic acid and tyrosol showed high correlation with the ORAC values. Considering the phenolic compounds by their chemical structure, flavanols were the group presenting higher correlation with all methods. Lower correlation was observed for benzoic and cinnamic acids (Fernández-Pachón *et al.*, 2006). In another study, protocatechuic acid, protocatechuic aldehyde, syringic acid, vanillin, and p-coumaric acid were the compounds that presented higher correlations with the antioxidant activity measured by electrochemical oxidation (Schwarz *et al.*, 2012).

Antioxidant activity in fortified wines is not necessarily correlated with the compounds which are present in the highest concentrations. For instance, caftaric acid is quantitatively the largest of the cinnamic acids but, nevertheless, it presents a weak correlation for all the methods involved (Fernández-Pachón *et al.*, 2006).

The antiradical capacity of Port wines was dependent on the type of Port. Higher antiradical activity was obtained for Ruby and LBV Port wines than for Tawny Port wines indicating that 'reductive' aging, with no

TABLE 1.4 Data from Literature Concerning Total Polyphenols, Anthocyanins and *in vitro* Antiradical Activity of Red Wines and Fortified Wines

Wines	Alcohol	Total Polyphenols (mg Gallic Acid/l)	Anthocyanins (mg/l)	Antiradical Activity DPPH/FRAP or ORAC	References
Portuguese red wines	11–12°	1012–5900		DPPH: 6.3–25.1 Trolox mM	Rodrigues (2011)
Spanish red wines	11–12°	1378–2360	4.6–92.9	DPPH: 1.03–2.45 mg/g DPPH: 5.22–17.4 Trolox mM ABTS: 3.06–11.14 Trolox mM ORAC: 4181–10800 μmol/l	Larrauri *et al.* (1999) Fernández-Pachón *et al.* (2004)
Red wine	12°	2371 921–1879	209 22–274	DPPH: 77 % inhibition	Bajčan *et al.* (2012) Tsanova-Savova *et al.* (2002)
Port ruby	19.5°		92–156 129	DPPH: 8.08 mg/ml	Pinho *et al.* (2012)
Port LBV	20°		32–78 58	DPPH: 6.79 mg/ml	Pinho *et al.* (2012)
Port tawny	19°		0–41 22	DPPH: 16.35 mg/ml	Pinho *et al.* (2012)
Sherry	19–21°	207–446		DPPH: 0.49-2.21 Trolox mM ABTS: 0.08-1.45 Trolox mM ORAC: 995–1782 μmol/l	Fernández-Pachón *et al.* (2004, 2006)
Young Sherry Old Sherry	20°			DPPH: 0.19 Trolox mM DPPH: 5.6 Trolox mM	Schwarz *et al.* (2012)
Madeira red	19°	1724–1936		DPPH 0.56–0.71 Trolox mM FRAP 3.45–3.86 QE mM	Paixão *et al.* (2008)
Madeira white	19°	282–770		DPPH 0.04–0.08 Trolox mM FRAP 0.44–0.67 QE mM	Paixão *et al.* (2008)
Moscatel	18°	1243		DPPH: 70.7% inhibition FRAP: 3098 mg Trolox eq ORAC: 10724 μmol/l	Feliciano *et al.* (2009)

LBV, Late Bottle Vintage; antioxidant activity was expressed as % inhibition of DPPH radical caused by a wine sample, amount of wine sample necessary to decrease by 50% the initial DPPH concentration (EC50), Quercetin Equivalents (QE) in mM or Trolox equivalent (Trolox) in mM antioxidant capacity.

exposure to air, increased antiradical activity as presented in Table 1.4 (Pinho *et al.*, 2012). In addition, LBV Port wines showed higher antiradical activity compared with Ruby Port wines, in spite of their lower amounts of anthocyanins. During the in-bottle aging period, LBV Port wine composition changes in a reducing environment and the oxidation–reduction potential decreases regularly until it reaches a minimum value that prevents oxidation reactions (Mateus and Freitas, 2001). The Tawny Port wines had lower antiradical activity regardless of whether they are reserve or indication of age wines. These wines also presented lower amounts of anthocyanins owing to their 'oxidative' aging in wooden barrels (Pinho *et al.*, 2012).

CONCLUSIONS

A high number of phenolic compounds and anthocyanins have been identified in fortified wines and different methods were used to evaluate their antioxidant activity, namely, polyphenolic compounds index, DPPH, ABTS, FRAP, and ORAC. There are also several studies associating the measurement of phenolic compounds in fortified wines and their antioxidant activity.

The color differences observed in different types of fortified wines are attributed to changes in the phenolic compounds extracted from the grapes occurring during vinification and maturation. Among these, anthocyanin degradation is of major importance to wine color and antioxidant activity.

However, antioxidant activity in fortified wines is not necessarily correlated with the compounds that present the highest concentrations. Aging conditions are a notable factor influencing antioxidant activity in these wines. In general, levels of phenolic compounds increase and anthocyanin content decreases with the duration of aging. Additionally, antioxidant activity is higher in fortified wines aged under 'reductive' conditions, with no exposure to air, than in fortified wines aged under 'oxidative' conditions such as wooden barrels. These aspects are important since classification within each fortified wine type is principally made as a function of aging.

References

Alonso, A.M., Castro, R., Rodríguez, M.C., Guillén, D.A., Barroso, C.G., 2004. Study of the antioxidant power of brandies and vinegars derived from Sherry wines and correlation with their content in polyphenols. Food Res. Int. 37, 715–721.

Andrade, P., Seabra, R., Ferreira, M., Ferreres, F., García-Viguera, C., 1998. Analysis of non-coloured phenolics in port wines by capillary zone electrophoresis. Influence of grape variety and ageing. Z. Leb. Unt. Forsch. A. 206, 161–164.

Bajčan, D., Čéryová, S., Tóth, T., 2012. Phenolic compounds and antioxidant activity of monovarietal red wines produced in Limbach region. JMBFS 1, 868–875.

Barroso, C.G., Rodríguez, M.C., Guillen, D.A., Pérez- Bustamante, J.A., 1996. Analysis of low molecular mass phenolic compounds, furfural and 5-hydroxymethylfurfural in Brandy de Jerez by high-performance liquid chromatography–diode array detection with direct injection. J. Chromatogr. A. 724, 125–129.

BOJA, 2012. Boletín Oficial de la Junta de Andalucía. 12 April 2012. p. 52.

Câmara, J.S., Alves, M.A., Marques, J.C., 2006. Development of headspace solid-phase microextraction–gas chromatography–mass spectrometry methodology for analysis of terpenoids in Madeira wines. Anal. Chim. Acta. 555, 191–200.

Castañeda-Ovando, A., Pacheco-Hernández, M.L., Páez-Hernández, M.E., Rodríquez, J.A., Galán-Vidal, C.A., 2009. Chemical studies of anthocyanins: A Review. Food Chem. 113, 859–871.

Esteves, V.I., Lima, S.S.F., Lima, D.L.D., Duarte, A.D., 2004. Using capillary electrophoresis for the determination of organic acids in Port wine. Anal. Chim. Acta. 513, 163–167.

Feliciano, R.P., Bravo, M.N., Pires, M.M., Serra, A.T., Duarte, C.M., Boas, L.V., Bronze, M.R., 2009. Phenolic Content and Antioxidant Activity of Moscatel Dessert Wines from the Setúbal Region in Portugal. Food Anal. Methods 2, 149–161.

Fernández-Pachón, M.S., Villaño, D., Garcia-Parrilla, M.C., Troncoso, A.M., 2004. Antioxidant activity of wines and relation with their polyphenolic composition. Anal. Chim. Acta. 513, 113–118.

Fernández-Pachón, M.S., Villaño, D., Troncoso, A.M., García-Parrilla, M.C., 2006. Determination of the phenolic composition of sherry and table white wines by liquid chromatography and their relation with antioxidant activity. Anal. Chim. Acta. 563, 101–108.

García-Viguera, C., Bakker, J., Bellworthy, S.J., Reader, H.P., Watkins, S.J., Bridle, P., 1997. The effect of some processing variables on non-coloured phenolic compounds in port wines. Z. Leb. Unt. Forsch. A. 205, 321–324.

He, J., Santos-Buelga, C., Mateus, N., Freitas, V., 2006. Isolation and quantification of oligomericpyrano anthocyanin-flavanol pigments from red wines by combination of column chromatographic techniques. J. Chromatogr. A. 1134, 215–225.

Heredia, F.J., Francia-Aricha, E.M., Rivas-Gonzalo, J.C., Vicario, I.M., Santos-Buelga, C., 1998. Chromatic characterization of anthocyanins from red grapes—I. pH effect. Food Chem. 63, 491–498.

Jackson, R.S., 2008. Specific and Distinctive Wine Styles. In: Wine Science: Principles and Applications, third ed. Academic Press, Canada, pp. 434–481.

Kähkonen, M.P., Heinonen, M., 2003. Antioxidant activity of anthocyanins and their aglycons. J. Agric. Food Chem. 51, 628–633.

La Torre, G.L., La Pera, L., Rando, R., Turco, Lo, DiBella, G., Saitta, M., Dugo, G., 2008. Classification of Marsala wines according to their polyphenol, carbohydrate and heavy metal levels using canonical discriminant analysis. Food Chem. 110, 729–734.

Larrauri, J.A., Sánchez-Moreno, C., Rupérez, P., Saura-Calixto, F., 1999. Free Radical Scavenging Capacity in the Aging of Selected Red Spanish Wines. J. Agric. Food Chem. 47, 1603–1606.

Mateus, N., Freitas, V., 2001. Evolution and Stability of Anthocyanin-Derived Pigments during Port Wine Aging. J. Agric. Food Chem. 49, 5217–5222.

Mateus, N., Machado, J.M., Freitas, V., 2002. Development changes of anthocyanins in Vitis vinifera grapes grown in the Douro Valley and concentration in respective wines. J. Sci. Food Agric. 82, 1689–1695.

Mateus, N., Oliveira, J., Pissarra, J., González-Paramás, A.M., Rivas-Gonzalo, J.C., Santos-Buelga, C., Silva, A.M.S., Freitas, V., 2006. A new vinylpyrano anthocyanin pigment occurring in aged red wine. Food Chem. 97, 689–695.

Oliveira, J., Santos-Buelga, C., Silva, A.M.S., Freitas, V., Mateus, N., 2006. Chromatic and structural features of blue anthocyanin-derived pigments present in Port wine. Anal. Chim. Acta. 563, 2–9.

Oliveira, C., Amaro, L.F., Pinho, O., Ferreira, I.M.P.L.V.O., 2010. Cooked blueberries: anthocyanin and anthocyanidins degradation and their radical-scavenging activity. J. Agric. Food Chem. 58, 9006–9012.

Paixão, N., Pereira, V., Marques, J.C., Câmara, J.C., 2008. Quantification of polyphenols with potential antioxidant properties in wines using reverse phase HPLC. J. Sep. Sci. 31, 2189–2198.

Pascual-Teresa, S., Moreno, D.A., Garcá-Viguera, C., 2010. Flavanols and Anthocyanins in Cardiovascular Health: A Review of Current Evidence. Int. J. Mol. Sci. 11, 1679–1703.

Pellegrini, N., Simonetti, P., Gardana, C., Brenna, O., Brighenti, F., Pietta, P., 2000. Polyphenol Content and Total Antioxidant Activity of Vini-Novelli (Young Red Wines). J. Agric. Food Chem. 48, 732–735.

Perestrelo, R., Albuquerque, F., Rocha, S.M., Câmara, J.S., 2011. Distinctive Characteristics of Madeira Wine Regarding Its Traditional Winemaking and Modern Analytical Methodologies. Adv. Food Nut. Res. 63, 207–249.

Perez-Magarino, S., González-San José, M.L., 2006. Polyphenols and colour variability of red wines made from grapes harvested at different ripeness grade. Food Chem. 96, 197–208.

Pinho, C., Melo, A., Mansilha, C., Ferreira, I.M.P.L.V.O., 2011. Optimization of Conditions for Anthocyanin Hydrolysis from Red Wine Using Response Surface Methodology (RSM). J. Agric. Food Chem. 59, 50–55.

Pinho, C., Couto, A.I., Valentão, P., Andrade, P., Ferreira, I.M.P.L.V.O., 2012. Assessing the anthocyanic composition of Port wines and musts and their free radical scavenging capacity. Food Chem. 131, 885–892.

Queiroz, F., Oliveira, C., Pinho, O., Ferreira, I.M.P.L.V.O., 2009. Degradation of anthocyanins and anthocyanidins in blueberry jams/stuffed fish. J. Agric. Food Chem. 57, 10712–10717.

Rentzsch, M., Schwarz, M., Winterhalter, P., 2007. Pyranoanthocyanins—an overview on structures, occurrence, and pathways of formation. Trends Food Sci. Technol. 18, 526–534.

Ribeiro de Lima, M.T., Waffo-Téguo, P., Teissedre, P.L., Pujolas, A., Vercauteren, J., Cabanis, J.C., Mérillon, J.M., 1999. Determination of Stilbenes (trans-Astringin, cis- and trans-Piceid, and cis- and trans-Resveratrol) in Portuguese Wines. J. Agric. Food Chem. 47, 2666–2670.

Rodrigues, J., 2011. Determinação da actividade antioxidante e composição fenólica de vinhos Portugeses e correlação com parâmeros de cor. Dissertação de Mestrado. Faculdade de Ciências e Tecnologia da Universidade Nova de Lisboa.

Romero, C., Bakker, J., 2000. Effect of storage temperature and pyruvate on kinetics of anthocyanin degradation, vitisin A derivative formation, and colour characteristics of model solutions. J. Agric. Food Chem. 48, 2135–2141.

Saunders, P., 2004. Wine Label Language, Firefly Books, pp. 176.

Schwarz, M., Rodríguez, M.C., Guillén, D.A., Barroso, C.G., 2012. Evolution of the colour, antioxidant activity and polyphenols in unusually aged Sherry wines. Food Chem. 133, 271–276.

Stevenson, T., 2005. The Sotheby's Wine Encyclopedia. Dorling Kindersley. ISBN 0-7566-1324-8. pp. 340–341.

Tsanova-Savova, S., Dimovw, S., Ribarova, F., 2002. Anthocyanins and Color Variables of Bulgarian Aged Red Wines. J. Food Comp. Anal 15, 647–654.

Tseng, K.C., Chang, H.M., Swi-Bea, W., 2006. Degradation kinetics of anthocyanin in ethanolic solutions. J. Food Process. Pres. 30, 503–514.

Vivar-Quintana, A.M., Santos-Buelga, C., Rivas-Gonzalo, J.C., 2002. Anthocyanin-derived pigments and colour of red wines. Anal. Chim. Acta. 458, 147–155.

Welch, C.R., Wu, Q., Simon, J., 2008. Recent advances in anthocyanin analysis and characterization. Cur. Anal. Chem. 4 (2), 75–101.

CHAPTER

2

Endogenous Antioxidants and Antioxidant Activities of Beers

Haifeng Zhao

College of Light Industry and Food Sciences, South China University of Technology, Guangzhou, China

CHAPTER POINTS

- Phenolic compounds, Maillard reaction products, and sulfites constitute the major endogenous antioxidants in beers, although other antioxidants such as reducing sugars, vitamins, carotenoids and chelating agents also exist in beer at very low levels.
- Endogenous antioxidants in beers originate from brewing raw materials or are formed during the brewing process; all of them make significant contributions to beer flavor stability and nutritional properties due to their antioxidant and antiradical properties, and other biological effects.
- The beneficial impacts of these antioxidant compounds on the stability of beer flavor have been demonstrated, but there has been a lack of consensus concerning the effectiveness of these compounds as potential antioxidants. The antioxidant or pro-oxidant activity of these compounds is determined partly by their content in beers.
- The content of endogenous antioxidants and antioxidant activities of beers are beer brand and type dependent. Generally, dark or brown beers contain higher levels of antioxidants and show higher antioxidant activity than pale or lager beers, because the former are brewed from worts with higher extracts or supplemented with some toast malts.
- Due to the different reaction mechanisms and antioxidant activity assays involved, various kinds of endogenous antioxidants make significantly different contributions to the antioxidant activities of beers.
- Optimization of the brewing process and screening of raw materials are considered as efficient measures to increase the antioxidant activities of beers, which is extremely important to beer flavor stability.
- Knowledge of the relationships between endogenous antioxidants and antioxidant activities of beers can be used to develop appropriate technologically innovative processes to control the flavor stability or to reinforce the nutritional properties of beer.

INTRODUCTION

Beer is one of the most widely consumed alcoholic beverages in the world for its fresh taste, low calories, and nutritional value. The main raw materials of beers are water, malt, hops, and brewer's yeast, and some beers also are brewed with some non-malted cereals, starch, or starch syrups as adjuncts. The production of beer involves extremely complicated processes and chemical and biochemical reactions. Thus, the final beer contains various compounds with antioxidant activity mainly originated from raw materials or formed during processing. The species and concentrations of these antioxidants in final beer varies, largely because brewing technology, raw materials, and yeast applied in the brewing can differ Slight changes in structure or conformation of these compounds can cause significant changes of the antioxidant activity, which alters the overall oxidative or flavor stability of beer. Nowadays, flavor stability has become

Processing and Impact on Antioxidants in Beverages
http://dx.doi.org/10.1016/B978-0-12-404738-9.00002-7

the most important factor in determining the shelf-life of packaged beer, and prolonging shelf-life by delaying flavor staling is one of the greatest challenges facing the brewer. Although the flavor stability of beer depends primarily on the oxygen content of the packaged beer, the brewing process and the raw materials used can influence the flavor stability. Therefore, attention is now increasingly shifting towards increasing the antioxidant activity of beer itself due to the fact that oxidative staling of beer is still noticeable even if the level of total packaged oxygen might be as low as 0.1 mg/l (Bamforth, 2000).

Antioxidants are generally thought to play a significant role in malting, mashing and brewing due to their ability to delay or prevent oxidation reactions and oxygen free radical reactions. Some synthetic antioxidants such as sulfites, formaldehyde, or ascorbate, can be added into the brewing process to improve beer flavor stability, but the effectiveness of these compounds is in doubt (Andersen *et al.*, 2000). In recent years, there has been a general trend toward minimizing the use of additives in beer production because of consumer demand and stiffening regulations. As a result, it is desirable to prevent oxidation by protection of endogenous antioxidants in beer and in its raw materials, or those formed during the processes of kilning, boiling, and fermentation (mainly Maillard reaction products and sulfites). These endogenous antioxidants can offer an overall antioxidant activity in the beer to increase its stability during storage. This issue – to keep or increase the antioxidant activity of beer so as to improve its flavor stability – is greatly recognized by consumers and brewers. Beer generally contains a range of antioxidants, such as phenolic compounds, Maillard reaction products (melanoidins and reductones), sulfites, thiols, non-fermentable reducing sugars, vitamins, carotenoids, and chelating agents. Among the antioxidants mentioned above, phenolic compounds, melanoidins, and sulfites are of particular interest to brewers because they are described as antioxidants that possess antioxidant and antiradical properties as well as other biological effects (Zhao *et al.*, 2013). The beneficial impacts of these antioxidant compounds on the stability of beer flavor have been demonstrated, whereas there has been a lack of consensus concerning the effectiveness of these compounds as potential antioxidants. For instance, results from electron spin resonance (ESR) lag phase studies showed that there were no significant effects of polyphenols on the formation of free radicals in beer during storage or in wort during brewing, while the author's group's previous study showed that phenolic compounds investigated made up 55.0–88.1% of the antioxidant activity of beer (Andersen *et al.*, 2000; Zhao *et al.*, 2010). Moreover, although the effects of sulfites on beer flavor stability evaluated by various assays have shown that sulfites stabilize the flavor in two ways, as antioxidants and as carbonyl scavengers in aldehyde–bisulfite adducts, while the efficiency of naturally produced bisulfites by yeast is controversial (Guido, 2005). In addition, antioxidant activity of melanoidins has been mentioned in previous studies (Papetti *et al.*, 2006), but melanoidins or their precursors may also exhibit pro-oxidative properties as indicated by their involvement in the oxidation of alcohols to aldehydes during beer storage (Martins *et al.*, 2001). Therefore, this overview aims to clarify the contributions of endogenous antioxidants to antioxidant activities of beers, which helps us to better understand beer flavor stability.

ENDOGENOUS ANTIOXIDANTS IN BEERS

Phenolic Compounds

Beer is generally considered to be one of the major sources of phenolic compounds, and the presence of these compounds contributes to colloidal, foam, flavor, color, and sensory properties of beer. Beers of various types and brands show similar phenolic profiles, but significant variations exist in the total and individual phenolics content due to various materials and brewing processes used (Table 2.1). The values of total phenolic content (TPC) in beers examined by the Folin–Ciocalteu assay usually exceeded 100 mg gallic acid equivalents (GAE)/l (Table 2.1), while the sums of individual phenolics content determined by high-performance liquid chromatography (HPLC) were in the range of 4.47 to 15.50 mg/l (Zhao *et al.*, 2010). The considerable difference in both assays for evaluating of the content of phenolic compounds is because the Folin–Ciocalteu method is not specific for phenolic compounds and does suffer interference from other compounds, and HPLC only reflects the sum content of limited or detectable phenolic species.

The phenolic compounds in beer mainly are phenolic acids and flavonoids (Zhao *et al.*, 2010; Piazzon *et al.*, 2010; Gorinstein *et al.*, 2000; McMurrough *et al.*, 1996). The main phenolic acids, flavanol and isoflavonoid, identified in beers are gallic and ferulic acids, (+)-catechin, and formononetin, respectively, but *p*-coumaric, caffeic, ferulic, sinapic, chlorogenic, *p*-hydroxybenzoic, and vanillic acids, (−)-epicatechin and their dimers, and the isoflavonoids of genistein, daidzein, and biochanin A have also been detected in beers (Table 2.1). Several research studies indicated that most dimers in beer were procyanidins B3 whilst most trimers were prodelphinidins (Callemien and Collin 2008; Callemien *et al.*, 2008). The brewing materials of malt and hop contain various phenolic acids and flavonoids, which are partially recovered in the final beer. Callemien and Collin (2010) reviewed the phenolic compounds in malts and hop, and found that the levels of *p*-coumaric, caffeic,

TABLE 2.1 Phenolic Compounds in Beers

	Phenolic Acids (mg/l)										
Beer Type	Gallic Acid	Protocatechuic Acid	Vanillic Acid	Caffeic Acid	Syringic Acid	*p*-Coumaric Acid	Ferulic Acid	4-Hydroxyphenylacetic Acid	Sinapic Acid	Chlorogenic Acid	Total phenolic content (TPC)
Lager	1.81–10.39[a]	0.02–1.30[a,c]	0.22–2.98[a,b,c]	0.08–1.22[a,b,c]	0.06–0.99[a,b]	0.01–1.12[a,b,c]	0.51–3.1[a,b,c]	0.27–0.61[b,c]	0.20–0.84[b,c]	0.21–0.26[c]	138.0–476.2[a,b,e,f,g]
Ale	—	—	0.61	0.14	0.20	0.58	1.54	0.53	0.29	—	280.10–563.0[b,e,h]
Pilsner	—	—	0.56	0.14	0.27	0.76	1.03	0.69	0.44	—	484[b]
Wheat	—	—	0.67	0.19	0.35	0.26	0.34	0.52	0.55	—	270.1–366.0[b,e]
Abbey	—	—	0.65	0.25	0.22	0.23	0.92	0.59	0.34	—	622[b]
Bock	—	—	0.81	0.17	0.22	0.72	2.00	0.81	0.45	—	875[b]
Dealcoholized	—	—	0.45	0.14	0.16	0.40	1.40	0.32	0.26	—	125.0–366[b,e]
Dark	—	—	—	—	—	—	—	—	—	—	333.6–600[e,f]

	Flavanoids										
	(+)-Catechin	(−)-Epicatechin	(−)-Gallocatechin	(−)-Epigallocatechin	Procyanidin Dimers B2	Procyanidin Dimers B3	Prodelphinidin B3	Genistein	Biochanin A	Daidzein	Formononetin
Lager	0.03–4.00[a,i]	0.02–0.73[a,i]	0.1[d,i]	0.1[d,i]	0.16[i]	0–3.6[i,j]	0–4.5[j]	0.05–1.82[k]	0.15–1.38[k]	0.08–0.65[k]	0.32–4.02[k]

[a]*Zhao* et al. *(2010)*
[b]*Piazzon* et al. *(2010)*
[c]*Nardini and Ghiselli (2004)*
[d]*Callemien* et al. *(2008)*
[e]*Gorjanović* et al.*(2010)*
[f]*Lugasi (2003)*
[g]*Martinez-Periñan* et al. *(2011)*
[h]*Granato* et al. *(2011)*
[i]*de Pascual-Teresa* et al. *(2000)*
[j]*Andersen* et al.*(2000)*
[k]*Lapcík* et al. *(1998)*.

ferulic, sinapic, and chlorogenic acids in malt reached ppm levels, while *p*-coumaric, caffeic, and ferulic acids in hop was more than 10 ppm in most cases. Most phenolic acids in malt or hop are esterified with starch and other polysaccharides by forming bridges or cross-links through hydrogen bonds, chelation, or covalent bonds. These phenolic acids bound with starch can be released by enzymatic hydrolysis of starch during mashing, which causes significantly higher levels of bound phenolic acids than those of free forms in raw materials and final beers. In addition, malt and hop also are an excellent source of catechins, proanthocyanidins and flavonoid oligomers. (+)-Catechin in hop dried cones or pellets can reach 2821 mg/kg (Callemien and Collin, 2010), while the major monomeric and dimeric flavan-3-ol in malts identified as (+)-catechin, prodelphinidin B3 and procyanidin B3 were found to be at values of 6-16 mg/kg, 99–175 mg/kg and 9–85 mg/kg, respectively (Dvorakova *et al.*, 2008).

The phenolic compounds content in beer is beer-type dependent. Piazzon *et al.* (2010) reported that the content of phenolic compounds increased in the order: dealcoholized < lager< pilsner < wheat < ale < abbey < bock. Vinson *et al.* (2003) also observed that total polyphenols in ales/porters/stouts beers were higher than those in lager and nonalcoholic beers. However, other research indicated that although higher TPC was found in dark beers, its difference between dark and lager beers was not statistically different (Lugasi, 2003). The higher phenolic compounds in dark, strong beers is because these beers are brewed from wort with higher extract content, while alcohol-free beers are usually brewed with lower original wort extract and inhibition of alcohol formation. Moreover, beers with higher original gravity also generally showed higher TPC (> 180 mg GAE/l), further indicating that phenolic compounds in beer mainly originate from barley malt and hops, although the brewing process also has some impact on phenolic compound content (Zhao *et al.*, 2010).

Melanoidins

Melanoidins are macromolecular, nitrogenous, and brown-colored final products of the Maillard reactions between reducing sugars and proteins or amino acids, which are formed during malt kilning, mashing and wort boiling, and are retained at least partially up to the final beer (Kuntcheva and Obretenov, 1996; Rivero *et al.*, 2005). Melanoidins are generally formed by cyclizations, dehydrations, retroaldolizations, rearrangements, isomerizations, and condensations of initial Maillard reaction products (Figure 2.1), but only very few structures have been identified completely (Martins *et al.*, 2001).

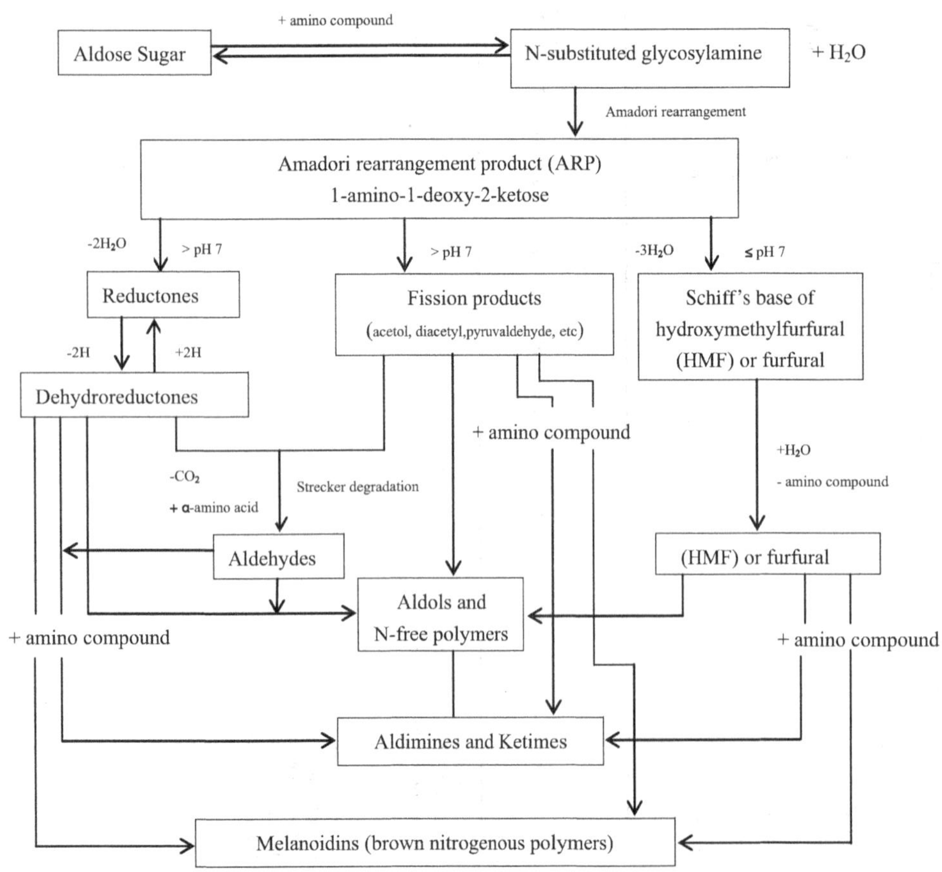

FIGURE 2.1 Maillard reaction scheme. *Adapted from Martins et al., (2001).*

Melanoidins contribute to the formation of color, aroma, and flavor, and also play an important role in stabilization of the foam and influence the oxidation–reduction processes of beer due to their strong antioxidant properties and other biological effects (Rivero *et al.*, 2005). The quantitative analysis of melanoidins in beers is very limited; a more than eight-fold difference in melanoidin content, varying from 1.64 mg/l to 13.71 mg/1, occurred between beers which can be ascribed to the different raw materials and brewing process used during beer brewing (Zhao *et al.*, 2013; Kuntcheva and Obretenov, 1996). The content of melanoidins in beers was influenced by the type of beers, and their content was found to be the highest in dark beer with values of 1.49 g/l, compared with to 0.61 g/l in blond beer and 0.58 g/l for alcohol-free beer (Rivero *et al.*, 2005). The main difference between the types of beers, especially for dark and blond beers, is that they are brewed from malt with diverse grades of toasting, although they are also largely influenced by the technological brewing factors. Indeed, malts contained more melanoidins than barley, and the melanoidins isolated from malt were observed as a group of compounds with high molecular weight (>3500 Da), brown color, and with a small percentage content of non-amino acidic nitrogen (1.12%) due to the relatively high polysaccharide and low protein content of barley grain and because of the specific reaction conditions of low water activity and lengthy roasting periods applied during the roasting process (Papetti *et al.*, 2006). Most of the molecular weight of such polymeric compounds isolated from food did not exceed 100,000 Da, while the molecular weight of melanoidins formed in roasted barley malt of was found to be up to 200,000 Da (Faist *et al.*, 2002). Moreover, significant differences in the distribution of the nitrogen content in different molecular weights of melanoidins isolated from barley were observed, and its content reached 30% in low-molecular-weight compounds, whereas only 3% was found in the high-molecular-weight polymers (30,000–60000 Da) (Papetti *et al.*, 2006).

Sulfites

Sulfites, one of important constituents of beer, are produced by yeast as by-products of the synthesis of sulfur-containing amino acids during fermentation and survive into the finished beer. The raw materials used in brewing may contain different amounts of sulfur dioxide but most of it disappears during wort boiling. Sulfites in beer are not only important antioxidants to prevent beer from oxidative staling, but they also act as important mask agents for stale flavor by reacting with carbonyl staling compounds in beer to form bisulfite–carbonyl adducts, which is important for beer flavor stability. The differences in brewer's yeast strain and brewing process lead to significant variations in sulfite content of final beers. The total concentration of SO_2 typically ranged from 1 to 30 mg/l (Illet, 1995). For example, the total SO_2 concentrations determined by the *p*-rosaniline method and the chronopotentiometric method in beers were found in the range of 1.15–37.32 mg/l and 0–4.1 mg/l, respectively (Zhao *et al.*, 2013; Dvořák *et al.*, 2006). Almeida *et al.* (2003) reported that the levels of free and total sulfur dioxide in beers determined by a proposed voltammetric methods were in the ranges of 0–3.0 mg/l and 8.1–14.3 mg/l, respectively.

The production of sulfites in beer is affected by yeast strain, fermentation conditions, and wort components. Saison *et al.* (2009) observed that sulfite production in beers appeared to be yeast strain dependent (Figure 2.2), but their content was quite low (2.44–3.91 mg/l) and their differences were small. Moreover, lager beers contained substantially higher levels of SO_2 than the ale because lager yeast strains produce more SO_2 than ale strains (Bushnell *et al.*, 2003). Over-production of sulfites by yeast can occur when the brewing yeast strain is restricted in growth during fermentation because endogenous sulfites in beer are produced from sulfates in the wort by the brewing yeast strain during synthesis of hydrogen sulfide. Thus, all conditions which promote yeast growth will reduce the amount of sulfites excreted in the medium. Guido *et al.* (2004) examined the effects of the physiological condition of the pitching yeast on sulfites production in beer, and fermentations with the lower yeast cell count secreted higher levels of sulfites (Figure 2.3). In recent decades, modification of characteristics of brewer's yeast by genetic engineering has been considered to be an efficient measure to control the accumulation of sulfites (Dequin, 2001). Korch *et al.* (1991) suggested that the overexpression of ATP sulfurylase (*MET3*) and APS kinase (*MET14*) in brewer's yeast could increase sulfites production (Figure 2.4). While Hansen

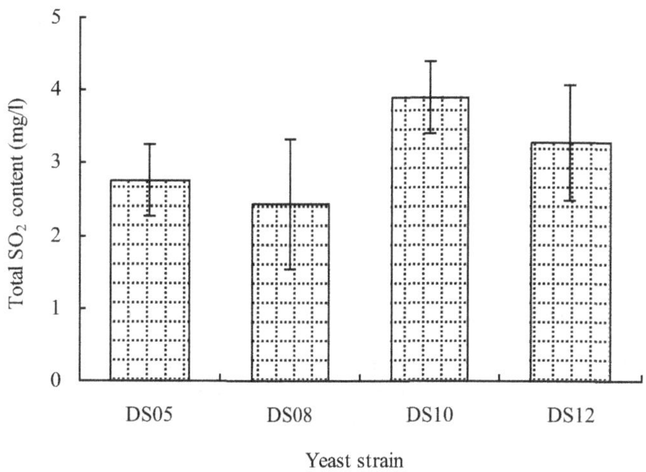

FIGURE 2.2 Influence of fermentation with different top-fermenting yeast strains on final total SO_2 content in beer. *Adapted from Saison et al. (2009).*

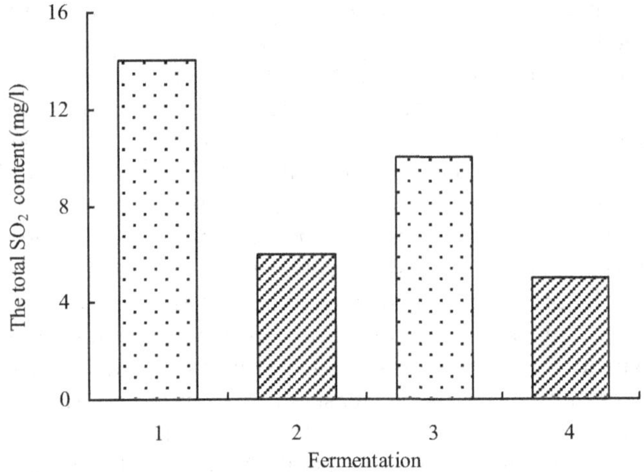

FIGURE 2.3 The effect of yeast physiological condition on total SO_2 content at the end of fermentation. Fermentation 1 and 3: higher vital pitching yeast; fermentation 2 and 4: lower vital pitching yeast. *Adapted from Guido et al. (2004).*

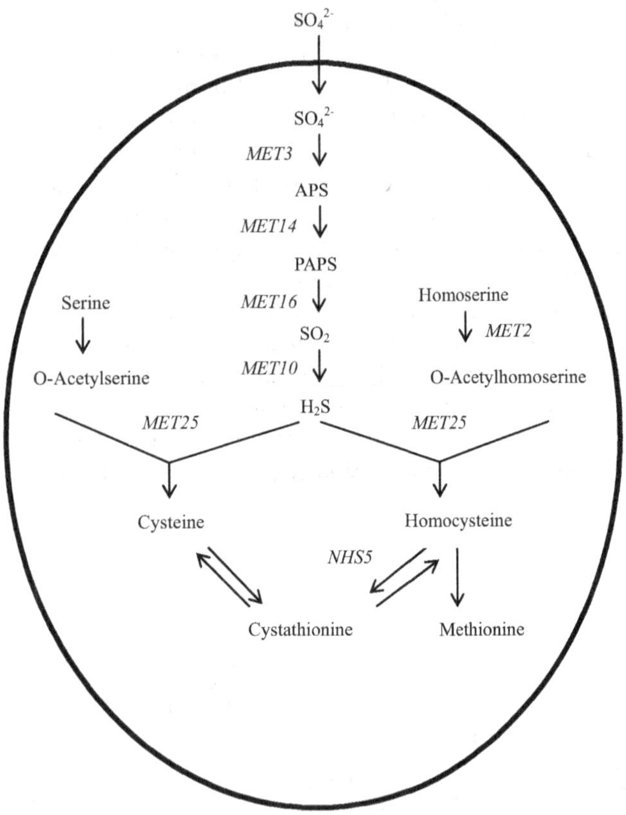

FIGURE 2.4 Pathways for sulfite production in *S. cerevisiae*. *APS*, adenosine 5′-phosphosulfate; *MET2*, serine acetyltransferase; *MET3*, ATP sulfurylase; *MET10*, sulfite reductase subunit; *MET14*, APS kinase; *MET16*, 3′-phospho adenosine 5′-phosphosulfate (PAPS) reductase; *MET25*, O-acetylhomoserine or O-acetylserine sulfhydrylase; *NHS5*, cystathionine β-synthase. *Adapted from Dequin, (2001).*

and Kielland-Brandt (1996) indicated that inactivation of serine acetyltransferase (*MET2*) in brewer's yeast could produce higher amounts of sulfites and hydrogen sulfide, and the inactivation of sulfite reductase subunit (*MET10*) was shown to be an efficient strategy for increasing sulfites formation and decreasing hydrogen sulfide production. However, it should be noted that the SO_2 level must be controlled at the end of beer production for human health and beer-quality reasons. A maximum level of total SO_2 in beer at 20 mg/l was established by the European Union and other countries. The SO_2 level in beer must also be controlled during storage because sulfites produced during fermentation influence the flavor staling of finished beer; a content of 8–9 mg/l of sulfites in packaged beer was considered to be the most appropriate for the flavor stability of beer (Narziss *et al.*, 1993).

ANTIOXIDANT ACTIVITIES OF BEERS

Antioxidant Activity Assays for Beers

Endogenous antioxidants present in beer are complex, and their activities and mechanisms largely depend on the composition and condition of the test system. The results are quite different even when an analysis of the same sample is carried out by each method since the antioxidant activity assays each rely on different chemical backgrounds and reaction mechanisms. Thus, many researchers have stressed the need to perform more than one type of antioxidant activity measurement to evaluate the antioxidant activity of beers. In the past few years, various *in vitro* methods have been investigated to estimate the antioxidant activity of beers or brewing raw materials. The varieties, principles, analytical performances and characteristics of methods for antioxidant activity evaluation have been reviewed by Pisoschi and Negulescu (2011) and Badarinath *et al.* (2010). These methods include 1,1-diphenyl-2-picrylhydrazyl (DPPH) radical scavenging activity assay, 2,2′-azinobis (3-ethylbenzothiazoline-6-sulfonic acid) diammonium salt radical cation scavenging activity (ABTS) assay, oxygen radical absorbing capacity (ORAC) assay, superoxide anion radical scavenging activity (SASA), ferric-reducing antioxidant power (FRAP) assay, total radical trapping antioxidant parameter (TRAP), lipid peroxidation inhibition capacity (LPIC) assay, cupric ion reducing antioxidant capacity (CUPRAC) assay, metal-chelating activity (MCA) assay, hydroxyl radical averting capacity (HORAC) assay, potassium ferricyanide reducing power (PFRAP) method, enhanced chemiluminescence (ECL), and ESR. The various analytical methods for antioxidant activity evaluation fall into distinct categories, which involve different reaction mechanisms, reaction conditions, and analytical techniques (Table 2.2). ORAC, LPIC, TRAP, ABTS, and SASA are classified as hydrogen atom transfer methods of monitoring competitive kinetic reactions,

TABLE 2.2 Antioxidant Activity Assays for Beers

Antioxidant Activity Assays	Type of Method	Principle of Method	End-Product Determination
ORAC	Hydrogen atom transfer	Antioxidant reaction with peroxyl radicals, induced by AAPH	Loss of fluorescence of fluorescein
LPIC	Hydrogen atom transfer	Antioxidant to inhibit peroxidation of lipids induced by a Fenton-like system	Colorimetry
TRAP	Hydrogen atom transfer	Antioxidant capacity to scavenge luminol-derived radicals, generated from AAPH decomposition	Chemiluminescence quenching
SASA	Hydrogen atom transfer	Antioxidant to scavenge superoxide anion radical formation by alkaline or enzyme	Colorimetry
ABTS	Hydrogen atom transfer	Antioxidant to scavenge an organic cation radical	Colorimetry
HORAC	Hydrogen atom transfer	Antioxidant capacity to quench OH radicals generated by a Co(II) based Fenton-like system	Loss of fluorescence of Fluorescein
CUPRAC	Electron transfer	Cu(II) reduction to Cu(I) by antioxidants	Colorimetry
FRAP	Electron transfer	Antioxidant reaction with a Fe(III) complex	Colorimetry
DPPH	Electron transfer	Antioxidant reaction with an organic radical	Colorimetry
PFRAP	Electron transfer	Potassium ferricyanide reduction by antioxidants and subsequent reaction of potassium ferrocyanide with Fe^{3+}	Colorimetry
MCA	Other	Chelating ferrous ions with ferrozine	Colorimetry
ECL	Other	Emission of light by a substance that has absorbed light or other electromagnetic radiation of a different wavelength	Recording of fluorescence excitation/emission spectra
ESR	Other	Free radicals by electron spin resonance	Recording the time of occurring of free radicals

AAPH, 2,2′-azinobis (2-amidinopropane) dihydrochloride; ABTS, 2,2′-azinobis (3-ethylbenzothiazoline-6-sulfonic acid) diammonium salt radical cation scavenging activity assay; CUPRAC, cupric ion reducing antioxidant capacity assay; DPPH, 1,1-diphenyl-2-picrylhydrazyl radical scavenging activity assay; ECL, enhanced chemiluminescence; ESR, electron spin resonance; FRAP, ferric-reducing antioxidant power assay; HORAC, hydroxyl radical averting capacity assay; LPIC, lipid peroxidation inhibition capacity assay; MCA, metal-chelating activity assay; ORAC, oxygen radical-absorbing capacity assay; PFRAP, potassium ferricyanide reducing power method; SASA, superoxide anion radical scavenging activity; TRAP, total radical trapping antioxidant parameter

while FRAP, DPPH, CUPRAC, HORAC, and PFRAP are grouped together as the electron transfer methods of a reduction reaction, and others such as MCA, ECL, and ESR are categorized as other methods. For routine evaluation of the antioxidant activity of beers or monitoring the brewing process with respect to antioxidant activity, spectrometry analytical techniques such as DPPH, ABTS, ORAC, FRAP, TRAP, etc. should be selected due to their better sensitivity, convenience, and short assay times.

Antioxidant Activities of Beers Evaluated by Different Assays

As mentioned above, beer contains numerous endogenous antioxidants, which endows beer with significant antioxidant activity. Zhao *et al.*, (2010) observed significant antioxidant activities in lager beers, suggesting that overall antioxidant activity evaluation results for beer samples using DPPH, ABTS, SASA, FRAP, MCA assays were consistent even though these assays involved different reaction mechanisms. Another study performed by the same group indicated that all 40 beers showed significant antioxidant activity evaluated by DPPH, ABTS, ORAC, FRAP, MCA assays, but clear discrimination among beer samples with significantly different antioxidant activity was achieved by principal component analysis (Zhao *et al.*, 2013). The antioxidant activity of 27 beers was tested by TRAP, TEAC, DPPH, FRAP, CUPRAC, and ORAC assays, and the results pointed out that all beers displayed antioxidant activities, although the values varied a lot with the sample, method, and standard (Tafulo *et al.*, 2010). Similarly, Piazzon *et al.* (2010), Granato *et al.* (2011), and Lugasi (2003) also found that all beers showed antioxidant activity measured with various assays, but a wide range of antioxidant activity was observed.

There are significant differences in antioxidant activity among different commercial beer types. Piazzon *et al.* (2010) reported that antioxidant activities decreased in the order: bock > abbey > ale > wheat > pilsner > lager > dealcoholized, and the antioxidant activities of bock beers were found to be about 3 times higher with respect to dealcoholized beers. In a comparative study, lager-type beers were found have lower antioxidant activity than the ale ones, and Portuguese beers showed slightly lower antioxidant activity then Belgian ones, regardless of the method or standard (Tafulo *et al.*, 2010). Granato *et al.* (2011) also examined the antioxidant activity of Brazilian lager and brown ale beers using ORAC and DPPH assays: higher antioxidant activity in brown ale beers was observed. However, another study performed by Lugasi (2003) indicated that there were no significant differences in antioxidant activity between lager and dark beers. It should also be noted that the differences in antioxidant activity across beer samples and types were due to the different parameters during brewing, such as the variety of barley and hop, the malting and mashing process, and the yeast fermentation employed.

Relationships between Endogenous Antioxidants and Antioxidant Activities of Beers

All endogenous antioxidants in beers may contribute to some extent to the overall antioxidant activities of beers although present in beer at very low levels. However, the real contributions of these antioxidants to antioxidant activities of beers have not been clarified yet.

Contributions of Phenolic Compounds to Antioxidant Activities of Beers

Most research has suggested that phenolic compounds are the most important antioxidants in beers. In addition to acting as chelating agents of metallic catalysts, phenolic compounds in beer have another indirect role in the antioxidant capacity due to their reversible interaction with sulfite (Illet, 1995). Our previous study observed that phenolic compounds including ferulic acid, syringic acid, (+)-catechin, caffeic acid, protocatechuic acid, and (−)-epicatechin contributed 55.0–88.1% of beer antioxidant activity evaluated by different assays. McMurrough *et al.* (1996) also found that partial removal of the polyphenol fraction by polyvinylpolypyrrolidone treatment diminished the reducing power by 9–38%. Piazzon *et al.* (2010) observed a strong correlation between FRAP values and total phenolic acids content of beers and a lack of correlation between FRAP values and the content of free phenolic acids, which indicated that conjugated forms of phenolic acids in beer were responsible for the antioxidant activity of beer. Granato *et al.* (2011) reported that the content of flavonoids and total phenolic compounds seemed to be strongly correlated with the antioxidant activity measured by ORAC and DPPH assays. TPC in the 40 commercial beers were also found to give strong correlations with DPPH, ABTS, ORAC, MCA, and FRAP assays tested, which strongly suggested that phenolic compounds were mainly responsible for the majority of the antioxidant activity of beer (Zhao *et al.*, 2013). However, some phenolic compounds can also act as pro-oxidants by transferring electrons to metal ions, and the levels of ferulic acid in beer were found to determine whether it is pro-oxidant or anti-oxidant (Walters *et al.*, 1997). Thus, some controversy concerning the relevance of polyphenolic antioxidants in beer has been reported in the literature. In an ESR lag phase study, phenolic compounds such as phenolic acids, catechin, epicatechin, and proanthocyanidin dimers had no effect on the formation of radicals, and they are neither antioxidants nor pro-oxidants (Andersen *et al.*, 2000). A Laccase–Sonogel–Carbon biosensor method proposed by Martinez-Periñan *et al.* (2011) for beer stability evaluation showed that the polyphenols cannot delay the formation of free radicals in beers, and polyphenols have an indirect role in beer aging by avoiding rapid SO_2 loss.

Contributions of Melanoidins to Antioxidant Activities of Beers

Maillard reaction products acting as antioxidants generally are involved in various mechanisms, and include: oxygen, reactive oxygen, peroxyl, or some specific stable radicals scavengers, reducing agents, and metal chelating agents. The overall effects of Maillard and caramelization products on the oxidative stability of beer are unknown. However, traditionally, the use of colored malt is known to improve the stability of the finished beer, and it has also been shown that more-highly colored beers retain a greater reducing power during storage (Coghe *et al.*, 2003). The final beer produced has a reducing power mainly dependent on melanoidin compounds and simple and polymerized polyphenols. Indeed, positive correlations between antioxidant activity and malt color also have been observed, which have been ascribed to the presence of Maillard components (Coghe *et al.*, 2003; Woffenden *et al.*, 2001). Moreover, the radical-scavenging and iron-chelating ability of melanoidins from dark beer has been found to be higher compared to melanoidins from pale beers (Coghe *et al.*, 2003; Morales *et al.*, 2005; Morales and Jiménez-Pérez, 2004). Furthermore, Woffenden *et al.* (2002) reported that correlations between levels of catechin or ferrulic acid and antioxidant activity are poor, suggesting that other components, such as polyphenols, oligomers, and Maillard reaction products also contribute to the total antioxidant activity of beer. Zhao *et al.* (2013) also indicated significant positive correlations between melanoidins content in beers and DPPH, FRAP, as well as MCA, and their content made an 8% contribution to ORAC of beers. But Rivero *et al.* (2005) concluded that the hydrogen-donating ability of beers is dependent on the total polyphenol levels but not on the total melanoidin content. Hydroxyl radical-scavenging activity of beer melanoidins was found to be unrelated to their color, suggesting that colored structures were not involved in such antioxidant effects (Morales, 2009).

Although melanoidins are known to act as antioxidants, some reports have also indicated a pro-oxidant activity. Pro-oxidative activity of melanoidins has mainly been established for low-molecular-weight melanoidins produced from model systems (Ames, 2001). A pro-oxidative effect of stout also has been observed in studies with stout added to lager beer analyzed by ESR spectroscopy (Nøddekær and Andersen 2007). Also, beers produced with dark malts kilned at high temperatures formed high concentrations of radicals, and also had a lower oxidative stability as compared to using paler malts (Cortés *et al.*, 2010). This might be because the pro-oxidative effect of Maillard reaction products was caused by reactions that were able to increase radical levels by other mechanisms than Fenton-catalysis since stout was found to decrease the level of radicals in a beer model system where radical formation was induced by the Fenton reaction (Nøddekær and Andersen, 2007).

Contributions of Sulfites to Antioxidant Activities of Beers

Sulfites formed in beer have been considered as one of the most effective antioxidants for beer flavor stability. Two main mechanisms for sulfites to control beer stabilization have been proposed: one is to inhibit beer oxidation during storage by acting as an antioxidant, the other is to combine with several carbonyls to produce sulfite adducts in beer and so mask staling flavor since the adducts are ionized and thus non-volatile. According to Andersen *et al.* (2000), sulfites were the most efficient naturally occurring antioxidants in beer, while thiols and ascorbic acid were shown to be pro-oxidants. Other authors obtained similar results studying flavor stability using ESR to detect free radicals, which supported the theory that phenolic compounds did not contribute to the antioxidant capacity but SO_2 did increase the lag time (Foster *et al.*, 1999). In another study performed by Martinez-Periñan *et al.* (2011), only SO_2 concentration showed relevant differences after the ageing process, thus, it was responsible for the antioxidant capacity of beers. Our study by stepwise regression analysis also indicated that total SO_2 content contributed 5% of the antioxidant activity of beer evaluated by reducing power assay. However, no significant correlations were found between SO_2 and all five antioxidant activity assays used in this study (Zhao *et al.* 2013). Moreover, the effects of sulfites on beer flavor stability evaluated by various assays have shown that sulfites in beer are not only one of the important antioxidants but also one of the important masking agents; however, the efficiency of naturally produced bisulfite by yeast was controversial (Guido, 2005).

Acknowledgments

I gratefully acknowledge the National Natural Science Foundation of China (No. 31000810), the Key Technology R&D Program of Guangdong Province (Nos. 2012A080107005 and 2011A020102001) and the Fundamental Research Funds for the Central Universities (No. 2012ZM0069) for their financial support.

References

Almeida, P.J., Rodrigues, J.A., Guido, L.F., Santos, J.R., Barros, A.A., Fogg, A.G., 2003. Voltammetric determination of free and total sulfur dioxide in beer. Electroanal 15, 587–590.

Ames, J.M., 2001. Melanoidins as pro- or antioxidants. Cerevisia 26, 210–216.

Andersen, M.L., Outtrup, H., Skibsted, L.H., 2000. Potential antioxidants in beer assessed by ESR spin trapping. J. Agric. Food Chem. 48, 3106–3111.

Badarinath, A.V., RAo, K.M., Chetty, C.M.S., Ramkanth, S., Rajan, T.V.S., Gnanaprakash, K., 2010. A Review on In-vitro Antioxidant Methods: Comparisions, Correlations and Considerations. Int. J. PharmTech Res. 2, 1276–1285.

Bamforth, C.W., 2000. Beer quality: oxidation. Brewer's Guardian 4, 31–34.

Bushnell, S.E., Guinard, J.X., Bamforth, C.W., 2003. Effects of sulfur dioxide and polyvinylpolypyrrolidone on the flavor stability of beer as measured by sensory and chemical analysis. J. Am. Soc. Brew. Chem. 61, 133–141.

Callemien, D., Collin, S., 2008. Use of RP-HPLC-ESI(–)-MS/MS to differentiate various proanthocyanidin isomers in lager beer extracts. J. Am. Soc. Brew. Chem. 66, 109–115.

Callemien, D., Collin, S., 2010. Structure, organoleptic properties, quantification methods, and stability of phenolic compounds in beer — a review. Food Rev. Int. 26, 1–84.

Callemien, D., Guyot, S., Collin, S., 2008. Use of thiolysis hyphenated to RP-HPLC-ESI(–)-MS/MS for the analysis of flavanoids of fresh lager beers. Food Chem. 110, 1012–1018.

Coghe, S., Vanderhaegen, B., Pelgrims, B., Basteyns, A.V., Delvaux, F.R., 2003. Characterization of dark specialty malts: New insights in color evaluation and pro- and antioxidative activity. J. Am. Soc. Brew. Chem. 61, 125–132.

Cortés, N., Kunz, T., Suárez, A.F., Hughes, P., Methner, F.J., 2010. Development and correlation between the organic radical concentration in different malt types and oxidative beer stability. J. Am. Soc. Brew. Chem. 68, 107–113.

de Pascual-Teresa, S., Santos-Buelga, C., Rivas-Gonzalo, J.C., 2000. Quantitative analysis of flavan-3-ols in Spanish foodstuffs and beverages. J. Agric. Food Chem. 48, 5331–5337.

Dequin, S., 2001. The potential of genetic engineering for improving brewing, wine-making and baking yeasts. Applied Microbiol. Biot. 56, 577–588.

Dvořák, J., Dostálek, P., Štěrba, K., Čejka, P., Kellner, V., Čulík, J., Beinrohr, E., 2006. Determination of total sulphur dioxide in beer samples by flow-through chronopotentiometry. J. Inst. Brew. 112, 308–313.

Dvorakova, M., Moreira, M.M., Dostalek, P., Skulilova, Z., Guido, L.F., Barros, A.A., 2008. Characterization of monomeric and oligomeric flavan-3-ols from barley and malt by liquid chromatography–ultraviolet detection–electrospray ionization mass spectrometry. J. Chromatogr. A. 1189, 398–405.

Faist, V., Lindenmeier, M., Geisler, C., Erbersdobler, H.F., Hofmann, T., 2002. Influence of molecular weight fractions isolated from roasted malt on the enzyme activities of NADPH–cytochrome c-reductase and glutathione-S-transferase in Caco-2 cells. J. Agric. Food Chem. 50, 602–606.

Foster, C., Schwieger, J., Narzib, L., Back, W., Uchida, M., Ono, M., Yanagi, K., 1999. Monatsschr. Brauwiss 5/6, 86–93.

Gorinstein, S., Caspi, A., Zemser, M., Trakhtenberg, S., 2000. Comparative contents of some phenolics in beer, red and white wines. Nutr. Res. 20, 131–139.

Gorjanović, S.Z., Novaković, M.M., Potkonjak, N.S., Leskošek-Cukalović, I., Sužnjević, D.Z., 2010. Application of a novel antioxidative assay in beer analysis and brewing process monitoring. J. Agric. Food Chem. 58, 744–751.

Granato, D., Branco, G.F., Faria, J.A.F., Cruzb, A.G., 2011. Characterization of Brazilian lager and brown ale beers based on color, phenolic compounds, and antioxidant activity using chemometrics. J. Sci. Food Agric. 91, 563–571.

Guido, L.F., 2005. How do sulfites help to control beer ageing. Cerevisia 30, 132–138.

Guido, L.F., Rodrigues, P.G., Rodrigues, J.A., Gonçalves, C.R., Barros, A.A., 2004. The impact of the physiological condition of the pitching yeast on beer flavour stability: an industrial approach. Food Chem. 87, 187–193.

Hansen, J., Kielland-Brandt, M., 1996. Inactivation of *MET10* in brewer's yeast specifically increases SO_2 formation during beer production. Nature Biotechnol. 14, 1587–1589.

Illet, D.R., 1995. Aspects of the analysis, role, and fate of sulphur dioxide in beer – A review. Tech. Q. Master Brew. Assoc. Am. 32, 213–221.

Korch, C., Moutain, H.A., Gyllang, H., Winge, M., Brehmer, P., 1991. A mechanism for sulphite production in beer and how to increase sulphite levels by recombinant genetics. In: Brewing Convention, European (Ed.), The proceedings of the 23rd Congress of the European Brewery Convention. IRL Press, Eynsham, pp. 201–208.

Kuntcheva, M.J., Obretenov, T.D., 1996. Isolation and characterization of melanoidins from beer. Z. Lebensm.Unters. F. A. 202, 238–243.

Lapcík, O., Hill, M., Hampl, R., Wähälä, K., Adlercreutz, H., 1998. Identification of isoflavonoids in beer. Steroids 63, 14–20.

Lugasi, A., 2003. Polyphenol content and antioxidant properties of beer. Acta. Aliment. 32, 181–192.

Martinez-Periñan, E., Hernández-Artiga, M.P., Palacios-Santander, J.M., ElKaoutita, M., Naranjo-Rodriguez, I., Bellido-Milla, D., 2011. Estimation of beer stability by sulphur dioxide and polyphenol determination. evaluation of a Laccase–Sonogel–Carbon biosensor. Food Chem. 127, 234–239.

Martins, S.I.F.S., Jongen, W.M.F., van Boekel, M.A.J.S., 2001. A review of Maillard reaction in food and implications to kinetic modelling. Trends Food Sci. Technol. 11, 364–373.

McMurrough, I., Madigan, D., Kelly, R.J., Smyth, M.R., 1996. The role of flavanoid polyphenols in beer stability. J. Am. Soc. Brew. Chem. 54, 141–148.

Morales, F.J., 2009. Non-specific hydroxyl radical scavenging properties of melanoidins from beer. In: Preedy, V.R. (Ed.), Beer in health and disease prevention. Elsevier Inc., San Diego, pp. 765–774.

Morales, F.J., Fernández-Fraguas, C., Jiménez-Pérez, S., 2005. Iron-binding ability of melanoidins from food and model systems. Food Chem. 90, 821–827.

Morales, F.J., Jiménez-Pérez, S., 2004. Peroxyl radical scavenging activity of melanoidins in aqueous systems. Eur. Food Res. Technol. 218, 515–520.

Nardini, M., Ghiselli, A., 2004. Determination of free and bound phenolic acids in beer. Food Chem. 84, 137–143.

Narziss, L., Miedaner, H., Graf, H., Eichhorn, P., Lustig, S., 1993. Technological approach to improve flavour stability. Tech. Q. Master Brew. Assoc. Am. 30, 48–53.

Nøddekær, T.V., Andersen, M.L., 2007. Effects of Maillard and caramelization products on oxidative reactions in lager beer. J. Am. Soc. Brew. Chem. 65, 15–20.

Papetti, A., Daglia, M., Aceti, C., Quaglia, M., Gregotti, C., Gazzani, G., 2006. Isolation of an in vitro and ex vivo antiradical melanoidin from roasted barley. J. Agric. Food Chem. 54, 1209–1216.

Piazzon, A., Forte, M., Nardini, M., 2010. Characterization of phenolic content and antioxidant activity of different beer types. J. Agric. Food Chem. 58, 10677–10683.

Pisoschi, A.M., Negulescu, G.P., 2011. Methods for total antioxidant activity determination: a review. Biochem. Anal. Biochem. 1, 1–10.

Rivero, D., Pérez-Magariño, S., González-Sanjosé, M.L., Valls-Belles, V., Codoñer, P., Muñiz, P., 2005. Inhibition of induced DNA oxidative damage by beers: Correlation with the content of polyphenols and melanoidins. J. Agric. Food Chem. 53, 3637–3642.

Saison, D., De Schutter, D.P., Overlaet-Michiels, W., Delvaux, F., Delvaux, F.R., 2009. Effect of fermentation conditions on staling indicators in beer. J. Am. Soc. Brew. Chem. 67, 222–228.

Tafulo, P.A.R., Queirós, R.B., Delerue-Matos, C.M., Sales, M.G.F., 2010. Control and comparison of the antioxidant capacity of beers. Food Res. Int. 43, 1702–1709.

Vinson, J.A., Mandarano, M., Hirst, M., Trevithick, J.R., Bose, P., 2003. Phenol antioxidant quantity and quality in foods: beers and the effect of two types of beer on an animal model of atherosclerosis. J. Agric. Food Chem. 51, 5528–5533.

Walters, M.T., Heasman, A.P., Hughes, P.S., 1997. Comparison of (+)-catechin and ferulic acid as natural antioxidants and their impact on beer flavor stability. 1. Forced-aging. J. Am. Soc. Brew. Chem. 55, 83–89.

Woffenden, H.M., Ames, J.M., Chandra, S., 2001. Relationships between antioxidant activity, color, and flavor compounds of crystal malt extracts. J. Agric. Food Chem. 49, 5524–5530.

Woffenden, H.M., Ames, J.A., Chandra, S., Anese, M., Nicoli, M.C., 2002. Effect of kilning on the antioxidant and pro-oxidant activities on pale malt. J. Agric. Food Chem. 50, 4925–4933.

Zhao, H., Chen, W., Lu, J., Zhao, M., 2010. Phenolic profiles and antioxidant activities of commercial beers. Food Chem. 119, 1150–1158.

Zhao, H., Li, H., Sun, G., Yang, B., Zhao, M., 2013. Assessment of endogenous antioxidative compounds and antioxidant activities of lager beers. J. Sci. Food Agric. 93, 910–917.

CHAPTER

3

Antioxidants in Coffee

Draženka Komes, Arijana Bušić

Department of Food Engineering, Faculty of Food Technology and Biotechnology, University of Zagreb, Zagreb, Croatia

CHAPTER POINTS

- Coffee beans are a rich source of biologically active compounds such as caffeine, chlorogenic acids, nicotinic acid, trigonelline, cafestol, and kahweol, which have significant potential as antioxidants.
- Antioxidant capacity of coffee brew is affected by the bioactive composition of green beans and post-harvest processing conditions (drying, storage, roasting, and grinding).
- Roasting process results in development of Maillard reaction products, induced by high temperatures and water loss, which are found to be powerful antioxidants.
- Overall antioxidant capacity of coffee brew can be attributed to the presence of remaining native and newly formed antioxidants during the roasting process.
- Progressive reduction of antioxidant capacity of coffee brew is related to the roasting degree of coffee beans, showing the highest capacity in medium-roasted coffee due to the balance between the degradation of phenolic compounds and generation of Maillard reaction products (melanoidins) during this process.
- Heterocyclic compounds present in coffee also contribute to antioxidant capacity of the brew, probably through the synergistic effect.
- Coffee by-products, such as coffee silverskin and spent coffee, have shown good potential as functional food ingredients.

INTRODUCTION

Coffee is a highly popular drink that is traditionally used to complement meals, as well as for hedonistic and psychostimulant purposes. It is estimated that throughout the world, 80% of the adult population consume coffee beverages (Sridevi *et al.*, 2011). Although coffee has been widely consumed for many centuries because of desirable aroma, newer epidemiological studies have shown numerous positive effects of regular coffee intake on human health due to its biochemical composition. Coffee is a rich source of antioxidants of the hidroxycinnamic acids family (caffeic, chlorogenic, coumaric, ferulic, and sinapic) and other biologically active compounds with significant antioxidant potential, such as caffeine, nicotinic acid, trigonelline, cafestol, and kahweol, as presented in Table 3.1 (Minamisawa *et al.*, 2004). In some countries, coffee is the source of two-thirds of the population's antioxidant nutrient intake (Sridevi *et al.*, 2011).

During processing, the antioxidant profile of coffee changes due to the degradation of native antioxidants and the formation of new ones. Thus, the antioxidant capacity of coffee is related to the presence of both natural constituents and compounds formed during processing (roasting) (Vignoli *et al.*, 2011). Low water activity and high temperatures favor the development of Maillard reactions (MR), and the formation of MR products between proteins and carbohydrates (Borrelli *et al.*, 2004).

Additionally, coffee is a source of minerals (such as magnesium) and low-molecular-weight volatile compounds, particularly heterocyclic compounds, obtained from the Maillard reaction, which have recently begun to receive much attention as antioxidants. More than 300 heterocyclic compounds, including pyrroles, oxazoles, furans, thiazoles, thiophenes, imidazoles, and pyrazines, were identified and reported in brewed coffee. Among them, pyrrols

Processing and Impact on Antioxidants in Beverages
http://dx.doi.org/10.1016/B978-0-12-404738-9.00003-9

and furans showed the strongest antioxidant activity, which was almost equal to that of α-tocopherol (Yanagimoto *et al.*, 2002).

It is interesting to note that by-products of coffee processing, such as coffee silverskin and spent coffee extract, have been recognized as potential new functional ingredients because of their content of soluble dietary fibers and marked antioxidant capacity (Borrelli *et al.*, 2004; Ramalakshmi *et al.*, 2009).

Due to the confirmed significance of antioxidants in human health, numerous methods for measuring antioxidant capacity have been developed, but so far a standardized procedure, applicable for all systems, has not been determined. The antioxidant potential of coffee has been mainly evaluated using several methods, such as the ferric reducing antioxidant power (FRAP) assay, the scavenging of ABTS [2,2-azino-bis(3-ethylbenzothiazoline-6-sulphonic acid)diammonium salt] and DPPH (2,2-diphenyl-1-picrylhydrazyl) radicals, and determination of total phenol content (Sánchez-González *et al.*, 2005).

PHENOLIC COMPOUNDS

Coffee is considered to be one of the richest sources of phenolics in the Western diet, contributing up to 350 mg per 7-oz cup (200 ml) (Higdon and Frei, 2006). The content of polyphenolic antioxidants in green coffee beans is influenced by the species and their origin, while in coffee brews it depends on the brewing procedure (Sánchez-González *et al.*, 2005). The most abundant class of phenolic compounds in coffee are chlorogenic acids (CGAs) which may account for up to 12% of the dry matter of green coffee beans (Ky *et al.*, 2001). In coffee trees, all biosynthetic pathways of CGA formation occur in the leaves (Colonna, 1986). CGAs are then exported to beans where they reach maximum accumulation when the fruits are green. During fruit maturation, the content of CGAs decreases. Thus, from their biosynthesis in leaves to their accumulation in beans at complete maturity, certain reactions occur that lead to quantitative variation in their content.

TABLE 3.1 Content of Major Antioxidants (% on dmb) Naturally Occuring in Green Beans of Arabica and Robusta Coffees and Additionally Formed during Roasting

	Green Coffee (%)		
Antioxidants	*C. arabica*	*C. canephora*	**Reference**
Chlorogenic acids	4.0–8.4	7.0–14.4	Farah and Donangelo (2006), Farah *et al.* (2005), Ky *et al.* (2001)
Caffeine	0.76–1.82	1.51–3.33	Ky *et al.* (2001), Anthony *et al.* (1993)
Trigonelline	0.88–2.76	0.75–3.42	Ky *et al.* (2001), Mazzafera (1991)
Nicotinic acid	3×10^{-6}	$2x10^{-6}$	Casal *et al.* (2000)
α-Tocopherol	2.2×10^{-4}–1.9×10^{-3}	3.4×10^{-4}–7.3×10^{-4}	Gonzales *et al.* (2001)
β-Tocopherol	4.3×10^{-3}–9.8×10^{-3}	1.1×10^{-3}–1.8×10^{-3}	
γ-Tocopherol	1×10^{-4}–5.3×10^{-4}	2.5×10^{-4}–6.2×10^{-4}	
Cafestol	0.27–1.1	0.15–0.37	Sridevi *et al.* (2011), Kitzberger *et al.* (2010)
Kahweol	0.11–0.67	<0.1	Sridevi *et al.* (2011), Kitzberger *et al.* (2010)
	Additionally Formed During Coffee Roasting (%)		
	C. arabica	*C. canephora*	
Chlorogenic acid lactones	0.40	0.42	Farah *et al.* (2005)
Melanoidins	25	25	Delgado-Andrade and Morales (2005)

An overview of the main antioxidants in *Coffea arabica* (Arabica coffee) and *Coffea canephora* (Robusta coffee) is shown. The presentation of results is adapted from the original sources for to consistency, taking into account the moisture content of 10–12% in primary processed green coffee, according to Jöet *et al.*, 2010. The values have been collected from various literature sources and expressed as percentage on dry matter basis (% on dmb).

CGAs are widely distributed in plants, and their content in green coffee is among the highest, with 45 different CGA compounds already identified (Clifford *et al.*, 2006), ranging from 4% to 14% (Farah and Donangelo, 2006). CGAs may be subdivided according to the nature and number of cinnamic substituents and their esterification position in the cyclohexane ring of quinic acid. The major class of CGAs in coffee is the caffeoylquinic acids (CQAs), with 5-caffeoylquinic acid (5-CQA) as the most represented isomer (Figure 3.1) (Farah *et al.*, 2005; Monteiro and Farah, 2012), comprising about 56–62% of total CGA content (Farah and Donangelo, 2006). The next most abundant CGA groups are the dicaffeoylquinic acids (diCQAs), accounting for about 15–20% of total CGAs in green coffee beans, and feruloylquinic acids (FQAs), accounting for 5–13% (Farah and Donangelo, 2006). Among others, *p*-coumaroylquinic acids (*p*-CoQAs) and caffeoylferuloylquinic acids (CFQAs) are represented. Recently, minor classes, such as diferuloylquinic acids (diFQAs), di-*p*-coumaroylquinic acids (di-*p*-CoQAs), dimethoxycinnamoylquinic acids, and others, which together constitute less than 1% of total CGA content, were identified (Clifford *et al.*, 2006). CGAs play an important role in the formation of roasted coffee flavor and have a marked influence in determining coffee brew quality (Farah *et al.*, 2006). Moreover, several beneficial health effects have been attributed to CGAs that may be largely explained by their potent antioxidant activity (Moreira *et al.*, 2005). The contribution of total CGAs to total antioxidant capacity of non-roasted coffee were estimated to be 69% and 56% for oxygen radical absorbance capacity (ORAC) and Trolox equivalent antioxidant capacity (TEAC) assays, respectively. Although the final cup content of coffee phenolics depends on starting material, roasting degree, and brewing method, heavy coffee consumers can approach dietary exposure levels of 1 g of CGAs and their phenolic derivatives per day (Clifford, 2000).

Due to their sensitivity to high temperatures during roasting, (200–250°C, 1.5–6 min) both phenolic and non-phenolic coffee constituents are partially degraded and/or bound to polymer structures, depending on

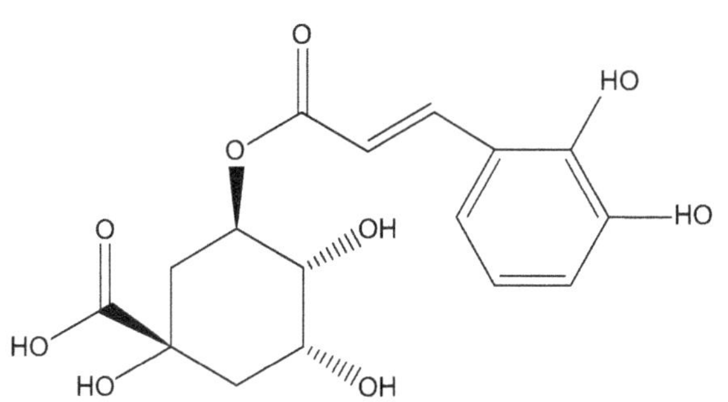

FIGURE 3.1 Chemical structure of 5-*O*-caffeoylquinic acid. *From Higdon and Frei (2006).*

roasting conditions. Therefore, the degradation of low-molecular-weight phenolic compounds and the formation of Maillard reaction products (melanoidins) are two major events during roasting of coffee beans (Farah *et al.*, 2005).

Non-roasted green coffee beans possess relatively higher antioxidant capacity than the corresponding roasted samples. The content of total CGA derivatives in light and medium roasted coffees ranges between 28 and 45 mg/g and 18 to 31 mg/g, respectively, representing 40–50% of the content in the original green coffee beans, making light- to medium-roasted coffee a significant source of dietary CGAs (Daglia *et al.*, 2000). Additionally, del Castillo *et al.* (2002) concluded that medium-roasted coffee exhibits the highest antioxidant capacity. By roasting at 205°C, the rate of total CGAs reduction in green coffee beans was 60.9% for light, 67.7% for medium, 88.8% for dark, and 96.5% for very dark roast of Arabica, and 59.7% for light, 76.4% for medium, 93.0% for dark, and 98.0% for very dark roast of Robusta (Fujioka and Shibamoto, 2008). High temperatures during the roasting process cause a breakage of the carbon–carbon bonds in CGA molecules, resulting in isomerization and degradation. Isomerization of CGAs takes place at the beginning of roasting, but it is possible that partial hydrolysis of diCQAs to monoester derivatives occurs in addition to isomerization (Leloup *et al.*, 1995). An initial increase in the total CGA content at 5 min of roasting could be a result of the loss of other compounds that are more sensitive to heat, causing a relative increase in the content of the remaining ones. Longer periods of roasting result in loss of total CGAs. Besides isomerization and degradation, other chemical transformations may occur, the dominant being dehydration of the quinic acid moiety and formation of a lactone ring (Scholz and Maier, 1990). The rate at which all these changes occur depends on both temperature and the amount of beans inside the roaster.

Recently, corresponding chlorogenic acid lactones (CGLs), which are formed during the roasting process, have received special attention due to their potential effects on brain function, independent of the pharmacological effects of caffeine (de Paulis *et al.*, 2004).

In addition to nine CGAs identified in green coffee samples, seven CGLs have been identified in roasted coffee. CGLs reached their maximum content approximately 7 min after the start of the roasting process, whereas the longer roasting time resulted in lower content of both CGAs and CGLs. The maximum content of total CGLs was 398 mg/100 g and 424 mg/100 g of coffee for *C. arabica* and *C. canephora*, respectively, which corresponded to 7.3% and 5.6% of the contents of total CGAs in green Arabica and Robusta samples, respectively. Most CGAs could not be detected after 11 min and only trace content of CGLs could be detected after 15 min of roasting. 3-Caffeoylquinic acid lactone (3–CQL) was the most abundant lactone in all samples, reaching its maximum amount at 7 min of roasting. The same pattern was observed for feruloylquinic acid lactone (FQL). The optimum degree of roasting to achieve a maximum content of lactones in coffee is light–medium roast, whereas darker roasts yield lower contents. Despite favorable structural configurations of CGAs, less than 10% of total CGAs and about 30% of possible precursors in green coffee were converted to lactones. Despite the higher content of 4-caffeoylquinic acid (4-CQA), as compared to 3-caffeoylquinic acid (3-CQA) in green beans, 3-CQL was the major lactone (Farah *et al.*, 2005).

MELANOIDINS

During roasting of coffee, the majority of phenolic compounds are destroyed or they may react with free radicals from the Maillard reaction and be incorporated in browning products (Djilas and Milic, 1994). Thus, the significantly higher oxygen scavenging properties of roasted coffee brews can be attributed to the Maillard reaction products (MRPs), especially melanoidins formed during the thermal treatment, although the remaining caffeic and chlorogenic acids and/or their degradation products may exhibit antioxidant capacity as well (Kroyer *et al.*, 1989).

Melanoidins are one of the major components of coffee brews, accounting for up to 25% of dry matter. They are responsible for the strong antioxidant properties and metal-chelating ability showed by coffee brews. Some authors have confirmed that this effect is due to their ability to break the radical chain by donation of hydrogen, effectivity as metal-chelating agents, capacity to reduce hydroperoxide to nonradical products, or to scavenge hydroxyl radicals (Delgado-Andrade and Morales, 2005).

Although there is not yet sufficient knowledge about their chemical structure, the mechanism of melanoidin formation is assumed to consist of an early stage and an advanced stage. In the early stage, the aldehydic moiety of sugars and amino groups react to produce a component which has the partial structure of $-CHOH-CO-CH_2NHR$, while in the advanced stage, various aldehydes, ketones, and furfurals are formed from the early-stage products, and these react with amino compounds to form high-molecular-weight melanoidins (Takenaka *et al.*, 2005).

High-molecular-weight polymerized melanoidins in coffee brews contribute to the antioxidant capacity less than the presence of low-molecular-weight MRPs. The loss of CGAs and possibly other native antioxidants in the coffee bean, combined with the generation of MRPs, attributed to high temperatures used in roasting, ultimately influence the composition and thus the antioxidant capacity of the final product (Delgado-Andrade and Morales, 2005).

Nicoli *et al.* (1997) evaluated the antioxidant capacity of coffee brews in relation to the roasting degree and found that all roasted coffee brews showed higher oxygen scavenging capacity than the green coffee brew. Progressive reduction in the antioxidant capacity of coffee brews was observed with the degree of roasting, showing the highest capacity in medium-roasted coffee. This was also in agreement with the results of del Castillo *et al.* (2002) and can be attributed to the balance between the degradation of phenolic compounds and the generation of MRPs during this process. In contrast, Duarte *et al.* (2005) found a higher antioxidant capacity for brews made from light-roasted coffee.

CAFFEINE

Caffeine (Figure 3.2) is an alkaloid present in varying amounts in brewed coffees (Higdon and Frei, 2006), mostly known for its specific effects on increased mental alertness, faster information processing, wakefulness, restlessness, reduction of fatigue, and delay in the need for sleep. It is also available in tablet form and is broadly used in medication and beverages (Harland, 2000). The content of caffeine in coffee brews depends on the type of bean, brewing strength, and roasting process. Generally, the same type of coffee bean brewed in the same manner may contain caffeine contents ranging from 130 to 282 mg/240 ml (Higdon and Frei, 2006), and more specifically, Arabica-brewed coffee may contain between 36 and 112 mg caffeine/100 ml, while in Robusta the content of caffeine may vary from 56 to 203 mg/100 ml (Oestreich-Janzen, 2010).

Regarding the green beans, the highest and the lowest caffeine contents are found in the samples of highest and lowest quality, respectively. The same tendency is maintained after roasting. Caffeine content of green coffee beans is within the range of 1–4% (on dry basis)

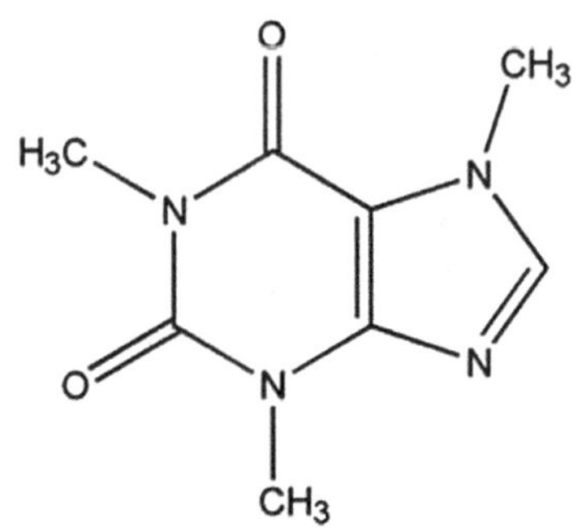

FIGURE 3.2 Chemical structure of caffeine (1, 3, 7-trimethylxanthine). *From Higdon and Frei (2006).*

(Mazzafera and Silvarolla, 2010), while the roasting causes a reduction in caffeine content of approximately 30%. Caffeine presents a distinct behavior for 200°C and 300°C roasting conditions, showing a sharper decrease for roasting at 300°C (Franca *et al.*, 2009). Since the solubility of this compound in water increases with temperature, the caffeine loss may be attributed to a drag by water vapor released during roasting. Previous studies have reported that caffeine was detected in the exhaust gas from roasting (Dutra *et al.*, 2001).

Caffeine and its catabolic products theobromine and xanthine exhibit both antioxidant and pro-oxidant properties. Therefore, caffeine and its metabolites may also contribute to the overall antioxidant and chemopreventive properties of caffeine-bearing beverages (Azam *et al.*, 2003). There is conflicting evidence regarding the contribution of caffeine to the antioxidant capacity of coffee brews. While Brezová *et al.* (2009) found a high antioxidant capacity of caffeic acid but not of caffeine, Vignoli *et al.* (2011) reported high correlations between antioxidant capacity of coffee brews and caffeine. Similarly, López-Galilea *et al.* (2007) reported a significant correlation between caffeine content and antioxidant capacity obtained by DPPH ($r = 0.83$) and redox potential ($r = 0.84$), depending on the brewing technique. In addition, Santini *et al.* (2011) also observed a direct relation between caffeine content of Arabica coffee brews and corresponding antioxidant capacity.

TRIGONELLINE

Trigonelline (1-methylpyridinium-3-carboxylate) is one of the major components of coffee beans, representing ~2% of the dry weight (Mazzafera, 1991). It is also one of the candidate compounds responsible for the bitter taste of the coffee brew. Trigonelline is thermally unstable and during roasting it is converted to nicotinic acid and other nitrogenous compounds that contribute to flavor (Mazzafera, 1991). During germination of coffee seeds, limited content of trigonelline is catabolized and used as a substrate for the synthesis of nitrogen-containing compounds (Shimizu and Mazzafera, 2000).

Trigonelline, demethylated to nicotinic acid, supplies 1–3 mg of nicotinic acid/240 ml of brewed coffee (Higdon and Frei, 2006; Stadler *et al.*, 2002). Trigonelline in coffee brews decreases with increasing roasting degree (Table 3.2). Casal *et al.* (2000) observed that trigonelline degradation increased after treatment at 200°C and only 5% of this compound remained in Arabica and 15% in Robusta coffee at 240°C. In contrast, nicotinic acid content increased ≥500% in both brews. The increase of nicotinic acid suggests that its formation occurs during roasting, particularly in the full city and French type, since the quantity of trigonelline in green coffee bean is very low. Proportionally to higher roasting temperature, the conversion to nicotinic acid is greater (Stadler *et al.*, 2002). Decomposition of trigonelline can be characterized by the content ratio of trigonelline to nicotinic acid for any roasted sample, because the nicotinic acid found in roasted coffee was essentially derived from the trigonelline in the green coffee (Minamisawa *et al.*, 2004).

TOCOPHEROLS

Tocopherols are a group of four lipid-soluble amphipathic molecules (α-, β-, γ-, δ-) that are exclusively synthesized by photosynthetic organisms. Collectively, they are an essential component of vitamin E (Gilliland *et al.*, 2006), which is known to be the most effective natural lipid-soluble antioxidant, protecting cell membranes from peroxyl radicals and mutagenic nitrogen oxide species (Gliszczyńka-Świgło and Sikorska, 2004). The presence of tocopherols in coffee oil was reported for the first time by Folstar *et al.* (1977).

Among tocopherols, two main tocopherols (α- and β-) were identified in Arabica and Robusta coffee beans, both green and roasted (Alves *et al.*, 2010). Ogawa *et al.* (1989) reported vitamin E mean content in coffee brew of $7 \pm 3\,\mu g/100\,ml$. The content of α-tocopherol in roasted coffees ranged between 7.55 μg/g and 33.54 μg/g, whereas in green coffees it was between 2.02 μg/g and 16.76 μg/g. In the case of β- and γ-tocopherols, remarkable differences between green and roasted samples were observed, with their contents being higher in roasted coffees. Thus, the mean values of β-tocopherol

TABLE 3.2 Content of Trigonelline and Nicotinic Acid (mg/kg on dmb) in Coffee Samples Depending on the Roasting Temperature

	Trigonelline		Nicotinic Acid	
	Arabica	*Robusta*	*Arabica*	*Robusta*
Green	8.91	6.32	0.03	0.02
140°C	8.47	6.37	0.06	0.06
160°C	8.31	5.86	0.08	0.05
180°C	8.29	5.78	0.06	0.04
200°C	7.80	5.43	0.07	0.06
220°C	5.57	4.20	0.13	0.06
240°C	0.49	0.97	0.17	0.13

The table shows a comparison of trigonelline and nicotinic acid content in Arabica and Robusta coffee samples, depending on different roasting temperature ranges (140–240°C) at constant time (15 min). Arabica coffee samples (*Coffea arabica*) originated from Brazil, while the Robusta coffee (*Coffea canephora* var. Robusta) originated from Ivory Coast.

The results are displayed as average values ± SD, expressed as mg/kg on dry matter basis.

The results are taken from Casal et al. *(2000).*

were evaluated to be 47.12 μg/g and 106.60 μg/g for green and roasted coffees, respectively. In the case of γ-tocopherol, its content varied between 2.63 μg/g in green samples and 70.99 μg/g in roasted coffee beans (Gonzales *et al.*, 2001). The increase in tocopherol content during the roasting process has been already described for some oilseeds (Yen, 1990), and was explained as the result of liberation of the combined tocopherols during the roasting process.

Recently, the usefulness of tocopherol content and profile as a marker of coffee adulteration with corn was also described by Jham *et al.* (2007).

HETEROCYCLIC COMPOUNDS PRODUCED BY MAILLARD REACTION

Characteristic volatile heterocyclic compounds found in brewed coffee extracts (pyrroles, furans, thiophenes, and thiazoles) exhibit certain levels of antioxidant capacity (Fuster *et al.*, 2000), although this activity is not as strong as that of the synthetic antioxidant butylated hydroxytoluene (BHT). Among them, pyrrols and furans showed the strongest activity. Conversely, most of the thiazoles and pyrazines, except for 2-methylpyrazine, did not produce oxidized products, which indicated that thiazoles and pyrazines are unable to scavenge hydroxyl radicals. However, because a tremendous number of these heterocyclic compounds are present in coffee, their combined activity might be comparable to those of known antioxidants (Yanagimoto *et al.*, 2002).

DITERPENES CAFESTOL AND KAHWEOL

Cafestol and kahweol are two diterpenes exclusively found in coffee (Silva *et al.*, 2012). Their structure is nearly identical, differing only in the degree of saturation of one conjugated bond to the furan ring (Figure 3.3).

Only a small fraction of cafestol and kahweol, about 0.7–3.5% (Kölling-Speer *et al.*, 1999), is present in coffee beans as free diterpene alcohols, while the majority occurs esterified with fatty acids in the form of diterpene esters. Kurzrock and Speer (2001) identified up to 14 different fatty acid esters linked to cafestol and kahweol, the most common being esters with palmitic acid (36–49%), followed by linoleic acid (22–30%), and several others present only in minor amounts, like oleic, stearic, and eicosanoic acid (Silva *et al.*, 2012).

Diterpenes exhibit good stability at high temperatures, but they can form dehydro- derivatives (dehydrocafestol and dehydrokahweol) in small amounts when roasting temperatures are raised. The presence of other decomposition products, such as kahweal, cafestal, isokahweol, and dehydroisokahweol, as well as alterations in the kahweol and cafestol esters, in a more severe roasting process is also reported (Speer and Kölling-Speer, 2006). Urgert *et al.* (1995) assessed the behavior of kahweol and cafestol in Arabica coffee subjected to a high degree of roasting and concluded that roasting did not reduce the content of these compounds. The dark-roasted samples presented lower cafestol content than the light- and medium-roasted ones, showing that the samples were affected differently by roasting. Campanha *et al.* (2010), working with three degrees of roasting for different blends of coffee, reported that increase of the roasting degree did not lead to a reduction in diterpene content (Table 3.3). Dias (2009) evaluated degradation of cafestol and kahweol (2–10 min, maximum temperature 230°C) and observed that the content of cafestol and kahweol remained stable, even though dehydro- derivatives occurred with more intense processes of roasting, due to the increased concentration of lipids during the roasting process. The influence of roasting on cafestol and kahweol indicated the significant impact of roasting methods on diterpene profiles in coffee beans. In Arabica coffee, the highest content of free forms of cafestol and kahweol were observed in light roast,

TABLE 3.3 Cafestol and Kahweol Content in Arabica (*C. arabica*) and Robusta (*C. canephora*) Coffees (mg/100 g) Subjected to Different Roasting Degrees (Light, Medium, Dark)

Roasting Degree	Arabica		Robusta	
	Cafestol	*Kahweol*	*Cafestol*	*Kahweol*
Light	463 ± 4	829 ± 16	163 ± 24	0
Medium	398 ± 46	744 ± 8	250 ± 8	0
Dark	420 ± 4	800 ± 28	242 ± 6	0

Both samples, originated from Brazil (Minas Gerais state). The results are presented as average values ± SD, expressed as mg/100 g.
The results are taken from Campanha et al. *(2010).*

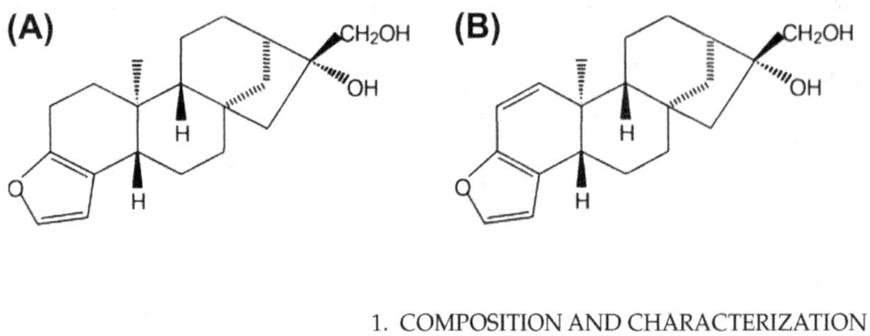

FIGURE 3.3 Chemical structures of cafestol (A) and kahweol (B). *From Kitzberger* et al. *(2010).*

followed by medium and high roasts. As the roasting temperature increased there was a significant decrease in both cafestol and kahweol profiles. Accordingly, in full city roast there was ~56% reduction in cafestol and 61% reduction in kahweol contents. A similar trend was noticed in Robusta beans, though the diterpene profiles were less compared to Arabica. The highest percentage of cafestol (363.3±8.0mg/100g) and kahweol (313±4.93mg/100mg), respectively, were found in light roast. The difference in cafestol content among different roasting degrees was insignificant. In Robusta full city roast, the cafestol and kahweol profiles showed ~44% and 10% increases, respectively, compared to full city roast of Arabica beans (Sridevi *et al.*, 2011).

INFLUENCE OF EXTRACTION PROCESS ON ANTIOXIDANT CAPACITY OF COFFEE RESIDUES

During coffee processing and preparation, some residues are generated. These coffee residues are frequently used in the industry as animal feed and fertilizers, but during the last few years other more environmentally friendy uses, such as biofuel production, have been proposed (Silva *et al.*, 1998). Furthermore, during the last few years, an increasing number of studies have shown the presence of phytochemicals related with health benefits in coffee residues (Borrelli *et al.*, 2004; Ramalakshmi *et al.*, 2009). In the case of coffee silverskin, some authors have demonstrated the high content of dietary fiber, as well as antioxidant capacity (Borrelli *et al.*, 2004; Napolitano *et al.*, 2007).

Also, in the residues obtained during the production of soluble coffee, antioxidant properties, attributed to phenolic and non-phenolic compounds, have been found (Ramalakshmi *et al.*, 2009). The results indicated that the filter coffeemaker is the most efficient extraction system for obtaining spent coffee extracts with high antioxidant capacity.

Finally, the application of defatting and lyophilization processes has the ability to increase the antioxidant capacity of spent coffee extracts, providing added value, and resulting in their preservation and development into spent coffee extract powders that can be used as an ingredient or additive in the food industry, with potential preservation and functional properties.

References

Alves, R.C., Soares, C., Casal, S., Fernandes, J.D., Beatriz, M., Oliveira, P.P., 2010. Acrylamide in espresso coffee: Influence of species, roast degree and brew length. Food Chem. 119, 929–934.

Anthony, F., Clifford, M.N., Noirot, M., 1993. Biochemical diversity in the genus *Coffea* L.: chlorogenic acids, caffeine and mozabioside contents. Genet. Resour. Crop Evol. 40, 61–70.

Azam, S., Hadi, N., Khan, N.U., Hadi, S.M., 2003. Antioxidant and prooxidant properties of caffeine, theobromine and xanthine. Med. Sci. Monit. 9, 325–330.

Borrelli, R.C., Esposito, F., Napolitano, A., Ritieni, A., Fogliano, V., 2004. Characterization of a new potential functional ingredient: coffee silverskin. J. Agric. Food Chem. 52, 1338–1343.

Brezová, V., Šlebodova, A., Staško, A., 2009. Coffee as a source of antioxidants: An EPR study. Food Chem. 114, 859–868.

Campanha, F.G., Dias, R.C.E., de Toledo Benassi, M., 2010. Discrimination of coffee species using kahweol and cafestol: effects of roasting and of defects. Coffee Sci. 5, 87–96.

Casal, S., Beatriz Oliveira, M., Ferreira, M.A., 2000. HPLC/diode array applied to the thermal degradation of trigonelline, nicotinic acid and caffeine in coffee. Food Chem. 68, 481–485.

Clifford, M.N., 2000. Chlorogenic acid and other cinnamates: nature, occurrence and dietary burden, absorption and metabolism. J. Sci. Food Agric. 80, 1033–1043.

Clifford, M.N., Knight, S., Sururu, B., Kuhnert, N., 2006. Characterization by LC-MSn of four new classes of chlorogenic acids in green coffee beans: dimetoxycinnamoyilquinic acids, diferuloylquinic acids, caffeoyl-dimethoxycinnamoylquinic acids and feruloyl-dimethoxycinnamoylquinic acids. J. Agric. Food Chem. 54, 1957–1969.

Colonna, J.P., 1986. Biosynthèse etre nouvellement de l'acide chlorogénique et des depsides voisins dans le genre. Café Cacao Thé. 30, 83–92, 247–258.

Daglia, M., Papetti, A., Gregotti, C., Berte, F., Gazzani, G., 2000. *In vitro* and *ex vivo* antihydroxyl radical activity of green and roasted coffee. J. Agric. Food Chem. 48, 1449–1454.

De Paulis, T., Commers, P., Farah, A., Zhao, J., McDonald, M.P., Galici, R., Martin, P.R., 2004. 4-Caffeoyl-1,5-quinide in roasted coffee inhibits [^{3}H]naloxone binding and reverses antinociceptive effects of morphine in mice. Psychopharmacology 176, 146–153.

Del Castillo, M.D., Ames, J.M., Gordon, M.H., 2002. Effect of roasting on the antioxidant activity of coffee brews. J. Agric. Food Chem. 50, 3698–3703.

Delgado-Andrade, C., Morales, F.J., 2005. Unraveling the contribution of melanoidins to the antioxidant activity of coffee brews. J. Agric. Food Chem. 53, 1403–1407.

Dias, R.C.E., 2009. Diterpenos em café: desenvolvimento de metodologia para quantificação, estudo da estabilidade com a torra e avaliação em cafés torrados, tecidos e frutos. Thesis (Doutorado em Ciência de Alimentos)—Universidade Estadual de Londrina, Londrina.

Djilas, S.M., Milic, B.L.J., 1994. Naturally occurring phenolic compounds as inhibitors of free radical formation in the Maillard reaction. In: Labuza, T.P., Reineccius, G.A., Monnier, V.M., Bayens, J.W. (Eds.), Maillard reaction in Chemistry, Food and Health. Royal Society of Chemistry Publisher, Cambridge, pp. 75–81.

Duarte, S.M.S., Abreu, C.M.P., Menezes, H.C., Santos, M.H., Gouvêa, C.M.C.P., 2005. Effect of processing and roasting on the antioxidant activity of coffee brews. Ciência Tecnol. Alime. 25, 387–393.

Dutra, E.R., Oliveira, L.S., Franca, A.S., Ferraz, V.P., Afonso, R.J.C.F., 2001. A preliminary study on the feasibility of using the composition of coffee roasting exhaust gas for the determination of the degree of roast. J. Food Eng. 47, 241–246.

Farah, A., de Paulis, T., Trugo, C.L., Martin, P.R., 2005. Effect of roasting on chlorogenic acid lactones in coffee. J. Agric. Food Chem. 53, 1505–1513.

Farah, A., Donangelo, C.M., 2006. Phenolic compounds in coffee. Braz. J. Plant Physiol. 18, 23–36.

Farah, A., Monteiro, M.C., Calado, V., Franca, A.S., Trugo, L.C., 2006. Correlation between cup quality and chemical attributes of Brazilian coffee. Food Chem. 98, 373–380.

Folstar, P., van der Plas, H.C., Pilnik, W., de Heus, J.G., 1977. Tocopherols in the unsaponifiable matter of coffee bean oil. J. Agric. Food Chem. 25, 283–285.

Franca, A.S., Oliveira, L.S., Oliveira, R.C.S., Agnesti, P.C.M., Augusti, R., 2009. A preliminary evaluation of the effect of processing temperature on coffee roasting degree assessment. J. Food Eng. 92, 345–352.

Fujioka, K., Shibamoto, T., 2008. Chlorogenic acid and caffeine contents in various commercial brewed coffees. Food Chem. 106, 217–221.

Fuster, M.D., Mitchell, A.E., Ochi, H., Shibamoto, T., 2000. Antioxidative activities of heterocyclic compounds formed in brewed coffee. J. Agric. Food Chem. 48, 5600–5603.

Gilliland, L.U., Magallanes-Lundback, M., Hemming, C., Supplee, A., Koornneef, M., Bentsink, L., DellaPenna, D., 2006. Genetic basis for natural variation in seed vitamin E levels in Arabidopsis thaliana. Proc. Nat. Acad. Sci. USA 103, 18834–18841.

Gliszczyńska-Świgło, A., Sikorska, E., 2004. Simple reversed-phase liquid chromatography method for determination of tocopherols in edible plant oils. J. Chromatogr. A. 1048, 195–198.

Gonzales, A.G., Pablos, F., Martín, M.J., Léon-Camacho, M., Valdenebro, M.S., 2001. HPLC analysis of tocopherols and triglycerides in coffee and their use as authentication parameters. Food Chem. 73, 93–101.

Harland, B.F., 2000. Caffeine and nutrition. Nutrition 16, 522–526.

Higdon, J.V., Frei, B., 2006. Coffee and health: A review of recent human research. Crit. Rev. Food Sci. Nutr. 46, 101–123.

Jham, A., Winkler, J.K., Berhow, M.A., Vaughn, S.F., 2007. γ-Tocopherol as a marker of Brazilian coffee (*Coffea arabica*) aduleration by corn. J. Agric. Food Chem. 55, 5995–5999.

Joët, T., Laffargue, A., Descroix, F., Doulbeau, S., Bertrand, B., de Kochko, A., Dussert, S., 2010. Influence of environmental factors, wet processing and their interactions on the biochemical composition of green Arabica coffee beans. Food Chem. 118, 30–36.

Kölling-Speer, I., Strohschneider, S., Speer, K., 1999. Determination of free diterpenes in green and roasted coffees. J. High Resolut. Chrom. 22, 43–46.

Kitzberger, C.S.G., Scholz, M.B.S., Pereira, L.F.P., Vieira, L.G.E., Sera, T., Silva, J.B.G.D., Benassi, M.T., 2010. Analysis of diterpenes in green and roasted coffee of *Coffea arabica* cultivars growing in the same edapho-climatic conditions. In: Proceedings of 23nd International Conference on Coffee Science. ASIC, Bali 110–117.

Kroyer, G.T., Kretschmer, L., Washuttl, J., 1989. Antioxidant properties of tea and coffee extracts. In: Proceedings of the 5th European Conference on Food Chemistry. Versailles, pp. 433–437.

Kurzrock, T., Speer, K., 2001. Diterpenes and diterpene esters in coffee. Food Rev. Int. 17, 433–450.

Ky, C.L., Louarn, J., Dussert, S., Guyot, B., Hamon, S., Noirot, M., 2001. Caffeine, trigonelline, chlorogenic acids and sucrose diversity in wild *Coffea arabica* L. and *C. canephora* P. accessions. Food Chem. 75, 223–230.

Leloup, V., Louvrier, A., Liardon, R., 1995. Degradation mechanisms of chlorogenic acids during roasting. In: Proceedings of the 16th international scientific colloquium on coffee. ASIC, Kyoto, pp. 192–198.

López-Galilea, I., de Peña, M.P., Cid, C., 2007. Correlation of selected constituents with the total antioxidant capacity of coffee beverages: Influence of the brewing procedure. J. Agric. Food Chem. 55, 6110–6117.

Mazzafera, P., 1991. Trigonelline in coffee. Phytochemistry 30, 2309–2310.

Mazzafera, P., Silvarolla, M.B., 2010. Caffeine content variation in single green Arabica coffee seeds. Seed Sci. Res. 20, 163–167.

Minamisawa, M., Yoshida, S., Takai, N., 2004. Determination of biologically active substances in roasted coffees using a diode-array HPLC system. Anal. Sci. 20, 325–328.

Monteiro, M.C., Farah, A., 2012. Chlorogenic acids in Brazilian *Coffea arabica* cultivars from various consecutive crops. Food Chem. 134, 611–614.

Moreira, D.P., Monteiro, M.C., Ribeiro-Alves, M., Donangelo, C.M., Trugo, L.C., 2005. Contribution of chlorogenic acids to the iron-reducing activity of coffee beverages. J. Agric. Food Chem. 53, 1399–1402.

Napolitano, A., Fogliano, V., Tafuri, A., Ritieni, A., 2007. Natural occurrence of ochratoxin and antioxidant activities of green and roasted coffees and corresponding byproducts. J. Agric. Food Chem. 55, 10499–10504.

Nicoli, M.C., Anese, M., Manzocco, L., Lerici, C., 1997. Antioxidants properties of coffee brews in relation to the roasting degree. Lebensm. Wiss. Technol. 30, 292–297.

Oestreich-Janzen, S., 2010. Chemistry of coffee. In: Mander, L. (Ed.), Comprehensive natural products II. Chemistry and biology. Elsevier Ltd., Oxford, pp. 1085–1113.

Ogawa, M., Kuroda, K., Kato, C., 1989. Preparation of montmorillonite-organic intercalation compounds by solid–solid reactions. Chem. Lett. 18, 1659–1662.

Ramalakshmi, K., Jagan Mohan Rao, L.J.M., Takano-Ishikawa, Y., Goto, M., 2009. Bioactivities of low-grade green coffee and spent coffee in different in vitro model systems. Food Chem. 115, 79–85.

Sánchez-González, I., Jiménez-Escrig, A., Saura-Calixto, F., 2005. *In vitro* antioxidant activity of coffees brewed using different procedures (Italian, espresso and filter). Food Chem. 90, 133–139.

Santini, A., Ferracane, R., Mikusová, P., Eged, S., Srobárová, A., Meca, G., Mañes, J., Ritieni, A., 2011. Influence of different coffee drink preparations on ochratoxin A content and evaluation of the antioxidant activity and caffeine variations. Food Control 22, 1240–1245.

Scholz, B.M., Maier, H.G., 1990. Isomers of quinic acid and quinide in roasted coffee. Lebensm. Unters. Forsch. 190, 132–134.

Shimizu, M.M., Mazzafera, P., 2000. A role for trigonelline during imbibition and germination of coffee seeds. Plant Biol. 2, 605–611.

Silva, J.A., Borges, N., Santos, A., Alves, A., 2012. Method validation for cafestol and kahweol quantification in coffee brews by HPLC-DAD. Food Anal. Meth. 5, 1404–1410.

Silva, M.A., Nebra, S.A., Machado, M.J., Sanchez, C.G., 1998. The use of biomass residues in the Brazilian soluble coffee industry. Biomass Bioenerg. 14, 457–467.

Speer, K., Kölling-Speer, I., 2006. The lipid fraction of the coffee bean. Braz. J. Plant Physiol. 18, 201–216.

Sridevi, V., Giridhar, P., Ravishankar, G.A., 2011. Evaluation of roasting and brewing effect on antinutritional diterpenes—cafestol and kahweol in coffee. Glob. J. Med. Res. 11, 16–22.

Stadler, R.H., Varga, N., Milo, C., Schilter, B., Vera, F.A., Welti, D.H., 2002. Alkylpyridiniums. 2. Isolation and quantification in roasted and ground coffees. J. Agric. Food Chem. 50, 1200–1206.

Takenaka, M., Sato, N., Asakawa, H., Wen, X., Murata, M., Homma, S., 2005. Characterization of a metal chelating substance in coffee. Biosci. Biotechnol. Biochem. 69, 26–30.

Urgert, R., van der Weg, G., Kosmeijer-Schuil, T.G., van de Bovenkamp, P., Hovenier, R., Katan, M.B., 1995. Levels of the cholesterol-elevating diterpenes cafestol and kahweol in various coffee brews. J. Agric. Food Chem. 43, 2167–2172.

Vignoli, J.A., Bassoli, D.G., Benassi, M.T., 2011. Antioxidant activity, polyphenols, caffeine and melanoidins in soluble coffee: The influence of processing conditions and raw material. Food Chem. 124, 863–868.

Yanagimoto, K., Lee, K.G., Ochi, H., Shibamoto, T., 2002. Antioxidative activity of heterocyclic compounds found in coffee volatiles produced by Maillard reaction. J. Agric. Food Chem. 50, 5480–5484.

Yen, G.C., 1990. Influence of seed roasting process on the changes in composition and quality of sesame (*Sesame indicum*) oil. J. Sci. Food Agric. 50, 563–570.

CHAPTER

4

Antioxidant Capacity of Green Tea (*Camellia sinensis*)

Richard S. Bruno, Joshua A. Bomser*, Mario G. Ferruzzi†*

*Department of Human Sciences, The Ohio State University, Columbus, OH, USA,
†Department of Food Science, Purdue University, West Lafayette, IN, USA

CHAPTER POINTS

- Green tea (*Camellia sinensis*) is rich in polyphenolic catechins.
- Epidemiological evidence supports the consumption of green tea for lowering the risks of cardiovascular disease (CVD) and nonalcoholic fatty liver disease (NAFLD).
- Green tea catechins have antioxidant and anti-inflammatory activities that mitigate oxidative stress responses implicated in CVD and NAFLD.
- The antioxidant activities of green tea catechins are likely to be mediated through indirect mechanisms given the poor bioavailability of dietary catechins.
- Anti-inflammatory activities of green tea, mediated in part by its antioxidant activities, reduce nuclear factor kappa B (NFκB) activation and NFκB-dependent pro-inflammatory responses.
- Green tea, when consumed at reasonable dietary intake levels, appears to be safe in humans.

INTRODUCTION

The tea plant (*Camellia sinensis*) is grown in more than 30 countries, and is the most commonly consumed prepared beverage worldwide. Even in the United States where tea is not a native crop, total tea imports are estimated to exceed 281 million pounds with sales up to $8.2 billion each year (www.teausa.com). Of the major tea varieties, green tea is minimally processed, compared to black tea, such that its high content of monomeric catechins (Figure 4.1) is preserved. Accumulating evidence from epidemiologic studies as well as controlled studies *in vitro* and *in vivo* have generated substantial support for the health benefits of green tea and its catechins in mitigating the risk of chronic disease. The health benefits of green tea are often attributed to the antioxidant and anti-inflammatory activities of its catechins that effectively regulate oxidative stress responses leading to chronic disease. Paradoxically, green tea catechins have relatively low bioavailability due to their instability in near-neutral pH environments, their high involvement in phase III efflux pathways that limit enterocyte absorption, and their rapid rate of biotransformation by phase II xenobiotic metabolism. This suggests that the direct free radical scavenging activities of green tea catechins observed *in vitro* may not fully represent the activities *in vivo,* such that the health benefits of green tea catechins are potentially mediated through indirect antioxidant activities. This chapter will discuss the indirect and direct antioxidant activities of green tea and its catechins. Specifically, it will describe their bioavailability and antioxidant activities within the framework of chronic disorders, such as obesity, cardiovascular disease (CVD), and nonalcoholic fatty liver disease (NAFLD), that are characterized by a pathogenesis of significant oxidative stress.

GREEN TEA AND ITS COMPOSITION

Green tea is a minimally processed product that is produced from freshly harvested leaves of the tea plant (*Camellia sinensis*) (Figure 4.2). Unlike other tea products, such as black and oolong teas, the tea leaves intended

Processing and Impact on Antioxidants in Beverages
http://dx.doi.org/10.1016/B978-0-12-404738-9.00004-0

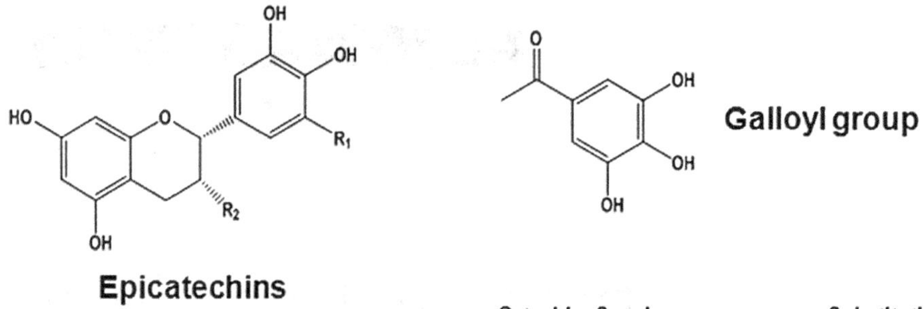

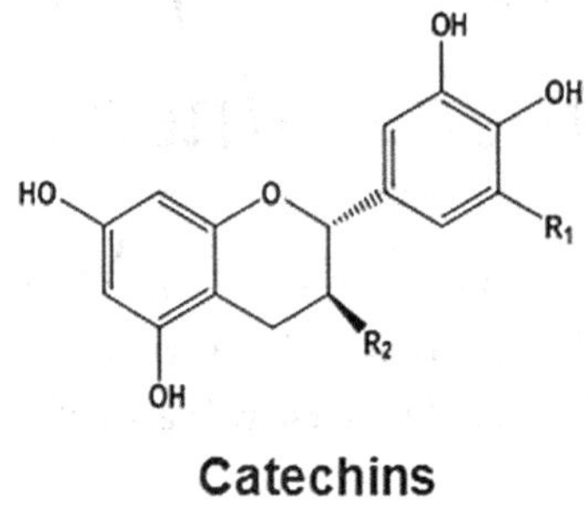

Catechin Species	Substitution	
	R_1	R_2
Epicatechin (EC)	H	H
Epicatechin gallate (ECG)	H	Galloyl
Epigallocatechin (EGC)	OH	H
Epigallocatechin gallate (EGCG)	OH	Galloyl
Catechin (C)	H	H
Catechin gallate (CG)	H	Galloyl
Gallocatechin (GC)	OH	H
Gallocatechin gallate (GCG)	OH	Galloyl

FIGURE 4.1 Chemical structures of the predominant catechins found in green tea. Green tea catechins include both catechin and epicatechin forms. Gallated catechins include epigallocatechin gallate (EGCG), gallocatechin gallate (GCG), epicatechin gallate (ECG), and catechin gallate (CG). The major non-gallated catechins are epicatechin (EC), catechin (C), epigallocatechin (EGC), and gallocatechin (GC).

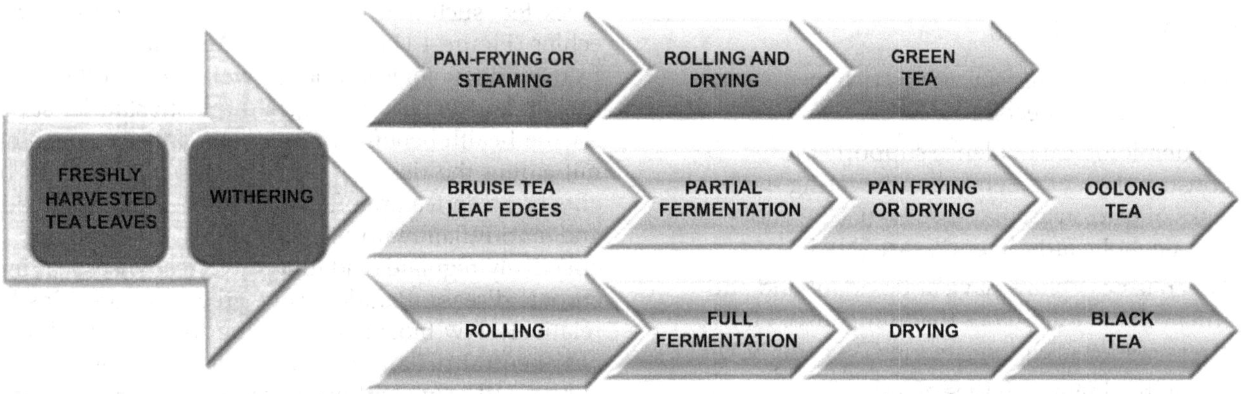

FIGURE 4.2 Post-harvest processing of green, black, and oolong teas. Common teas, including green tea, oolong tea, and black tea, are derived from the same tea plant (*Camellia sinensis*). Freshly harvested tea leaves are withered, and those directed for the production of green tea are then immediately fried or steamed, and subsequently rolled and dried. In contrast, oolong and black tea are fermented prior to drying. Steaming or frying in combination with the lack of fermentation of tea leaves preserves the naturally occurring catechins that are otherwise oxidized during the production of oolong and black tea.

for green tea are harvested, immediately withered, and steamed or pan fried to rapidly inactivate polyphenol oxidase and preserve its fresh character and monomeric polyphenol profile (Frei and Higdon, 2003). Thus, these timely post-harvest processes preserve the catechins that would have been otherwise oxidized. In fact, catechins account for approximately one-third of the dry weight of the tea leaves, with the eight major green tea catechins being epigallocatechin gallate (EGCG), gallocatechin gallate (GCG), epicatechin gallate (ECG), catechin gallate (CG), epigallocatechin (EGC), gallocatechin (GC), epicatechin (EC) and catechein (C) (Figure 4.1) (Neilson and Ferruzzi, 2011). Of these compounds, EGCG, EGC, ECG, and EC are the most common in tea leaves and brewed tea products (Graham, 1992). EGCG is the most abundant representing ~50% of the total catechin content (Figure 4.3) and exhibits the greatest antioxidant activity. In contrast, partially fermented oolong tea and fully fermented black tea have high levels of theaflavins and thearubigins, but a total catechin content that can be more than 50% lower than that of green tea. Regardless of the catechin quantity and distribution, green, black, and oolong teas have similar caffeine content with 40–50 mg of caffeine per freshly brewed cup of tea.

BIOAVAILABILITY OF GREEN TEA CATECHINS

The potential bioactivity of green tea catechins *in vivo* is contingent upon their bioavailability and rates of biotransformation and elimination. Oral administration of green tea catechins to rats results in their intestinal absorption and subsequent delivery to the liver via the

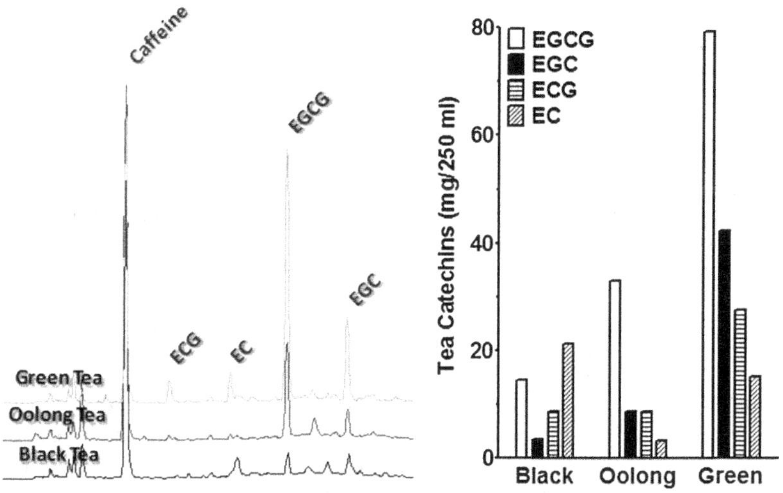

FIGURE 4.3 Analysis and distribution of catechins and caffeine found in green, black, and oolong teas. HPLC analysis of freshly brewed teas in 250 mL of boiled water shows that green tea is especially high in epigallocatechin gallate (EGCG) compared to black and oolong teas. Green tea also contains appreciably higher levels of epigallocatechin (EGC), epicatechin gallate (ECG), and epicatechin (EC). In total, the catechin content of freshly brewed green tea is more than 3-times higher than that of black or oolong tea.

portal vein (Okushio *et al.*, 1996). The pharmacokinetics of green tea catechins have received considerable study resulting in the understanding that: (1) catechin concentrations in the circulation peak at 1–5 μM within 2–4 h following oral ingestion of tea or purified catechins; (2) plasma half-lives of catechins are relatively short, ranging from 0.75 to 4.0 h following green tea ingestion; (3) those catechins that are absorbed undergo extensive biotransformation through phase II xenobiotic metabolism resulting in numerous catechin metabolites that are glucuronidated, sulfated, and/or methylated; and (4) efflux transport proteins such as p-glycoprotein and multidrug resistance proteins (MRP)-1 and -2 limit enterocyte retention and/or apical-to-basolateral transport of catechins (Feng, 2006). For example, biotransformation of EGCG and EGC by catechol-*O*-methyltransferase (COMT) results in the formation of 4′-*O*-methyl-(−)-EGC and 4″-*O*-methyl-(−)-EGCG or 4′,4″-*O*-dimethyl-(−)-EGCG (Lambert *et al.*, 2007). However, the extent to which methylation is a major route of catechin biotransformation is unclear since COMT activity in liver cytosolic fractions from rats is greater than those from humans. Catechin methylation is also mediated in a concentration-dependent manner such that dimethylated EGCG is the major product when EGCG is at low concentrations whereas monomethylated metabolites of EGCG are more abundant when EGCG is administered at higher doses. Unsurprisingly, mice co-administered EGCG with an inhibitor of glucuronidation had greater bioavailability of EGCG by increasing its intestinal absorption (Lambert *et al.*, 2004). Similarly, studies *in vitro* that co-treat cells with chemical inhibitors of MRP along with EGCG demonstrate greater uptake and retention of EGCG (Hong *et al.*, 2003). These studies emphasize that phase II and III xenobiotic metabolism occurring in the small intestines substantially regulate catechin bioavailability.

Further complicating the bioavailability of catechins is their susceptibility to auto-oxidation at near-neutral pH (Neilson and Ferruzzi, 2011). This implicates not only low-acid foods/beverages, but also the human intestines as a potential site of conversion of monomers to dimerized products. Indeed, simulated gastric and intestinal digestion of EGCG, EGC, and ECG results in substantial degradation of these compounds. For example, degradation of EGCG resulted in the formation of the auto-oxidation homodimers theasinensins A and D. Collectively, the oral bioavailability of green tea catechins is rather low resulting from a combination of instability during gastrointestinal digestion, relatively poor absorption, and rapid metabolism and elimination. This has resulted in controversy over whether green tea catechins from dietary sources can exert antioxidant activity *in vivo* similar to the direct reactive oxygen/nitrogen species scavenging activities observed *in vitro*. In addition, studies are currently underway that are considering the health-promoting bioactivities of specific catechin metabolites that result from phase II xenobiotic metabolism. It should be noted that the unique acidic environment of the stomach may provide stabilization of tea catechins *in vivo* and promote their uptake and/or bioactivity in the gastric epithelium. Interestingly, epidemiological evidence also suggests that tea consumption is correlated with reduced risk of gastric cancer. The stabilization, uptake, and antioxidant/anti-inflammatory effects of green tea and its catechins during the gastric phase of digestion are not well understood but are currently the focus of ongoing research efforts.

IN VITRO ANTIOXIDANT ACTIVITIES OF CATECHINS

The propensity of green tea catechins to exert antioxidant activity by chemically reducing an oxidative species (i.e., free radical) is related to their one-electron reduction potential. Compounds having lower electron

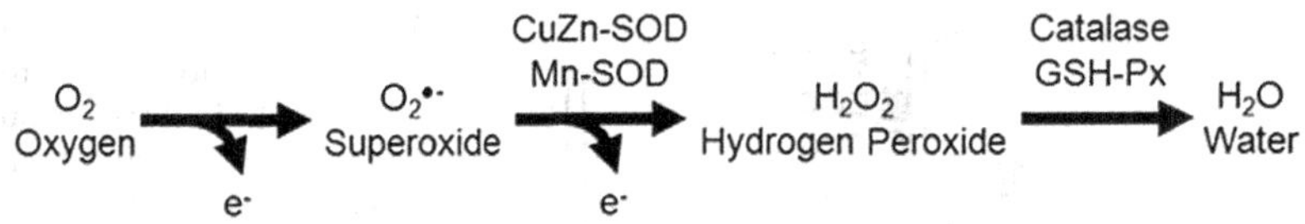

FIGURE 4.4 Detoxification of reactive oxygen species. Aerobic organisms are exposed to oxygen, which becomes oxidized to superoxide in response an initiating event of oxidative stress. Green tea protects against oxidative stress, in part, by upregulating the activities of superoxide dismutases (Cu/Zn- and Mn-SOD) that detoxify superoxide to hydrogen peroxide. Subsequently, green tea-mediated increases in the activities of catalase and glutathione peroxidase (GSH-Px) result in the detoxification of hydrogen peroxide.

reduction potentials are more likely to participate in hydrogen or electron exchange, thereby resulting in reduction of the reactive species and simultaneous oxidation of the reducing agent. Green tea catechins (Figure 4.1) have reduction potentials in the range of 430–570 mV suggesting that they have antioxidant activities similar to α-tocopherol (480 mV), but somewhat less than that of ascorbic acid (280 mV) (Jovanovic *et al.*, 1996, 1998; Frei and Higdon, 2003).

Studies *in vitro* have provided clear evidence that green tea catechins protect against reactive oxygen species (ROS). Studies *in vitro* taking advantage of a superoxide-generating system showed that green tea extract readily scavenged superoxide at a high rate (Nakagawa and Yokozawa, 2002). Direct comparison of purified green tea catechins in this free radical generating system indicated that catechins having a galloyl moeity have higher scavenging activities than those lacking this structure. This implies that EGC, GC, EGCG and ECG (Figure 4.1) have greater antioxidant scavenging against ROS in comparison to the non-gallated forms (C, EC). Likewise, studies utilizing electron spin resonance technology demonstrated that green tea catechins also scavenge peroxyl radicals as well as singlet oxygen (Guo *et al.*, 1999).

Green tea catechins also effectively scavenge reactive nitrogen species (RNS) (Paquay *et al.*, 2000). *In vitro*, catechins were shown to scavenge nitric oxide and peroxynitrite (Haenen *et al.*, 1997; Haenen and Bast, 1999). Moreover, green tea, containing 4.8-times more catechins than black tea, was 5-times more effective than black tea in scavenging nitric oxide (Paquay *et al.*, 2000). A separate line of evidence also indicates that green tea catechins chelate redox-active transition metals, such as free iron and copper, supporting that these polyphenols may mitigate metal-catalyzed ROS generation (Frei and Higdon, 2003; Higdon and Frei, 2003). Taken together, green tea has the potential to protect against an array of ROS and RNS, and these effects are mediated, in part, through its polyphenolic catechins.

Contrasting these antioxidant activities are studies showing that EGCG exerts pro-oxidant activities at high doses. Indeed, in studies, mice that were provided a single high dose of EGCG intragastrically were shown to exhibit hepatotoxicity characterized by substantially higher levels of the liver injury marker plasma alanine aminotransferase, moderate to severe hepatic necrosis, and a dramatically lower survival rate (Lambert *et al.*, 2010). Livers from these mice also had higher levels of lipid peroxidation markers and pro-inflammatory cytokines. Thus, physiologically relevant intakes of green tea or its catechins are seemingly beneficial to health outcomes with respect to mitigating oxidative stress. In contrast, supraphysiological doses that bypass tight homeostatic controls otherwise limiting catechin bioavailability at the intestinal level readily trigger oxidative stress and inflammatory responses that increase the risk of morbidity and mortality.

PHYSIOLOGIC ANTIOXIDANT ACTIVITIES

Despite catechins having similar reduction potentials to other dietary antioxidants, and clear *in vitro* evidence showing their direct scavenging of ROS/RNS, the extent to which green tea catechins exhibit direct antioxidant activity *in vivo* remains controversial. This is largely related to the lack of analytical methods available to detect and accurately quantify oxidized catechin forms from biological matrices. Because of these limitations, studies *in vivo* have primarily focused on indirect antioxidant activities of green tea catechins. Studies in animal models suggest that green tea protects against oxidative stress-mediated antioxidant depletion, including liver and erythrocyte glutathione (Skrzydlewska *et al.*, 2002; Oz *et al.*, 2005) and circulating α-tocopherol (Nanjo *et al.*, 1993). Antioxidant activities of green tea catechins also recycle α-tocopheroxyl radicals back to their reduced α-tocopherol form (Zhou *et al.*, 2005), which is an important mechanism for maintaining vitamin E status (Bruno *et al.*, 2006). Antioxidant activities of green tea catechins also lower oxidative stress by increasing activities of enzymatic antioxidants, which lowers the magnitude of ROS by enhancing their detoxification (Figure 4.4). For example, green tea increases the activities of copper/zinc- and manganese superoxide dismutase found in the cytosol and mitochondria, respectively, which facilitates the detoxification of superoxide radical to hydrogen peroxide (Park *et al.*, 2011). Green tea also increases catalase and glutathione peroxidase, which function to degrade hydrogen peroxide to water (Agarwal *et al.*, 1993; Park *et al.*, 2011). These protective effects are thought to occur by upregulating nuclear factor erythoid 2-related factor 2 (Nrf2) (Shen *et al.*, 2005), a cytoprotective transcription factor regulating gene expression for antioxidant defenses and xenobiotic metabolism.

PROTECTIVE ASPECTS OF GREEN TEA IN OBESITY AND RELATED DISORDERS

Green tea and its catechins exert anti-obesity activities that promote weight loss, and also protect against oxidative and metabolic insults that otherwise contribute to obesity-related disorders. The role of green tea in weight management stems largely from work in rodents because of the relative ease of controlling critical variables in these models in comparison to human populations. Green tea and its catechins may protect against obesity by reducing intestinal lipid absorption (Koo and Noh, 2007). In humans, a beverage containing green tea catechins, caffeine, and calcium increased 24 h energy expenditure (Rudelle *et al.*, 2007). This is complemented by studies in rodents showing that green tea or its catechins lower body mass without affecting energy intake, and that decreases in body mass are attributed to lower adiposity (Masterjohn and Bruno, 2012). Green tea or its catechins also decrease body mass by increasing β-oxidation via the upregulation of the expression of genes from this energy-producing pathway and reduction of the expression of lipogenic genes. These effects of green tea are likely, at least in part, to reduce oxidative stress secondary to decreases in obesity.

Epidemiological evidence from Japan suggests that ≥5 cups per day of green tea lowers the risk of CVD-related mortality by 26% and that ≥10 cups each day may protect against livery injury (Imai and Nakachi, 1995; Kuriyama *et al.*, 2006). Although these health benefits remain unclear, obese mice provided with green tea extract are protected from liver steatosis by lowering lipogenic gene expression concurrent with decreases in hepatic lipid peroxidation and increases in antioxidant defenses (Park *et al.*, 2011). Green tea supplementation also reduces 4-hydroxynonenal in livers from obese mice having NAFLD in association with lowering the activity of nicotinamide adenine dinucleotide phosphate (NADPH) oxidase and expression of inducible nitric oxide synthase (iNOS) (Chung *et al.*, 2012). Thus, green tea-dependent reductions in lipid peroxidation, an early event implicated in both CVD and NAFLD, are likely mediated through antioxidant activities that increase ROS detoxification, anti-inflammatory activities that suppress ROS/RNS generation, and its hypolipidemic activities that reduce the availability of oxidizable substrate. Similar to these effects in rodents, chronic ingestion of green tea in humans with metabolic syndrome was shown to reduce circulating levels of malondialdehyde and hydroxynonenal (Basu *et al.*, 2010). However, future work is needed to determine whether green tea reduces lipid peroxidation in humans through direct or indirect antioxidant activities.

Anti-inflammatory activities of green tea have received considerable attention because of its role in regulating oxidative stress. Supplementation of green tea to rats fed a high-fat diet lowered hepatic and adipose nuclear factor kappa B (NFκB) binding activity and expression of pro-inflammatory cytokines by decreasing the phosphorylation of IκBα (Park *et al.*, 2012). The mechanism by which green tea inhibits NFκB activation is not fully understood and is likely mediated through overlapping mechanisms that converge at the phosphorylation of IκBα. Antioxidant activities of green tea that increase hepatic glutathione appear to suppress NFκB activation (Park *et al.*, 2012). Because green tea increases mRNA expression of glutamate cysteine ligase (GCL) (Park *et al.*, 2011), a Nrf2-dependent gene, catechins likely promote Nrf2-dependent antioxidant activities that attenuate ROS, and in turn, lower NFκB activation. It is also likely that the protective effects occur extracellularly by reducing toll-like receptor-4 (TLR4) signaling that otherwise increases NFκB activation (Byun *et al.*, 2012). That green tea reduces TLR4-dependent inflammation suggests the possibility that catechins reduce inflammation by inhibiting bacterial endotoxin translocation from the small intestine, potentially by preventing intestinal tight junction barrier dysfunction (Watson *et al.*, 2004).

SAFETY OF GREEN TEA

Several dozen anecdotal and case reports of idiosyncratic hepatotoxicity in humans have questioned the safety of green tea (Masterjohn and Bruno, 2012). While studies in mice demonstrated that single administration of EGCG at the equivalent of ~20 cups of green tea in humans results in hepatotoxicity (Lambert *et al.*, 2010), the findings of epidemiologic studies suggest that green tea at ≥5–10 cups/day is associated with lower risks of liver-related morbidity and CVD-related mortality (Imai and Nakachi, 1995; Kuriyama *et al.*, 2006). Thus, it is possible that substantial inter-individual differences exist for catechin metabolism and/or that only supraphysiological doses are hepatotoxic. Moreover, because most case reports have been linked to dietary supplements, and dechallenge/rechallenge evidence is only provided in relatively fewer reports, product contamination remains a possibility to explain the acute hepatotoxicity of green tea (Masterjohn and Bruno, 2012). In contrast, intervention studies in otherwise healthy overweight men indicated that dietary supplementation of green tea polyphenols at 714 mg/day for 3 weeks did not affect numerous clinical chemistries for liver and kidney function (Frank *et al.*, 2009). In addition, although a systemic review performed by the United States Pharmacopeia was unable to identify a tolerable upper limit for safety, it concluded that green tea and green tea extracts are safe when used and manufactured appropriately (Sarma *et al.*, 2008).

CONCLUSIONS

The 2010 Dietary Guidelines for Americans recommend an increase in the consumption of plant-based foods that are nutrient dense, low in energy, and rich in bioactive compounds in order to reduce the risk of chronic disease. Although a beverage, green tea is clearly a functional food rich in bioactive polyphenolic catechins that protect against chronic disorders that are characterized by high levels of inflammation and oxidative stress. Although much of the available evidence regarding the beneficial effects of green tea has been provided by studies *in vitro* and in experimental model systems, controlled studies in humans are accumulating. In some cases, these clinical interventions are providing experimental evidence that directly support the profound health relations observed in epidemiological studies. More work is clearly needed to identify optimal intake levels and define safe strategies that optimize catechin bioavailability and bioefficacy. In addition, further work is warranted to consider the potential bioactivity of catechin forms that evade absorptive processes and accumulate in the lower gastrointestinal tract. The accumulation of catechins, their metabolites, and products of oxidation in the colon have the potential to favorably augment colonic health. This will certainly be an active area of investigation in the coming years as considerable technological advances have made it possible to more fully characterize and profile the microbiome in response to dietary bioactive components.

Acknowledgments

This work was supported by state and federal funds appropriated to The Ohio State University Ohio Agricultural Research and Development Center (to RSB and JAB), and grants from the USDI-NRI to RSB (CSREES 2007-02303) and the Food Innovation Center at The Ohio State University (to RSB and JAB).

References

Agarwal, R., Katiyar, S.K., Khan, S.G., Mukhtar, H., 1993. Protection against ultraviolet B radiation-induced effects in the skin of SKH-1 hairless mice by a polyphenolic fraction isolated from green tea. Photochem. Photobiol. 58, 695–700.

Basu, A., Sanchez, K., Leyva, M.J., Wu, M., Betts, N.M., Aston, C.E., Lyons, T.J., 2010. Green tea supplementation affects body weight, lipids, and lipid peroxidation in obese subjects with metabolic syndrome. J. Am. Coll. Nutr. 29, 31–40.

Bruno, R.S., Leonard, S.W., Atkinson, J., Montine, T.J., Ramakrishnan, R., Bray, T.M., Traber, M.G., 2006. Faster plasma vitamin E disappearance in smokers is normalized by vitamin C supplementation. Free Radic. Biol. Med. 40, 689–697.

Byun, E.B., Choi, H.G., Sung, N.Y., Byun, E.H., 2012. Green tea polyphenol epigallocatechin-3-gallate inhibits TLR4 signaling through the 67-kDa laminin receptor on lipopolysaccharide-stimulated dendritic cells. Biochem. Biophys. Res. Commun. 426, 480–485.

Chung, M.Y., Park, H.J., Manautou, J.E., Koo, S.I., Bruno, R.S., 2012. Green tea extract protects against nonalcoholic steatohepatitis in ob/ob mice by decreasing oxidative and nitrative stress responses induced by proinflammatory enzymes. J. Nutr. Biochem. 23, 361–367.

Feng, W.Y., 2006. Metabolism of green tea catechins: an overview. Curr. Drug. Metab. 7, 755–809.

Frank, J., George, T.W., Lodge, J.K., Rodriguez-Mateos, A.M., Spencer, J.P., Minihane, A.M., Rimbach, G., 2009. Daily consumption of an aqueous green tea extract supplement does not impair liver function or alter cardiovascular disease risk biomarkers in healthy men. J. Nutr. 139, 58–62.

Frei, B., Higdon, J.V., 2003. Antioxidant activity of tea polyphenols in vivo: evidence from animal studies. J. Nutr. 133, 3275S–3284S.

Graham, H.N., 1992. Green tea composition, consumption, and polyphenol chemistry. Prev. Med. 21, 334–350.

Guo, Q., Zhao, B., Shen, S., Hou, J., Hu, J., Xin, W., 1999. ESR study on the structure-antioxidant activity relationship of tea catechins and their epimers. Biochim. Biophys. Acta. 1427, 13–23.

Haenen, G.R., Bast, A., 1999. Nitric oxide radical scavenging of flavonoids. Methods Enzymol. 301, 490–503.

Haenen, G.R., Paquay, J.B., Korthouwer, R.E., Bast, A., 1997. Peroxynitrite scavenging by flavonoids. Biochem. Biophys. Res. Commun. 236, 591–593.

Higdon, J.V., Frei, B., 2003. Tea catechins and polyphenols: health effects, metabolism, and antioxidant functions. Crit. Rev. Food Sci. Nutr. 43, 89–143.

Hong, J., Lambert, J.D., Lee, S.H., Sinko, P.J., Yang, C.S., 2003. Involvement of multidrug resistance-associated proteins in regulating cellular levels of (−)-epigallocatechin-3-gallate and its methyl metabolites. Biochem. Biophys. Res. Commun. 310, 222–227.

Imai, K., Nakachi, K., 1995. Cross sectional study of effects of drinking green tea on cardiovascular and liver diseases. BMJ 310, 693–696.

Jovanovic, S.V., Steenken, S., Hara, Y., Simic, M.G., 1996. Reduction potentials of flavonoid and model phenoxyl radicals. Which ring in flavonoids is responsible for antioxidant activity? J. Chem. Soc. Perkin Trans. 2: Phys. Organ. Chem., 2497–2504.

Jovanovic, S.V., Steenken, S., Simic, M.G., Hara, Y., 1998. Antioxidant properties of flavonoids: reduction potentials and electron transfer reactions of flavonoid radicals. Antioxid. Health Dis. 7, 137–161.

Koo, S.I., Noh, S.K., 2007. Green tea as inhibitor of the intestinal absorption of lipids: potential mechanism for its lipid-lowering effect. J. Nutr. Biochem. 18, 179–183.

Kuriyama, S., Shimazu, T., Ohmori, K., Kikuchi, N., Nakaya, N., Nishino, Y., Tsubono, Y., Tsuji, I., 2006. Green tea consumption and mortality due to cardiovascular disease, cancer, and all causes in Japan: the Ohsaki study. JAMA 296, 1255–1265.

Lambert, J.D., Hong, J., Kim, D.H., Mishin, V.M., Yang, C.S., 2004. Piperine enhances the bioavailability of the tea polyphenol (−)-epigallocatechin-3-gallate in mice. J. Nutr. 134, 1948–1952.

Lambert, J.D., Kennett, M.J., Sang, S., Reuhl, K.R., Ju, J., Yang, C.S., 2010. Hepatotoxicity of high oral dose (−)-epigallocatechin-3-gallate in mice. Food Chem. Toxicol. 48, 409–416.

Lambert, J.D., Sang, S., Yang, C.S., 2007. Biotransformation of green tea polyphenols and the biological activities of those metabolites. Mol. Pharm. 4, 819–825.

Masterjohn, C., Bruno, R.S., 2012. Therapeutic potential of green tea in nonalcoholic fatty liver disease. Nutr. Rev. 70, 41–56.

Nakagawa, T., Yokozawa, T., 2002. Direct scavenging of nitric oxide and superoxide by green tea. Food Chem. Toxicol. 40, 1745–1750.

Nanjo, F., Honda, M., Okushio, K., Matsumoto, N., Ishigaki, F., Ishigami, T., Hara, Y., 1993. Effects of dietary tea catechins on alpha-tocopherol levels, lipid peroxidation, and erythrocyte deformability in rats fed on high palm oil and perilla oil diets. Biol. Pharm. Bull. 16, 1156–1159.

Neilson, A.P., Ferruzzi, M.G., 2011. Influence of formulation and processing on absorption and metabolism of flavan-3-ols from tea and cocoa. Annu. Rev. Food Sci. Technol. 2, 125–151.

Okushio, K., Matsumoto, N., Kohri, T., Suzuki, M., Nanjo, F., Hara, Y., 1996. Absorption of tea catechins into rat portal vein. Biol. Pharm. Bull. 19, 326–329.

Oz, H.S., Chen, T.S., McClain, C.J., de Villiers, W.J., 2005. Antioxidants as novel therapy in a murine model of colitis. J. Nutr. Biochem. 16, 297–304.

Paquay, J.B., Haenen, G.R., Stender, G., Wiseman, S.A., Tijburg, L.B., Bast, A., 2000. Protection against nitric oxide toxicity by tea. J. Agric. Food Chem. 48, 5768–5772.

Park, H.J., DiNatale, D.A., Chung, M.Y., Park, Y.K., Lee, J.Y., Koo, S.I., O'Connor, M., Manautou, J.E., Bruno, R.S., 2011. Green tea extract attenuates hepatic steatosis by decreasing adipose lipogenesis and enhancing hepatic antioxidant defenses in ob/ob mice. J. Nutr. Biochem. 22, 393–400.

Park, H.J., Lee, J.Y., Chung, M.Y., Park, Y.K., Bower, A.M., Koo, S.I., Giardina, C., Bruno, R.S., 2012. Green tea extract suppresses NFkappaB activation and inflammatory responses in diet-induced obese rats with nonalcoholic steatohepatitis. J. Nutr. 142, 57–63.

Rudelle, S., Ferruzzi, M.G., Cristiani, I., Moulin, J., Mace, K., Acheson, K.J., Tappy, L., 2007. Effect of a thermogenic beverage on 24-hour energy metabolism in humans. Obes. (Silver Spring) 15, 349–355.

Sarma, D.N., Barrett, M.L., Chavez, M.L., Gardiner, P., Ko, R., Mahady, G.B., Marles, R.J., Pellicore, L.S., Giancaspro, G.I., Low Dog, T., 2008. Safety of green tea extracts: a systematic review by the US Pharmacopeia. Drug Saf. 31, 469–484.

Shen, G., Xu, C., Hu, R., Jain, M.R., Nair, S., Lin, W., Yang, C.S., Chan, J.Y., Kong, A.N., 2005. Comparison of (−)-epigallocatechin-3-gallate elicited liver and small intestine gene expression profiles between C57BL/6J mice and C57BL/6J/Nrf2 (−/−) mice. Pharm. Res. 22, 1805–1820.

Skrzydlewska, E., Ostrowska, J., Stankiewicz, A., Farbiszewski, R., 2002. Green tea as a potent antioxidant in alcohol intoxication. Addict. Biol. 7, 307–314.

Watson, J.L., Ansari, S., Cameron, H., Wang, A., Akhtar, M., McKay, D.M., 2004. Green tea polyphenol (−)-epigallocatechin gallate blocks epithelial barrier dysfunction provoked by IFN-gamma but not by IL-4. Am. J. Physiol. Gastrointest Liver Physiol. 287, G954–961.

Zhou, B., Wu, L.M., Yang, L., Liu, Z.L., 2005. Evidence for alpha-tocopherol regeneration reaction of green tea polyphenols in SDS micelles. Free Radic. Biol. Med. 38, 78–84.

CHAPTER

5

Antioxidant Capacities of Herbal Infusions

Sha Li, Shu-Ke Li*, Hua-Bin Li*, Xiang-Rong Xu†, Gui-Fang Deng*, Dong-Ping Xu**

*Guangdong Provincial Key Laboratory of Food, Nutrition and Health, Department of Nutrition, School of Public Health, Sun Yat-Sen University, Guangzhou, China, †Key Laboratory of Marine Bio-resources Sustainable Utilization, South China Sea Institute of Oceanology, Chinese Academy of Sciences, Guangzhou, China

CHAPTER POINTS

- Oxidative stress plays an important role in many chronic and degenerative diseases, such as cardiovascular diseases, cancer, and aging.
- Dietary supplements of antioxidants are very important to enhance the body's antioxidant defenses.
- Natural antioxidants may come from fruits, vegetables, and beverages.
- Herbal infusions are known to be safe because they have been drunk as beverages for thousands of years.
- Some herbal infusions possess high antioxidant capacities, and could be important dietary sources of natural antioxidants for prevention of diseases caused by oxidative stress.
- The antioxidant components in some herbal infusions have been identified, and their antioxidant mechanism should be studied further.

INTRODUCTION

The oxidative stress caused by reactive oxygen species plays an important role in many chronic and degenerative diseases, such as cardiovascular diseases, cancer, aging, and diabetes mellitus. Dietary supplements of antioxidants have become a popular way to enhance the body's antioxidant defenses. Natural antioxidants may come from vegetables, fruits, and beverages. Herbal infusions have been drunk as beverages for thousands of years. One special kind of herbal infusion is called cool tea ('Liang cha' in Chinese), which originated from South China. The cool tea is made from different kinds of herbs, and its spread has been from South China to the whole of China, and from China to about 20 countries around the world, such as the USA, Canada, UK, France and Germany. The cool tea has the efficacies of clearing away heat, detoxification, dewetting, moistening lungs, and stopping thirst (Fu *et al.*, 2011). Antioxidant capacities of many herbal infusions have been evaluated in the literature, and it has been found that some herbal infusions possess high antioxidant capacities, and could be important dietary sources of natural antioxidants for the prevention of diseases caused by oxidative stress (Fu *et al.*, 2011). In this chapter, antioxidant capacities and main antioxidant components of some herbal infusions are summarized.

ANTIOXIDANT CAPACITIES OF HERBAL INFUSIONS

Lemon verbena is used as a primary ingredient in infusions and nonalcoholic drinks because of its good sensorial attributes. The antioxidant activity of lemon verbena infusion as well as the thermal stability of its antioxidant activity and the content of polyphenolic compounds was studied (Abderrahim *et al.*, 2011). The values reflecting the antioxidant activity of lemon verbena infusion, including antioxidant activity index, fast scavenging rate against 2,2-diphenyl-1-picrylhydrazyl, Trolox-equivalent antioxidant capacity (TEAC), and hydroxyl radical scavenging, are higher than those of many herbal infusions and antioxidant drinks estimated from reported data. In addition, the slope lag time and

Processing and Impact on Antioxidants in Beverages
http://dx.doi.org/10.1016/B978-0-12-404738-9.00005-2

specific oxyradical antioxidant capacity values of lemon verbena infusion are comparable to those of a commercial antioxidant drink based on green tea. In another study, four different oak leaves were analyzed as potential nutraceutical beverages (Rocha-Guzman *et al.*, 2012). The phenolic composition, antioxidant capacity, and sensory preferences of leaf infusions from *Quercus resinosa*, *Q. sideroxyla*, *Q. eduadii*, and *Q. durifolia* were investigated. Herbal infusions from oak leaves were evaluated for total polyphenol content (TPC), total flavonoid content (TFC), high-performance liquid chromatography (HPLC) analysis, TEAC, oxygen radical absorbance capacity (ORAC), soluble solids, pH, color, and consumer preference analysis. *Q. resinosa* leaves infusions showed the highest TPC, TEAC, and ORAC values but they attained the lowest preference score. *Quercus* leaf infusions with higher contents of gallic acid and catechins showed the best antioxidant capacity but lower consumer preference.

Small centaury (*Centaurium erythraea* Rafin.) is an herbal species that has long been used in traditional medicine due to its digestive, stomachic, tonic, depurative, sedative, and antipyretic properties, and contains considerable amounts of polyphenolic compounds, namely xanthones and phenolic acids as the main constituents. The ability of small centaury infusion to act as a scavenger of the reactive oxygen species hydroxyl radical and hypochlorous acid was studied and compared with that of green tea (*Camellia sinensis* L.). Hydroxyl radical was generated in the presence of Fe^{3+}-EDTA, ascorbate and H_2O_2 (Fenton system) and monitored by evaluating hydroxyl radical-induced deoxyribose degradation. The reactivity towards hypochlorous acid was determined by measuring the inhibition of hypochlorous acid-induced 5-thio-2-nitrobenzoic acid oxidation to 5,5′-dithiobis(2-nitrobenzoic acid). The results demonstrated that small centaury infusion exhibited interesting antioxidant properties, expressed both by its capacity to effectively scavenge hydroxyl radical and hypochlorous acid, although with a lower activity against the second than that observed for green tea (Valentao *et al.*, 2003). In another study, the total antioxidant capacity of the aqueous extracts of some endemic herbs—prepared as infusions by steeping these herbs in hot water—was assayed with bis(neocuproine) copper(II) chloride, also known as the cupric-ion-reducing antioxidant capacity (CUPRAC) reagent, which was easily accessible, rapid, stable, and responsive to both hydrophilic and lipophilic antioxidants. The highest antioxidant capacities of some herbal teas available on the Turkish market were observed for scarlet pimpernel (*Anagallis arvensis*), sweet basil (*Ocimum basilicum*), green tea (*Camellia sinensis*), and lemon balm (*Melissa officinalis*), in the order of 1.63, 1.18, 1.07, and 0.99 mmol Trolox equivalent/g, respectively (Apak *et al.*, 2006).

Hot water infusions (teas) of air-dried petals of 12 rose cultivars were assayed for antioxidant activity, total phenols, and total anthocyanins contents. Their composition was analyzed by HPLC. Rose petal teas from different cultivars exhibited scavenging capacity toward 2,2′-azino-bis-(3-ethylbenzothiazoline)-6-sulfonate cation radical ($ABTS^{\bullet+}$) ranging between 712.7 and 1770.7 μM Trolox equivalents (TE) per gram of dry petals, as compared with 1227.6 μM TE/g dry weight in the green tea. The range of total phenol content in rose teas was 50.7 to 119.5 mg gallic acid equivalents (GAE) per gram of dry matter, as compared with 62.1 mg GAE/g dry weight in the green tea. The rose teas were rich in free gallic acid. The highest values of antioxidant activity, total phenols, and gallic acid contents were found in the cultivars San Francisco, Katharina Zeimet, and Mercedes, and in the essential-oil-bearing rose *Rosa damascena*. The correlation coefficients between antioxidant activity and the contents of total phenols and of gallic acid in various rose cultivars, were 0.79 and 0.81, respectively. No clear relationship between anthocyanin level and radical-scavenging activity was revealed. Teas from different rose cultivars significantly differed in their sensory properties. The dried rose petals may be used for preparing antioxidant-rich caffeine-free beverages, either separately or in combination with other herbal materials (Vinokur *et al.*, 2006). In a study, the antioxidant capacity of six herbal tea infusions (*Mentha piperita*, *Erythroxylum coca*, *Rosa moschata*, *Tilia* spp, *Plantago major*, and *Aloysia citriodora*) was measured by ORAC-type methodologies employing fluorescein (ORAC-FL) and pyrogallol red (ORAC-PGR) as target compounds. Probe consumption profiles were widely different, with neat induction times when FL was employed as target. Relative ORAC values of different herbal infusions depended upon the test molecule employed. Relative ORAC-PGR values followed the order: *Rosa moschata* > *Mentha piperita* > *Tilia* sp > *Plantago major* > *Aloysia citriodora* > *Erythroxylum coca*, while for ORAC-FL the order was: *Mentha piperita* > *Aloysia citriodora* > *Erythroxylum coca* > *Rosa moschata* > *Tilia* sp > *Plantago major* (Poblete *et al.*, 2009). In another study, 12 plants used traditionally in ethno-medicine from the South American Andes were investigated. Infusions were compared in terms of their free radical-scavenger properties, using four different methods: AAPH-induced hemolysis, TEAC-DPPH, ferric reducing antioxidant power (FRAP), and TEAC-crocin. All herbal teas showed strong antioxidant capacities, and the results obtained with FRAP and TEAC-DPPH were very similar. When compared at a dilution of 0.03%, *Parastrephia lucida*, *Parastrephia lepidophylla*, *Senecio nutans*, *Azorella compacta*, and *Baccharis tola*, showed a protective effect against oxidative damage on human erythrocytes greater than that of Trolox ($p < 0.001$). Upon correlating the polyphenolic contents with the antioxidant capacity, a positive association for all plants (TEAC-DPPH *versus*

GAE; $r^2=0.7437$) was found. Interestingly, there was a better correlation between polyphenolic contents and the protective effect against APPH-induced haemolytic activity (Rojo *et al.*, 2009).

Lippia citriodora is an herbal species which contains several flavonoids and phenolic acids. In a study, the superoxide radical, hydroxyl radical, and hypochlorous-acid-scavenging activities of *L. citriodora* infusion were examined. Superoxide radical was generated either in an enzymatic or in a chemical system, and scavenging ability was assessed by the inhibition of nitroblue tetrazolium reduction. Hydroxyl radical was generated by the reaction of an iron–EDTA complex in the presence of ascorbic acid, and was assayed by evaluating deoxyribose degradation. Hypochlorous acid scavenging activity was tested by measuring the inhibition of 5-thio-2-nitrobenzoic acid oxidation. The results demonstrated that this infusion had a potent superoxide radical scavenging activity and moderate scavenging activities of hydroxyl radical and hypochlorous acid (Valentao *et al.*, 2002). *Lippia javanica* and *Lippia scaberrima* are used as herbal remedies and are commercially traded as health teas in southern Africa under the brands 'Mosukujane' and 'Musukudu', respectively. A study evaluated the relationship between the presence of phenolic compounds and the antioxidant activities of infusions prepared from four *Lippia* species (*L. javanica, L. scaberrima, L. rehmannii* and *L. wilmsii*) indigenous to South Africa. Of the four indigenous species, infusions of *L. javanica* and *L. wilmsii* exhibited the highest antioxidant activities (EC_{50}: 358 and 525 μg/ml, respectively) and contained the most phenolic compounds (14.8 and 14.5 mg/ml, respectively) (Shikanga *et al.*, 2010). *Lippia multiflora*, a perennial, aromatic shrub commonly known as bush tea has recently been identified as an African plant with high commercial potential due to its medicinal properties. The plant material was subjected to steam pasteurization to improve its microbial quality. The major compounds of *L. multiflora* herbal infusion, i.e., the phenylethanoid glycosides, verbascoside, isoverbascoside, nuomioside A and isonuomioside A, and the flavone, luteolin-7-O–glucuronide, were quantified by HPLC. Verbascoside was the most abundant phenylethanoid glycoside. The phenylethanoid glycosides are of interest due to their pharmacological properties. Liquid chromatography tandem-mass spectrometry was used to tentatively identify the compounds. The on-line DPPH (2,2′-diphenyl-1-picrylhydrazyl radical) scavenging assay (reaction time = 0.45 s) applied to the infusion in 'quantitative' mode, showed the relative order of activity: isonuomioside A > isoverbascoside > verbascoside > nuomioside A. In the microplate assay (reaction time = 2 h), isoverbascoside and verbascoside had similar activities. Both compounds were less active in the latter assay than the well-known flavan-3-ol antioxidant, (−)-epigallocatechin gallate, but more active than caffeic acid and an ester, rosmarinic acid. Steam pasteurization of *L. multiflora* leaves at maximum exposure (150 s at ca. 99°C) for improved microbial quality did not decrease the soluble solids content, phenolic content, and antioxidant activities of the infusion compared to the untreated control ($p<0.05$). The phenylethanoid glycoside content of the water-soluble solids was as high as 15%, underscoring the potential of *L. multiflora* extract as a functional ingredient (Arthur *et al.*, 2011).

Hot water extracts of the herbal tea, rooibos, are increasingly used as an ingredient in ready-to-drink beverages and a variety of food products. The extent of variation in total polyphenol, aspalathin, orientin, and isoorientin contents, as well as total antioxidant capacity (TAC) of hot water extract of fermented rooibos was determined and compared with that of the hot-water-soluble solids of infusions, prepared similarly to a cup of tea. Extract preparation from a large number of individual rooibos production batches ($n=74$) partly simulated industrial processing, while infusions were prepared from a subset of samples ($n=20$). Based on the total polyphenol and aspalathin contents, rooibos extract and infusion were equivalent when compared on a soluble-solids basis. The isoorientin and orientin contents of the soluble solids of the infusion were slightly higher than those of the extract. The TAC of the soluble solids of the infusion, measured with the ORAC assay, was slightly higher than that of the extract, while the opposite was observed for the TAC, measured with the DPPH radical scavenging assay (Joubert and de Beer, 2012). In another study, 15 Thai herbal teas in comparison with teas of *Camellia sinensis* were investigated for their antioxidant properties in correlation with their total phenolics, flavonoids, and non-flavonoids contents. Significant differences were observed among the tea infusions. Only stevia and sappan herbal teas had primary antioxidant capacities comparable to *C. sinensis*. However, purple velvet, mulberry, and false mallow herbal teas were exceptionally stronger in metal-chelating capacity than the *C. sinensis* teas. Principal component analysis showed that total phenolics, particularly flavonoids, highly correlated with primary antioxidants. Cluster analysis showed that the properties of stevia and sappan herbal teas were similar to green, black, and oolong teas. Some common, but rarely mentioned, Thai herbal teas could be choices of interest for healthy beverages and could be new dietary sources for bioactive compounds (Deetae *et al.*, 2012).

An *in vitro* digestion/Caco-2 cell culture model was used to assess the stability and bioavailability of phenolic compounds in aqueous extracts of four herbal infusions traditionally used as functional drinks in Portugal. Alterations in antioxidant power were monitored by $ABTS^{\bullet+}$, whereas the profile of phenolic compounds was ascertained by HPLC with diode-array detection (HPLC-DAD). The bioavailability of rutin—an important flavonoid present in such extracts, and thus a representative of those compounds—was assessed across Caco-2 cells via LC-MS/MS. The results indicated that several antioxidant compounds are not affected by the (simulated) digestive process, whereas others undergo decreases in their activity throughout said process; a few have their antioxidant capacity hampered under stomach-like conditions. It was also found that rutin can be transported across the Caco-2 cell barrier (Giao *et al.*, 2012).

Infusions (herbal teas) and decoctions are used frequently to administer oral doses of herbs. Although some herbs are used as single ingredients, they are often prepared as mixtures, as reported by numerous ethnobotanical surveys. A study was carried out to identify the different types of interaction (synergistic, additive and antagonistic effects) that may be found in the antioxidant activities of preparations from mixtures of the popular herbs *Aloysia citrodora* (lemon verbena), *Foeniculum vulgare* (fennel), and *Mentha spicata* (spearmint). Herbs were prepared using traditional methods, and the effects after different periods of storage, up to 120 days, were also evaluated. Antioxidant activity was evaluated using DPPH radical scavenging activity, reducing power, and inhibition of lipid peroxidation by the β-carotene-linoleate system and the thiobarbituric acid reactive substances (TBARS) assay. Known antioxidant compounds such as total phenolics, flavonoids, ascorbic acid, and reducing sugars were also determined. Spearmint was found to possess the greatest antioxidant activity and the highest flavonoid content. The most potent antioxidant activity was found in combinations of different herbs, suggesting synergistic effects (Guimaraes *et al.*, 2011).

Total phenolic content of 28 kinds of herbal infusions made in China were measured by the Folin–Ciocalteu method, and their antioxidant capacities were evaluated using FRAP and TEAC assays. Generally, these beverages had high antioxidant capacities and total phenolic contents, and could be important dietary sources of antioxidant phenolics for prevention of diseases caused by oxidative stress (Fu *et al.*, 2011). In another study, antioxidant capacities of the infusions from 32 medicinal plants were measured by FRAP and TEAC assays, and total phenolic contents of the extracts were determined using the Folin–Ciocalteu method. The infusion of *Rhodiola sacra* Fu showed the highest antioxidant activity among the infusions from the 32 plants tested, followed by the infusion of *Polygonum multiflorum* Thunb stem, and the infusion of *P. multiflorum* Thunb root (Li *et al.*, 2007).

It is very difficult to summarize and compare antioxidant capacities of herbal infusions because different extraction and evaluation methods were used in the literature. The antioxidant capacities and total phenolic contents of some herbal infusions are summarized in Tables 5.1–5.3, which employed the same evaluation methods (Fu *et al.*, 2011; Li *et al.*, 2007), and correlation between FRAP values and TEAC values as well as correlation between the antioxidant capacities and total phenolic content are shown in Figures 5.1–5.3. A significant correlation between FRAP and TEAC values (Figure 5.1) suggested that antioxidant components in these beverages were capable of reducing oxidants and scavenging free radicals. In addition, the high correlation between antioxidant capacities and total phenolic contents (Figures 5.2 and 5.3) indicated that phenolic compounds could be one of the main components responsible for antioxidant activities of these beverages.

ANTIOXIDANT COMPONENTS IN HERBAL INFUSIONS

Herbal infusions often contain many different kinds of natural antioxidants. As an important category of phytochemicals, phenolic compounds universally exist in plants. They have attracted increasing attention as potential agents for preventing and treating many oxidative stress-related diseases. The fermented leaves and stems of *Cyclopia intermedia* are used to brew honeybush tea, an herbal tea indigenous to South Africa. The plant is also used to manufacture a sweet herbal infusion used for its restorative properties such as soothing coughs and alleviating bronchial complaints including tuberculosis, pneumonia, and catarrh. It is claimed to have low tannin content and no caffeine and contains various antioxidants. The leaves and stems of *C. intermedia* contain tyrosol, 2-{4-[O-α-apiofuranosyl-(1″→6′)-β-D-glucopyranosyloxy] phenyl}ethanol, 4-[O-α-apiofuranosyl-(1″→2′)-β-D-glucopyranosyloxy] benzaldehyde, five glycosylated flavonols, two isoflavones, four flavanones, two isoflavones, and two flavones. Because flavonoids are presumed to contribute significantly toward the scavenging effects of active oxygen species, the results indicated that the tentative claimed health-promoting properties may be attributed to the presence of the phenolics in *C. intermedia* (Kamara *et al.*, 2003). *Cyclopia subternata* is one of the 24 *Cyclopia* species that are used to brew honeybush tea, a unique South African herbal beverage with a pleasant taste and flavor. It contains various antioxidants, is very low tannin content, and has no caffeine. Many health properties are associated with regular consumption of the tea. Honeybush infusion has been noted as a tonic for colds and influenza, catarrh, and pulmonic tuberculosis, and is becoming well known for its effectiveness in alleviating menopausal symptoms in women. The unfermented leaves of *C. subternata* contain pinitol, shikimic acid,

TABLE 5.1 Antioxidant Capacities (Ferric Reducing Antioxidant Power, FRAP) of Herbal Infusions

Sample Number	Name of Sample	FRAP Values (mmol Fe(II)/l)
1	*Rhodiola sacra* Fu	26.575 ± 0.240
2	*Polygonum multiflorum* Thunb. (Stem)	24.895 ± 0.100
3	*Polygonum multiflorum* Thunb. (Root)	17.165 ± 0.085
4	*Dipsacus japonicus* Miq.	8.515 ± 0.015
5	*Epimedium brevicomum* Maxim.	8.275 ± 0.135
6	*Paeonia lactiflora* pall.	5.870 ± 0.255
7	*Ligustrum lucidum* Ait.	6.230 ± 0.305
8	*Cynomorium songaricum* Rupr.	8.335 ± 0.185
9	*Rehmannia glutihosa* Libosch.	8.115 ± 0.065
10	*Eclipta prostrata* L.	6.585 ± 0.045
11	*Cistanche deserticola* Y. C. Ma	3.885 ± 0.395
12	*Psoralea corylifolia* L.	1.385 ± 0.025
13	*Astragalus complanatus* R. Br.	3.345 ± 0.070
14	*Polygonatum sibiricum* Redoute	2.500 ± 0.025
15	*Morus alba* L.	2.635 ± 0.020
16	*Curculigo orchioides* Gaerten.	2.600 ± 0.035
17	*Cuscuta chinensis* Lam.	2.040 ± 0.075
18	*Angelica sinensis* (Oliv.) Diels	1.760 ± 0.020
19	*Eucommia ulmoides* Oliv.	1.965 ± 0.050
20	*Glycyrrhiza uralensis* Fisch.	0.575 ± 0.010
21	*Lycium barbarum* L.	1.060 ± 0.015
22	*Alpinia oxyphylla* Mig.	1.050 ± 0.020
23	*Dendrobium nobile* LindL.	0.880 ± 0.005
24	*Morinda officinalis* How	0.945 ± 0.015
25	*Astragalus membranaceus* (Fisch.) Bge.	0.610 ± 0.020
26	*Asparagus cochinchinensis* (Lour.) Merr	0.460 ± 0.010
27	*Atractylodes macrocephala* Koidz.	0.870 ± 0.010
28	*Lilium brownii* F. E. Brown	0.360 ± 0.005
29	*Ophiopogon japonicus* Ker-GawL	0.375 ± 0.010
30	*Dioscorea opposita* Thunb.	0.355 ± 0.005
31	*Tremella fuciformis* Berk.	0.245 ± 0.010
32	*Polygonatum odoratum* (Mill.) Druce	0.095 ± 0.003
33	Ping An Tang xue li ju hua cha	4687 ± 208
34	Ping An Tang mao geng zhu zhe shui	1003 ± 24
35	Ping An Tang shen ju cha	7234 ± 212
36	Ping An Tang luo han guo wu hua cha	5452 ± 88
37	Ping An Tang suan mei tang	506 ± 11
38	Ping An Tang huo ma ren	1504 ± 101
39	Ping An Tang li yan cha	30581 ± 1285
40	Ping An Tang shi gan cha	26314 ± 63
41	Qing Xin Tang ju hua xue li cha	2722 ± 116
42	Qing Xin Tang suan mei tang	3114 ± 168
43	Qing Xin Tang mao geng zhu zhe shui	3162 ± 116
44	Qing Xin Tang luo han guo wu hua cha	5246 ± 266
45	Qing Xin Tang gan mao cha	10382 ± 845
46	Qing Xin Tang zhi ke hua tan tang	11697 ± 777
47	Qing Xin Tang hou zheng tang	12490 ± 615
48	Qing Xin Tang jiang huo wang	25454 ± 1175
49	Qing Xin Tang er shi si wei	13252 ± 225
50	Deng lao liang cha	8341 ± 322
51	Wang lao ji (ting zhuang)	3508 ± 39
52	Wang lao ji (he zhuang)	3764 ± 151
53	Qing liang cha (he zhuang)	1550 ± 40
54	Nian ci an run (qing xing lü se ting zhuang)	812 ± 16
55	Nian ci an run (chun cui hong se ting zhuang)	392 ± 14
56	Bao Qing Tang xue li ju hua cha	1218 ± 28
57	Pan Gao Shou liang cha	1159 ± 20
58	Er shi si wei	1785 ± 55
59	Bai Yun Shan liang cha	1510 ± 20
60	Ben cao mi liang cha	4279 ± 82

p-coumaric acid, 4-glucosyltyrosol, epigallocatechin, gallate, orobol, hesperedin, narirutin, eriocitrin, a glycosylated flavan, luteolin, 5-deoxyluteolin, scolymoside, mangiferin, and C-6-glucosylkaempferol (Kamara *et al.*, 2004).

Burrito tea originates from the leaves of *Wendita calysina*, an indigenous Paraguayan plant, which is commonly consumed in South America and in western countries. Phytochemical investigation of this species has led to the isolation of 14 constituents, among

TABLE 5.2 Antioxidant Capacities (Trolox Equivalent Antioxidant Capacity, TEAC) of Herbal Infusions

Sample Number	Name of Sample	TEAC Values (mmol Trolox/l)
1	*Rhodiola sacra* Fu	24.245 ± 0.255
2	*Polygonum multiflorum* Thunb. (Stem)	24.700 ± 0.235
3	*Polygonum multiflorum* Thunb. (Root)	12.895 ± 0.185
4	*Dipsacus japonicus* Miq.	3.915 ± 0.050
5	*Epimedium brevicomum* Maxim.	4.330 ± 0.050
6	*Paeonia lactiflora* pall.	4.130 ± 0.085
7	*Ligustrum lucidum* Ait.	4.490 ± 0.055
8	*Cynomorium songaricum* Rupr.	3.900 ± 0.120
9	*Rehmannia glutihosa* Libosch.	4.215 ± 0.095
10	*Eclipta prostrata* L.	4.140 ± 0.010
11	*Cistanche deserticola* Y. C. Ma	1.405 ± 0.015
12	*Psoralea corylifolia* L.	2.075 ± 0.040
13	*Astragalus complanatus* R. Br.	0.580 ± 0.002
14	*Polygonatum sibiricum* Redoute	2.195 ± 0.010
15	*Morus alba* L.	2.355 ± 0.080
16	*Curculigo orchioides* Gaerten.	1.350 ± 0.002
17	*Cuscuta chinensis* Lam.	1.125 ± 0.010
18	*Angelica sinensis* (Oliv.) Diels	1.365 ± 0.010
19	*Eucommia ulmoides* Oliv.	1.190 ± 0.004
20	*Glycyrrhiza uralensis* Fisch.	1.050 ± 0.020
21	*Lycium barbarum* L.	0.535 ± 0.004
22	*Alpinia oxyphylla* Mig.	0.810 ± 0.015
23	*Dendrobium nobile* LindL.	0.875 ± 0.015
24	*Morinda officinalis* How	0.460 ± 0.005
25	*Astragalus membranaceus* (Fisch.) Bge.	0.430 ± 0.001
26	*Asparagus cochinchinensis* (Lour.) Merr	0.475 ± 0.020
27	*Atractylodes macrocephala* Koidz.	0.340 ± 0.004
28	*Lilium brownii* F. E. Brown	0.300 ± 0.002
29	*Ophiopogon japonicus* Ker-GawL	0.245 ± 0.004
30	*Dioscorea opposita* Thunb.	0.140 ± 0.002
31	*Tremella fuciformis* Berk.	0.155 ± 0.002
32	*Polygonatum odoratum* (Mill.) Druce	0.075 ± 0.004
33	Ping An Tang xue li ju hua cha	2988 ± 177
34	Ping An Tang mao geng zhu zhe shui	613 ± 15

TABLE 5.2 Antioxidant Capacities (Trolox Equivalent Antioxidant Capacity, TEAC) of Herbal Infusions—cont'd

Sample Number	Name of Sample	TEAC Values (mmol Trolox/l)
35	Ping An Tang shen ju cha	504 ± 7
36	Ping An Tang luo han guo wu hua cha	3529 ± 60
37	Ping An Tang suan mei tang	2673 ± 94
38	Ping An Tang huo ma ren	386 ± 7
39	Ping An Tang li yan cha	19296 ± 692
40	Ping An Tang shi gan cha	16269 ± 230
41	Qing Xin Tang ju hua xue li cha	2040 ± 85
42	Qing Xin Tang suan mei tang	2176 ± 64
43	Qing Xin Tang mao geng zhu zhe shui	2480 ± 74
44	Qing Xin Tang luo han guo wu hua cha	3669 ± 36
45	Qing Xin Tang gan mao cha	6188 ± 238
46	Qing Xin Tang zhi ke hua tan tang	6695 ± 114
47	Qing Xin Tang hou zheng tang	6499 ± 46
48	Qing Xin Tang jiang huo wang	6474 ± 19
49	Qing Xin Tang er shi si wei	6310 ± 321
50	Deng lao liang cha	3438 ± 76
51	Wang lao ji (ting zhuang)	2083 ± 85
52	Wang lao ji (he zhuang)	2348 ± 13
53	Qing liang cha (he zhuang)	637 ± 4
54	Nian ci an run (qing xing lü se ting zhuang)	428 ± 16
55	Nian ci an run (chun cui hong se ting zhuang)	250 ± 6
56	Bao Qing Tang xue li ju hua cha	445 ± 18
57	Pan Gao Shou liang cha	537 ± 13
58	Er shi si wei	825 ± 13
59	Bai Yun Shan liang cha	446 ± 9
60	Ben cao mi liang cha	2351 ± 93

them two new flavanonols, dihydroquercetagetin and 3,5,6,7,4′-pentahydroxyflavanonol, in addition to several known methoxyflavones, methoxyflavonols, phenylethanoid glycosides, and benzoic acid derivatives. Quantitative determination of phenolic constituents from burrito water infusions was also performed by HPLC-UV-DAD (Piccinelli *et al.*, 2004). In a study, the profile of caffeoylquinic, feruloylquinic, and dicaffeoylquinic acids in 20 different Brazilian herbal infusions was established using a reversed-phase HPLC method. The

TABLE 5.3 Total Phenolic Contents (Gallic Acid Equivalent, GAE) of Herbal Infusions

Sample Number	Name of Sample	Total Phenolic Contents (g GAE/l)
1	*Rhodiola sacra* Fu	2.540 ± 0.069
2	*Polygonum multiflorum* Thunb. (Stem)	2.116 ± 0.059
3	*Polygonum multiflorum* Thunb. (Root)	1.696 ± 0.031
4	*Dipsacus japonicus* Miq.	1.607 ± 0.355
5	*Epimedium brevicomum* Maxim.	1.186 ± 0.014
6	*Paeonia lactiflora* pall.	0.946 ± 0.011
7	*Ligustrum lucidum* Ait.	1.054 ± 0.006
8	*Cynomorium songaricum* Rupr.	0.846 ± 0.014
9	*Rehmannia glutihosa* Libosch.	0.948 ± 0.030
10	*Eclipta prostrata* L.	0.810 ± 0.010
11	*Cistanche deserticola* Y. C. Ma	0.698 ± 0.006
12	*Psoralea corylifolia* L.	0.835 ± 0.008
13	*Astragalus complanatus* R. Br.	0.814 ± 0.028
14	*Polygonatum sibiricum* Redoute	0.626 ± 0.005
15	*Morus alba* L.	0.538 ± 0.003
16	*Curculigo orchioides* Gaerten.	0.512 ± 0.005
17	*Cuscuta chinensis* Lam.	0.455 ± 0.008
18	*Angelica sinensis* (Oliv.) Diels	0.376 ± 0.012
19	*Eucommia ulmoides* Oliv.	0.290 ± 0.002
20	*Glycyrrhiza uralensis* Fisch.	0.416 ± 0.004
21	*Lycium barbarum* L.	0.264 ± 0.003
22	*Alpinia oxyphylla* Mig.	0.205 ± 0.002
23	*Dendrobium nobile* LindL.	0.240 ± 0.004
24	*Morinda officinalis* How	0.198 ± 0.006
25	*Astragalus membranaceus* (Fisch.) Bge.	0.159 ± 0.002
26	*Asparagus cochinchinensis* (Lour.) Merr	0.150 ± 0.004
27	*Atractylodes macrocephala* Koidz.	0.184 ± 0.005
28	*Lilium brownii* F. E. Brown	0.154 ± 0.003
29	*Ophiopogon japonicus* Ker-GawL	0.129 ± 0.001
30	*Dioscorea opposita* Thunb.	0.124 ± 0.003
31	*Tremella fuciformis* Berk.	0.118 ± 0.001
32	*Polygonatum odoratum* (Mill.) Druce	0.125 ± 0.002
33	Ping An Tang xue li ju hua cha	406 ± 14
34	Ping An Tang mao geng zhu zhe shui	74 ± 3

TABLE 5.3 Total Phenolic Contents (Gallic Acid Equivalent, GAE) of Herbal Infusions—cont'd

Sample Number	Name of Sample	Total Phenolic Contents (g GAE/l)
35	Ping An Tang shen ju cha	53 ± 1
36	Ping An Tang luo han guo wu hua cha	201 ± 9
37	Ping An Tang suan mei tang	392 ± 9
38	Ping An Tang huo ma ren	128 ± 5
39	Ping An Tang li yan cha	1395 ± 68
40	Ping An Tang shi gan cha	1192 ± 11
41	Qing Xin Tang ju hua xue li cha	249 ± 4
42	Qing Xin Tang suan mei tang	390 ± 3
43	Qing Xin Tang mao geng zhu zhe shui	347 ± 8
44	Qing Xin Tang luo han guo wu hua cha	572 ± 5
45	Qing Xin Tang gan mao cha	844 ± 13
46	Qing Xin Tang zhi ke hua tan tang	909 ± 37
47	Qing Xin Tang hou zheng tang	875 ± 19
48	Qing Xin Tang jiang huo wang	1028 ± 55
49	Qing Xin Tang er shi si wei	1007 ± 39
50	Deng lao liang cha	443 ± 13
51	Wang lao ji (ting zhuang)	147 ± 2
52	Wang lao ji (he zhuang)	148 ± 1
53	Qing liang cha (he zhuang)	72 ± 1
54	Nian ci an run (qing xing lü se ting zhuang)	56 ± 1
55	Nian ci an run (chun cui hong se ting zhuang)	32 ± 1
56	Bao Qing Tang xue li ju hua cha	70 ± 1
57	Pan Gao Shou liang cha	68 ± 1
58	Er shi si wei	99 ± 1
59	Bai Yun Shan liang cha	114 ± 4
60	Ben cao mi liang cha	162 ± 5

3-caffeoylquinic acid was found in 75% of the infusions (18–5200 mg/kg), while the 5-caffeoylquinic acid was found in 80% of them (34–9990 mg/kg). The highest concentrations of these isomers were found in 'Cha mate tostado'. The 4-caffeoylquinic acid appeared in 55% of the herbs (23–1600 mg/kg), with 'Chapeu-de-couro' and 'Carqueja' being the richer ones. The 4-feruloylquinic acid was found in 35% of the infusions (51–1800 mg/kg) and the 5-feruloylquinic acid was detected in 45%

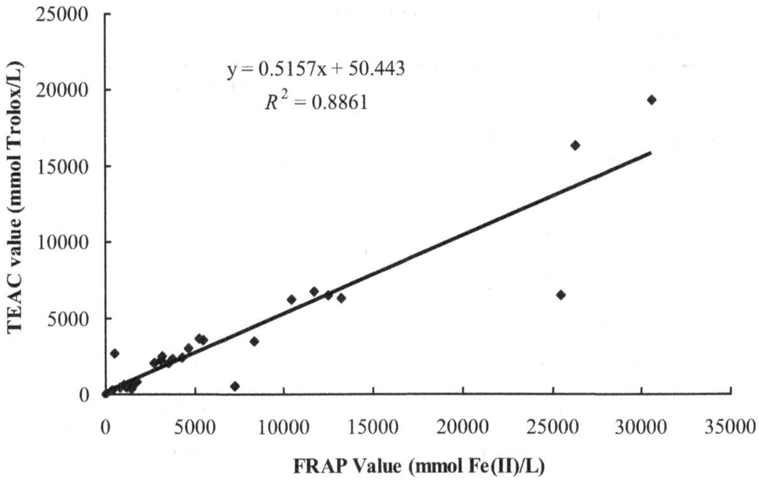

FIGURE 5.1 Correlation between ferric reducing antioxidant power (FRAP) values and Trolox equivalent antioxidant capacity (TEAC) values.

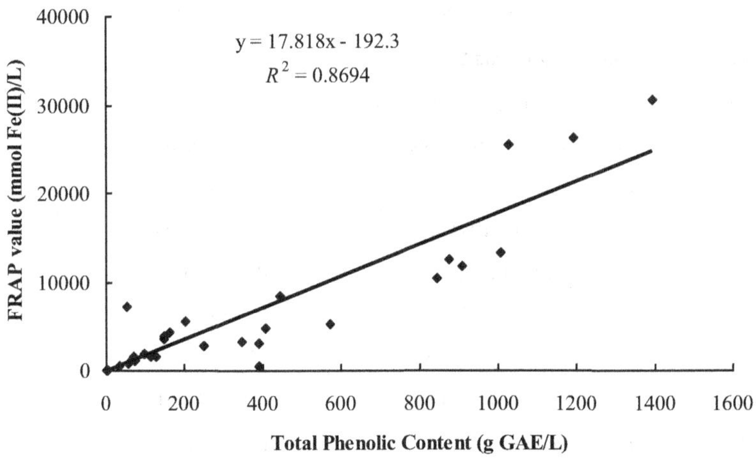

FIGURE 5.2 Correlation between the antioxidant capacities (FRAP values) and total phenolic content. GAE, gallic acid equivalents.

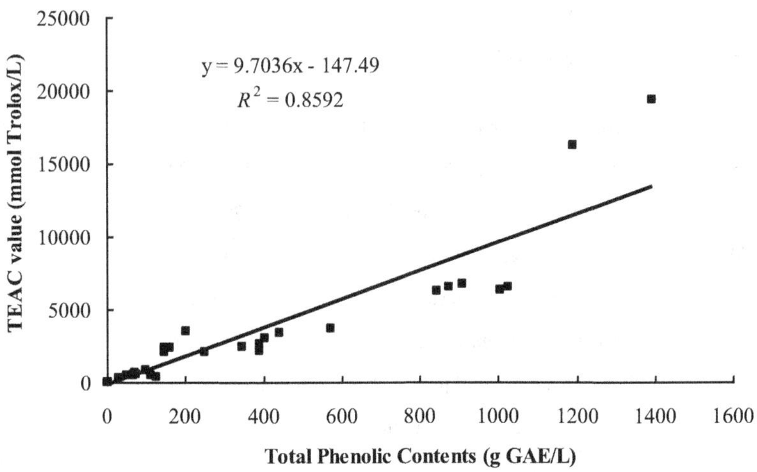

FIGURE 5.3 Correlation between the antioxidant capacities (TEAC values) and total phenolic content. GAE, gallic acid equivalents.

of them (39–1300 mg/kg). In spite of the low concentrations (30–560 mg/kg), the 3,5-dicaffeoylquinic acid appeared in 50% of the herbs. The distribution of the other dicaffeoylquinic isomers in the infusions was poor. The 3,4-dicaffeoylquinic acid was detected in 20% of them (190–1700 mg/kg), while only 15% of the herbs comprised the 4,5-dicaffeoylquinic isomer. Considerable amounts of this last isomer were found just in 'Salvia' (1100 ± 270 mg/kg) (da Silva *et al.*, 2006). In another study, chromatographic techniques (HPLC and HPTLC) were used for qualitative and quantitative determination of eriocitrin, luteolin 7-O-rutinoside, luteolin 7-O-β-glucuronide, lithospermic acid, rosmarinic acid, and methyl rosmarinate, together with other known compounds in commercial herbal teas from the Lamiaceae family: peppermint leaf (*Menthae piperitae* folium), melissa leaf (*Melissae* folium), and sage leaf (*Salviae officinalis* folium). The investigated herbal teas delivered polyphenols in high amounts, up to 182.2 mg for the infusion of one peppermint tea bag (Fecka and Turek, 2007). In addition, verbena and lemon verbena aqueous preparations were investigated for their constituents, especially polyphenols by HPLC/DAD/ESI/MS analysis because they are used worldwide as herbal teas. The main class of compounds of these plants were phenylpropanoids (from 16 to 120 mg/g of dried extract), and verbascoside was the most abundant in all the preparations up to 97% of the total phenylpropanoids. Iridoids, hastatoside, and verbenalin together with flavonoids, mono- and di-glucuronidic derivatives of luteolin and apigenin were also found (Bilia *et al.*, 2008).

Plectranthus barbatus, known as 'falso boldo' in Brazil, is used in herbal tea or cooked as a vegetable. Infusions and decoctions of leaves from *P. barbatus* were analyzed for their antioxidant activities. The decoction showed high antioxidant activity (IC_{50} = 45.8 ± 0.5 μg of dry extract/ml in the DPPH test; IC_{50} = 69.8 ± 3.1 μg of dry extract/ml in the β-carotene-linoleic acid test). Rosmarinic acid, scutellarein 4′-methyl ether 7-O-glucuronide and (16S)-coleon E were the main constituents identified (Fale *et al.*, 2009). In another study, phenolic compounds in an aqueous infusion of leaves of *Ficus deltoidea* (Moraceae), a well-known herbal tea in Malaysia, were analyzed by HPLC coupled to photodiode array and fluorescence detectors and an electrospray ionization tandem mass spectrometer. Following chromatography of extracts on a reversed-phase C_{12} column, 25 flavonoids were characterized and/or tentatively identified with the main constituents flavan-3-ol monomers, proanthocyanidins, and C-linked flavone glycosides. The proanthocyanidins were dimers and trimers comprising (epi)catechin and (epi)afzelechin units. No higher-molecular-weight proanthocyanidin polymers were detected. The antioxidant activity of *F. deltoidea* extract was analyzed using HPLC with online antioxidant detection. This revealed that 85% of the total antioxidant activity of the aqueous *F. deltoidea* infusion was attributable to the flavan-3-ol monomers and the proanthocyanidins (Omar *et al.*, 2011). In addition, 'Pichi' or 'pichi romero' (*Fabiana imbricata* R. et. P., Solanaceae) is a Chilean plant used as a tea in the Andean regions of Chile and Argentina. The phenolic constituents identified in the tea were chlorogenic acid (3-O-caffeoylquinic acid), p-hydroxyacetophenone, scopoletin, and quercetin derivatives. The glycosides were mainly glucosides from p-hydroxyacetophenone and scopoletin while di- and tri-glycosides from quercetin were the main flavonoids. The content of the main phenolic compounds in the tea (g/100 g lyophilized infusion) was 0.8–1.9% for scopoletin, 0.4–6.2% for p-hydroxyacetophenone, and 2.1–4.3% for rutin, respectively. The health-promoting properties reported for this herbal tea can be associated with the presence of several phenolics with known antioxidant, diuretic, and anti-inflammatory activities (Quispe *et al.*, 2012).

In order to calculate the dietary exposure to rooibos herbal tea flavonoids and phenolic acids, representative content values for the principal phenolic compounds and total antioxidant capacity of fermented rooibos infusion, taking into account variation caused by production seasons (2009, 2010, and 2011) and quality grades (A, B, C, and D), were determined for samples (n = 114) from different geographical areas and producers. The major phenolic constituents were isoorientin and orientin (> 10 mg/l), with quercetin-3-O-robinobioside, phenylpyruvic acid glucoside, and aspalathin present at > 5 mg/l. Isovitexin, vitexin, and hyperoside were present at < 3 mg/l. Rutin, ferulic acid, and isoquercitrin were present at < 2 mg/l. Nothofagin was present at < 1 mg/l. Only traces of luteolin-7-O-glucoside and the aglycones quercetin, luteolin, and chrysoeriol were present. Substantial variation was observed in the individual content values of the phenolic compounds and total antioxidant capacity within production seasons and quality grades (Joubert *et al.*, 2012).

References

Abderrahim, F., Estrella, S., Susin, C., Arribas, S.M., Gonzalez, M.C., Condezo-Hoyos, L., 2011. The antioxidant activity and thermal stability of lemon verbena (*Aloysia triphylla*) infusion. J. Med. Food 14, 517–527.

Apak, R., Guclu, K., Ozyurek, M., Karademir, S.E., Ercag, E., 2006. The cupric ion reducing antioxidant capacity and polyphenolic content of some herbal teas. Int. J. Food Sci. Nutr. 57, 292–304.

Arthur, H., Joubert, E., De Beer, D., Malherbe, C., Witthuhn, R.C., 2011. Phenylethanoid glycosides as major antioxidants in *Lippia multiflora* herbal infusion and their stability during steam pasteurisation of plant material. Food Chem. 127, 581–588.

Bilia, A.R., Giomi, M., Innocenti, M., Gallori, S., Vincieri, F.F., 2008. HPLC-DAD-ESI-MS analysis of the constituents of aqueous preparations of verbena and lemon verbena and evaluation of the antioxidant activity. J. Pharm. Biomed. Anal. 46, 463–470.

da Silva, A.I., Santana, C.S., Pivato, S.C.L., De Maria, C.A.B., Moreira, R.F.A., 2006. Chlorogenic acid profile of commercial Brazilian herbal infusions. Sci. Alimen. 26, 173–180.

Deetae, P., Parichanon, P., Trakunleewatthana, P., Chanseetis, C., Lertsiri, S., 2012. Antioxidant and anti-glycation properties of Thai herbal teas in comparison with conventional teas. Food Chem. 133, 953–959.

Fale, P.L., Borges, C., Madeira, P.J.A., Ascensao, L., Araujo, M.E.M., Florencio, M.H., Serralheiro, M.L.M., 2009. Rosmarinic acid, scutellarein 4 ′-methyl ether 7-O-glucuronide and (16S)-coleon E are the main compounds responsible for the antiacetylcholinesterase and antioxidant activity in herbal tea of *Plectranthus barbatus* ("falso boldo. Food Chem. 114, 798–805.

Fecka, I., Turek, S., 2007. Determination of water-soluble polyphenolic compounds in commercial herbal teas from Lamiaceae: Peppermint, melissa, and sage. J. Agric. Food Chem. 55, 10908–10917.

Fu, L., Xu, B.T., Gan, R.Y., Zhang, Y., Xu, X.R., Xia, E.Q., Li, H.B., 2011. Total phenolic contents and antioxidant capacities of herbal and tea infusions. Int. J. Mol. Sci. 12, 2112–2124.

Giao, M.S., Gomes, S., Madureira, A.R., Faria, A., Pestana, D., Calhau, C., Pintado, M.E., Azevedo, I., Malcata, F.X., 2012. Effect of in vitro digestion upon the antioxidant capacity of aqueous extracts of *Agrimonia eupatoria, Rubus idaeus, Salvia* sp and Satureja montana. Food Chem. 131, 761–767.

Guimaraes, R., Barros, L., Carvalho, A.M., Ferreira, I.C.F.R., 2011. Infusions and decoctions of mixed herbs used in folk medicine: Synergism in antioxidant potential. Phytother. Res. 25, 1209–1214.

Joubert, E., de Beer, D., 2012. Phenolic content and antioxidant activity of rooibos food ingredient extracts. J. Food Compos. Anal. 27, 45–51.

Joubert, E., Beelders, T., de Beer, D., Malherbe, C.J., de Villiers, A.J., Sigge, G.O., 2012. Variation in phenolic content and antioxidant activity of fermented rooibos herbal tea infusions: Role of production season and quality grade. J. Agric. Food Chem. 60, S9171–S9179.

Kamara, B.I., Brandt, E.V., Ferreira, D., Joubert, E., 2003. Polyphenols from honeybush tea (*Cyclopia intermedia*). J. Agric. Food Chem. 51, 3874–3879.

Kamara, B.I., Brand, D.J., Brandt, E.V., Joubert, E., 2004. Phenolic metabolites from honeybush tea (*Cyclopia subternata*). J. Agric. Food Chem. 52, 5391–5395.

Li, H.B., Jiang, Y., Wong, C.C., Cheng, K.W., Chen, F., 2007. Evaluation of two extraction methods for the extraction of antioxidants from medicinal plants. Anal. Bioanal. Chem. 388, 483–488.

Omar, M.H., Mullen, W., Crozier, A., 2011. Identification of proanthocyanidin dimers and trimers, flavone C-glycosides, and antioxidants in *Ficus deltoidea*, a Malaysian herbal tea. J. Agric. Food Chem. 59, 1363–1369.

Quispe, C., Viveros-Valdez, E., Schmeda-Hirschmann, G., 2012. Phenolic constituents of the Chilean herbal tea *Fabiana imbricata* R. et P. Plant Food Human Nutr. 67, 242–246.

Piccinelli, A.L., De Simone, F., Passi, S., Rastrelli, L., 2004. Phenolic constituents and antioxidant activity of *Wendita caiysina* leaves (Burrito), a folk Paraguayan tea. J. Agric. Food Chem. 52, 5863–5868.

Poblete, A., Lopez-Alarcon, C., Lissi, E., Campos, A.M., 2009. Oxygen radical antioxidant capacity (ORAC) values of herbal teas obtained employing different methodologies can provide complementary data. J. Chilean Chem. Soc. 54, 154–157.

Rocha-Guzman, N.E., Medina-Medrano, J.R., Gallegos-Infante, J.A., Gonzalez-Laredo, R.F., Ramos-Gomez, M., Reynoso-Camacho, R., Guzman-Maldonado, H., Gonzalez-Herrera, S.M., 2012. Chemical evaluation, antioxidant capacity, and consumer acceptance of several oak infusions. J. Food Sci. 77, 162–166.

Rojo, L.E., Benites, J., Lopez, J., Rojas, M., Diaz, P., Ordonez, J., Pastene, E., 2009. Antioxidant capacity and polyphenolic content of twelve traditionally used herbal medicinal infusions from the South American Andes. Bol. Latin. Carib. Plant. Med. Arom. 8, 498–508.

Shikanga, E.A., Combrinck, S., Regnier, T., 2010. South African *Lippia* herbal infusions: Total phenolic content, antioxidant and antibacterial activities. South Afric. J. Botany 76, 567–571.

Valentao, P., Fernandes, E., Carvalho, F., Andrade, P.B., Seabra, R.M., Bastos, M.D., 2002. Studies on the antioxidant activity of *Lippia citriodora* infusion: Scavenging effect on superoxide radical, hydroxyl radical and hypochlorous acid. Biol. Pharm. Bull. 25, 1324–1327.

Valentao, P., Fernandes, E., Carvalho, F., Andrade, P.B., Seabra, R.M., Bastos, M.L., 2003. Hydroxyl radical and hypochlorous acid scavenging activity of small centaury (*Centaurium erythraea*) infusion: A comparative study with green tea (*Camellia sinensis*). Phytomedicine 10, 517–522.

Vinokur, Y., Rodov, V., Reznick, N., Goldman, G., Horev, B., Umiel, N., Friedman, H., 2006. Rose petal tea as an antioxidant rich beverage: Cultivar effects. J. Food Sci. 71, 42–47.

CHAPTER

6

Antioxidant Capacity of Soft Drinks

Oreste V. Brenna

Department of Food, Environmental and Nutritional Sciences, University of Milan, Milan, Italy

CHAPTER POINTS

- Beverages are necessary for human life.
- Billions of liters of soft drinks are consumed annually around the world.
- They can be categorized into two main groups containing, respectively, caramel and caffeine or diluted fruit juice.
- Their antioxidant properties depend on their composition, mainly in the case of fruit-juice-containing soft drinks.
- There are positive and negative aspects related to the consumption of soft drinks.
- Positive aspects include refreshing and appealing features to quench thirst, and consumption of some antioxidant compounds
- Negative aspects include an excessive consumption of calories in the case of sugary soft drinks and damage to tooth integrity caused by their acidity.

INTRODUCTION

Human beings are mostly made of water (67%). Virtually all biochemical reactions in our bodies require an aqueous medium. It follows that fluid intake in our diet is essential. Most of the water needed to satisfy our daily needs is taken in with beverages, of which the most available (at least in the Western world) is drinking water. Water is also taken in with liquid foods such as milk (87% water) and, without realizing it, with the so-called solid foods, which contain considerable amounts of water, ranging from 70 to 95% in the case of fruit and vegetables, to 45–75% in meat, and also in butter (13%). Over time and in different cultures, different ingredients have been added to improve the taste of water and to make its consumption more appealing.

In Far East civilizations, water is the principal ingredient of leaf infusions, such as tea, while in pre-Columbian America it is the main ingredient of beverages made from roasted seeds, such as coffee and cocoa. In the Mediterranean region, water has mainly been used for the preparation of fermented beverages (beer) as well as to dilute the highly tannic wines of the old days. (In ancient times, wine was obtained by soaking the grapes with their stems, rich in tannins, while today, the grapes are destalked, thus obtaining a less tannic wine.) In the latter case, the addition of a red wine rich in tannins also assumed the function of sterilizing the water source (i.e, by reaction with microorganisms and proteinaceous compounds, if present). This function in the first two examples was obtained by boiling the water to prepare infusions or extracts.

ORIGIN OF SOFT DRINKS

The addition of one or more components to drinking water has led in recent times to the creation of hundreds of beverages with different properties. For example, a more refreshing drink can be obtained with the simple addition of a few drops of lemon juice. Other changes are aimed to make drinking water more appealing to the taste and, more recently, beneficial to the health of the consumer.

These beverages can be roughly divided into two large groups: beverages made from extracts of leaves or seeds (other than tea or coffee) with artificial colorings, and beverages based on fruit juices. These two groups share the general addition of carbon dioxide to the beverage, which contributes to their thirst-quenching effect: they can all be therefore considered under the denomination of 'soft drinks.' They have a very important market

Processing and Impact on Antioxidants in Beverages
http://dx.doi.org/10.1016/B978-0-12-404738-9.00006-4

share that grew by 2.3% in 2011, and it is forecast that this will increase on average by 3% per year over the next 5 years (Carbonated Soft Drinks, 2012). All of these beverages are also characterized by the absence of ethyl alcohol, and they can be freely consumed by children and drivers.

In many cases, however, soft drinks contain caffeine, a central nervous system and metabolic stimulant derived from the kola (cola) nut extract, which is added as a flavoring agent, even if the amount which is usually present is less than that which is found in tea and coffee, with the exception of energy drinks. In cola-type beverages, caffeine is considered as 'generally recognized as safe' (GRAS) up to a maximum use level of 0.02% (CFR, 2004). Due to the bitter flavor and astringency caused by the presence of both caffeine and a rather significant amount of procyanidins (Atawodi *et al.*, 2007), sugar is added to the drinks containing kola nut extract.

The health aspect related to the excessive intake of sugars has resulted in the creation of 'sugar free' synthetically sweetened soft drinks, and their encouraged consumption, especially by children, whose problem of early obesity is becoming more important in all Western countries. This effort seems to have produced some results, since a recent paper reports that children's consumption of beverages sweetened with low-calorie or no-calorie sweeteners has more than doubled over the past decade in the United States, moving from 6.1% in 1999–2000 to 12.5% in 2007–2008, together with a smaller but significant increment in the adult consumption (Sylvetsky *et al.*, 2012).

Another division can be made along the main characterizing components between soft drinks containing caramel, fruit juice, or caffeine. This division, however, is only indicative, because several soft drinks that can be considered under more than one of these aspects are commercially available.

CARAMEL-CONTAINING SOFT DRINKS

Color is the first characteristic perceived by the consumer, if the physical form of the product, solid or liquid, is not considered. It is therefore clear that a beverage can be more pleasing to the eye if it has a color that characterizes it.

The first cola-based soft drinks could occur in different colors depending on the quantity of added extract which, however, were not chemically stable over time. It was then realized that 'adding color' to the beverage with an artificial dye would give a constant quality that would characterize the drink itself.

If we consider at first the caramel-containing soft drinks, they are characterized by the presence of an artificial coloring, caramel, particularly Class IV Caramel (mainly Sulfite Ammonia Process Caramel Color), as well as small amounts of 'natural extracts'.

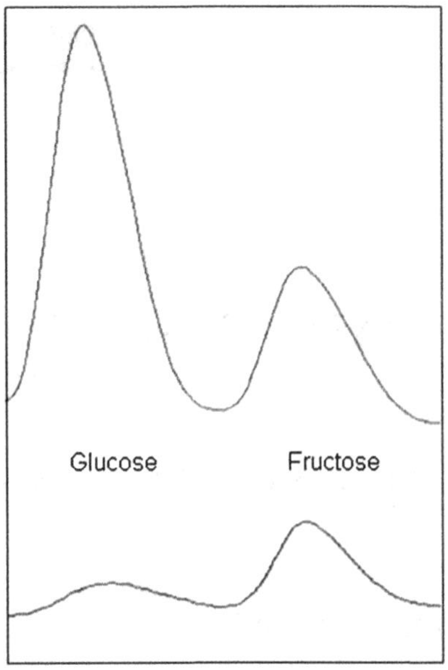

FIGURE 6.1 Main differences between sugary and diet colas. High-performance liquid chromatography (HPLC) analysis of sugars present in normal cola A (upper trace) and in diet cola F (lower trace) are shown. A small amount of sugars can be present in diet colas due to caramel.

The IV Caramel color is obtained by the controlled heating of carbohydrates in the presence of both ammonia and sulfite-containing compounds (Licht *et al.*, 1992).

In Europe this artificial color is identified as caramel E150d and the EU Scientific Committee for Food (SCF) and the Joint FAO/WHO Expert Committee on Food Additives (JECFA) had established for E150d an acceptable daily intake (ADI) ranging from 160 to 200 mg/kg body weight per day. This value is related to the fact that E150d contains a chemical derivative, 4-methyl imidazole (4-MeI), which has been shown to possibly exhibit carcinogenic properties (Moon and Shibamoto, 2011). Common brands of cola drinks colored with caramel may contain highly variable amounts of 4-MeI, from traces to 0.6 mg/l (Schlee *et al.*, 2012).

However, based on all available data, the European Food Safety Authority (EFSA) Panel on Food Additives and Nutrient Sources added to Food (ANS) concluded that all the four caramel colors admitted, E150 a, b, c, and d, are neither genotoxic, nor carcinogenic, and that there is no evidence to show they have any adverse effects on human reproduction or child development.

Based on their similar properties, the ANS Panel has established an ADI of 300 mg/kg body weight per day applicable to all four colors; in the case of one of them, caramel E150c, the Panel has set a more restrictive ADI of 100 mg/kg body weight per day (EFSA, 2011).

Caramel may still contain a small part of the sugars resulting from those employed for its preparation, as can be seen in Figure 6.1, which shows part of the high-performance liquid chromatography (HPLC) trace

TABLE 6.1 Glucose and Fructose Content in Some Colas

	Glucose (g/l)		Fructose (g/l)	
Sample	*Mean*	*DS*	*Mean*	*DS*
Cola A	55.08	0.02	48.81	0.38
Cola B	48.32	3.38	32.71	3.56
Cola C	57.11	1.20	49.06	2.34
Cola D	0.16	0.01	0.29	0.01
Cola E	0.05	0.01	nd	
Cola F	0.06	0.01	nd	
Cola G	77.19	1.60	37.19	0.94
Cola H	59.14	0.56	51.57	0.24
Cola I	61.23	0.88	37.42	0.80
Chinotto A	62.17	0.25	49.34	0.22
Chinotto B	71.14	4.05	40.65	1.27

The amounts of sugars present in different analyzed cola drinks are shown. A,B,C were normal colas; D,E,F were light colas; G and H were colas without caffeine and I was cola with lemon flavor added. Two samples of chinotto drinks were also analyzed. nd, not detected.
Adapted from Brenna et al. *(2009).*

TABLE 6.2 5-(Hydroxymethyl)-2-Furfural (HMF) and Antioxidant Capacity of Some Soft Drinks Containing Caramel

	HMF (mg/l)		GAE (mg/l)		FRAP (TE mM)	
Sample	*Mean*	*DS*	*Mean*	*DS*	*Mean*	*DS*
Cola A	1.05	0.01	98.47	4.39	0.30	0.03
Cola B	1.36	0.01	80.62	5.27	0.26	0.03
Cola C	2.01	0.04	79.29	0.13	0.24	0.02
Cola D	1.36	0.01	57.54	0.75	0.25	0.02
Cola E	0.84	0.01	59.31	2.76	0.25	0.02
Cola F	1.63	0.01	56.38	1.88	0.24	0.01
Cola G	1.33	0.04	72.10	2.51	0.25	0.02
Cola H	2.01	0.04	87.99	6.66	0.27	0.03
Cola I	0.81	0.03	72.19	2.13	0.32	0.02
Chinotto A	79.57	0.14	213.81	10.05	1.44	0.02
Chinotto B	1.85	0.01	136.03	4.77	0.78	0.01

HMF content, reducing power measured with the Folin–Ciocalteu reagent; gallic acid equivalent (GAE); ferric reducing antioxidant power (FRAP), as Trolox® equivalent (TE), in the tested soft drinks. Chinotto drinks contain more antioxidant then colas, due to their polyphenol content.

obtained for a light cola we analyzed (Brenna, *et al.*, 2009): In the sample, a small amount of sugars was found and, therefore, even if minimal, the labels actually should report a caloric value related to the amount of glucose and fructose found in them (Table 6.1).

ANTIOXIDANT ACTIVITY OF CARAMEL-CONTAINING SOFT DRINKS

Caramel also contains compounds with reducing activity, such as 5-(hydroxymethyl)-2-furfural (HMF, Table 6.2) and many other compounds arising from the reaction of sugars and amino compounds present in the initial mixture, like the browning derivatives from the early stages of the Maillard reaction, which give the typical color to caramel. Similar compounds are formed in all heating steps when reducing sugars and amino groups are present together and are known as melanoidins (Wang *et al.*, 2011). The antioxidant activity of caramel-based soft drinks is therefore probably not due only to the minimal amount of natural extracts present, but also to the mixture of chemical compounds which are present in the added coloring agent.

Besides the phenolics present in the kola nut extract (Atawodi *et al.*, 2007), there is another important natural compound possessing an antioxidant activity, related in particular to its excellent •OH scavenging activity (León-Carmona and Galano, 2011). This compound is not a polyphenol but an alkaloid, caffeine, the concentration of which was evaluated for a wide range of soft drinks and found to vary from 90 to 170 mg/l (Chou and Bell, 2007).

Table 6.2 shows some other characteristics of several cola- and caramel-containing soft drinks, along with two varieties of chinotto drink. The latter is a quite popular product in Italy, almost the only country where this fruit is grown. It can be seen that the first group of colas, containing sugars, and the second group, containing non-caloric sweeteners, are noticeably different for their total reducing substances content, when determined using the Folin–Ciocalteu reagent (Singleton and Rossi, 1965). Since the caffeine present in all drinks does not react with this reagent, as confirmed by the values obtained for samples G and H which contain no caffeine, the difference must be caused by some other compound absent in sugar-free colas. This difference was not shown when drinks were analyzed with the ferric reducing antioxidant power (FRAP) test, i.e., following the procedure of Benzie and Strain (1996), with some modifications (Brenna *et al.*, 2009).

It can be seen that the samples of chinotto drink, which, in addition to caramel, contain polyphenols, mainly flavonoids extracted from the chinotto fruit (Barreca *et al.*, 2009), are found to have values significantly higher both with the Folin–Ciocalteu reagent and the FRAP test, compared with those found for the colas. This aspect leads to confirmation of the importance of polyphenols, whose presence is therefore decisive for the evaluation of the antioxidant activity of the soft-drinks made from fruit juices.

TABLE 6.3 A List of Soft Drinks Marketed as 'Fanta' Varieties

Aloe Vera Muscat, Apple, Apple Grape, Apple Kiwi, Apple Peach, Apple Vanilla, Apricot, Banana, Banana Fermented Milk, Berry, Berry Blackcurrant, Berry Cherry, Berry Orange, Birch Beer, Bitter Herbal, Bitter Orange, Bitter Water, Blackcurrant, Blackcurrant Blueberry Raspberry, Blackcurrant Lemon, Blueberry, Blueberry Salacider, Bubble Gum, Cherry, Citron, Citrus Blend, Club Soda, Coconut Pineapple, Cranberry, Cranberry White Grape, Floral Lemon, Fruit Punch, Fruit Punch Orange, Ginger Ale, Granadilla, Grape, Grape Lime Strawberry, Grape Pineapple, Grapefruit, Grapefruit Lemon Lime, Grapefruit Lime, Grenadine, Guaraná, Herbal, Honey Lemon, Kiwi Strawberry, Lemon, Lemon Lime, Lemon Lime Mint, Lemon Passionfruit, Lime, Lime Peach, Lychee, Lychee Soursop, Mandarin Orange, Mandarin Orange Mint, Manderin Tangerine, Mango, Mango Melon, Mango Orange, Mango Passionfruit, Mango Pineapple, Melon, Melon Orange, Melon Vanilla Cream, Mint Orange, Mint Raspberry Strawberry, Orange, Orange Passionfruit, Orange Peach, Orange Tangerine, Orange Vanilla, Passionfruit, Peach, Peach Strawberry, Pineapple, Raspberry, Root Beer, Sangria, Strawberry, Strawberry Cream, Strawberry Milk, Strawberry Pineapple, Tamarind, Tangerine, Tropical Fruit Punch, Tutti-Fruti, Unflavored, Vanilla Cream, Watermelon, Wild Berry Lemon.

The different compositions influence the antioxidant power of a particular brand of soft drink, being linked to the amount of antioxidant compounds present in the original fruit juice.

SOFT DRINKS CONTAINING FRUIT JUICE

Fruits are a source of many important compounds necessary to our health, in particular vitamins and salts. Like vegetables, fruits are a rich source of many different compounds belonging to the class of polyphenols, which have been shown to have antioxidant activity, and to act as dietary protective agents against some pathologies such as cardiovascular diseases and certain forms of cancer (Frankel, 1999).

Besides their consumption as fresh fruits, they are processed by the food industry with many procedures to many different products, with the aim of obtaining a prolonged shelf-life. Fruit juices, for example, are prepared from fresh fruits and usually stored after a concentration procedure, in order to reduce to a minimum changes caused by time and temperature. Concentrated fruit juices are then employed to prepare fruit drinks by suitable dilution.

A simple look at the market of these drinks shows that there are hundreds of formulations containing one or more different juices, so it is extremely difficult to give a special assessment. It is clear, however, that the values of antioxidant activity and content in their different phenolic compounds can be extremely variable: for example, it is possible to find, though not everywhere available to the consumer, about 100 soft drinks which are included under the trade name of 'Fanta' (Table 6.3). Most of them are present only locally, like Fanta Chinotto in Italy. In fact, Fanta composition changes from one country to another, depending on the different flavors preferred in different countries and the presence of different juices from which soft drinks are prepared.

ANTIOXIDANT ACTIVITY OF FRUIT-JUICE-BASED SOFT DRINKS

The antioxidant capacity of these soft drink mixes depends on the kind of fruit juice employed and on its percentage present in the drink. The juice concentration can be very different: for example, in most countries, orange juice concentration is 5%, while in Italy it is currently 12% and recently it has been proposed that the orange juice concentration be increased to 20%.

The healthy features offered by the presence of polyphenolic compounds of a certain number of soft drinks could be explained from the data reported in some papers on unaltered fruit juices (Seeram *et al.*, 2008). Besides the fact that soft drinks contain less juice because the juice is diluted, little more can be determined unless the exact recipe of the different juice blend used as a basis is known.

Since the polyphenolic compounds present depend on the type of fruit from which the concentrate was prepared, in general, it might be assumed that the soft drinks based on concentrated colored fruit juices should have a higher content of active antioxidants belonging to the anthocyanins.

In addition to the antioxidant power deriving from the content of the particular polyphenols present in the fruit used for the preparation of the concentrated juice, the presence of ascorbic acid (vitamin C) should not be forgotten.

In fact, in excess of the amount naturally present in the fruit, during fruit processing more ascorbic acid is normally added (ranging from 700 to 1000 mg/kg), with the aim of protecting from the non-enzymatic browning caused by the processing of the raw material. Of course, like all reducing agents, its presence strongly influences the response to the Folin–Ciocalteu test.

Regarding the influence of the presence of ascorbic acid, Fiore *et al.* (2005) reported that no differences were detected for ascorbic acid, which was found to range from 38.9 to 89.0 mg per 100 ml of juice (Table 6.4), both for pure red orange juice and sterilized juice containing various amounts of concentrated juice (from 25 to 35%). This means that a content of roughly 300 mg/l of ascorbic acid can be assumed to be present in a soft drink containing 12% fruit juice and this amount is well in excess of the possible content of natural polyphenolic compounds.

TABLE 6.4 The Influence of Basic Components of a Fruit Juice on the Antioxidant Power of its Derived Soft Drink

Sample	% Juice	Vitamin C (mg/100 ml)	Total Anthocyanins (mg/l)
REF 1	100	81.1	75.5
REF 2	100	75.5	97.7
REF 3	100	62.6	44.7
REF 4	100	46.8	43.0
REF 5	100	70.5	72.0
REF 6	100	83.7	62.2
REF 7	100	38.4	67.2
LC 1	35.2	54.7	0.9
LC 2	35.2	60.0	0.4
LC 3	25.2	89.0	2.0
LC 4	30.3	75.8	3.3
LC 5	26.3	75.5	5.4

Vitamin C and total anthocyanins content in different samples of pure pasteurized and refrigerated red orange juice (REF) and of commercial concentrated sterilized juice are shown. The quality and composition of the fruit juices used are of critical importance to the derived soft drink.
Adapted from Fiore et al. *(2005).*

Conversely, a huge difference was revealed regarding the presence of anthocyanins in the two groups. This clearly shows that, besides ascorbic acid, a colorant has been added to the concentrated juice, leading as a consequence to a rather insignificant content of natural anthocyanin in a derived soft drink.

Furthermore it is known that anthocyanins, which usually give the red coloring to soft drinks, are not chemically stable under the conditions of the sterilization process normally employed which leads to the degradation of anthocyanins and mainly of those that possess the higher antioxidant activity, such as delphinidin and cyanidin derivatives (Kirca and Cemeroglu, 2003).

CAFFEINE-CONTAINING SOFT DRINKS

As previously stated, in cola-based soft drinks, the only important natural component with antioxidant action that these drinks offer is caffeine. A sub-family of very particular soft drinks, in which caffeine is the main component, is based on the seed extract of guaraná (*Paullinia cupana* Kunth). These seeds contain significant amounts of caffeine (up to 3.4% w/w) and a lesser amount of catechin and epicatechin, respectively, of the order of 0.7 and 0.9% w/w.

Figure 6.2 shows a typical HPLC chromatogram of an aqueous guaraná seed extract. Minor amounts of theobromine and theophylline (Meurer-Grimes *et al.*, 1998) and of some procyanidins are also present (Yamaguti-Sasaki *et al.*, 2007). The presence of caffeine in much higher amounts in respect to these polyphenols, which are involved in different ways in exercising their antioxidant activity, could explain the health qualities that are ascribed to guaraná soft drink consumption. The drink, however, is also colored with caramel.

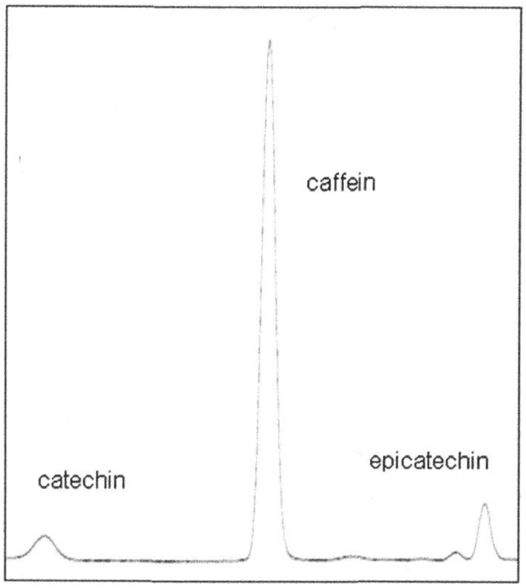

FIGURE 6.2 HPLC trace of a guaraná seeds extract. Caffeine is the main compound of the guaraná seed extract; monomeric catechins are also present.

Guaraná drink is widely consumed, especially in Brazil where it has almost assumed the status of a national drink, competing with cola soft drinks. More than three trillion cans were consumed in the year 2008, perhaps not only because guaraná is a typical local tree, originating from the Amazonian basin, but also for its real or supposed aphrodisiac properties (Smith and Atroch, 2010).

Even in this case, the production has led to the offer on the market of many beverages where other more or less 'secret' components are added to the extract of guaraná seeds.

In Brazil, guaraná berry drinks are available in various flavors, in particular a recent formulation contains both the natural flavor of guaraná and a natural extract from açaí (*Euterpe oleracea*), a palm tree that yields drupes very rich in anthocyanins (Dos Santos *et al.*, 2008).

In the group of soft drinks containing guaraná we can include the so-called 'energy drinks'. In general, in addition to significant quantities of caffeine (up to 0.04% w/v), numerous other components are included, such as taurine, glucuronolactone, amino acids, and vitamins of the B group, as well as salts and sugars. This is the reason why it is important to establish a limit range for the values of these components, so that we can distinguish whether they are to be considered simply as beverages or as liquid dietary supplements.

A RETURN TO THE PAST?

The latest trend in the world of soft drinks seems to be the return its origins, adding 'flavors' to water, such as the addition of lemon juice. In fact, concentrated drinks named with the generic designation of 'water flavor enhancers' have recently been made available to consumers. Generally they are zero-calorie liquid beverage enhancers that allow consumers to mix flavor drops into their water according to the desired concentration.

Many varieties already exist, some of which contain the main components already present in normal soft drinks, such as caffeine, guaraná extract, ginseng extracts; taurine and B vitamins can also be present. Their labels show that they 'contain natural flavors.' In this case, their popularity could be a problem due to the possible risk of consumer misuse, especially in the case of young people.

CONCLUSION

Soft drinks are a group of non alcoholic beverages useful to quench our thirst in a pleasant way. As well as containing some useful compounds with antioxidant properties, it is necessary however to consider some of their drawbacks, related to their composition.

Regular sugared soft drinks can be considered as a primary source of added sugar to the diet, and they have been linked to obesity and diabetes (de Koning *et al.*, 2011). Some other health problems have been reported to be related to the presence of advanced glycation end-products that give their color to the caramel added to the cola soft drinks and potentially increase insulin resistance (Nseir *et al.*, 2010).

Sugar consumption with soft drinks can be overcome, by replacing with 'diet' versions containing artificial sweeteners, but finding a substitute for caramel colorant seems to be difficult.

As a further general problem reported for the consumers of soft drinks, their erosive potential must be taken in account. This problem seems to stem from their rather low pH value (de Carvalho Sales-Peres *et al.*, 2007) due to the presence of phosphoric and/or citric acid.

References

Atawodi, S.E., Pfundstein, B., Haubner, R., Spiegelhalder, B., Bartsch, H., Owen, R.W., 2007. Content of Polyphenolic Compounds in the Nigerian Stimulants *Cola nitida* ssp. *alba*, *Cola nitida* ssp. *rubra* A. Chev, and *Cola acuminata* Schott & Endl and Their Antioxidant Capacity. J. Agric. Food Chem. 55, 9824–9828.

Barreca, D., Bellocco, E., Caristi, C., Leuzzi, U., Gattuso, G., 2009. Flavonoid Composition and Antioxidant Activity of Juices from Chinotto (*Citrus x myrtifolia* Raf.) Fruits at Different Ripening Stages. J. Agric. Food Chem. 58, 3031–3036.

Benzie, I.F.F., Strain, J.J., 1996. The ferric reducing agent of plasma (FRAP) as a measure of "antioxidant power": The FRAP assay. Anal. Biochem. 239, 70–76.

Brenna, O.V., Ceppi, E.L.M., Giovanelli, G., 2009. Antioxidant capacity of some caramel-containing soft drinks. Food Chem. 115, 119–123.

Carbonated Soft Drinks: Global Industry Guide Published by: Datamonitor (Product Code: 860CD620-EA2C-42B0-B07E-323B87672C90) Published on: 01 March 2012.

Chou, K.H., Bell, L.N., 2007. Caffeine Content of Prepackaged National-Brand and Private-Label Carbonated Beverages. J. Food Sci. 72, C337–342.

EFSA, 2011. http://www.efsa.europa.eu/en/press/news/ans110308.htm.

de Carvalho Sales-Peres, S.H., Magalhães, A.C., de Andrade Moreira Machado, M.A., Buzalaf, M.A., 2007. Evaluation of The Erosive Potential of Soft Drinks. Eur. J. Dent. 1, 10–13.

de Koning, L., Malik, V.S., Rimm, E.B., Willett, W.C., Hu, F.B., 2011. Sugar-sweetened and artificially sweetened beverage consumption and risk of type 2 diabetes in men. Am. J. Clin. Nutr. 93, 1321–1327.

Dos Santos, G.M., Maia, G.A., De Sousa, P.H.M., Da Costa, J.M.C., De Figueiredo, R.W., Do Prado, G.M., 2008. Correlação entre atividade antioxidante e compostos bioativos de polpas comerciais de açaí (*Euterpe oleracea* Mart. Arch. Latinoam. de Nutr. 58, 187–192.

Fiore, A., La Fauci, L., Cervellati, R., Guerra, M.C., Speroni, E., Costa, S., Galvano, G., De Lorenzo, A., Bacchelli, V., Fogliano, V., Galvano, F., 2005. Antioxidant activity of pasteurized and sterilized red orange juices. Mol. Nutr. Food Res. 49, 1129–1135.

Frankel, E.N., 1999. Natural phenolic antioxidants and their impact on health. In: Packer, L. (Ed.), Antioxidant food supplements in human health. Academic Press, London, pp. 385–392.

Kirca, A., Cemeroglu, B., 2003. Degradation kinetics of anthocyanins in blood orange juice and concentrate. Food Chem. 81, 583–587.

León-Carmona, J.R., Galano, A., 2011. Is Caffeine a Good Scavenger of Oxygenated Free Radicals? J. Phys. Chem. B. 115, 4538–4546.

Licht, B.H., Shaw, K., Smith, C., Mendoza, M., Orr, J., Myers, D.V., 1992. Characterization of caramel colour IV. Fd. Chem. Toxic. 30, 365–373.

Meurer-Grimes, B., Berkov, A., Beck, H., 1998. Theobromine, Theophylline, and Caffeine in 42 Samples and Products of Guarana (*Paullinia Cupana*, Sapindaceae). Econ. Bot. 52, 293–301.

Moon, J.K., Shibamoto, T., 2011. Formation of carcinogenic 4(5)-methylimidazole in Maillard reaction systems. J. Agric. Food Chem. 59, 615–618.

Nseir, W., Nassar, F., Assy, N., 2010. Soft drinks consumption and nonalcoholic fatty liver disease. World J. Gastroenterol. 16, 2579–2588.

Schlee, C., Markova, M., Schrank, J., Laplagne, F., Schneider, R., Dirk W. Lachenmeier, D.W., 2012. Determination of 2-methylimidazole, 4-methylimidazole and 2-acetyl-4-(1,2,3,4-tetrahydroxybutyl)imidazole in caramel colours and cola using LC/MS/MS. J. Chromatogr. B Analyt. Technol. Biomed. Life Sci. Available online 23 October 2012.

Seeram, N.P., Aviram, M., Zhang, Y.., Henning, S.M., Feng, L., Dreher, M., Heber, D., 2008. Comparison of Antioxidant Potency of Commonly Consumed Polyphenol-Rich Beverages in the United States. J. Agric. Food Chem. 56, 1415–1422.

Singleton, V.L., Rossi, J.A., 1965. Colorimetry of total phenolics with phosphomolybdic-phosphotungstic acid reagents. Am. J. Enol. Vitic. 16, 144–158.

Smith, N., Atroch, A.L., 2010. Guaraná's Journey from Regional Tonic to Aphrodisiac and Global Energy Drink. Evid. Based Complement. Alter. Med. 7, 279–282.

Sylvetsky, A.C., Welsh, J.A., Brown, R.J., Vos, M.B., 2012. Low-calorie sweetener consumption is increasing in the United States. Am. J. Clin. Nutr. 96, 640–646.

US Code of Federal Regulations (CFR), 2004. Title 21 section 182.20.

Wang, H.Y., Qian, H., Yao, W.R., 2011. Melanoidins produced by the Maillard reaction: Structure and biological activity. Food Chem. 128, 573–584.

Yamaguti-Sasaki, E., Ito, L.A., Canteli, V.C.D., Ushirobira, T.M.A., Ueda-Nakamura, T., Filho, B.P.D., Nakamura, C.V., Palazzo de Mello, J.C., 2007. Antioxidant Capacity and In Vitro Prevention of Dental Plaque Formation by Extracts and Condensed Tannins of Paullinia cupana. Molecules 12, 1950–1963.

SECTION 2

Effects of Production and Processing

CHAPTER

7

Antioxidants in Wine during Fermentation

Wessel du Toit, Anita Oberholster*[†]

*Department of Viticulture and Oenology, Stellenbosch University, South Africa,
[†]Department of Viticulture and Enology, University of California, Davis, CA, USA

INTRODUCTION

White and red wine production entails grape harvesting, juice processing, and alcoholic fermentation. These steps in the wine-production process can all influence the levels of antioxidants found in grape juice and wine. Interest in recent years in the ability of humans to absorb and metabolize wine antioxidants such as phenolics has sparked renewed interest in processes that may influence or even increase antioxidants in wine. Two main types of antioxidants in wine will be discussed in the following section, namely phenolic compounds and glutathione.

GRAPE COMPOSITION

The grape berry comprises of the skin, pulp, and seeds. Grape berry skin consists of 6 to 10 cell wall layers. The outer layer of the skin is the wax-like cuticle, which makes a grape berry waterproof. The main components of the skin are the colored compounds (red and yellow pigments), phenolics, and aromatic compounds, as well as potassium and other minerals. Below the skin layer is the pulp that makes up most of the berry volume. The pulp consists of large vacuoles that contain the grape juice, which gets released during crushing of the grapes when the grape skins and vacuoles are broken down. Grape juice consists of 70–80% water with many dissolved solids. The main solutes are sugars (mainly glucose and fructose) and organic acids, such as tartaric, malic, citric, and small amounts of amino acids. Tartaric and malic acid accounts for over 90% of the organic acids present. The seeds are located in the center of the flesh and contain mainly phenolics also commonly referred to as tannins. Following sugars and organic acids, phenolics are the main constituents present in grapes. Phenolics are major contributors to the flavor and color of both grapes and wine, and play an important role in wine quality, aging potential, and other sensory characteristics. They are also the main antioxidants present in grapes and wine.

Total phenol distribution in grape pulp, juice, skins, and seeds have been estimated to be 1, 5, 50, and 44%, respectively (Singleton and Esau, 1969). These estimates will be influenced by the extractability of the phenolics in the different matrixes. Phenolic compounds are mainly responsible for the differences observed between white and red grapes and wine, and are extracted from grape skins and seeds during wine-making. The concentration of phenols can be influenced by grape variety, viticultural practices, wine-making processes, and transformations that can take place during wine ageing (Ribéreau-Gayon *et al.*, 2006). White wines are customarily produced with little skin and seed contact, resulting in lower phenol content in white wines (100–250 mg/l gallic acid equivalents (GAE)) compared to red wines (1000–3500 mg/l GAE), which are mostly produced with extensive skin and seed contact (Ribéreau-Gayon *et al.*, 2006). Grapes also contain nitrogenous compounds such as amino acids, peptides, and proteins. Nitrogen-containing compounds serve as nutrients for yeast during fermentation, and proteins mostly precipitate as the alcohol content of the wines increases. Many volatile odorous compounds are present in the wine. They can be grape and fermentation derived. Grape berry skin and pulp contain significant amounts of pectic compounds that, together with other constituents such as cellulose, hemicellulose, and lignin, contribute to the structure of cell walls (Vidal *et al.*, 2001). Many macromolecules such as proteins and polysaccharides will be eliminated by clarification and stabilization treatments of the wine (Ribéreau-Gayon *et al.*, 2006).

Processing and Impact on Antioxidants in Beverages
http://dx.doi.org/10.1016/B978-0-12-404738-9.00007-6

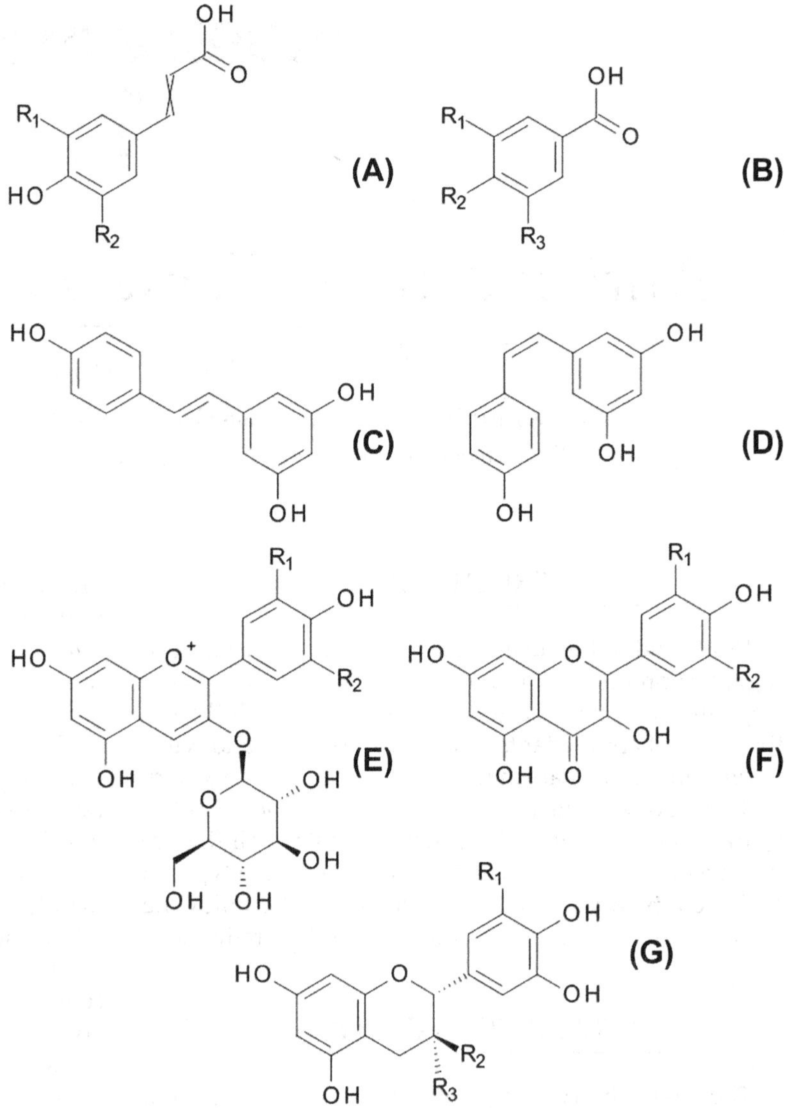

FIGURE 7.1 A, hydroxycinnamic acid (R_1=H, R_2=H, *p*-coumaric acid; R_1=H, R_2=OH, caffeic acid; R_1=H, R_2=OCH_3, ferulic acid); B, benzoic acid (R_1=OH, R_2=OH, R_3=OH, gallic acid); C, *trans*-resveratrol; D, *cis*-resveratrol; E, anthocyanin (R_1=OH, R_2=H, cyanidin-3-gluc; R_1=OCH_3, R_2=H, delphinidin-3-gluc; R_1=OH, R_2=OCH_3, petunidin-3-gluc; R_1=OCH_3, R_2=OCH_3; malvidin-3-gluc); F, flavonol (R_1=H, R_2=H, kaempferol; R_1=OH, R_2=H, quercetin; R_1=OH, R_2=OH, myricetin); G, flavan-3-ol (R_1=H, R_2=OH, R_3=H, (+)-catechin; R_1=OH, R_2=OH, R_3=H, (+)-gallocatechin; R_1=H, R_2=H, R_3=OH, (−)-epicatechin; R_1=OH, R_2=H, R_3=OH, (−)-epigallocatechin).

ANTIOXIDANTS IN WHITE AND RED GRAPES

It is important to briefly discuss antioxidant levels in white and red grapes, as these will be reflected in the corresponding wines as well (Du Toit and Visagie, 2012). The phenolics present in grapes are well known and can be divided into two main groups: the flavonoids and non-flavonoids. The non-flavonoids consist mostly of the hydroxycinnamic (HCA) and benzoic acids (gallic acid). Resveratrol is present in grape skin and is another important compound, although a stilbene, mainly extracted during red wine fermentation. Resveratrol and its isomers and oligomers have been extensively investigated due to their potential health benefits including antioxidant activities (Frankel *et al.* 1995). The main flavonoid compounds present in grapes and wine are flavonols, anthocyanins, flavan-3-ols and their polymerized products (Adams, 2006) (Figure 7.1). Flavonols are the yellow pigments present in the skins of both white and red grapes, while anthocyanins are the red pigments present in red grape skins.

Several cinnamic acids are present in grapes and wine. In grapes they are usually present esterified with particularly tartaric acid. The main HCA found in grapes are tartaric esters of caffeic, *p*-coumaric, and ferulic acids, yielding *-cis* and *-trans* forms of caftaric, coutaric, and fertaric acids, respectively (Vrhovšek, 1998). The non-flavonoids are more important in white than red grapes, due to the higher concentration of flavonoids in red wines from grape seed and skin extraction. Flavonols are present in grapes as glycosides, which are hydrolyzed during fermentation, resulting in the increasing presence of the aclycones in wine. The main flavonols identified are quercetin, isorhamnetin, kaempferol, and myricetin, and their glycosides. Anthocyanins are present in grapes and wine as monoglucosides and their acetylated esters with caftaric, *p*-coumaric, and acetic acid. Their diglucoside forms have been identified in trace amounts in *Vitis vinifera*, although larger quantities have been found in other species (Ribéreau-Gayon *et al.*, 2006). Anthocyanins are more stable in their glycosylated forms. Flavan-3-ols can be present as monomers, oligomers, and polymers in red wine, also known as proanthocyanidins or condensed tannins (Ribéreau-Gayon *et al.*, 2006; Adams, 2006).

The composition of skin and seed tannins have been determined by several researchers. In general, tannins in skins are larger and less galloylated and have the extra complexity of prodelphinidins, meaning they contain epi(gallo)catechin units. These compounds are mostly found in the hypodermal layers of the skin as well as the area between the parenchyma of the seed between the cuticle and the hard seed coating (Adams, 2006). The stems of grapes also contain significant amounts of phenolic compounds, such as flavanols, tannins, and dihydroquercitin 3-rhamnoside (Souquet *et al.*, 2000).

The most important phenolic compounds found in white grapes and often white wine as well are the hydroxycinnamates (HCA) (Vrhovšek, 1998). Total HCA concentrations up to 281 mg/l have been found in white grapes (Vrhovšek, 1998), with Singleton *et al.*, (1986) reporting that caftaric acid levels can also differ dramatically between cultivars, ranging from 40 to 400 mg/l. Catechin and caffeic acid levels in juices are normally lower, with levels of 0.7–8.7 mg/l and 0.5–3.7 mg/l, respectively, being found (Fracassetti *et al.*, 2011).

Flavan-3-ols and tannins are formed early on in berry development via the shikimate pathway and change little from veraison (Adams, 2006; Ribérau-Gayon *et al.*, 2006). However, their concentration decreases, especially in the skin, due to berry growth. Anthocyanin synthesis starts from veraison (berry softening) until ripeness. Flavonols have two distinct phases of synthesis: the first around flowering with the other during ripening of the grape berry (Downey *et al.* 2003). HCA are synthesized in the early stages of berry formation, while levels decrease in the early stages of ripening, but then seem to stabilize at the later stages of ripening (Suklje *et al.*, 2012). Although most phenolics have antioxidant activities, studies have shown that anthocyanins, followed by flavonols, phenolic acids, and resveratrol contribute the most to the antioxidant potential of grape juice and wine (Frankel *et al.*, 1995; Sánchez-Moreno *et al.*, 1999). Makris *et al.* (2003) found that antioxidant potential was strongly associated with the total phenol content of white wine as well as the hydroxycinnamic acids, but poorly associated with the total non-hydroxycinnamates ((+)-catechin, (−)-epicatechin, gallic acid).

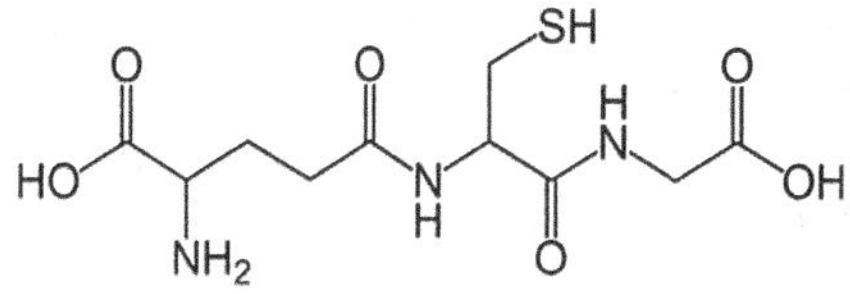

FIGURE 7.2 Glutathione.

Glutathione (GSH, Figure 7.2), an important sulfur-containing tripeptide, consisting of glutamic acid, cysteine, and glycine are also produced during grape ripening. γ-Glutamylcysteine synthetase produces γ-glutamylcysteine from L-glutamate and L-cysteine. GSH synthetase then catalyzes the addition of glycine to form GSH. GSH can occur in white grapes at a few milligrams to more than 100 mg/kg (Kritzinger *et al.*, 2013a), depending on the cultivar and ripening stage. Certain authors reported GSH concentrations in ripening grapes to increase up to about 16 degree Brix (Okuda and Yokotsuka, 1999), while others found levels to increase up to commercial harvest at around 25 degree Brix (Suklje *et al.*, 2012). GSH seems to be produced in the leaves, but whether the berry can also produce this antioxidant is unclear. Higher concentrations of available nitrogen also lead to increased GSH levels in nitrogen-deficient vines (Liyanage and Adams, 1992). GSH concentrations in commercial South African white grape juices ranged from 1.1 mg/l to 42.3 mg/l (Fracassetti *et al.*, 2011), which correlates well with an average value of 12.5 mg/l found in 28 young Slovenian wines (Janěs *et al.*, 2010).

It has been found that red and white wines have higher antioxidant potentials than their corresponding juices (Sánchez-Moreno *et al.*, 1999; Talcott and Lee, 2002). Differences were not attributed to polyphenolic content but rather to additional chemical constituents, synergistic responses, or other chemical changes that occurred during fermentation (Sánchez-Moreno *et al.*, 1999).

ANTIOXIDANTS IN WHITE GRAPE JUICE DURING PROCESSING

During white wine production the solid part such as the skins, pips and stems are often removed before alcoholic fermentation. This is done to reduce the occurrence of bitter- and astringent-tasting flavanols and tannins (Brossaud *et al.*, 2001) from these solid parts in the resulting wines. White grapes are thus normally crushed and then pressed to extract the juice. However, this process can also influence the antioxidant composition of the juice and the resulting wines. During crushing and de-stemming, the aim is to separate the berry from its cuticle and then tear the berry open to facilitate juice extraction. During this process extraction of phenolic compounds from the solid parts into the juice may already occur, depending on the type of crusher and de-stemmer (Ribéreau-Gayon *et al.*, 2006). The skins can then be left in contact with the juice to extract certain flavor compounds, such as terpenes and methoxypyrazines, of which high concentrations can be found in the skins (Marais, 1998). This process of skin contact normally lasts a few hours to one day and is conducted at lower temperatures to prevent spontaneous oxidation from occurring and to lower the rate of enzymatic oxidation. During skin contact, antioxidants such as phenolic compounds are then also extracted into the juice. Singleton and Trousdale (1983) reported that skin contact increases hydroxycinnamates, flavonoids, and gallic acid. Other authors also reported an increase in phenolic compounds with skin contact, such as hydroxycinnamic acids, flavonoids, benzoic acids, resveratrol, and total phenolics with skin contact (Cejudo-Bastante *et al.*, 2011). Higher skin contact temperatures (0°C *versus* 20°C) in the presence of SO_2 also led to higher levels of total phenols and flavonoids in Sauvignon Blanc wines (Marais, 1998), with periods as short as 1 h also affecting the concentration of certain phenolic compounds in the juice (Maggu *et al.*, 2007). These increased concentrations of phenolic-derived compounds in white juices increase their resulting wine's free radical scavenging capacity. A skin contact period of 2 h resulted in the same inhibition of copper ion-induced low-density lipoprotein (LDL) oxidation of the resulting Chardonnay wines as those produced from the free run juice fraction. However, prolonged skin contact time increased this capacity, with an 18-h skin contact time leading to a similar inhibition to that of red wine (Fuhrman *et al.*, 2001).

The following step in white wine production is to separate the juice from the skins by pressing. This leads to different fractions of juice being obtained. The first fraction of juice obtained from the press (about 55% of the volume) is the so-called free run fraction, where no pressure is applied to the skins. This juice fraction is often considered to be the best-quality juice, due to it containing the lowest concentration of bitter-tasting flavonoids. Pressure is then applied to skins in the press to extract most of the remaining juice. This obviously leads to higher extraction of phenolic compounds, especially flavonoids such as catechin, epicatechin, tannins, and flavonols. HCA, catechins, and proanthocyanidin concentrations have been reported to increase up to threefold from the free run fraction to heavy pressing at 1.8 bar under certain conditions (Mattivi *et al.*, 2012). However, oxidation that may occur may also drastically influence the antioxidant composition of the juice during pressing and the subsequent settling process (Du Toit *et al.*, 2006).

The disruption of the grape berry during crushing and pressing with the accompanying phenolic extraction and exposure to oxygen can induce enzymatic oxidation of white juice. Enzymatic oxidation in white juice is normally due to oxidation enzymes such as tyrosinase, cresolase, and catechol oxidase, which occurs in healthy grapes, while laccase occurrs in *Botrytis*-infected grapes. The rate of oxidation between these two types of oxidation enzymes does not differ dramatically, but laccase tends to be more resistant to the inhibitory effect of sulfur dioxide (SO_2) (Du Toit *et al.*, 2006). The main phenolic substrate for oxidation by these oxidation enzymes are cinnamic acid derivatives, especially caftaric acid. The later compound is oxidized to its corresponding quinone by these oxidation enzymes. The antioxidant GSH, if present in sufficient amounts, can then reduce caftaric acid to the so-called Grape Reaction Product (GRP). GSH, with a mercapto group, has a nucleophilic center to substitute into the electrophilic ring of the quinone, thus leading to regeneration of the vicinal dihydroxy ring of the oxidized phenolic compound (Du Toit *et al.*, 2006). GRP can be further oxidized by laccase and if sufficient GSH still exists in the must be reduced again to GRP2. This regeneration of the oxidized quinones thus buffer the must against excessive browning, and the GSH to cinnamic acid ratio has been proposed as an indication of the browning potential of a must (Du Toit *et al.*, 2006).

However, large amounts of oxygen, which comes into contact with the must during pressing and settling of the juice in the absence of SO_2, can lead to a rapid decline in especially caftaric acid and GSH levels. Such decreases are often also reflected in the corresponding wines (Coetzee and Du Toit, 2012; Coetzee *et al.*, 2013). Other phenolic compounds, such as catechin may also decrease, but not at the same rate than caftaric acid (Coetzee *et al.*, 2013). This led to the technique of juice treatment called hyperoxidation of must, where large amounts of oxygen are added to the must before alcoholic fermentation. The reason for this is to remove phenolic compounds that may lead to oxidative browning at a later stage of wine-making or during wine ageing. Research has shown that this decreases the phenolic composition of the must and wines to a large extent, but the quality of the resulting wines is not always improved (Du Toit *et al.*, 2006). The reason for this is not clear, but it is thought that volatile thiols, which can impart favorable tropical aromas to white wine, may undergo transformation due to Michael addition with the *o*-quinone leading to loss in aroma (Coetzee *et al.*, 2013) amongst others.

However, the presence of sufficient SO_2 and a reductive environment during pressing can preserve these phenolic compounds and GSH to a large extent, with levels of caftaric acid being similar in must where 0.5 mg/l or 4 mg/l O_2 has been added in the presence of 60 mg/l SO_2 (Coetzee *et al.*, 2013). During a pressing cycle, most the SO_2 runs off with the free run juice, leaving the press juice with low levels of this preservative. Press juice also contains higher levels of phenolics, especially flavonoid derived, which makes press juices more sensitive to oxidative browning and leading to further increases in GRP formation (Maggu *et al*, 2007). This led to the development of so-called reductive presses, where the oxygen level in the press is drastically lowered before the pressing cycle. Under these reductive conditions, higher levels of GSH and caftaric acid can be obtained after pressing (Du Toit *et al.*, 2006; Mattivi *et al.*, 2012). Little information is available on the effect of settling on antioxidants in wine, but some unpublished data from our laboratory found levels to decrease slightly with settling.

ANTIOXIDANTS IN WHITE WINE DURING ALCOHOLIC FERMENTATION

Alcoholic fermentation is the process where yeast transforms fructose and glucose in grape juice to mainly ethanol, CO_2, and heat. A wide range of other compounds are also being produced during this process, but this review will only focus on antioxidants. It seems that variations in phenolic compounds in white juice are also reflected in the corresponding white wines. Practices leading to lower or higher phenolic concentrations in the juice also yield wines with similar levels of these compounds, as can be seen in Table 7.1. During alcoholic fermentation, yeast also displays enzymatic activities which release the caffeic acid moiety from caftaric acid, leading to higher levels of the latter compound in the resulting wines (Winetech, 2006).

The other important antioxidant for which the concentration can change during alcoholic fermentation is GSH. Contradicting data exist on the levels of GSH during fermentation, with levels reported to either increase or decrease (Kritzinger *et al.*, 2013a). Recently Kritzinger *et al.* (2013b) reported on an exhaustive study consisting of a variety of yeast strains in four different grapes juices. In this study it was found that GSH levels normally decrease rapidly after the onset of fermentation, then increase at the middle to latter stages of fermentation and, in some instances, again decrease towards the end of fermentation. This variation depended on the juice, as well as the yeast strain used. In general, GSH levels were lower or in some cases similar at the end of fermentation than in the corresponding juice, which correlated with Du Toit *et al.*, 2006.

GSH can be taken up by the yeast from the juice with a GSH transporter (Hgt1p) (Bourbouloux *et al.*, 2000). Once in the cell, it is used in many stress-response mechanisms such as oxidative stress, the detoxification of heavy metals, and nitrogen starvation. Later increases in GSH concentrations in the must may be due to active GSH intracellular production and secretion by the yeast. Such secreted GSH may then again be taken up by the yeast, leading to fluctuations observed during alcoholic fermentation. The exact mechanism of GSH export is not known, but a novel GSH exchanger, Gex1, located at both the vacuolar and plasma membrane was recently identified in *Saccharomyces cerevisiae* and seems to be involved (Kritzinger *et al.*, 2013b).

In an effort to increase GSH levels in wine, glutathione-enriched inactive dry yeast preparations are currently available, containing different levels of GSH (Tirelli *et al.*, 2010). These preparations are normally added at the beginning or before the middle of alcoholic fermentation. It seems that the usage of some of these preparations can increase the GSH level in the wines by a few mg/l, if added at the correct stage. The exact mechanisms of how these preparations lead to higher GSH levels are not known, but it is doubtful whether the direct addition GSH to the must in these preparations leads to increased levels (Kritzinger *et al.*, 2012). These

TABLE 7.1 Levels of Phenolics (mg/l) in a South African Sauvignon Blanc Juice and Corresponding Wine which Received Different Levels of Oxygen and SO_2 (Winetech, 2006)

Treatment	(+)-Catechin	(−)-Epicatechin	*Trans*- Caftaric Acid	*Trans*-Coutaric Acid	GRP	*Trans*-Fertaric Acid	Caffeic Acid
JUICE							
A	3.52^{a}	13.21^{b}	51.06^{c}	0.45^{b}	0.80^{a}	4.31^{b}	0.00^{a}
B	2.94^{a}	10.89^{a}	35.82^{b}	0.34^{a}	2.48^{b}	2.67^{a}	0.00^{a}
C	3.12^{a}	11.67^{a}	28.45^{a}	0.33^{a}	10.81^{c}	2.60^{a}	0.49^{b}
WINE							
A	4.95^{ab}	13.72$_{b}$	41.42^{b}	0.35^{b}	0.27^{a}	3.29^{b}	0.57^{a}
B	4.69^{a}	12.84^{a}	37.50^{b}	0.38^{b}	1.19^{b}	2.77^{ab}	0.61^{a}
C	5.27^{b}	13.35^{ab}	22.39^{a}	0.29^{a}	7.77^{c}	2.48^{a}	0.54^{a}

(A) SO_2 (60 mg/l) and ascorbic acid (50 mg/l) added, dissolved O_2 levels in the juice: <0.3 mg/l. (B) SO_2 (60 mg/l) and ascorbic acid (50 mg/l) added, dissolved O_2 levels in the juice: 1–1.5 mg/l. (C) No SO_2 or ascorbic acid added, dissolved O_2 levels in the juice: 3–4 mg/l.
Means in columns followed by the same letter are not significantly different by Duncan's multiple range test, $p = 0.05$

TABLE 7.2 Factors Affecting Antioxidant Levels in White Wine

Factor	Effect	Source
PHENOLICS		
Ripening degree of grapes	Increase	Suklje *et al.* (2012)
Cultivars	Varying level	Vrhovšek (1998)
De-stemming of grapes	Decrease	Souquet *et al.* (2000)
Skin contact	Increase	Cejudo-Bastante *et al.* (2011)
Pressing of grape skins	Increase	Mattivi *et al.* (2012)
Oxidation of juice	Decrease	Coetzee *et al.* (2013)
SO_2 addition to juice	Stabilize levels	Marais (1998)
GLUTATHIONE		
Ripening degree of grapes	Increase	Suklje *et al.* (2012)
Oxidation of juice	Decrease	Du Toit *et al.* (2006)
Pressing of grape skins	Decrease	Maggu *et al.* (2007)
SO_2 addition to juice	Stabilize levels	Coetzee *et al.* (2013)
Yeast strain used for fermentation	Increase or decrease	Kritzinger *et al.* (2013b)
Addition of glutathione-enriched inactive dry yeast preparation	Increase	Kritzinger *et al.* (2013a)

preparations contain different levels of amino acids, which could lead to increased GSH synthesis by the yeast, but this needs further investigation.

It is thus clear that a number of factors can affect the antioxidant levels in white juice and ultimately the resulting wine, which are summarized in Table 7.2.

ANTIOXIDANTS IN RED GRAPE JUICE DURING PROCESSING

During red wine production, grapes are mostly crushed and de-stemmed prior to fermentation. In contrast to white wine production, the juice is fermented in contact with the skins and seeds and, in fact, winemaking techniques have been optimized to maximize extraction especially from the skins. Sulfur dioxide is added after crushing to protect must against oxidation before the start of fermentation. Studies have shown that SO_2 has only a small effect on phenol extraction (Sacchi *et al.*, 2005). Heatherbell *et al.* (1996) observed increased extraction at lower temperatures during cold soak with the use of higher levels of SO_2. Cold soak is a treatment applied especially to Pinot Noir to increase extraction particularly of anthocyanins and other compounds that will increase color expression in wines, a major quality indicator. Cold soak treatment entails the cooling of must to temperatures of usually 10–15°C, for several days before fermentation. The reasoning is that that aqueous extraction improves wine color. However, based on the known effects of alcohol and temperature on extraction of anthocyanins and tannins, improved extraction would not be expected. The results obtained with Pinot Noir, Tempranillo and Mencia grapes consistently show that a cold soak alone has either no effect or a negative effect on the phenolic composition (Feuillat 1996; Puertas *et al.*, 2008; Vazques, 2010). Another practice called saignée or juice runoff, refers to the removal of juice before fermentation, thus increasing the skin to juice ratio. Since the anthocyanins and tannins extracted during fermentation are found in the skins and seeds, this practice should theoretically increase their concentration in the finished wine. If phenol extraction were strictly solubility limited, the effect would be the opposite as there would be less liquid to dissolve the phenolics. Generally the former is mostly observed (Gerbaux 1993; Puertas *et al.*, 2008). In contrast Gawel *et al.* (2001) found no difference between a control Syrah wine and Syrah wines made from must with 10 and 20% runoff after 6 months of aging. This raises questions regarding the long-term value of these treatments. Carbonic maceration is the process where whole berries or clusters are held under a carbon dioxide atmosphere and partial fermentation occurs. After a week or two, the grapes are pressed and the juice is inoculated to complete fermentation. The practice usually produces lighter fruitier wines. Pellegrini *et al.* (2000) found that although the average phenol content of wine produced by carbonic maceration was lower than the standard fermentation wines, the antioxidant potential was similar.

ANTIOXIDANTS IN RED WINE DURING ALCOHOLIC FERMENTATION

Color is the most important quality contributor in red wine. Both anthocyanins originating from grape skins and polymeric pigments contribute to wine color. Once the grapes are crushed, polymeric pigments begin to form and their concentrations continue to increase with time (Nagel and Wulf, 1979). Several studies have shown that anthocyanin extraction reaches a maximum early on during fermentation followed by a decrease (Nagel and Wulf, 1979; Gao *et al.*, 1997; Reichwage, 2012). This decrease is due to either reabsorption on to the grape skins or yeast cell walls, or as a result of polymerization reactions and the formation of polymeric pigments (Bakker *et al.*, 1998; Gao *et al.*, 1997). Phenolic compounds have a significant influence on red wine quality, resulting in the development of many wine-making techniques to increase the extraction of these compounds during vinification (Cheynier and Ricardo Da Silva, 1991). The extraction of grape skin and seed tannin increase with increasing alcohol content, sulfur dioxide, temperature, and skin contact time, which increases the solubility of tannin in wine (Girard *et al.*, 1997, 2001). During a traditional fermentation of the skins, the alcohol content, carbon dioxide, and sulfur dioxide, together with the heat of fermentation, increase the permeability of these cells and membranes. Other treatments that destroy the cell membranes, such as pectolytic enzymes may also increase release of these pigments. The purity of the enzyme preparation is important, as the presence of *B*-glucosidases can result in the hydrolysis of the sugar moiety from anthocyanins, which will make it unstable, resulting in color loss. There are many studies that support an improvement in color extraction (Sacchi *et al.*, 2005; Vasquez, 2010). In some cases no increase in anthocyanin content were found but a simultaneous increase in polymeric pigments were observed (Parley *et al.*, 2001; Wightman *et al.*, 1997). Thus, the lack of anthocyanin increase could be the result of polymerization reactions.

Some researchers found little difference in anthocyanin content with increasing fermentation temperature (15–30°C), although the total phenol content increased with higher fermentation temperatures (Girard *et al.*, 1997, 2001). Others found an increase in color and anthocyanin content with increases in fermentation temperature (Reynolds *et al.*, 2001). Sacchi *et al.* (2005) stated that the increased extraction of phenolics was the result of increased permeability of the hypodermal cells and solubility of the phenolics in the wine solution. Netzel *et al.* (2003) found that the combination of mash heating (heating at 65°C, cooling down for 24 h, 30°C) with fermentation with skin contact resulted in wines with the highest antioxidant capacity and phenolic content, in comparison with treatments using either treatment.

During red wine fermentation a dense cap forms on top of the liquid due to carbon dioxide liberated during fermentation causing the pomace to rise. It is generally accepted that temperatures attained in the cap can be significantly higher than those measured in the liquid portion. Heat is readily removed from the liquid portion but accumulates in the cap which is a poor conductor of heat. Schmid *et al.* (2009) found a temperature gradient of more than 12°C between the cap and the liquid. We found that pump-overs of at least one volume of wine are efficient at dispersing this temperature difference (Reichwage, 2012). Different cap management techniques are used to mix the cap and liquid. In general it has been found that different techniques give variable variety-dependent results (Sacchi *et al.*, 2005). Research at the LEED Platinum winery at UC Davis showed that different pump-over volumes of wine did not significantly influence phenolic extraction but did impact yeast fermentation kinetics and thus fermentation times (Reichwage, 2012). The use of micro-oxygenation or the controlled sparging of wine with oxygen has also become popular during red wine production. This promotes polymerization reactions, resulting in the softening of grape tannins and stabilization of color (Gómez-Plaza and Cano-López, 2011). However, when applied during fermentation the aim is mostly to improve the health of the yeast.

During standard wine-making practices, wines are pressed to separate the fermented juice from the skins at dryness. However, in some cases the practice of extended maceration is employed, which is the practice of prolonged skin contact after the must has fermented to dryness. According to current knowledge regarding anthocyanin and tannin extraction, extended maceration would be expected to increase tannin, but not anthocyanin content. This fact is generally supported by the literature (Sacchi *et al.*, 2005). Yokotsuka *et al.* (2000) found that extraction of phenolics increases up to 36 days of skin contact time. Gambuti *et al.* (2004) found that in general the maximum extractions of (+)-catechin, (−)-epicatechin, and *trans*-resveratrol were reached after 12 days' skin contact time. Red wines are often fined to remove undesirable substances in wine. The main purpose of using a protein-based fining agent such as egg albumin or gelatin in particular is to soften the wine and reduce astringency (Oberholster *et al.*, 2013). Yildirim (2011) demonstrated that wines were mostly negatively affected by fining with egg albumin and gelatin. Fining with a combination of gelatin and Kieselsol resulted in the highest total phenol content and antioxidant potential among the fining agents tested (gelatin, gelatin and Kieselsol, egg white, isinglass, PVPP, bentonite). During the first year of storage, rapid changes take place in the color composition of the wine. The purple–red color of a young red wine changes to orange–red with aging. These changes occur when the monomeric pigments are replaced with oligomeric forms which are more stable through direct and indirect polymerization reactions (Cheynier *et al.*, 2006). In general, studies have shown that although vinification practices have small influences on phenolic content and thus antioxidant potential (Table 7.3), changes in the phenolic composition of wines with storage and aging have a larger effect on antioxidant potential (Talcott and Lee, 2002).

CONCLUSION

The antioxidant capacity of both white and red grapes and wines can differ and depends mostly on the cultivar, ripening stage, and wine-making techniques used. Wine-making practices such as skin contact, oxidation control, using more press juice, and enzyme additions may all influence the phenolic and thus antioxidant capacity of a wine. However, how these steps influence

TABLE 7.3 Factors Affecting Antioxidant Levels in Red Wine

Factor	Effect	Source
PHENOLICS		
Cold soak	None or decrease	Feuillat (1996), Puertas *et al.* (2008), Vazquez *et al.* (2010)
Saignée or juice runoff	Increase	Puertas *et al.* (2008)
Carbonic maceration	Decrease	Pellegrini *et al.* (2000)
Heating with skin contact	Increase	Netzel *et al.* (2003)
Increase fermentation temperature	Increase	Reynolds *et al.* (2001)
Extended maceration	Increase	Sacchi *et al.* (2005), Yokotsuka *et al.* (2000)
Fining	Decrease	Yildirim (2011)

the antioxidant capacity of both white and red wines are not fully understood yet and needs further attention. Advances in analytical chemistry should help to overcome these unknowns, as well as assess whether differences observed after alcoholic fermentation are also reflected in wine that underwent maturation, which is normally the end product that the consumer purchases.

References

Adams, D.O., 2006. Phenolics and Ripening in Grape Berries. Am. J. Enol. Vitic. 57 (3), 249–256.

Bakker, J., Bridle, P., Bellworthy, S., Garcia-Viguera, C., Reader, H., Watkins, S., 1998. Effect of Sulphur Dioxide and Must Extraction on Colour, Phenolic Composition and Sensory Quality of Red Table Wine. J. Sci. Food Agric. 78, 297–307.

Brossaud, F., Cheynier, V., Noble, A.C., 2001. Bitterness and astringency of grape and wine polyphenols. Austr. J. Grape Wine Res. 7, 33–39.

Bourbouloux, A., Shahi, P., Chakladar, A., Delrot, S., Bachhawat, A.K., 2000. Hgt1p, a high affinity glutathione transporter from the yeast Saccharomyces cerevisiae. J. Biol. Chem. 275, 13259–13265.

Cejudo-Bastante, M.J., Castro-Vazquez, L., Hermosín-Gutierrez, I., Perez-Coello, M.S., 2011. Combined Effects of Prefermentative Skin Maceration and Oxygen Addition of Must on Color-Related Phenolics, Volatile Composition, and Sensory Characteristics of Airen White Wine. J. Agric. Food Chem. 59, 12171–12182.

Cheynier, V., Dueñas-Paton, M., Salas, E., Maury, C., Souquet, J.-M., Sarni-Manchado, P., Fulcrand, H., 2006. Structure and Properties of Wine Pigments and Tannins. Am. J. Enol. Vitic. 57 (3), 298–305.

Cheynier, V., Ricardo da Silva, J., 1991. Oxidation of Grape Procyanidins in Model Solutions Containing trans-Caffeoyltartaric Acid and Polyphenol Oxidase. J. Agric. Food Chem. 39, 1047–1049.

Coetzee, C., du Toit, W.J., 2012. A comprehensive review on Sauvignon blanc aroma with a focus on certain positive volatile thiols. Food Res. Int. 45, 287–298.

Coetzee, C., Lisjak, K., Nicolau, L., Kilmartin, P., du Toit, W.J., 2013. Oxygen and sulphur dioxide additions to Sauvignon blanc must: effect on must and wine composition. Frag. Flavour J. 28, 155–167.

Downey, M.O., Harvey, J.S., Robinson, S.P., 2003. Synthesis of flavonols and expression of flavonol synthase genes in the developing grape berries of Shiraz and Chardonnay (Vitis vinifera L.). Aust. J. Grape Wine Res. 9 (2), 110–121.

Du Toit, W.J., Lisjak, K., Stander, M.A., Prevoo, D., 2007. Using LC-MSMS to assess glutathione levels in South African white grape juices and wines made with different levels of oxygen. J. Agric. Food Chem. 55, 2765–2769.

Du Toit, W.J., Marais, J., Pretorius, I.S., du Toit, M., 2006. Oxygen in must and wine: A review. South Afr. J. Enol. Viticul. 27, 76–94.

Du Toit, W.J., Visagie, M., 2012. Correlations between South African Red Grape and Wine Colour and Phenolic Composition: Comparing the Glories, Iland and Bovine Serum Albumin Tannin Precipitation Methods. South Afr. J. Enol. Viticul. 33, 33–41.

Feuillat, M., 1996. Vinification du Pinot noir en Bourgogne par maceration prefermentaire a froid. Rev. Oenol. 82, 29–31.

Fracassetti, D., Lawrence, N., Tredoux, A.G.J., Tirelli, A., Nieuwoudt, H.H., Du Toit, W.J., 2011. Quantification of glutathione, catechin and caffeic acid in grape juice and wine by a novel ultra-performance liquid chromatography method. Food Chem. 128, 1136–1142.

Frankel, E.N., Waterhouse, A.L., Teissedre, P.L., 1995. Principal phenolic phytochemicals in selected California wines and their antioxidant activity in inhibiting oxidation of human low-density lipoproteins. J. Agri. Food Chem., 43,890–43,894.

Fuhrman, B., Volkova, N., Surasko, A., Aviram, M., 2001. White Wine with Red Wine-like Properties: Increased Extraction of Grape Skin Polyphenols Improves the Antioxidant Capacity of the Derived White Wine. J. Agric. Food Chem. 49 3164(3168.

Gambuti, A., Strollo, D., Ugliano, M., Lecce, L., Moio, L., 2004. trans-Resveratrol, Quercetin, (+)-Catechin, and (−)-Epicatechin Content in South Italian Monovarietal Wines: Relationship with Maceration Time and Marc Pressing during Winemaking. J. Agric. Food Chem. 52, 5747–5751.

Gao, L., Girard, B., Mazza, G., Reynolds, A.G., 1997. Changes in anthocyanins and color characteristics of Pinot noir wines during different vinification processes. J. Agric. Food Chem. 45 (6), 2003–2008.

Gawel, R., Iland, P., Leske, P., Dunn, C., 2001. Compositional and sensory differences in Syrah wines following juice run-off prior to fermentation. J. Wine Res. 12, 5–18.

Gerbaux, V., 1993. Etude de quelques conditions de cuvaison susceptibles d''augmenter la composition polyphénolique des vins de Pinot noir. Rev Oenol. 69, 15–18.

Girard, B., Kopp, T.G., Reynolds, A.G., Cliff, M., 1997. Influence of vinification treatments on aroma constituents and sensory descriptors of Pinot noir wines. Am. J. Enol. Vitic. 48 (2), 198–206.

Girard, B., Yuksel, D., Cliff, M.A., Delaquis, P., Reynolds, A.G., 2001. Vinification effects on the sensory, colour and GC profiles of Pinot noir wines from British Columbia. Food Res. Int. 34 (6), 483–499.

Gómez-Plaza, E., Cano-López, M., 2011. A review on micro-oxygenation of red wines: Claims, benefits and underlaying chemistry. Food Chem. 125, 1131–1140.

Heatherbell, D., Dicey, S., Goldsworthy, S., Vanhanen, L., 1996. Effect of cold maceration on the composition, color, and flavor of Pinot noir wine. In: Henick-Kling, T. (Ed.), Proceedings of the Fourth International Symposium on Cool Climate Enology and Viticulture. New York State Agricultural Experiment Station, Geneva, pp. 10–17.

Janěs, L., Lisjak, K., Vanzo, A., 2010. Determination of glutathione content in grape juice and wine by high-performance liquid chromatography with fluorescence detection. Anali. Chem. Acta 674, 239–242.

Kritzinger, E.C., Bauer, F.F., du Toit, W.J., 2013a. Role of glutathione in wine: a review. J. Agric. Food Chem. 61, 269–277.

Kritzinger, E., Stander, M., Du Toit, W.J., 2012. Assessment of glutathione levels in model solution and grape ferments supplemented with glutathione-enriched inactive dry yeast preparations using a novel UPLC-MS/MS method. Food Addit. Contam. Part A: Chem. Anal. Control Expo. Risk Assess. 30, 80–92.

Kritzinger, E., Bauer, F.F., Du Toit, W.J., 2013b. Influence of yeast strain, extended les contact and nitrogen supplementation on glutathione concentrations in wine. Aust. J. Grape Wine Res. 19, 161–170.

Liyanage, C., Adams, D.O., 1992. Glutathione content of grape leaves and berries at the onset of ripening. Am. J. Enol. Vitic. 43, 304.

Makris, D.P., Psarra, E., Kallithraka, S., Kefalas, P., 2003. The effect of polyphenolic composition as related to antioxidant capacity in white wines. Food Res. Int. 36, 805–814.

Maggu, M., Winz, R., Kilmartin, P.A., Trought, M.C.T., Nicolau, L., 2007. Effect of skin contact and pressure on the composition of Sauvignon blanc must. J. Agric. Food Chem. 55, 10281–10288.

Marais, J., 1998. Effect of grape temperature, oxidation and skin contact on Sauvignon blanc juice and wine composition and wine quality. South Afr. J. Enol. Viticul. 19, 10–16.

Mattivi, F., Fedrizzi, B., Zenato, A., Tiefenthaler, P., Tempesta, S., 2012. Development of reliable analytical tools for evaluating the influence of reductive winemaking on the quality of Lugana wines. Anali. Chem. Acta 732, 194–202.

Nagel, C.W., Wulf, L.W., 1979. Changes in the anthocyanins, flavonoids and hydroxycinnamic acid esters during fermentation and aging of Merlot and Cabernet Sauvignon. Am. J. Enol. Vitic. 30, 111–116.

Netzel, M., Strass, G., Bitsch, I., Könitz, R., Christmann, M., Bitsch, R., 2003. Effect of grape processing on selected antioxidant phenolics in red wine. J. Food Eng. 56, 223–228.

Oberholster, A., Carstens, L.M., Du Toit, W.J., 2013. Investigation of the effect of gelatine, egg albumin and cross-flow microfiltration on the phenolic composition of Pinotage wine. Food Chem. 138, 1275–1281.

Okuda, T., Yokotsuka, K., 1999. Levels of glutathione and activities of related enzymes during ripening of Koshu and Cabernet Sauvignon grapes and during winemaking. Am. J. Enol. Vitic. 50, 264–270.

Parley, A., Vanhanen, L., Heatherbell, D., 2001. Effects of pre-fermentation enzyme maceration on extraction and colour stability in Pinot Noir wine. flavonol synthase genes in the developing grape berries of Shiraz and Chardonnay (Vitis vinifera L.). Aust. J. Grape Wine Res. 7 (3), 146–152.

Pelligrini, N., Simonetti, P., Gardana, C., Brenna, O., Brighenti, F., OPietta, P., 2000. Polyphenol content and total antioxidant activity of vini novelli (young red wines). J. Agric. Food Chem. 48, 732–735.

Puertas, B., Guerrero, R.F., Jurado, M.S., Jimenez, M.J., Cantos-Villar, E., 2008. Evaluation of Alternative Winemaking Processes for Red Wine Color Enhancement. Food Sci. Technol. Int. 14, 21–27.

Reichwage, M., 2012. Managing the Phenolic Profile of Cabernet Sauvignon through the Monitoring of Cap Temperature. M.Sc. Thesis. University of California, Davis, Davis.

Reynolds, A., Cliff, M., Girard, B., Kopp, T.G., 2001. Influence of fermentation temperature on composition and sensory properties of Semillon and Shiraz wines. Am. J. Enol. Vit. 52, 235–240.

Ribéreau-Gayon, P., Glories, Y., Maujean, A., Dubourdieu, D., 2006. Handbook of Enology: The Chemistry of Wine and Stabilization and Treatments, second ed. John Wiley & Sons Ltd., West, Sussex, England.

Sacchi, K.L., Bisson, L.F., Adams, D.O., 2005. A review of the effect of winemaking techniques on phenolic extraction in red wines. Am. J. Enol. Vitic. 56 (3), 197–206.

Sánchez-Moreno, C., Larrauri, J.A., Saura-Calixto, F., 1999. Free radical scavenging capacity of selected red, rose and white wines. J. Sci. Food Agric. 79 (10), 1301–1304.

Schmid, F., Schadt, J., Jiranek, V., Block, D.E., 2009. Formation of temperature gradients in large- and small-scale red wine fermentations during cap management. Aust. J. Grape Wine Res. 15 (3), 249–255.

Singleton, V.L., Esau, P., 1969. Phenolic substances in grapes and wine, and their significance. Academic Press, New York.

Singleton, V.K., Trousdale, E., 1986. White Wine Phenolics: Varietal and Processing Differences as Shown by HPLC. Am. J. Enol. Vitic. 34, 27–34.

Singleton, V.L., Zaya, J., Trousdale, E.K., 1986. Caftaric and coutaric acids in fruit of Vitis. Phytochem. 25, 2127–2133.

Suklje, A., Lisjak, K., Cesnik, H.B., Janes, L., Vanzo, A., Du Toit, W.J., Coetzee, Z., Deloire, A., 2012. Classification of grape berries according to diameter and total soluble solids to study the effect of light and temperature on methoxypyrazines, glutathione and hydroxycinnamates evolution during ripening of Sauvignon blanc. J. Agric. Food Chem. 60, 9454–9461.

Souquet, J.M., Labarbe, B., Le Guernevé, C., Cheynier, V., Moutounet, M., 2000. Phenolic composition of grape stems. J. Agric. Food Chem. 48, 1076–1080.

Talcott, S.T., Lee, J.H., 2002. Ellagic acid and flavonoid antioxidant content of Muscadine wine and juice. J. Agric. Food Chem. 50, 3186–3192.

Tirelli, A., Fracassetti, D., De Noni, I., 2010. Determination of reduced cysteine in oenological cell wall fractions of Saccharomyces cerevisiae. J. Agric. Food Chem. 58, 4565–4570.

Vazquez, E.S., Segade, S.R., Fernandez, I.O., 2010. Effect of the winemaking technique on phenolic composition and chromatic characteristics in young red wines. Eur. Food Res. Technol. 231 (5), 789–802.

Vidal, S., Williams, P., O'Neill, M.A., Pellerin, P., 2001. Polysaccharides from grape berry cell walls. Part I: tissue distribution and structural characterization of the pectic polysaccharides. Carbohydr. Polym. 45, 315–323.

Vrhovšek, U., 1998. Extraction of Hydroxycinnamoyltartaric Acids from Berries of Different Grape Varieties. J. Agric. Food Chem. 46, 4203–4208.

Wightman, J.D., Price, S.F., Watson, B.T., Wrolstad, R.E., 1997. Some effects of processing enzymes on anthocyanins and phenolics in Pinot noir and Cabernet Sauvignon wines. Am. J. Enol. Vitic. 48 (1), 39–48.

Winetech technical report, 2006. Oxygen management during the winemaking process.

Yildirim, H.K., 2011. Effects of Fining Agents on Antioxidant Capacity of Red Wines. The Institute of Brewing & Distilling. 117 (1), 55–60.

Yokotsuka, K., Sato, M., Ueno, N., Singleton, V.L., 2000. Colour and Sensory Characteristics of Merlot Red Wines Caused by Prolonged Pomace Contact. J. Wine Res. 11 (1), 7–18.

CHAPTER

8

Effects of Aging on the Antioxidant Capacity of Red Wines

Eduardo Lissi[*], *Ana Maria Campos*[*], *Cristian Calderón*[*], *Sergio Lobato*[*], *Camilo López-Alarcón*[†]

[*]Facultad de Química y Biología, Universidad de Santiago de Chile, Santiago, Chile, [†]Pontificia Universidad Católica de Chile, Santiago, Chile

CHAPTER POINTS

- The antioxidant activity (AOA) of wines can be modulated by vinification procedures.
- A large number of factors influence the AOA of red wines.
- The effect of barrel and/or bottle aging is very difficult to predict and depends on the selected samples of wines.
- The retail price of wines is independent or even inversely correlated to their total phenolic content and/or antioxidant activity.
- It would be useful if AOA values of wines at the moment of bottling were indicated to consumers.

INTRODUCTION

Wine-making techniques and bottle aging are important factors influencing phenolic and polyphenolic levels in wine (Lachman *et al.*, 2009). Since these compounds confer to wines their free-radical-removing characteristics, it can be expected that this property, frequently related to the health-beneficial effects of wines, can also be modulated by the oenological procedures and/or aging. Red wines contain wood- and yeast-derived phenolic compounds in addition to those extracted from grapes, particularly from their skins. These compounds are transferred to the wine during maintenance together with the grape juice at the first stage of fermentation. This close contact leads, in red wines, to ca. 10-times higher concentrations of phenols and polyphenols than in white wines. This difference in phenol and polyphenol content is usually reflected in the antioxidant activity (AOA) of red and white wines. However, the AOA of white wines can be improved by increasing the efficiency of phenol and polyphenol extraction processes during the wine-making stages (Fuhrman *et al.*, 2001; Landrault *et al.*, 2001).

At the time of drinking of a particular bottle of wine, there are noticeable differences in phenol concentrations among different wines and even among the same wines produced in different vintages. This fact is ascribed to the multiple origins of phenols present in wines, which directly influence both the amounts of individual phenols and the total amount of phenols present in a given wine bottle. This implies that the antioxidant capacity of a wine, an important property closely related to the total amount of phenols, varies with grape varieties, vintages, weather, and wine-making procedures. Also, the possible extraction/removal/reactions of phenols during aging could significantly modify their phenol content and AOA in a fashion that is very difficult to predict. In fact, simple but relevant questions such as "Is it possible to improve the antioxidant capacity of a wine during aging?" or "Is there any relationship between the acceptance (and price) of a red wine and its antioxidant capacity?" have been only partially addressed with a wide variety of apparently contradictory answers.

FACTORS AFFECTING THE PHENOL CONTENT AND *IN VITRO* ANTIOXIDANT ACTIVITY OF RED WINES

Phenols and polyphenols contained within the skin, seeds, and flesh of black grapes are extracted into red

Processing and Impact on Antioxidants in Beverages
http://dx.doi.org/10.1016/B978-0-12-404738-9.00008-8

wines during the processes of vinification. The processes of viticulture and vinification determine the content and profile of phenolic compounds, and hence the antioxidant capacity of the resulting young wine. Vineyard factors such as grape variety and ripening, quality, climate, geographical origin, soil characteristics, and disease pressure affect the accumulation of phenolic compounds in grapes (Burns *et al.*, 2001). During vinification, the length of skin contact, temperature, and presence of seeds, stems, and enzymes affect the extraction of the grapes' phenolics into the must. Therefore, systematization of results is difficult due to the large number of factors that influence the concentration and main characteristics of the phenols. Furthermore, the approaches employed to evaluate the AOA of wine samples varies among published works, making a meaningful comparison of the results almost impossible. Examples of techniques employed in the evaluation of red-wine AOA are collected in Table 8.1.

As Table 8.1 shows, a large number of methodologies have been employed to quantify the total AOA of wines and several of them only evaluate the amount of phenols in wines, irrespective of their reactivity towards free radicals and their capacity to act as antioxidants. In this sense, good correlations are expected, and generally found (Simonetti *et al.*, 1997; Poblete *et al.*, 2009; Bernardi *et al.*, 2008; Kondrashov *et al.*, 2009; Alarcón *et al.*, 2008) between AOA indexes, measured by 1,1-diphenyl-2-picrylhydrazyl (DPPH), 2,2′-azino-bis(3-ethylbenzthiazoline-6-sulfonic acid (ABTS), electron paramagnetic resonance (EPR) and oxygen radical absorbance capacity (ORAC) using fluorescein (ORAC-Fl) methods, and total phenol content (measured by Folin's assay). Furthermore, relative ratios such as ORAC using pyrogallol red (ORAC-PGR)/ORAC-Fl (López-Alarcón *et al.*, 2011), total antioxidant reactivity (TAR)/total reactive antioxidant potential (TRAP), DPPH initial rate/DPPH total bleaching (Campos *et al.*, 2012), can be considered as a measure of the average reactivity of the antioxidants present in the tested sample. A similar intent to introduce indexes that measure the average 'quantity' and 'quality' of the antioxidants present in a complex sample is based on measurement of the amount

TABLE 8.1 Experimental Methodologies Employed in the Evaluation of the AOA of Red Wines

Methology	Measurement	Wine Property	References
Folin	Phenol titrations	Amount of phenol groups	Monagas *et al.* (2006), De Beer *et al.* (2005)
ABTS radical bleaching	Consumption after a single (long) time	Amount of antioxidants	Simonetti *et al.*, (1997), Campos and Lissi (1996)
DPPH radical	Consumption at a fixed time	Amount of antioxidants	Campos *et al.* (2012)
	Consumption initial rate	Antioxidants amount and reactivity	Campos *et al.* (2012)
	Time and concentration to afford 50% of protection	Antioxidants amount and reactivity	Larrauri *et al.* (1999)
Luminol CL induced by AAPH	Induction time (TRAP)	Amount of antioxidants	Perez *et al.* (2000)
	Instantaneous decrease after wine addition	Antioxidants amount and reactivity	Campos *et al.* (1997)
ORAC	Fluorescein protection	Amounts of AO	Dávalos *et al.* (2004)
	PGR protection	Antioxidants amount and reactivity.	López-Alarcón *et al.* (2011)
Cu(II) promoted LDL oxidation	Induction time	Amounts of AO and Cu(II) chelants	Hotzer *et al.* (2005)
FRAP	Fe(III) reduction at a fixed time	Amounts of reductants	Kondrashov *et al.* (2009)
Fremy's salt	EPR signal reduction	Amounts of reductants	Burns *et al.* (2001)
EPR	Nitroxides reduction	Amounts of reductants	
Rat brain homogenate	MDA decrease	?	Echeverry *et al.* (2005)
Chain oxidation of methyl linoleate	Rate of oxygen consumption	Amounts of chain breaking compounds	Roginsky *et al.* (2006)

AAPH, 2,2′-azobis(2-amidino-propane) dihydrochloride; ABTS, 2,2′-azino-bis(3-ethylbenzthiazoline-6-sulfonic acid; AO, antioxidant; CL, chemiluminescence; DPPH, 1,1-diphenyl-2-picrylhydrazyl; EPR, electron paramagnetic resonance; FRAP, ferric reducing antioxidant power; LDL, low-density lipoprotein; MDA, malondialdehyde; ORAC, oxygen radical absorbance capacity; PGR, pyrogallol red; TRAP, total reactive antioxidant potential.

of sample needed to bleach 50% of DPPH and the time required to achieve this goal (Larrauri *et al.*, 1999; Sanchez-Moreno *et al.*, 1998)

EFFECTS OF PRODUCTION AND PROCESSING

Several studies have been devoted to establishing how the amount of phenols and the must AOA evolves during vinification. Netzel *et al.* (2003) compared the effects of different vinification techniques on the AOA (DPPH assay) of red wines. A combination of fermentation on skin and mash heating leads to significant increases in the AOA of the three varietals considered. Lachman *et al.* (2007) followed the amount of antioxidants (DPPH and ABTS methodologies) during the wine-making process. Samples were collected every week during 21 weeks until bottling, confirming the impact of the oenological practices on the AOA indexes. In particular, it was established that addition of SO_2 produces a sharp increase in AOA in the sixth to seventh week of the process.

Burns *et al.* (2001) have evaluated the extraction of phenolics (Folin's assay) and changes in AOA (measured by the reduction of Fremy's salt) for four wines. Some of the results obtained are summarized in Table 8.2.

The data given in Table 8.2 show, for the standard vinification process, an increase in phenol extraction (from grapes) during the first days of vinification. Conversely,

TABLE 8.2 Changes in Total Phenolic Index (TPI) and AOA (Reduction of Fremy's Salt) During Vinification of Four Chilean Wines

Wine	Day	TPI*	AOA**
Cabernet Sauvignon, basic quality	0	2.8	0.12
	4	8.4	1.61
	9	8.2	2.01
Merlot, basic quality, thermovinified	0	6.5	1.84
	4	8.4	1.93
	9	7.0	1.89
Cabernet Sauvignon, Reserva quality	0	2.1	n.d.
	4	8.7	2.37
	8	10.5	3.64
Merlot, varietal quality	0	3.4	0.03
	4	11.4	2.75
	1	11.4	2.95

n.d., not detected.
**TPI expressed in mM gallic acid equivalents.*
***AOA given as 10^{21} radicals reduced by 1 liter of wine.*
Adapted from Burns et al. *(2001).*

during thermovinification an early increase is evident attributable to the high temperature of the process. Furthermore, a very close relationship was observed between total polyphenolic index (TPI) and the AOA of each wine during vinification. However, it must be noted that AOA indexes measured after 9 days are considerably smaller than those of finished wines, whose values range between 4.5 and 9.3×10^{21} radicals reduced per liter of wine. Such a difference is not observed for the TPI.

Villano *et al.* (2006) analyzed the influence of grape varietal and AOA evaluation methodology on the changes elicited during maceration and clarification. These authors analyzed during 14 days the variation of TPI and antioxidants capacities measured by the β-phycoerytrin-based ORAC technique, the prompt and slow consumption of ABTS-derived radicals, and DPPH consumption at long reaction times. Results were obtained in 27 monovarietal red wines, including eight Cabernet Sauvignon samples. A traditional maceration with SO_2 addition was performed, without addition of protolytic enzymes. At day zero of maceration, the wine was mainly composed of phenolic acids from the grapes pulp similar to those of white wines. During the following 14 days of maceration, the TPI and AOA increased progressively. However, the temporal profiles depended on the varietal and the methodology employed to evaluate the AOA. This means that the relation between the different AOA indexes are qualitatively but not quantitatively correlated. A relevant result is that extraction from Cabernet Sauvignon samples is almost complete after 4 days of maceration.

Campodónico *et al.* (1998) measured the total phenols and AOA (ABTS method) of a Cabernet Sauvignon wine during the early stages of the wine-making process. Both indexes strongly correlate and present a rapid increase during the early stages of primary fermentation. Leaving the samples for 3 months after the secondary fermentation or extending maceration in contact with the skins for two days prior to pressing does not significantly increase the measured values.

Ginjom *et al.* (2010, 2011) have measured the total phenols and the antioxidant activities (DPPH and ORAC-Fl) of two Cabernet Sauvignon and one Shiraz red wine samples at all stages of vinification: crush, after the primary (5–16 days) and malolactic fermentations (45–51), post-oaking (430–470 days, young wines), and post-bottling (young wines plus 3 months' bottling). The authors reported, in the three samples, a fair correlation between total phenols and AOA at all stages of the process. Similarly, they observed a strong increase in phenols during the primary fermentation and small changes during the other stages of the process. In particular, it is observed that ORAC-Fl values tend to decrease during oak aging in the three samples considered.

Different viticulture, vinification, and maturation processes of the red wine variety Tannat were studied by Echeverry *et al.* (2005). AOA (measured by the protection in the autoxidation of rat-brain homogenates) correlated with total polyphenol content at 1, 6, and 12 months post-fermentation. Both the amount of phenols and the AOA of the samples steadily increased with time. In all wines studied, changes in field treatments such as etephon application, the timing of the application, and variations in wine-making techniques such as the use of *Saccharomyces cerevisiae* var. *ellipsoideus* or a 'color extractor' pectinolytic enzyme were the most effective in increasing AOA of the wines.

EFFECT OF BARREL AND/OR BOTTLE AGING ON THE PHENOL CONTENT AND ANTIOXIDANT ACTIVITY

It is well recognized that barrel and bottle aging of red wines improves their 'quality' and acceptance by the consumer. However, it is not clear how aging modifies the antioxidant capacity of the beverage. Some data are presented in Table 8.3 and emphasize the wide range of behaviors elicited by different samples. In particular for bottle aging, reported values imply:

1. Increases in both TPI and AOA;
2. Decreases in both TPI and AOA;
3. Decrease in TPI and increase in AOA;
4. Increase in AOA (DPPH assay) and decrease in FRAP index;
5. Increase in AOA during the first years of bottling and decrease at longer times.

Furthermore it is important to note that, in bottle aging, besides the length period, other factors (temperature control, illumination levels, bottles position, cork type, etc.) can be relevant in determining the magnitude of the observed changes (Gómez-Plaza *et al.*, 2000). No systematic studies of these factors have been reported.

TABLE 8.3 Aging and Total Phenols and Free Radical Removal Capacity of Red Wines

Aging Procedure	Conclusion	Comment	References
In bottle	No statistically significant difference between young and 1-year-old wines	AOA measured by DPPH method	Giovanelli (2005)
3 and 12 months in bottles	Mencía and Brancellao red wines increase their AOA during bottle aging	AOA measured by DPPH consumption correlates with TPI	Alén-Ruiz *et al.* (2009)
6, 12,18, 24 and 30 months in bottle	Cabernet Sauvignon wines from two regions increase their AOA (ca. 50%) and decrease the FRAP index to a half	AOA measured by FRAP assay and DPPH consumption	Poiana *et al.* (2008)
Up to 12 months of aging at controlled temperatures	Several grape cultivars. Decrease in TPI (up to 16.5%) and AOA (up to 34%). Similar results with ABTS and DPPH	TPI measured by Folin's and AOA by ABTS and DPPH.	De Beer *et al.* (2005)
Four months' aging in oak barrels	AOA values higher for oak-aged wines (35.8–63.8 mM Trolox equivalents). Young wines: 30.8–40.7 mM	ORAC-Fl. AOA values influenced by grapes variety and vintage, but not by oak origin	Dávalos *et al.* (2003)
1.5 , 3, 7, 9,12, 19.5, and 26 months in bottle	TPI show complex temporal profile with minimum values after 12 months (Tempranillo, Graciano, and Cabernet Sauvignon)	TPI measured by Folin's	Monagas *et al.* (2006)
Aging in bottles	Montepulciano d'Abruzzo, vintages 1973, 1995, and 1996 produced without SO_2 addition. Large decrease in TPI and AOA with bottle aging time	Folin's and DPPH	Manzocco *et al.* (1998)
Aging in bottles	Young and typical Italian wines. Aged wines had higher TPI (28%) and lower AOA (36%)	Folin's and DPPH	Pellegrini *et al.* (2000)
Aging in bottles	21 Californian red wines from 1 to 28 years old. AOA does not correlate with age	Methyl linoleate oxidation measured by oxygen consumption	Roginsky *et al.* (2006)
Aging in bottles	7 months in the dark at 20°C more than 50% decrease in total phenols with minor changes in AOA	Folin's and DPPH	Zafrilla *et al.* (2003)
Aging in tanks	Decrease in phenols depends on the position of the tanks	Folin's	Catania *et al.* (2011)

ABTS, 2,2′-azino-bis(3-ethylbenzthiazoline-6-sulfonic acid; AOA, antioxidant activity; DPPH, 1,1-diphenyl-2-picrylhydrazyl; FRAP, ferric reducing antioxidant power; ORAC-Fl, oxygen radical absorbance capacity method using fluorescein; TPI, total phenolic index.

Similarly, in barrel aging factors such as the oak origin and the shape of the barrel could influence changes in TPI and AOA

We have evaluated TPI and ORAC-PGR indexes in wines of 27 Cabernet Sauvignon red wines from successive vintages. We did not observe any tendency in either of the indexes. The average values of the ratio of the indexes for the two years were:

6. Folin's: average ratio $=0.99 \pm 0.23$, minimum 0.74, maximum 1.3
7. ORAC-PGR: average ratio $=1.00 \pm 0.08$, minimum 0.86, maximum 1.11

These data imply that there are no clear tendencies with bottle aging for TPI (total phenolic index) and ORAC-PGR (quantity and reactivity of the phenols). In fact, ORAC-PGR values (given in Figure 8.1) do not correlate significantly with the vintage year ($p=0.45$).

De Beer *et al.* (2005) evaluated the effect of bottle aging on the AOA of several red wine varietals employing ABTS- and DPPH-based assays. The final total phenol content of the wines decreased between 10.2 % and 16.5 % after 12 month storage, almost independently of the storage temperature (0 to 15°C). This result was attributed to the oxidation of phenols and to anthocyanins condensations. Similarly, the AOA measured employing ABTS radical cation and DPPH decreased by as much as 34.5% during the storage.

Wines change during barrel aging partly by incorporating new substances released from the wood and partly by oxidation, condensation, and polymerization reactions of wine components mediated by the oxygen permeating through the staves (Sanchez-Iglesias *et al.*, 2009). In order to mimic these conditions micro-oxygenation techniques have been developed in which small controlled doses of oxygen are added to fermenting wines. In fact, best results are obtained when oxygen is added at the end of alcohol fermentation. Irrespective of micro-oxygenation, TPI of wine samples remain almost constant during 12 months' aging in oak barrels.

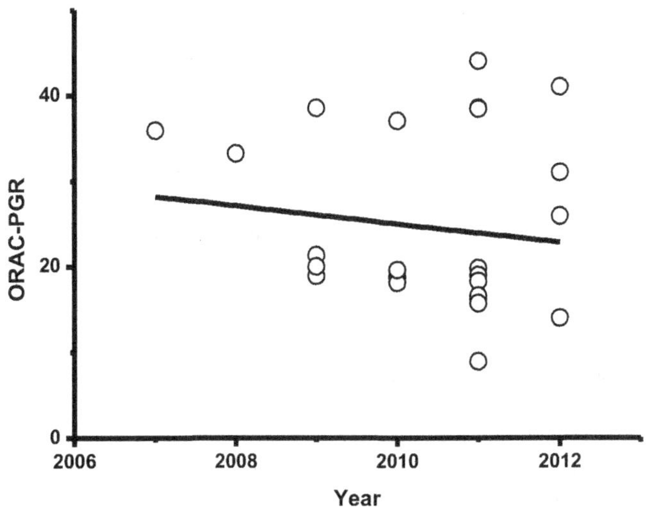

FIGURE 8.1 ORAC-PGR (ORAC-pyrogallol red) values plotted as a function of the vintage year for 27 Cabernet Sauvignon Chilean wines (unpublished results). ORAC-PGR given as mM Trolox equivalents.

Cabernet Sauvignon wines from the vintage 1995 (7 years old) showed higher antioxidant activity (ORAC-Fl) than either younger wines (vintage 2002) or older wines (vintages 1992 and 1989). Similar results were obtained for Tempranillo wines. The authors concluded that there is not a direct relationship between age and AOA. Thus, it might be an optimum aging time that could vary with grape variety (Dávalos *et al.*, 2003, 2004). Representative results for Cabernet Sauvignon wines are given in Table 8.4 (Dávalos *et al.*, 2004). It is interesting to note that during oak aging (1 year) of a Sangiovese wine a steady increase in the phenol content was observed (Folin's method). Conversely, the levels of antioxidants (chemiluminescence and Crocin bleaching) presented a different behavior, without clear tendencies with time or significant fluctuations (Girotti *et al.*, 2002).

A study of the relationship among AOA values of French red wines (assessed by ABTS methodology) and vintage year showed that the AOA of the wines decreases from ca. 20 mM (Vintage 1999) to 13.4 mM for vintages 1995–1991. This would imply that bottle aging decreases the AOA of the wines. Furthermore, given the strong correlation among total phenolic content and AOA (ABTS) ($r=0.959$ in this work) a decrease in total phenolic content with age can be expected (Landrault *et al.*, 2001).

Spanish 'young' (to be consumed during the vintage year), 'crianza' (aging at least 6 months in oak and 18 months in bottles), and 'reserve' (aging at least 12 months in oak and 24 months in bottles) were compared in their capacity to bleach DPPH. Three of four tested wines decrease their AOA when moving from the young to the crianza categories. No further changes were observed for the reserve wines.

Eight young Italian wines (*vini novelli*) prepared by carbonic maceration and aimed to be consumed within 3

TABLE 8.4 Effect of Four Month's Oak Aging on Total Phenols and ORAC-Fl Values

Vintage	Oak Aged	ORAC-Fl*	Total Phenols**
2000	No	34.7	1428
2002	Yes	45.7	2001
2000	Yes	41.6	2074
1995	Yes	63.8	2350
1992	Yes	54.1	1960
1989	Yes	53.5	2025

**ORAC-FL (ORAC-fluorescein) is expressed as mM Trolox equivalents.*
***Total phenolic content represents mg of equivalents of gallic acid/liter.*
Adapted from Dávalos et al. *(2004).*

months from their wine-making were evaluated for their TPI and AOA (ABTS) and the results compared to those of traditional Italian wines (mean age 3 years) (Pellegrini *et al.*, 2000). Young and traditional wines do not show significant differences in their TPI. Conversely, AOA of fresh wines was 26.7 % higher ($p=0.015$) than Italian traditional wines.

AOA and TPI of 21 Californian red wines from 1 to 28 years old were measured after large periods of aging (Roginsky *et al.*, 2006). Both properties were not correlated to ageing. In fact, since samples with different ages correspond to different vintages, the different values measured were more dependent on the grape variety, climate conditions and the particular viticulture and wine-making practices rather than on age.

Okuda *et al.* (2002) investigated the relationship between the concentration of phenols in Japanese Cabernet Sauvignon red wines and their radical scavenging activities (AOA) after long bottle-aging periods. Total phenols and flavonoid concentrations and the total AOA of the wines decreased during a period of about 10 years of storage, with a close correlation between the concentrations of total phenols or total flavonoids with values of AOA index.

EFFECT OF BARREL AND BOTTLE AGING ON THE LEVEL OF POTENTIALLY VALUABLE PHENOLS AND POLYPHENOLS

The effect of aging on the concentrations of specific compounds has been also studied, in particular for resveratrol due to the relevance that it could have on the beneficial effects of moderate red wine ingestion. The results obtained emphasize the variability of results. For example, Abril *et al.* (2005) have carried out a study of resveratrol content in Aragón red wines. In this work, they compare resveratrol concentrations in young and aged wines (matured in barrels), concluding that there were appreciable reductions of *trans-*, *cis-* and total resveratrol concentrations in aged wines in the Somontano and Calatayud designation of origin. Similar results have been reported by other authors (Jeandet *et al.*, 1995; Lamuela-Raventós *et al.*, 1995). However, in other wines these concentrations were slightly higher. These different results were attributed to the possible different climatic conditions of each vintage.

Anthocyanin levels generally decrease with aging. The effect of bottle aging (up to 26 months) of red wines on phenol concentrations has been evaluated in Tempranillo, Graciano, and Cabernet Saugvignon wines by Monagas *et al.* (2006). A relevant result was a noticeable decrease in total anthocyanins in the three varieties (43% in Tempranillo, 65% in Graciano, and 66% in Cabernet Saugvignon wines).

RELATIONSHIP BETWEEN ANTIOXIDANT CAPACITY AND ACCEPTANCE (PRICE) OF RED WINES

There are very few studies in which the relationship between antioxidant capacities of wines and their retail price is discussed. Campos and Lissi (1996), after analyzing a reduced number of Chilean red wines concluded that no relationship was evidenced between the price of the wine and its antioxidant capacity. However, the antioxidant capacity was assessed from the consumption of the radical derived from ABTS, a procedure that determines the amount of phenols and not their antioxidant capacity.

Twenty-three Cabernet Sauvignon wines from the Mudgee region and 32 Shiraz wines from the Hunter Valley region were analyzed for phenolic content and antioxidant activity (Yoo *et al.*, 2011). DPPH antioxidant activity varied considerably, both within and between varieties. Total phenols, and antioxidant activity, were correlated with price and vintage. Wines showed positive and significant correlations for total phenols and antioxidant activity. However, there was a negative and significant correlation between price and antioxidant activity for Cabernet Sauvignon wines. It was concluded that the price (acceptance) of studied wines is not reflective of their health functionality, based on antioxidant activities.

Granato *et al.* (2011) measured the AOA (ORAC-Fl and DPPH consumption) of 71 South American red wines from five varietals. They found a clear correlation between both indexes ($p=0.012$, $r=0.648$) and between them and total phenolics evaluated by Folin's (Folin/DPPH $r=0.59$; Folin/ORAC $=0.61$). The retail price was correlated to the

TABLE 8.5 Total Phenols and Antioxidant Capacity of Red Wines and Acceptance

Study	Conclusions
Evaluation of phenols in grapes and wines from three retail price wine categories (US$ 6–8, US$ 28–30, US$ 150–160)	No difference was observed in total phenols of the three price categories (Cáceres *et al.*, 2012)
Evaluation of phenols and AOA	Price inversely correlated with AOA (Yoo *et al.*, 2011) and catechin (Granato *et al.*, 2011)
Evaluation of phenols, AOA, retail price, and sensory quality	Retail price correlates with the sensory quality (Granato *et al.*, 2011) but not with TPI or AOA
Phenolic content and retail price	Retail prices inversely correlated with catechin concentration (Faustino *et al.*, 2003)
Phenolic, AOA, and retail price, 20 Cabernet Sauvignon Chilean wines	Prices uncorrelated to TPI ($r^2=0.2$) and to ORAC-PGR index ($r^2=0.08$) Unpublished

AOA, antioxidant activity; TPI, total phenolic index.

sensory quality of the wine but not to its phenolic content. In agreement with these results it has been reported that, for a series of Merlot wines from Canada, Chile, and the United States, wine prices were inversely correlated with catechin concentration (Faustino *et al.*, 2003).

The conclusions of these studies are summarized in Table 8.5. It is clear that the levels of potentially beneficial antioxidants are not reflected in the retail price of wines.

Acknowledgments

This work was supported by FONDECYT (Project no. 1100659) and DICYT (Universidad de Santiago de Chile).

References

Abril, M., Negueruela, A., Pérez, C., Juan, T., Estopañán, G., 2005. Preliminary study of resveratrol content in Aragón red and rose wines. Food Chem. 92, 729–736.

Alarcón, E., Campos, A., Edwards, A.M., Lissi, E., López-Alarcón, C., 2008. Antioxidant capacity of herbal infusions and tea extracts: A comparison of ORAC-Fluorescein and ORAC-pyrogallol red methodologies. Food Chem. 107, 1114–1119.

Alén-Ruiz, F., García-Falcón, M.., Pérez-Lamela, M., Martínez-Carballo, E., Simal-Gándara, J., 2009. Influence of mayor polyphenols on antioxidant activity in Mencía and Brancellao red wines. Food Chem. 113, 53–60.

Burns, J., Garner, P., Mathews, D., Duthie, G., Lean, M., Crozier, A., 2001. Extraction of phenolics and changes in antioxidant activity of red wines during vinification. J. Agric. Food Chem. 49, 5797–5808.

De Beer, D., Boubert, E., Gelderblom, W., Manley, M., 2005. Changes in the phenolic composition and antioxidant activity of Pinotage, Cabernet Sauvignon, Chardonay and Chenimblanc wines during bottle ageing. S. Afr. J. Enol. Vitic. 26, 6–15.

Bernardi, A., López-Alarcón, C., Aspee, A., Rech, S., von Poser, G., Bridi, R., Dutra Filho, C., Lissi, E., 2008. Antioxidant activity in Southern Brazil hypericum species. J. Chil. Chem. Soc. 53, 1658–1662.

Cáceres, A., Peña-Neira, P., Galvez, A., Obreque-Slier, E., López-Solís, R., Canals, J., 2012. Phenolic composition of grapes and wines from CV. Cabernet Sauvignon produced in Chile and their relation on commercial value. J. Agric. Food Chem. 60, 8694–8702.

Campodónico, P., Barbieri, E., Pizarro, M., Sotomayor, C., Lissi, E., 1998. A comparison between total phenol content of wines and their TRAP values measured by the bleaching of ABTS radical cations. Bol. Soc. Chil. Quim. 43, 281–285.

Campos, A., Escobar, J., Lissi, E., 1997. The total reactive antioxidant potential (TRAP) and total antioxidant reactivity (TAR) of Ilex paraguayensis extracts and red wine.J. Braz. Chem. Soc. 7, 43–49.

Campos, A., Lissi, E., 1996. Total antioxidant potential of Chilean wines. Nutr. Res. 16, 385–389.

Campos, A., Duran, N., López-Alarcón, C., Lissi, E., 2012. Kinetic and stoichiometric evaluation of free radical scavengers activities based on DPPH consumption. J. Chil. Chem. Soc. 57, 1199–1202.

Catania, P., Pipitone, F., Vallone, M., 2011. Comparison between two types of winemaking tanks for the production of quality wines CV. Nero D'Avola. J. Agric. Eng. 2, 19–24.

Dávalos, A., Bartolomé, B., Suberviola, J., Gómez-Córdoba, C., 2003. ORAC-Fluorescein as a model for evaluating antioxidant activity of wines. Pol. J. Food Nutr. Sci. 53, 133–136.

Dávalos, A., Gómez-Cordovés, C., Bartolomé, B., 2004. Extending applicability of the ORAC-Fluorescence Assay. J. Agric. Food Chem. 52, 48–54.

Echeverry, C., Ferreira, M., Reyes-Parada, M., Abin-Carriquiry, J., Blasina, F., González Nieves, G., Dajas, F., 2005. Changes in antioxidant capacity of Tannat red wines during early maturation. J. Food Eng. 69, 147–154.

Faustino, R.S., Sobrattee, S., Edel, A.L., Pierce, G.N., 2003. Comparative analysis of the phenolic content of selected Chilean, Canadian and American merlot red wines. Mol. Cell Biochem. 249, 11–19.

Fuhrman, B., Volkova, N., Suraski, A., Aviran, M., 2001. White wine with red wine-like properties: Increased extraction of grape skin polypenols improves the antioxidant capacity of the derived white wine. J. Agric. Food Chem. 49, 1164–1168.

Ginjom, I., D'Arcy, B., Caffin, N., Gidley, M., 2010. Phenolic content and antioxidant activity of major Australian red wines throughout the wine making process. J. Agric. Food Chem. 58, 10133–10142.

Ginjom, I., D'Arcy, B., Caffin, N., Gidley, M., 2011. Phenolic compounds profiles in selected Queensland red wines at all stages of the wine making process. Food Chem. 125, 823–834.

Giovanelli, G., 2005. Evaluation of the antioxidant activity of red wines in relation to their phenolic content. Italian J. Food Sci. 17, 381–393.

Girotti, S., Botelli, L., Budin, R., Arfelli, C., 2002. Comparison of analytical methods in determining total antioxidant capacity in red wine. Anal. Lett. 35, 747–758.

Gómez-Plaza, E., Gil-Muñoz, R., López-Roca, J., Martínez, A., 2000. Color and phenolic compounds of a Young red wine. Influence of wine-making techniques, storage temperature and length of storage time. J. Agric. Food Chem. 48, 736–741.

Granato, D., Katayanam, F., de Castro, U., 2011. Phenolic composition of South American red wines classified according to their antioxidant activity, retail price and sensory quality. Food Chem. 129, 366–373.

Hotzer, K.A., Henriquez, C., Pino, E., Miranda-Rottmann, S., Aspillaga, A., Leighton, F., Lissi, E., 2005. Antioxidant and pro-oxidant effect of red wine and its fractions on Cu(II) induced LDL oxidation evaluated by absorbance and chemiluminescence measurements. Free Radic. Res. 39, 175–183.

Jeandet, P., Bessis, R., Sbaghi, M., Meunier, P., Trollat, P., 1995. Resveratrol content of wines of different ages: relationship with fungal disease pressure in the vineyard. Am. J. Enol. Viticult. 46, 1–4.

Kondrashov, A., Sevcik, R., Benakova, H., Kostirova, M., Stipek, S., 2009. The key role of grape variety for antioxidant capacity of red wine. Eur. e-J. Clin. Nutrit. Metab. 4, e41–e46.

Lachman, J., Sulc, M., Schilla, M., 2007. Comparison of the total antioxidant status of Bohemian wines during the wine-making process. Food Chem. 103, 802–807.

Lachman, J., Sulc, M., Faitova, K., Pivex, V., 2009. Major factors influencing antioxidant contents and antioxidant activity in grapes and wines. Int. J. Wine Res. 1, 101–121.

Landrault, N., Poucheret, P., Ravel, P., Gasc, F., Cros, G., Teissedre, P.-L., 2001. Antioxidant capacities and phenolics levels of French wines from different varieties and vintages. J. Agric. Food Chem. 49, 3341–3348.

Lamuela-Raventós, R., Romero-Pérez, A., Waterhouse, A., de la Torre-Boronat, R., 1995. Direct HPLC analysis of cis and trans resveratrol and piceid isomers in Spanish red Vitisvinifera wines. J. Agric. Food Chem. 43, 281–283.

Larrauri, J., Sanchez-Moreno, C., Rupérez, P., Saura-Calixto, F., 1999. Free radical scavenging capacity in the aging of selected red Spanish wines. J. Agric. Food Chem. 47, 1603–1606.

López-Alarcón, C., Ortiz, R., Benavides, J., Mura, E., Lissi, E., 2011. Use of the ORAC-PGR/ORAC Fl ratio to assess the quality of antioxidants in Chilean wines. J. Chil. Chem. Soc. 56, 764–767.

Manzocco, L., Mastrocola, D., Nicoli, M., 1998. Chain-breaking and oxygen scavenging properties of wine as affected by some technological procedures. Food Res. Int. 31, 673–678.

Monagas, M., Gómez-Cordovés, C., Bartolomé, B., 2006. Evolution of the phenolic content of red wines from Vitis Vinífera during ageing in bottle. Food Chem. 94, 405–412.

Netzel, M., Strass, G., Bitsch, I., Könitz, R., Christmann, M., Bitsch, R., 2003. Effect of grape processing on selected antioxidant phenolics in red wine. J. Food Eng. 56, 223–228.

Okuda, T., Takayanagi, T., Sato, M., Yokotsuta, K., 2002. Changes in radical scavenger activity of Japanese cavernet sauvignon red wines during aging. J. Wine Res. 13, 93–100.

Pellegrini, N., Simonetti, P., Gardana, C., Brenna, O., Brighenti, F., Pietta, P., 2000. Polyphenol content and total antioxidant activity of *Vini novelli* (young red wines). J. Agric. Food Chem. 48, 732–735.

Perez, D., Leighton, F., Aspée, A., Aliaga, C., Lissi, E., 2000. A comparison of methods employed to evaluate antioxidant capabilities. Biol. Res. 33, 71–77.

Poblete, A., López-Alarcón, C., Lissi, E., Campos, A., 2009. Oxygen radical antioxidant capacity (ORAC) values of herbal teas obtained employing different methodologies can provide complementary data. J. Chil. Chem. Soc. 54, 154–157.

Poiana, M., Dobrei, A., Stoin, D., Ghita, A., 2008. The influence of viticultural region and the ageing process on the color structure and antioxidant profile of Cabernet Sauvignon red wines. J. Food Agric. Environm. 6, 104–108.

Roginsky, V., De Beer, D., Harbertson, J., Kilmartin, P., Barsukova, T., Adams, D., 2006. The antioxidant activity of californian red wines does not correlate with wine age. J. Agric. Food Chem. 86, 834–840.

Sanchez-Iglesias, M., González-San, M., Perez-Magariño, S., Ortega-Heras, M., González-Huerta, C., 2009. Effect of micro-oxygenation and Wood type on the phenolic composition and color of an aged red wine. J. Agric. Food Chem. 57, 11499–11509.

Sanchez-Moreno, C., Larrauri, J., Saura-Calixto, F., 1998. A procedure to measure de antiradical efficiency of polyphenols. J. Sci. Food Agric 76, 270–276.

Simonetti, P., Pietta, P., Testolin, G., 1997. Polyphenol content and total antioxidant potential of selected Italian wines. J. Agric. Food Chem. 45, 1152–1155.

Villano, D., Fernández-Pachón, M., Troncoso, A., García-Parrilla, M., 2006. Influence of enological practices on the antioxidant activity of wines. Food Chem. 95, 394–404.

Yoo, Y., Prenzer, P., Saliba, A., Ryan, D., 2011. Assessment of some Australian red wines for price, phenolic content, antioxidant activity, and vintage in relation to functional food prospects. J. Food Sci. 76, 1355–1360.

Zafrilla, P., Morillas, J., Mulero, J., Cayuela, J., Martinez-Cachá, A., Pardo, F., López, J.M., 2003. Changes during storage in conventional and ecological wines: Phenolic content and antioxidant activity. J. Agric. Food Chem. 51, 4694–4700.

CHAPTER

9

Effects of Varieties and Growing Conditions on Antioxidant Capacity of Coffee

Draženka Komes, Aleksandra Vojvodić

Department of Food Engineering, Faculty of Food Technology and Biotechnology, University of Zagreb, Zagreb, Croatia

CHAPTER POINTS

- Chemical composition of green coffee beans is mainly genetically defined.
- Robusta green coffees (*Coffea canephora* species) are known to contain higher contents of chlorogenic acids and caffeine, whereas Arabica coffees (*Coffea arabica* species) contain higher contents of trigonelline and coffee diterpenes kahweol and cafestol.
- Environmental characteristics of coffee-growing areas and growing practices significantly influence the content of chlorogenic acids, caffeine, trigonelline, and coffee diterpenes, explaining the annual deviations among the same cultivars grown in different regions, and differences caused by origin in general.
- Post-harvest handling of green coffee (dry, semi-dry, and wet methods) may result in variations in coffee bean composition, reflected especially in the content of water-soluble compounds.
- Storage of green coffee beans is mainly evaluated from the microbiological standpoint and visual appearance but could be applied to the content of antioxidants, chlorogenic acids in particular, due to their involvement in unwanted enzymatic changes and susceptibility to deterioration.
- In order to asses and compare the 'antioxidant quality' of green coffees, besides well-chosen representative samples and analytical methods, a defined set of environmental parameters should also be included.

INTRODUCTION

At least 66 species of genus *Coffea L.* have been identified so far, and two of these varieties are economically and commercially important: *C. arabica L.* (Arabica coffee) and *C. canephora* Pierre ex Froehn (Robusta coffee) (Söndahl and van der Vossen, 2005), accounting for 90% and 9% of world's coffee production, respectively (Briandet *et al.*, 1996). Both coffee varieties differ not only from a botanical standpoint but also in terms of chemical composition. Green coffee beans of both species contain a wide range of different chemical compounds that affect the quality of final coffee products (El-Abassy *et al.*, 2011). Those differences are recognized commercially, and Robusta usually sells at prices 20–25% lower than Arabica, which is more appreciated by the consumers and generally considered to be of better quality (Esteban-Diez *et al.*, 2004). Due to this price difference, there is a tendency to adulterate Arabica with Robusta green coffee, which consequently leads to the improvement of Robusta coffee quality. Therefore, as far as mislabeling or eventual fraudulence are concerned, it is important to identify the bean variety (El-Abassy *et al.*, 2011). The green beans of *C. arabica* and *C. canephora* have different colors, shapes, and sizes, but after roasting and grinding, these species can not be visually distinguished, and therefore it is necessary to use chemical parameters for their discrimination. Several components can be used as indicators or discriminators of Arabica and Robusta species: water-soluble compounds (caffeine, trigonelline, nicotinic acid and chlorogenic acids) and fat-soluble compounds (kahweol and cafestol) (De Souza and Benassi, 2012). The content of these compounds generally depends not only on the species but also on the geographical origin, presence of defective beans, and finally, roasting degree. Additionally, it is reported that also metal content, sterolic profile,

Processing and Impact on Antioxidants in Beverages
http://dx.doi.org/10.1016/B978-0-12-404738-9.00009-X

fatty acid profile, tocopherols and triglycerides could be used as suitable chemical descriptors of a certain coffee variety (Esteban-Diez *et al.*, 2004).

Distinction between Arabica and Robusta coffees is mostly made through the evaluation of their chemical composition by means of the formation of specific aroma compounds which define coffee cup quality. Many of the compounds that act like precursors in aroma development may also have other significant contributions. In light of this, both species, *C. arabica* and *C. canephora,* represent a rich source of biologically active compounds, among which antioxidants especially have received a lot of scientific interest in recent years, focused on the antioxidant potential of coffee as a raw material and consequently as a beverage. Despite several reports on the antioxidant capacity of coffee brew from roasted coffee, the antioxidant capacity of green coffee in particular is not well studied.

INFLUENCE OF COFFEE VARIETIES ON THE CONTENT OF CHLOROGENIC ACIDS

The evaluation of antioxidant properties of coffee beans is of interest primarily in order to find a promising new source of natural antioxidants for functional foods and nutraceuticals, promoting the utilization of green coffee for uses other than as a beverage (Madhava *et al.*, 2008).

Due to the fact that green coffee is known as one of the main food sources of chlorogenic acids (CGAs), with contents being equalled or exceeded only by the green leaves of *Ilex paraguariensis* (maté) (Clifford, 1999), and supported by their bioactive contribution, many studies have been designed with the aim of setting CGAs as a distinctive parameter for *Coffea* species and their cultivars, thus adding value to the quality of a particular coffee variety. Content of CGAs in different coffee varieties and cultivars is presented in Table 9.1. Generally, Robusta green coffees contain a higher content of CGAs than the *C. arabica* species (Ky *et al.*, 2001; Farah *et al.*, 2005), with values ranging from 7.0% to 14.4% for *C. canephora,* and from 4.0% to 8.4% for *C. arabica,* both expressed on a dry weight basis (dwb), with some hybrids presenting the intermediate contents (Farah and Donangelo, 2006). Non-flavonoid compounds, which include the CGAs, were found to account for more than 50% of the total phenol content in both Arabica and Robusta cultivars (Table 9.2) (Hečimović *et al.*, 2011). As it generally applies for CGAs, the content of individual CGAs subgroups and isomers varies between the varieties and cultivars (Table 9.1). Caffeoylquinic acids (CQAs) are the predominant CGA subgroup, with the extent of contribution relative to the species and cultivars, but maintained at high levels such as approximately 80% and 76% of total CGA content in some *C. arabica* and *C. canephora* cultivars, respectively (Farah *et al.*, 2005). Dicaffeoylquinic acids (diCQAs) are reported to be represented at higher levels in Robusta than in Arabica coffees, while the content of feruloylquinic acids (FQAs) is not significantly different between Arabica and Robusta coffees (Farah *et al.*, 2005). In the case of minor CGA components, caffeoylferuloylquinic acids (CFQAs) content was found to be four-fold higher in *C. canephora* than in *C. arabica* samples (Perrone *et al.*, 2008), which may be due to genetic factors rather than agricultural practices,

TABLE 9.1 Content of Chlorogenic Acids (% dmb) in Green Coffee Beans of Different *Coffea* Varieties

	Chlorogenic Acids (% dmb)				
Coffee Samples	**CQAs**	**FQAs**	**diCQAs**	**Total CGAs**	**References**
Wild *C. arabica* (average)	3.26	0.19	0.60	4.10	Ky *et al.* (2001)
C. arabica (Angola)	4.30	0.57	1.23	6.10	Correia *et al.* (1995)
C. arabica (Angola)	4.84	0.28	0.53	5.65	Correia *et al.* (1995)
C. arabica (Angola)	5.67	0.79	1.39	7.85	Correia *et al.* (1995)
C. arabica cv. Burbon (Brazil)	4.20	0.28	0.77	5.25	Farah *et al.* (2005)
C. arabica cv. Longberry (Ethiopia)	4.60	0.29	0.84	5.73	Farah *et al.* (2005)
Wild *C. canephora*	7.66	1.43	2.31	11.3	Ky *et al.* (2001)
C. canephora cv. Robusta (Angola)	3.43	0.54	1.20	6.08	Correia *et al.* (1995)
C. canephora cv. Robusta (Anogola)	4.97	0.75	1.46	7.18	Correia *et al.* (1995)
C. canephora cv. Robusta (Uganda)	5.77	0.47	1.34	7.58	Farah *et al.* (2005)

The results represent average values of total chlorogenic acids (CGAs) and their individual subgroups, such as caffeoylquinic acids (CQAs), feruloylquinic acids (FQAs), dicaffeoylquinic acids (diCQAs), expressed as percentage on a dry matter basis (% dmb), in different *Coffea* varieties. The results are collected from various literature sources and are adapted from the original source for consistency.

TABLE 9.2 Total Phenol Content (mg GAE/g), Total Flavonoid Content (mg GAE/g) and Total Flavan-3-ol Content [mg (+)-catechin/g] of Green Coffee Beans of *C. arabica* and *C. canephora*

Coffee Samples	Total Phenols (mg GAE/g)	Total Flavonoids (mg GAE/g)	Total Flavan-3-ols [mg (+)-catechin/g]
C. arabica			
Minas	30.67 ± 0.23	11.08 ± 0.83	0.20 ± 0.08
Cioccolatato	21.00 ± 0.71	9.67 ± 0.12	0.17 ± 0.00
C. canephora			
Vietnam	26.08 ± 0.59	6.41 ± 0.12	0.39 ± 0.14
Cherry	38.83 ± 0.59	18.83 ± 0.24	0.57 ± 0.23

Presented results of total phenols and phenolic subgroups are average values ± SD, expressed as mg of gallic acid equivalents (GAE) per g of green coffee, for total phenols and total flavonoids, while for flavan-3-ols as mg of (+)-catechin per g of green coffee. Coffee samples were assembled of two varieties of *C. arabica* (Minas and Cioccolatato) and two varieties of *C. canephora* (Vietnam and Cherry).
The results are excerpted from Hečimović et al. *(2011) and their presentation is adapted from the original source to form a single display.*

supported by the correlation between the pattern of minor CGA-like compounds and the geographical origin (Correia *et al.*, 1995). Large variations in CGA content and distribution of their isomers, along with qualitative differences (presence and absence of some isomers) in green coffee beans, make the CGA content a potential criterion of coffee genotype selection (Correia *et al.*, 1995; Ky *et al.*, 2001). Significant differences in the content of CGAs may also be observed among cultivars of the same *Coffea* varieties. This can be illustrated with the example of the contents of total CQAs and diCQAs in Arabica green coffees, which were higher by 8.9% and 8.4%, respectively, in Longberry cultivar originated from Ethiopia than in Bourbon cultivar, originated from Brazil (Farah *et al.*, 2005). In addition, among the Robusta coffees, Conillon cultivar was reported to contain higher CGA content when compared to the values reported for Robusta cultivar (Correia *et al.*, 1995; Farah and Donangelo, 2006). Conversely, between *C. arabica* cultivars originated from Brazil (Guaxupé, Minas Gerais), no significant deviations in CGA contents were observed, complemented also with similar pattern of CGA distribution among Arabica coffees (Monteiro and Farah, 2012). The small variations in CGA content of coffee beans from ripe fruits though may be observed due to weather and agricultural practices (Farah and Donangelo, 2006).

The large range of CQA, diCQA, and FQA content in genus *Coffea* and their implications on quality of coffee brews emphasize the importance of CGA inbreeding programs, opening the possibilities for the enhancement of CGA profiles in favor of the best composition, including also the antioxidant outcome. Among coffee trees of less commercially interesting *Coffea* species, CGA content in green coffee beans varies greatly between the species, from 0.14% dry matter basis (dmb) in *Coffea rhamnifolia* Bridson to 9.90% dmb in *C. sessilifora* Bridson. The content of CQAs ranges from 61% of total CGAs in *C. mouloundou* to 100% in *C. farafanganensis* Leroy (Anthony *et al.*, 1993). In a study by Ky *et al.* (1999) on quantitative inheritance of CGAs in hybrids obtained by crossing *C. pseudozanguebariae* Bridson and *C. liberica* var. Dewevrei, the variations in CGA content among resulting hybrids was affected by genetic attributors at 78% in the best cases, thereby showing the relative importance of environmental effects and interactions. The latter would have been probably more important in crosses of less genetically differentiated species. It is implied that CGAs are sensitive to all types of environmental changes and consequently, interactions should exist when genotypes are growing in different regions. Quantitative inheritance parameters for caffeine, CGAs, sucrose, and trigonelline are known in coffee trees (Barre *et al.*, 1998; Ky *et al.*, 1999, 2001), therefore suggesting a considerable potential for biochemical improvement of *Coffea* species.

INFLUENCE OF ENVIRONMENTAL FACTORS ON THE CONTENT OF CHLOROGENIC ACIDS

Distinctions by a specific compound are not easy to estimate since they may arise not only from genetic predispositions of a certain coffee variety or cultivar, but also from various environmental factors (Da Silva *et al.*, 2005; Farah and Donangelo, 2006; Joët *et al.*, 2010). These effects include inter-annual climatic variations during bean maturation, and also maturity variations at harvest and bean-drying conditions after harvest. Severe weather conditions, such as cold, high visible light, and water stress conditions tend to increase the content of phenolic compounds not only in the coffee plant and seed but also in other plants. In growing practice it is known that higher growing altitudes improve coffee quality in general (Guyot *et al.*, 1996; Decazy *et al.*, 2003; Avelino *et al.*, 2005). However, the results obtained for the influence of altitude on the chemical composition are rather inconsistent. For instance, chlorogenic acids were reported to accumulate with an increase in altitude (Avelino *et al.*, 2005; Bertrand *et al.*, 2006), and also to remain stable (Guyot *et al.*, 1996). Joët *et al.* (2010) emphasized that these correlations may be modified by soil–climatic conditions as well as agricultural management that may also have varied at the sampling sites of previous studies. In addition, the authors evaluated the chemical composition of 16 experimental plots of *C. arabica* cv. Laurina cultivated on Reunion Island. The results showed very high climatic variations within the growing region, especially a very broad temperature gradient, which was highly correlated with altitude. Total CGA content was found not to have any significant correlation with any considered climatic variable, although within CGAs, individual isomers exhibited either a positive or a negative correlation with temperature, indicating that temperature may act directly on routing towards different isomers within the CGA metabolic pathway, without affecting the final CGA content. Influence of climatic changes in the same area during a certain period was reported by Monteiro and Farah (2012), who found that extraneous factors, such as rainfall and temperature extremes, can significantly influence the yield of CGAs in a particular cultivar grown on the same location throughout several years. By comparing *C. arabica* cultivars, cvs. Mundo Novo, Bourbon, Red Catuai, and Yellow Catuai, originated from Brasil (Guaxupé, Minas Gerais) throughout four or five annual crops, no significant variations were found between CGA content, also maintaining their relative ratio within a year. Conversely, annual crops of a specific cultivar were found to be significantly different in terms of total CGA content and individual isomer content. These results suggest that the CGA content can be significantly affected by yearly environmental changes, in contrast to the impact of cultivar differences. Due to the natural fluctuations in the content of phenolic compounds among annual coffee crops, a defined set of environmental conditions, as well as a carefully selected representative sample (including several crops), is required in order to distinguish characteristics of different cultivars. Apart from climate and soil composition, agricultural practices may also contribute to the changes in CGA content and distribution (Camacho-Cristóbal, *et al.*, 2002; Farah *et al.*, 2005; Farah and Donangelo, 2006; Joët *et al.*, 2010; Monteiro and Farah, 2012). For example, fertilization with zinc was able to increase CGA content in Arabica coffee (Perrone, *et al.*, 2008). The use of nitrogen-rich fertilizers (Malta *et al.*, 2003) and situations of boron deficiency (Camacho-Cristóbal *et al.*, 2002) have also been shown to increase the content of total CGAs in coffee beans and other plants.

From a broader aspect, considering the impact of various environmental factors, either isolated or combined, it is obvious that the importance of geographical origin in general, with all its environmental attributes, certainly can not be diminished in the assessment of 'antioxidant quality' of green coffee beans. To illustrate this on a small scale, two *C. arabica* cultivars (var. Caturra and var. Colombia) grown in two coffee-growing regions in Colombia, Cauca and Nariño, were screened for chemical and sensory differences and correlated with environmental parameters such as altitude and average annual rainfall. The regional differences in environmental and product quality characteristics between the regions of Cauca and Nariño were generally significant and not random, but had a clearly expressed spatial structure. Moreover, the spatial structure in the data of both the environmental and the product quality characteristics was correlated. The approach presents climate and topography information as the main predictors for regional coffee quality. Regarding the latter case, in Cauca, both cultivars (cv. Colombia and cv. Caturra) were grown in very similar production environments of the region, while the production environments were very different in Nariño, causing the differences in coffee bean composition (Oberthür *et al.*, 2011). In addition, significant differences between *C. arabica* cultivars, namely Caturra and Catuai, grown in two *terroirs* of Costa Rica (Santa María de Dota and Orosi) were observed. Chlorogenic acid content was found to be more abundant in Santa María de Dota coffees, probably due to the higher altitude of the growing terrain, thus suggesting a meaningful impact of origin on the chemical composition (Avelino *et al.*, 2005).

INFLUENCE OF VARIETIES AND ENVIRONMENTAL FACTORS ON THE CONTENT OF OTHER ANTIOXIDANT CONSTITUENTS IN COFFEE

Apart from chlorogenic acids, other compounds present in green coffee beans, such as caffeine, trigonelline, diterpenes, and tocopherols, contribute to the 'antioxidant quality' of coffee both as a raw material as well as a final product. These compounds are also being considered as distinctive factors between the species and cultivars of *Coffea* genus, thus adding to the contribution of chlorogenic acids in the relation between raw material quality, in terms of chemical composition, and possible health-beneficial effects of the product.

Caffeine

Besides CGAs, caffeine is probably the most interesting compound to consider with regard to its physiological effects and corresponding merit to coffee popularity. In green coffee beans, caffeine content varies markedly, both between and within the species (Campa *et al.*, 2005), as presented in Table 9.3. For instance, *C. canephora* (Robusta coffee), *C. brevipes* and *C. stenophylla* are coffee species with the highest caffeine content, while *C. pseudozanguebariae* and *C. humblotiana* are almost caffeine-free. A significant difference in caffeine content of green coffee beans of *C. arabica* from Brazil and *C. canephora* var. Robusta from Ivory Coast was observed, promoting Robusta coffee as richer in caffeine, with a value of 20.84 mg/kg dmb, in contrast to 12.36 mg/kg dmb evaluated for Arabica

TABLE 9.3 Caffeine Content (% dmb) in Green Coffee Beans of Different *Coffea* Varieties

Coffee samples		Origin	Caffeine content (% dmb)
C. arabica cv.	Burbon vermelho	Brazil	1.26
	Purpurescens	Tanzania	0.76
	Laurina	Tanzania	0.79
	Burbon amarelo	Tanzania	1.49
	Burbon amarelo	Brazil	1.54
	Caturra	Tanzania	1.59
	Mundo novo	Brazil	1.57
	Mundo novo	Brazil	1.59
	Catuai vermelho	Brazil	1.82
	Catuai vermelho	Brazil	1.53
	Catuai vermelho	Brazil	1.28
C. canephora Pierre		Côte-d'Ivore	3.19
	Ugandae	Tanzania	3.02
	Robusta	Tanzania	2.29
	Robusta	Brazil	2.48
	Kouillou	Tanzania	2.22
C. brevipes Hiern		Côte-d'Ivore	2.16
		Côte-d'Ivore	1.93
C. stenophylla G. Don		Côte-d'Ivore	2.64
		Brazil	1.53
		Kew*	1.50
C. pseudozanguebariae Bridson		Côte-d'Ivore	0.02
C. rhamnifolia (Chiov.) Bridson		Kew*	0.06

**Kew—Royal Botanic Garden, UK.*
Presented results are average values of caffeine content in green coffee beans of various *Coffea* varieties, expressed as percentage on a dry matter basis (% dmb).
The results are excerpt fromed Anthony et al. *(1993).*

TABLE 9.4 Trigonelline Content (% dwb in Green Coffee Beans) of Different *Coffea* Varieties

Coffee Samples		Trigonelline Content (% dwb)
C. arabica cv.	Bourbon Red	2.06
	Catuai Red	2.48
	Caturra Red	2.23
	Ibaré	1.97
	Laurina	1.52
	Mundo Novo Red	2.13
C. liberica cv. *Dewevrei*		1.37
C. canephora cv. *Guarini*		3.08
C. stanophylla		1.89
C. salvatrix		1.21
C. liberica		1.01
C. racemosa		2.31
C. brevipes		2.23

Presented results are average values of trigonelline content, expressed as percentage on a dry weight basis (% dwb) in different *Coffea* varieties.
The results are excerpted from Mazzafera (1991).

coffee (Casal *et al.*, 2000). This ratio was maintained during coffee processing, specifically roasting, corresponding to the outcome of caffeine content in brews of Arabica and Robusta coffees. Additionally, in a large sample of *C. arabica* and *C. canephora* specimens, the determined average caffeine content showed prevalence in *C. canephora* samples, ranging from 1.51–3.33% dmb, in contrast to 0.9–1.62% dmb evaluated for *C. arabica* samples (Ky *et al.*, 2001). The Robusta dominance in caffeine was confirmed as well with high-performance liquid chromatography (HPLC) analysis (Hečimović *et al.*, 2011) of green beans of two Robusta and Arabica samples.

As far as other factors apart from genetic distinctions are considered, caffeine was significantly affected by the differences in environmental characteristics of Costa Rica *terroirs* Orosi and Santa María de Dota, where the higher altitude of Santa María de Dota *terroir* favored the higher caffeine content (Avelino *et al.*, 2005). Conversely, significant correlation of caffeine content with climatic variables such as altitude, temperature range, and rainfall in *C. arabica* cv. Laurina, grown on Reunion Island was not found (Joët *et al.*, 2010). With the exception of some isolated cases, it can generally be observed that caffeine content in coffee beans may be susceptible to geographical origin and environmental descriptors of coffee-growing areas, since in cultivated species, caffeine accumulation in coffee beans depends on the period between flowering and ripening of the coffee cherry (Campa *et al.*, 2005). Antioxidant capacity of green coffee beans, evaluated as ferric reducing antioxidant power (FRAP) and Trolox equivalent antioxidant capacity (TEAC) has shown higher potential of Robusta coffees (Hečimović *et al.*, 2011), correlating well with chlorogenic acids and caffeine contents (more abundant in Robustas) and implying their dominant antioxidant contribution.

Trigonelline

Trigonelline, as well as other coffee components, is dependent on the species (Mazzafera, 1991) (Table 9.4). In contrast to caffeine content, Arabica species were found to be more abundant in trigonelline (Casal *et al.*, 2000), ranging from 0.88% to 1.77% dmb, while in *C. canephora* species, trigonelline was found to range from 0.75% to 1.24% dmb, all determined on a large sample of green coffees (Ky *et al.*, 2001). This suggests the relevance of trigonelline, as well as of caffeine, as a discriminating factor for Robustas and Arabicas (Casal *et al.*, 2000). Among coffees (cvs. Catura and Catuai of *C. arabica* sp.) grown in different parts of Costa Rica, inter-regional variations in trigonelline content were low, whereas between the two *terroirs*, the difference was significant. Coffees grown in a region characterized as a very wet zone, under Atlantic influence, and with higher altitude, exhibited higher values of trigonelline (Avelino *et al.*, 2005).

Kahweol and Cafestol

Kahweol and cafestol are diterpenes present in coffee beans, recognized for their antioxidant potential and additionally characterized as potential discriminators of the species *Coffea arabica* and *Coffea canephora* (Campanha *et al.*, 2010). Their high thermal stability favors the latter's use in coffees subjected to different degrees of roasting in commercial coffee blends. Among the species, both diterpenes are represented in higher contents in *C. arabica* than in *C. canephora* (Speer and Kölling-Speer, 2006), with cafestol varying between 0.27% and 0.67% of dry matter (dm) and 0.15–0.37% of dm, respectively (Sridevi *et al.*, 2011), while kahweol is reported to be almost exclusive for Arabica coffee beans (Campanha *et al.*, 2010; Dias *et al.*, 2010), ranging from 0.11% to 0.35% of dm (Sridevi *et al.*, 2011) and detected only at trace levels in *C. canephora* (Speer and Kölling-Speer, 2006). Besides these two main diterpenes, 16-O-methylcafestol, which is found only in the Robusta coffee variety, is defined as a specific marker according to the German industrial standard DIN 10779 (1999).

Observing the contents of kahweol and cafestol esters in green coffee beans of different origin, besides defaults by species, geographic origin was also set up as a significant source of variations (Kurzrock and Speer, 2001). Regarding different cultivars, a great variability in the composition of diterpenes was observed for both green and roasted coffees (Table 9.5), but there is little information on the influence of genetic variability. Comparing the *C. arabica* cultivars grown under the same edapho-climatic conditions, the obtained differences are mainly attributed to particular characteristics of each cultivar. However, no significant differences in total diterpene content were observed between the cultivars (Kitzberger *et al.*, 2010).

TABLE 9.5 Contents of Kahweol and Cafestol (mg/100 g dwb) of Green Coffee Beans of Different *C. arabica* cultivars

Diterpenes Content in green Coffee Beans (mg/100 g dwb)			
Coffee Samples	**Kahweol**	**Cafestol**	**Total Diterpenes**
C. arabica cv.			
Catuaí	604±8	371±6	975±13
Icatu	501±37	433±18	934±56
IPR 100	328±32	892±59	1221±91
IPR 102	356±34	605±52	960±86
IPR 106	325±14	986±46	1312±60

Presented results are average values ± SD, expressed as mg/100 g on a dry weight basis (dwb) of green coffee. Analyzed samples were different cultivars of *C. Arabica*.
The results are excerpted from Kitzberger et al. *(2010).*

Tocopherols

Tocopherols in coffee beans are lipophilic constituents present in the coffee oil, which possess antioxidant properties (Abidi and Mounts, 1997). The influence of coffee species, varieties, and environment on their content is, to our knowledge, the least reported topic within this context. Nevertheless, the content of α-, β-, and γ-tocopherols varies between Arabicas and Robustas, exhibiting generally higher values in green Arabica beans (Gonzáles *et al.*, 2001). Furthermore, tocopherol content significantly differs according to the origin, namely the country from which both Arabica and Robusta coffees originate. Taking into account coffee processing in a general way, Arabica roasted beans are known to contain higher tocopherols content than the Robusta ones, especially of β-tocopherol, which is associated with higher lipid content.

INFLUENCE OF PRIMARY PROCESSING AND STORAGE ON COFFEE ANTIOXIDANTS

Many characteristics of coffee beans that later determine the properties and quality of the brewed coffee, including its antioxidant potential, are not only dependent on the species (e.g., *C. arabica* or *C. canephora*, including different cultivars), cultivation conditions of the plants (shade, pruning, fertilization, soil, altitude, sun exposure, rainfall, and temperature), degree of ripeness at harvest and harvesting method, but also on the processing method (dry or wet) (Belay *et al.*, 2008). Post-harvest handling is a primary step in coffee processing, having as a result pronounced effects on the chemical composition of coffee beans, especially in water-soluble components like sugars, caffeine, trigonelline, and chlorogenic acids (Smith, 1985). After harvesting of the fruits, green coffee seeds are processed by one of three different methods known as dry, wet, and semi-dry (Teixeira *et al.*, 1995). The methods aim to remove the pericarp of the coffee cherry and to dry the raw coffee seed to a moisture content of 10–12%

of fresh weight (Joët *et al.*, 2010), resulting in 'natural' ('unwashed'), 'washed' ('parchment'), and 'pulped natural' coffee according to the applied method (dry, wet, and semi-dry), respectively (Duarte *et al.*, 2010). Dry-processed coffees are generally characterized by more body, whereas in most *terroirs*, wet-processed coffees have a better aroma, generally resulting in higher acceptance (Selmar *et al.*, 2002). Literature data linked to the varietal dependences on the preservation of antioxidant compounds in coffee in relation to the post-harvest processing show that no significant differences were observed in CGA content when semi-dry and wet methods were applied, comparing the cultivars of *C. arabica* (Red Cataui, Rubi, Yellow Burbon and Topázio) (Duarte *et al.*, 2010). Additionaly, in the case of *C. arabica* cv. Laurina, CGAs were only relatively affected by wet processing (in comparison to fresh seeds), reflected in significant change of two minor isomers, di-3,5-CQA and di-4,5-CQA, as well as CQA and diCQA ratio increase (Joët *et al.*, 2010). In contrast, higher content of CGAs in wet-processed Robusta beans in comparison to dry-processed beans has been reported (Duarte *et al.*, 2010), which can be interpreted as a relative increase due to the losses of other water-soluble components by leaching and fermentation in the wet method, in addition to slight degradation of such components in dry-processed seeds exposed to heat from the sun (Table 9.6). Accordingly, significant differences in CGA content can be observed in Arabica/Robusta hybrids with respect to wet and semi-dry methods, which may be attributed to their genetic background and the changes in cell-wall constitution (Toci and Farah, 2008). Regarding other constituents with antioxidant properties, caffeine was not affected by the post-harvest handling due to its thermal stability (Leloup *et al.*, 2004). Trigonelline showed an increase in wet-processed beans when compared to the semi-dried, being in accordance with the effect of relative increase, above described for CGAs in Robusta coffee (Table 9.6) (Duarte *et al.*, 2010). Regardless of the post-harvest treatment, wet, dry or semi-dry, freshly processed coffee beans remain viable and exhibit active metabolic processes (Bytof *et al.*, 2005) and are therefore susceptible to biochemical changes.

Presenting the last stage in green coffee beans production, storage can strongly influence their commercialization and therefore is considered as one of the most important factors for maintaining final product quality and securing the best market price for the producer (Ribeiro *et al.*, 2011). Traditionally, green coffee beans have been stored in jute sacks for some practical reasons, such as adaptability to small-scale commerce and ease of sampling for lot inspections. During shipping, coffee beans stored in jute sacks are often exposed to potentially harmful variations in ambient conditions, including fluctuations in temperature and relative humidity, due to packaging permeability to water and gases. This is also applicable to warehouse storage without ambient control. Characterization of coffee deterioration and quality loss in permeable packaging includes undesirable changes in the physico-chemical composition of the beans as a result of an increase in water content and higher metabolic activity (Vilela *et al.*, 2000). Injury to the cellular membrane systems are accompanied by a subsequent loss of permeability control and by the leakage of solutes (Malta *et al.*, 2005). Additionally, the phenomenon of whitening is also observed during inadequate storage of green coffee beans. To our knowledge, deterioration with respect to antioxidant compounds has not been reported, but some general guidelines for quality preservation might be adopted in that sense as well. Since preservation of product quality over longer periods of storage secures a longer sales period for growers and guarantees better prices, a viable alternative had been proposed in the form of hermetically sealed systems that permit atmospheric modification or control (Nobre *et al.*, 2007). Plastic sacks impermeable to CO_2 have recently entered the market in response to the growing demand

TABLE 9.6 Contents of Caffeine and Trigonelline (mg/100 g dwb) of Semi-Dried and Wet Processed Green Coffee Beans of *C. Arabica* Cultivars

	Semi-Dry Method (mg/100 g dwb)			Wet Method (mg/100 g dwb)		
Samples	**5-CQA**	**Caffeine**	**Trigonelline**	**5-CQA**	**Caffeine**	**Trigonelline**
Coffea arabica						
cv. Yellow Burbon	3305.0±74.0	1.13±0.02	0.90±0.01	3410.2±17.5	1.05±0.02	0.89±0.01
cv. Red Catuai	3226.0±41.2	1.13±0.02	0.78±0.04	3256.8±41.6	1.26±0.02	0.92±0.01
cv. Rubi	3191.5±54.2	1.21±0.01	0.74±0.01	3682.6±38.8	1.19±0.02	0.80±0.01
cv. Topázio	3251.2±35.4	1.43±0.01	0.92±0.01	3162.8±19.0	1.16±0.01	0.85±0.02

Presented results are average values±SD, expressed as mg/100g on a dry weight basis (dwb), of 5-caffeoylquinic acid (5-CQA), caffeine, and trigonelline contents in *C. arabica* cultivars, processed by two different primary processing methods.
The results are excerpted from Duarte et al. *(2010).*

for new methods of storage for coffee beans. However, because of the elevated cost, this method has not been viable for many producers (Borém *et al.*, 2008). Hermetically sealed containers, such as hermetic big bags, have proved technically feasibile to maintain coffee quality over a prolonged storage period (Ribeiro *et al.*, 2011). They ensured stable water content for secure storage (10–11%) and color over a 12-month storage period, providing a visually desirable product that retained most of its native quality.

References

Abidi, S.L., Mounts, T.L., 1997. Reversed-phase high.performance liquid chromatographic separations of tocopherols. J. Chromatogr. A. 782, 25–32.

Anthony, F., Clifford, M.N., Noirot, M., 1993. Biochemical diversity in the genus *Coffea* L.: chlorogenic acids, caffeine and mozabioside contents. Genet. Resour. Crop Evol. 40, 61–70.

Avelino, J., Barboza, B., Araya, J.C., Fonesca, C., Davrieux, F., Guyot, B., Cilas, C., 2005. Effects of slope, exposure, altitude and yield on coffee quality in two altitude terroirs of Costa Rica, Orosi, and Santa María de Dota. J. Sci. Food Agric. 85, 1869–1876.

Barre, P., Akaffou, S., Louarn, J., Charrier, A., Hamon, S., Noirot, M., 1998. Inheritance of caffeine and heteroside contents in an interspecific cross between cultivated species *Coffea liberica* var. "dewevrei" and a wild species caffeine-free C. pseudozanguebariae. Theor. Appl. Genet. 96, 306–311.

Belay, A., Ture, K., Redi, M., Asfaw, A., 2008. Measurement of caffeine in coffee beans with UV/vis spectrometer. Food Chem. 108, 310–315.

Bertrand, B., Vaast, P., Alpizar, E., Etienne, H., Davrieux, F., Charmetant, P., 2006. Comparison of bean biochemical composition and beverage quality of Arabica hybrids involving Sudanese–Ethiopian origins with traditional varieties at various elevations in Central America. Tree Physiol. 26, 1239–1248.

Borém, F.M., Nobre, G.W., Fernandes, S.M., Pereira, R.G.F.A., Oliveira, P.D., 2008. Avaliação sensorial do café cereja descascado, armazenado sob atmosfera artificial e convencional. Ciênc. Agrotec. 32, 1724–1729.

Briandet, R., Kemsley, E.K., Wilson, R.H., 1996. Discrimination of Arabica and Robusta in instant coffee by Fourier transform infrared spectroscopy and chemometrics. J. Agric. Food Chem. 44, 170–174.

Bytof, G., Knopp, S.E., Schieberle, P., Teutsch, I., Selmar, D., 2005. Influence of processing on the generation of gamma-aminobutyric acid in green coffee beans. Eur. Food Res. Technol. 220, 245–250.

Camacho-Cristóbal, J.J., Anzelotti, D., González-Fontes, A., 2002. Changes in phenolic metabolism of tobacco plants during short-term boron deficiency. Plant Physiol. Biochem. 40, 997–1002.

Campa, C., Doulbeau, S., Dussert, S., Hamon, S., Noirot, M., 2005. Diversity in bean caffeine content among wild *Coffea* species: Evidence of a discontinuous distribution. Food Chem. 91 663–637.

Campanha, F.G., Dias, R.C.E., Benassi, M, T., 2010. Discrimination of coffee species using kahweol and cafestol: effects of roasting and of defects. Coffee Sci. 44, 87–96.

Casal, S., Oliveira, M.B., Ferreira, M.A., 2000. HPLC/diode array applied to the thermal degradation of trigonelline, nicotinic acid and caffeine in coffee. Food Chem. 68, 481–485.

Clifford, M.N., 1999. Chlorogenic acids and other cinnamates. Nature, occurrence and dietary burden. J. Sci. Food Agric. 79, 362–372.

Correia, A.M.N.G., Leitão, M.C.A., Clifford, M.N., 1995. Caffeoyiltyrosine and Angola II as characteristic markers for Angolan robusta coffees. Food Chem. 53, 309–313.

Da Silva, E., Mazzafera, P., Brunini, O., Sakai, E., Arruda, F.B., Mattoso, L.H.C., Carvalho, C.R.L., Pires, R.C.M., 2005. The influence of weter management and environmental conditions on the chemical composition and beveage quality of coffee beans. Braz. J. Plant Physiol. 17, 229–238.

De Souza, R.M.N., Benassi, M.T., 2012. Discrimination of commercial roasted coffee according to chemical composition. J. Braz. Chem. Soc. 23, 1347–1354.

Decazy, F., Avelino, J., Guyot, B., Perriot, J.I., Pineda, C., Cilas, C., 2003. Quality of different Honduran coffees in relation to several environments. J. Food Sci. 68, 2356–2361.

Dias, R.C.E., Campanha, F.G., Vieira, L.G.E., Pereira, L.P., Pot, D., Marraccini, P., Benassi, M.T., 2010. Evaluation of kahweol and cafestol in coffee tissues and roasted coffee by a new High-Performance Liquid Chromatography methodology. J. Agric. Food Chem. 58, 88–93.

DIN 10779, 1999. Bestimmung des Gehaltesan 16-O-Methylcafestol in Röstkaffee - HPLC-Verfahren, German Institute for Standardization.

Duarte, G.S., Pereira, A.A., Farah, A., 2010. Chlorogenic acids and other relevant compounds in Brazilian coffee processed by semi-dry and wet post-harvesting methods. Food Chem. 118, 851–855.

El-Abassy, R.M., Donfack, P., Matery, A., 2011. Discrimination between Arabica and Robusta green coffee using visible and micro Raman spectroscopy and chemometric analysis. Food Chem. 126, 1443–1448.

Esteban-Díez, I., Gonzáles-Sáiz, J.M., Pizarro, C., 2004. An evaluation of orthogonal signal correction methods for the characterization of Arabica and Robusta coffee varieties by NIRS. Anal. Chem. Acta. 514, 57–67.

Farah, A., de Paulis, T., Trugo, L.C., Martin, P., 2005. Effect of Roasting on the formation of chlorogenic acid lactones in coffee. J. Agric. Food Chem. 53, 1505–1513.

Farah, A., Donangelo, C.M., 2006. Phenolic compounds in coffee. Braz. J. Plant Physiol. 18, 23–36.

Gonzáles, A.G., Pablos, F., Martín, M.J., León-Camacho, M., Valdenebro, M.S., 2001. HPLC analysis of tocopherols and triglycerides in coffee and their use as authentication parameters. Food Chem. 73, 93–101.

Guyot, B., Gueule, D., Manez, J.C., Peirrot, J.J., Giron, J., Vilain, J., 1996. Influence de l'altitude et de de l'ombrage des cafés Arabica. Plantat Rech. 3, 272–280.

Hečimović, I., Belščak-Cvitanović, A., Horžić, D., Komes, D., 2011. Comparative study of polyphenols and caffeine in different coffee varieties affected by the degree of roasting. Food Chem. 129, 991–1000.

Joët, T., Laffargue, A., Descroix, F., Doulbeau, S., Bertrand, B., de Kochko, A., Dussert, S., 2010. Influence of environmental factors, wet processing and their interactions on the biochemical composition of green Arabica coffee beans. Food Chem. 118, 30–36.

Kitzberger, C.S.G., Scholz, M.B.S., Pereira, L.F.P., Vieira, L.G.E., Sera, T., Silva, J.B.G.D., Benassi, M.T., 2010. Analysis of diterpenes in green and roasted coffee of *Coffea arabica* cultivars growing in the same edapho-climatic conditions. In: Proceedings of 23rd International Conference on Coffee Science. ASIC, Bali, pp. 110–117.

Kurzrock, T., Speer, K., 2001. Diterpenes and diterpene esters in coffee. Food Rev. Int. 17, 433–450.

Ky, C.-L., Louarn, J., Dussert, S., Guyot, B., Hamon, S., Noirot, M., 2001. Caffeine, trigonelline, chlorogenic acids and sucrose diversity in wild *Coffea arabica* and *C. canephora* P. accessions. Food Chem. 75, 223–230.

Ky, C.-L., Louarn, J., Guyot, B., Charrier, A., Hamon, S., Noirot, M., 1999. Relations between inheritance of chlorogenic acid and contents in an interspecific cross between *Coffea pseudozanguebariae* and *Coffea liberica* var "dewevrei". Theor. Appl. Genet. 98, 628–637.

Leloup, V., Cancel, C., Liardon, R., Rytz, A., Pithon, A., 2004. Impact of wet and dry process on green composition and sensory characteristics. In: Proceedings of 20th International Conference on Coffee Science. ASIC, Bangalore, pp. 93–100.

Madhava Naidu, M., Sulochanamma, G., Sampathu, S.R., Srinivas, P., 2008. Studies on extraction and antioxidant potential of green coffee. Food Chem. 107, 377–384.

Malta, M.R., Nogueira, F.D., Gimarães, P.T.G., 2003. Composição química, produção e qualidade do café fertilizado com diferentes fontes e doses de nitrogênio. Ciênc. Agrotec. 27, 1246–1252.

Malta, M.R., Pereira, G.F.A., Chagas, S.J., de, R., 2005. Potassium leaching and electric conductivity of grain coffee (*Coffea arabica* L.) exudate: some factors that may affect these evaluations. Cienc. Agrotecnol. 29, 1015–1020.

Mazzafera, P., 1991. Trigonelline in coffee. Phytochemistry 30, 2309–2310.

Monteiro, M.C., Farah, A., 2012. Chlorogenic acids in Brazilian Coffea arabica cultivars from various consecutive crops. Food Chem. 134, 611–614.

Nobre, G.W., Borém, F.M., Fernandes, S.M., Pereira, R.G.F.A., 2007. Chemical alterations of parchment coffee beans during storage. Coffee Sci. 2, 1–9.

Oberthür, T., Läderach, P., Posada, H., Fisher, M.J., Sampler, L.F., Illera, J., Collet, L., Moreno, E., Alarcón, R., Villegas, A., Usma, H., Perez, C., Jarvis, A., 2011. Regional relationships between inherent coffee quality and growing environment for denomination of origin labels in Nariño and Cauca, Colombia. Food Policy 36, 783–794.

Perrone, D., Farah, A., Donangelo, C.M., de Paulis, T., Martin, P., 2008. Comprehensive analysis of major and minor chlorogenic acids and lactones in economically relevant Brazilian coffee cultivars. Food Chem. 106, 859–867.

Ribeiro, F.C., Borém, F.M., Giomo, G.S., De Lima, R.R., Malta, M.R., Figueirdo, L.P., 2011. Storage of green coffee in hermetic packaging injected with CO_2. J. Stored Prod. Res. 47, 341–348.

Selmar, D., Bytof, G., Knopp, S., 2002. New aspects of coffee processing: The relation between seed germination and coffee quality. In: Proceedings of 19th International Conference on Coffee Science, ASIC, Trieste, pp. 19.

Smith, A.W., 1985. Agricultural practices. In: Clarke, R.J., Macrae, R. (Eds.), Coffee chemistry. Elsevier, Applied Science, Amsterdam, pp. 18–23.

Söndahl, M.R., van der Vossen, H.A.M., 2005. Origin, production and botany. In: Illy, A., Viani, R. (Eds.), Espresso coffee: The Chemistry of Quality. Elsevier, Academic Press, San Diego, USA, p. 22.

Speer, K., Kölling-Speer, I., 2006. The lipid fraction of the coffee bean. Braz. J. Plant Physiol. 18, 201–216.

Sridevi, B.V., Giridhar, P., Ravishankar, G.A., 2011. Evaluation of roasting and brewing effect on antinutritional diterpenes-cafestol and kahweol in coffee. Glob. J. Med. Res. 11, 17–22.

Teixeira, A.A., Brando, C.H.J., Thomaziello, R.A., Teixiera, R., 1995. In: Illy, A., Viani, R. (Eds.), Espresso coffee: The science of quality. Elsevier, Academic Press, Italy, pp. 91–95.

Toci, A.T., Farah, A., 2008. Volatile compounds as potential defective coffee seeds' markers. Food Chem. 52, 385–387.

Vilela, E.R., Chandra, P.K., Oliveira, G.A., 2000. Effect of temperature and relative humidity on discoloration of coffee beans. Rev. Bras. Armazenamento 25, 31–37.

CHAPTER

10

Effects of Preparation Techniques on the Antioxidant Capacity of Coffee Brews

Draženka Komes, Ana Belščak-Cvitanović

Department of Food Engineering, Faculty of Food Technology and Biotechnology, University of Zagreb, Pierottijeva, Zagreb, Croatia

CHAPTER POINTS

- Contradictory results regarding the antioxidant capacity and polyphenolic compound content exist in differently prepared coffee brews.
- The majority of studies indicate that caffeine content is higher in boiled and instant coffee brews than in espresso and filter brews.
- Decaffeinated coffee brews exhibit lower antioxidant capacity than regular brews.
- Contents of cafestol and tocopherols are the highest in boiled and Turkish coffee brews, and the lowest in filtered and espresso brews.
- Addition of milk to coffee brews, regardless of the milk content, decreases their antioxidant capacity when compared to plain brews.
- Antioxidant potential and caffeine content of instant cappuccino brews is in the range of decaffeinated and filtered coffee.

INTRODUCTION

Coffee brews are increasingly popular worldwide due to their unique flavor and psychostimulant purposes. Some sensory properties of coffee beans are attributed to volatile substances developed during roasting and brewing, which in turn result in a variety of choices for preparing beverages (Lewis, 2004). How coffee is consumed is a matter of individual preference, offer and availability, and socio-cultural habits (Nestle, 2007). In industrialized countries, specialized coffee establishments are as visible as restaurants and fast-food chains. In these affluent societies, in addition to a great variety of coffee products, there is a growing industry in coffee-making machinery suitable for homes and offices. Coffee brews are prepared in a wide variety of formats including Scandinavian–type boiled coffee, drip filtered coffee, instant or soluble coffee, espresso, Turkish and Mocha. The preparation technique has a significant influence on the taste, aroma and composition of coffee brews. The process of coffee brew preparation, although fundamentally simple for the consumer (mixing hot water and ground roasted coffee), is complicated to predict because all process variations may affect the product's characteristics (Oosterveld *et al.*, 2003).

The coffee-brewing technique is generally not taken into consideration or taken for granted by consumers, despite the large variations of beverage chemical composition, physico-chemical and sensory properties, which are dependent on the selected preparation methodologies. Brewing process accompanied by other factors, such as origin or variety of coffee beans, blending, roasting degree, and grinding is essential for antioxidant composition and health properties of a coffee brew. Even though the brewing time is given by the coffee-brewing technique, the knowledge of extraction behavior of the main coffee antioxidants during this time might elucidate the technological factors with major impact on antioxidant extraction and enable the production of coffee brews with higher antioxidant capacity (Ludwig *et al.*, 2012).

The contact of water with roasted coffee solids is the crucial step for producing a coffee beverage. Ground roasted coffee can be set in contact with water resorting to three different extraction techniques: decoction, infusion, or pressure methods (Petracco, 2001). Boiled coffee (very popular in Nordic countries) and Turkish/Greek coffee (served in all Mediterranean countries) are prepared

Processing and Impact on Antioxidants in Beverages
http://dx.doi.org/10.1016/B978-0-12-404738-9.00010-6

by decoction methods. Among different brewing techniques, filter coffee (drip filter) is the most widely used coffee brew obtained by the infusion method, whereas espresso coffee is the most appreciated coffee brew produced by the pressure method. According to Navarini *et al.* (2004), espresso brews also exhibited higher sensory preference when compared to other coffee brews. In drip filtration methods, water at 92–96°C flows through a hardly compressed ground coffee bed and the extract drips from the brewing chamber into the pot. Turbulence in the brewing chamber prevents water from becoming saturated (Lingle, 1996). In pressure methods, water at approximately 9 bars and 88–92°C is forced to go through coffee grounds compacted in a small brewing chamber (coffee cake). Also, rapid brewing time and fine particle size are necessary (Lingle, 1996). Instant coffee is produced by water extraction under high temperatures (180°C) followed by concentration in evaporators, and lastly dehydration (using freeze-drying or spray drying) (Noguiera and Trugo, 2003). Around 17% of the green coffee produced worldwide is used to make soluble (instant) coffee.

Within each class of brew preparation method, individual population groups consume coffee in a range of formats, e.g., 37–88°C (Lee *et al.*, 2003), 0–80% milk (Lee and O'Mahony, 2002), 0–16 g of sugar, 25–880 ml in volume (Hsu and Hung, 2005), with foamed milk, cream, ice, flavorings, brew adjuncts, or co-adjuncts (Fisk *et al.*, 2011). Caffeine is the most well-known psychoactive substance in coffee, appreciated for its stimulating effect; however, the strong pharmacological effects of caffeine have led to consumers' demand for caffeine-free coffee beverages. Different processes have been developed to eliminate caffeine from the coffee bean, with the method using only water being preferred, but solvents or supercritical carbon dioxide may also be used. Almost all of the methods of decaffeinating coffee consist of soaking the beans in water to dissolve the caffeine, extracting the caffeine with either a solvent or activated carbon, and then re-soaking the beans in the decaffeinated water to reabsorb the flavor compounds that were removed in the initial extraction. However, all brands of decaffeinated coffee still contain very low residual caffeine (Nestle, 2007).

As compared to other beverages, coffee stands out for its significant antioxidant potential. According to some authors, antioxidant capacity of soluble and espresso coffee are higher than those of red wine and green tea (Pellegrini *et al.*, 2003). The antioxidant capacity of coffee brew is attributed to both natural antioxidants, like phenolic acids (Farah and Donangelo, 2006), caffeine (Casal *et al.*, 2000), diterpens cafestol and kahweol (Zhang *et al.*, 2012), and tocopherols (Speer and Kölling-Speer, 2006), as well as roasting-induced antioxidants, like melanoidins and other Maillard reaction products (MRPs) (Nunes and Coimbra, 2007). The content of all bioactive compounds in coffee brews is affected primarily by the brewing procedure (Sánchez-González *et al.*, 2005).

EFFECTS OF PREPARATION TECHNIQUES ON THE ANTIOXIDANT CAPACITY OF COFFEE BREWS

The antioxidant capacity of coffee beverages has been assessed by several studies, using different methods (Pellegrini *et al.*, 2003; Daglia *et al.*, 2000; del Castillo *et al.*, 2002; Nicoli *et al.*, 1997a, b). Coffee has been reported to be a better contributor to human antioxidant defense than fruits, vegetables, and wine (Svilaas *et al.*, 2004) and the major source of dietary phenolic antioxidants in several countries, such as Denmark, USA, Brazil and Japan (Farah, 2009). It has been established that coffee has relatively high antioxidant activity, compared to other beverages: three-fold higher than that of red wine and five-fold higher than that of green tea (Pellegrini *et al.*, 2003). However, as indicated previously, the content of bioactive compounds and antioxidant capacity of different coffee brews is mostly affected by the coffee brew preparation method, which points to a need for a detailed characterization and comparison of the antioxidant potential of various coffee brews.

Table 10.1 provides the comparison of antioxidant capacities (assessed by various *in vitro* assays) of differently prepared coffee brews obtained in several studies. According to the results of Ludwig *et al.* (2012), espresso coffee brews prepared from both Arabica and Robusta coffees exhibited higher antioxidant capacity in comparison to filter coffees, but showed a decrease in antioxidant capacity with brewing time. These results demonstrate that the compounds responsible for the antioxidant capacity of an espresso coffee brew are mainly extracted at the beginning of the brewing process and afterwards are diluted. Similar results were found by Alves *et al.* (2010), who determined antioxidant capacity of espresso coffees with different brew lengths/volumes [varying from 20 ml or less ('short') to 50 ml or more ('long')] using DPPH (2,2-diphenyl-1-picrylhydrazyl) assay.

Regarding the effect of brewing time, Minamisawa *et al.* (2004) also determined that filter coffee showed different antioxidant extraction behaviors depending on the brewing duration (being different in two coffee samples; Arabica coffee originating from Guatemala and Robusta coffee originating from Vietnam), in comparison to espresso coffee. According to their results, Arabica filter coffee exhibited the highest antioxidant capacity in the initial (0–75 s) and final fractions (300–375 s), while the middle fractions (150–225 s) generated lower antioxidant capacity. Additionally, Robusta filter coffee exhibited the

TABLE 10.1 Antioxidant Capacity of Coffee Brews Prepared by Different Brewing Techniques and Assessed by Three Different *In Vitro* Assays

	ABTS	DPPH	FRAP	Reference
FILTER				
	< 50 μmol Trolox/ml	< 60 μmol Trolox/ml	—	Ludwig *et al.* (2012)
	7.75–12.29 μmol Trolox/ml	—	—	Parras *et al.* (2007)
	968–1695 μg Trolox/serving	—	2527–2801 μg Trolox/serving	Sánchez-González *et al.* (2005)
	14.37 μmol Trolox/ml	—	36.56 μmol Fe(II)/ml	Niseteo *et al.* (2012)
American	5.867 ± 1.54 μmol Trolox/ml	—		Santini *et al.* (2011)
	—	—	39.6–120.3 μmol Fe(II) /ml*	Moreira *et al.* (2005)
ESPRESSO				
	< 130 μmol Trolox/ml	< 20 μmol Trolox/ml	—	Ludwig *et al.* (2012)
	6.97–11.42 μmol Trolox/ml	—	—	Parras *et al.* (2007)
	15.23–39.89 μmol Trolox/ml	—	27.81–122.35 μmol Fe(II)/ml	Niseteo *et al.* (2012)
	636–829 μg Trolox/serving	1401–1716 μg Trolox/serving	—	Sánchez-González *et al.* (2005)
	39.149 ± 4.88 μmol Trolox/ml	—	—	Santini *et al.* (2011)
INSTANT				
	453–454 μg Trolox/serving	—	1012–1015 μg Trolox/serving	Sánchez-González *et al.* (2005)
	41.04–49.34 μmol Trolox/ml	—	235.90–244.82 μmol Fe(II)/ml	Niseteo *et al.* (2012)
	—	—	105.8–162.2 μmol Fe(II)/ml	Moreira *et al.* (2005)
TURKISH				
	54.068 ± 7.62 μmol Trolox/ml	—	—	Santini *et al.* (2011)
	37.73–37.98 μmol Trolox/ml	—	82.76–87.40 μmol Fe(II)/ml	Niseteo *et al.* (2012)
ITALIAN OR MOCCHA				
	8.77–12.82 μmol Trolox/ml	—	—	Parras *et al.* (2007)
	10.303 ± 1.76 μmol Trolox/ml	—	—	Santini *et al.* (2011)
	479–673 μg Trolox/serving	—	561–1319 μg Trolox/serving	Sánchez-González *et al.* (2005)
NEAPOLITAN				
	56.968 ± 8.13 μmol Trolox/ml	—	—	Santini *et al.* (2011)

**Coffee brew prepared by pouring with boiling water, stirring and filtering through filter paper.*
Comparison of antioxidant capacity of coffee brews prepared by different brewing techniques and assessed by ABTS [scavenging assay of 2,2-azino-bis(3-ethylbenzthiazoline-6-sulphonic acid) diammonium salt radical], DPPH (scavenging assay of 2,2-diphenyl-1-picrylhydrazyl radical) and FRAP (ferric reducing antioxidant power) assays, performed by different authors. The displayed results are provided as ranges of molar concentrations (equivalents) for a specific assay. The results are adapted from the original source for consistency.

lowest value already in the first, initial fraction, which was attributed to higher water absorption in Robusta coffee that leads to a longer wetting stage. The increase in antioxidant capacity in the final fractions of filter coffee brews could be due to the water pressure decrease that induces a slower flow and a longer contact time between water and ground coffee. However, due to the lowest volumes of the final fractions, their contribution to the antioxidant capacity of the overall coffee brew was limited (~9% and ~14%, respectively).

Parras *et al.* (2007) evaluated the antioxidant capacity of coffees (Arabica and Robusta) from 12 different points of origin and two decaffeinated coffees from Colombia and Brazil prepared by three commonly used procedures (espresso, filter, and Italian). According to their results, all analyzed coffees were very effective as scavengers of lipoperoxyl and $OH^{\bullet}$ radicals. Regarding the effects of preparation technique, the results showed no significant differences between the three brewing techniques (espresso, filter, and Italian). The filtered coffee showed a greater capacity to react with H_2O_2 than the Italian and espresso coffees. All the coffee samples improved the oxidative stability of butter (Rancimat test), espresso and Italian coffee providing greater protection than the filtered brews. However, filter brews produced stronger antioxidant capacity during storage than espresso and Italian brews. Similar to these findings, Sánchez-González *et al.* (2005) observed a high correlation between the polyphenolic content and ferric reducing power (FRAP) and radical scavenging capacity (ABTS) of different types of coffee (filter > espresso > freeze dried (instant) > Italian or Mocha coffee). These results are, however, in contrast with the ones obtained by Niseteo *et al.* (2012), who determined the highest antioxidant capacity in instant coffees, followed by Turkish/boiled and espresso coffees, while filter coffee exhibited the lowest content of polyphenols and antioxidant capacity.

Additionally, Carlsen *et al.* (2010) determined that among 54 different types of prepared coffee brews procured from 16 different manufacturers, the variations in antioxidant capacity determined by FRAP assay were large, ranging from a minimum of 0.89 mmol/100 g for one type of brewed coffee with milk to 16.33 mmol/100 g for a double-espresso coffee, the highest antioxidant value of all beverages analyzed in the study (tea, coffee, cocoa). When summarizing the obtained results, it becomes obvious that no general conclusion can be deduced regarding the coffee brew with the best antioxidant properties. As displayed, numerous factors such as coffee species, origin, particle size, brewing equipment and parameters, even the quality of water, affect the antioxidant capacity of the final brew.

EFFECTS OF PREPARATION TECHNIQUES ON THE CONTENT OF POLYPHENOLIC COMPOUNDS

Polyphenols are the most abundant antioxidant compounds of coffee brews. The polyphenolic fraction of coffee brews is mainly composed of chlorogenic acids (CGAs), with caffeoylquinic acids (CQAs) being the most abundant chlorogenic acid class. These compounds are powerful antioxidants (Daglia *et al.*, 2004) to which several positive physiological actions have been attributed both *in vitro* and *in vivo*, such as chemopreventive anticancer properties (Belkaid *et al.*, 2006).

Regarding the effect of coffee brewing techniques on the content of polyphenolic compounds, Niseteo *et al.* (2012) determined the following decreasing order of total polyphenolic compounds (TPC) (as well as of antioxidant capacity) among differently prepared coffee beverages: instant coffees > Turkish/boiled > espresso coffees > filter coffee (Table 10.2). This is not surprising, since according to the production process of instant coffee, which involves extraction and concentration of water-soluble or dispersible compounds (including polyphenols), instant coffee contains higher content of these substances when compared to other types of coffee. However these results are not in accordance with the results of Sánchez-González *et al.* (2005) who observed the highest TPC of filter coffee brew, followed by espresso coffee brew. Lower total phenol and total flavonoid contents of filter coffee could be due to the filter used during the brew preparation. Mišík *et al.* (2010) studied the effect of paper-filtered coffee on oxidative DNA damage and found that filtration of coffee brews led to removal of some bioactive compounds. Additionally, decaffeinated Turkish/Greek and instant coffee brews showed lower TPC than their regular coffee counterparts (Niseteo *et al.*, 2012), confirming previously published results by Fujioka and Shibamoto (2008).

Although Ludwig *et al.* (2012) determined a higher content of both total polyphenols and total CGAs in espresso coffee brews, Niseteo *et al.* (2012) observed a higher content of total CGAs in filter coffee than in espresso (despite the higher TPC of espresso brew). A similar discrepancy can also be observed in the results of Moreira *et al.* (2005), who found higher total CGAs in filter coffee brews, rather than in instant ones, which is in contrast to the results of Niseteo *et al.* (2012) (Table 10.2).

The esterification of an additional caffeic acid moiety in dicaffeoylquinic acids (diCQAs) increases the number of hydroxyl groups, which might favor the retention of these compounds by interaction with melanoidins or other polymeric compounds. The diCQAs also exhibit weaker polarity when compared to the CQAs, which might explain the slower release of these compounds during extraction with water (Kroll *et al.*, 2003). Blumberg *et al.* (2010) studied the influence of hot water percolation on the concentration of monocaffeoylquinic acids and chlorogenic acid lactones and reported that dicaffeoylquinic acid lactones were extracted rather more slowly in comparison to the monocaffeoylquinic ones. The delay in caffeoylquinic acid extraction might be attributed to the longer wetting stage observed in Robusta (Vietnam) coffee, which might facilitate the hydrolysis of caffeoylquinic acids bound to melanoidins inducing their release during advanced stages of filter coffee brewing (Lingle, 1996). Ludwig *et al.* (2012) determined higher extraction of CQAs and diCQAs per gram of coffee in filter coffee brews than in espresso ones, which is in agreement with the findings of Pérez-Martínez *et al.* (2010). This may be due to the technological differences between espresso and filter coffeemaker. Although the high water pressure applied in an espresso

TABLE 10.2 Content of Total Polyphenols and Chlorogenic Acids in Coffee Brews Prepared by Different Brewing Techniques

	Total Polyphenols (Folin–Ciocalteu Assay)	Σ Chlorogenic Acids	Reference
FILTER			
	2.97 mg GAE/ml	1.24 mg/ml	Niseteo *et al.* (2012)
	<5.0 mg GAE/ml	0.41–1.57 mg/ml	Ludwig *et al.* (2012)
	0.20–0.31 g GAE/serving	—	Sánchez-González *et al.* (2005)
	—	1.36–25.29 mg/ml*	Moreira *et al.* (2005)
	—	0.15–0.47 mg/ml	Fujioka and Shibamoto (2008)
ESPRESSO			
	4.0 < TPC < 14.0 mg GAE/ml	0.50–0.93 mg/ml	Niseteo *et al.* (2012)
	<15.0 mg GAE/ml	0.43–4.31 mg/ml	Ludwig *et al.* (2012)
	0.15–0.22 g GAE/serving	—	Sánchez-González *et al.* (2005)
INSTANT			
	15.0 < TPC < 18.0 mg GAE/ml	2.30–2.52 mg/ml	Niseteo *et al.* (2012)
	0.12 g GAE/serving	—	Sánchez-González *et al.* (2005)
	—	0.79–3.26 mg/ml	Moreira *et al.* (2005)
TURKISH			
	7.0 < TPC < 9.0 mg GAE/ml	1.54–2.01 mg/ml	Niseteo *et al.* (2012)
Boiled 10 min	0.25–0.44 mg GAE/ml	—	Sachetti *et al.* (2009)
ITALIAN OR MOCCHA			
	0.11–0.17 g GAE/serving	—	Sánchez-González *et al.* (2005)

**Coffee brew prepared by pouring with boiling water, stirring and filtering through filter paper.*

Comparison of total phenol content and the sum of chlorogenic acid derivatives of coffee brews prepared by different brewing techniques. The displayed results are deduced from several studies and are provided as ranges of concentrations (mg/ml or serving) for each category. The results are adapted from the original source for consistency. GAE, gallic acid equivalents; TPC, total polyphenolic compounds.

coffeemaker favors the extraction process, the short contact time between water and coffee grounds, the high coffee/water ratio and the limited space in coffee cake does not allow equilibrium to be reached (Petracco, 2005). In contrast, longer time and turbulences in the extraction chamber of the filter coffeemaker allow the water to be in immediate contact with the coffee to extract additional compounds when it has not become so saturated with dissolved material. Thus, both technological factors might favor the extraction of CQAs and diCQAs, free and bound with melanoidins. In fact, turbulences are considered, after time and temperature, as the third most important factor in filter coffee brewing (Lingle, 1996). Less turbulences during sequential coffee percolation could also be the reason why Blumberg *et al.* (2010) found that monocaffeoylquinic acid and monocaffeoyl and dicaffeoyl quinide extraction behaviors were more similar to those of espresso coffee fractions than filter ones, i.e., higher extraction in the first fractions and slower release of dicaffeoyl quinides.

EFFECTS OF PREPARATION TECHNIQUES ON THE CONTENT OF MELANOIDINS

Apart from chlorogenic acids (CGAs), other antioxidants attributed to high antioxidant capacity of coffee brews are Maillard reaction products (MRPs) formed during coffee roasting. Sacchetti *et al.* (2009) evaluated the radical scavenging activity of coffee brews obtained from different types of coffee as a function of roasting degree and the relative contribution of the phenolic fraction and non-phenolic fraction to the overall radical scavenging activity. The authors observed that the relative contribution of non-phenolic fraction to the overall radical scavenging activity of the brew was much lower than that of the phenolic fraction (Sacchetti *et al.*, 2009), indicating that the MRPs formed during roasting do not contribute significantly to the antioxidant capacity of coffee brews. However, Sánchez-González *et al.* (2005) found that antioxidant activity increased significantly (by 34%) after 4h of heating the coffee pot, since in the USA and Northern Europe, the pot containing coffee is usually kept hot (85°C) for several hours. The cause of this increase was explained as the consequence of formation of Maillard reaction products, due to the heating process. These findings are in compliance with the ones by Niseteo *et al.* (2012) who also determined that reheating of prepared coffee beverage to the boiling state significantly increased the content of CGAs and caffeine when compared to the initially prepared brews.

EFFECTS OF PREPARATION TECHNIQUES ON THE CONTENT OF CAFFEINE

Owing to its broad application in the food and pharmaceutical industries, caffeine is the most well known and pharmacologically studied component of coffee (Harland, 2000). The caffeine content of different coffee preparations was evaluated by numerous authors over the years. Several differences exist in the preparation techniques of coffee, which may influence caffeine consumption, especially the coffee grounds to water ratio and the volume of coffee prepared. Smaller volumes of coffee yield less caffeine than larger volumes, while the extraction of caffeine is dependent on the time of brewing. Also, the difference in caffeine content between finely and coarsely ground coffee is expected because the former has larger surface areas, allowing greater caffeine extraction. Larger extent of grinding results in more total solids being extracted from the coffee grounds. According to the study by Bell *et al.* (1996), the use of more coffee grounds, larger extents of grinding, and larger volumes of coffee prepared yielded higher caffeine contents in filtered coffee. Caffeine contents per 177ml (6oz) of coffee ranged from 50 to 143mg, depending upon the mode of preparation. Domestic grinding, which yielded less distinctive grinding patterns, had caffeine contents similar to that of store-ground coffee. According to their results, boiled coffee produced similar or higher caffeine contents than filtered coffee, depending upon the length of boiling time (Table 10.3). Phan *et al.* (2012) also determined that higher temperature (100°C) and longer brewing time (5min) are favorable for obtaining higher caffeine content from ground and roasted coffee.

Regarding the effect of brewing technique on the caffeine content of different coffee brews, Ludwig *et al.* (2012) evidenced a higher content of caffeine in both Arabica and Robusta espresso coffees (ranging from 0.39 to 5.75 mg/ml), when compared to filter coffees (ranging from 0.36 to 1.58mg/ml). According to the results of Santini *et al.* (2011) caffeine content was the highest in 'Moka' coffee (5.4mg/ml), followed by espresso (3.33mg/ml), Turkish (1.94mg/ml), and Neapolitan coffee (1.89mg/ml), while the lowest value was observed for the American coffee that had a caffeine content of 0.44mg/ml. Lau *et al.* (1992) studied the presence of this bioactive compound in different coffee samples divided in instant and regular coffees. In the instant coffee samples the authors determined a content of caffeine varying from 2.16 to 4.30mg/ml. Instant coffee brew contained the highest content of caffeine according to Niseteo *et al.* (2012), followed by Turkish and filtered coffee brews and espresso as the lowest-caffeine-containing coffee brew. The same authors also evidenced the presence of caffeine in all decaffeinated coffee brews, which was previously confirmed by Fujioka and Shibamoto (2008) (Table 10.3).

As indicated above, it has been well established that the quantity of caffeine extracted from coffee powder during the coffee-making procedure depends on the method used, while the antioxidant capacity is strongly dependent on the roasting of the coffee beans and is retained in the brewed coffee, with respect to the procedure of extraction used in the coffee making (Fujioka and Shibamoto, 2008). Additionally, several studies suggested that caffeine might be a good contributor to the antioxidant capacity or reducing power of differently prepared coffee brews (Vignoli *et al.*, 2011; López-Galilea *et al.*, 2007; Santini *et al.*, 2011). The presence of caffeine might partially explain the higher antioxidant capacity of Vietnam (*C. canephora*) coffee brews, that could not be attributed to the main chlorogenic acids found in lower content in these coffee brews. López-Galilea *et al.* (2007) also observed higher antioxidant capacity but lower 5-CQA amounts in brews prepared with Robusta coffee or torrefacto blends.

The removal of caffeine from coffee brews negatively affected their antioxidant capacity. Decaffeination of espresso coffee led to a decrease in total antioxidant capacity values of ~25–30%, likely due to the antioxidant capacity of caffeine (Shi *et al.*, 1991). Table 10.4 displays the results of antioxidant capacity of espresso and decaffeinated espresso determined using FRAP and ABTS assays. Parras *et al.* (2007) evaluated the antioxidant capacity of coffee from different origins in brews prepared by different procedures (espresso, Italian, and brewed) and found higher Trolox equivalent antioxidant capacity (TEAC) (determined by the ABTS assay) values for regular products than for decaffeinated products. The decrease of antioxidant capacity in decaffeinated coffee brews (when compared to their regular counterparts) was also observed by Moreira *et al.* (2005) and Niseteo *et al.* (2012), although Sánchez-González *et al.* (2005) observed somewhat higher antioxidant capacity of decaffeinated filter and espresso brews (Table 10.4). However, it can generally be concluded that the decaffeination process of coffee impairs the antioxidant properties of the prepared coffee brews, regardless of the brewing technique, to an extent that reaches up to 26.2% and 28.1% lower antioxidant capacity of decaffeinated

TABLE 10.3 Caffeine Content (mg/ml) of Regular and Decaffeinated Coffee Brews Prepared by Different Brewing Techniques

	Caffeine	Decaffeinated Brews	Reference
FILTER			
	1.0 < caffeine > 1.5 mg/ml	—	Niseteo *et al.* (2012)
	0.36–1.58 mg/ml	—	Ludwig *et al.* (2012)
	0.28 < caffeine > 0.45 mg/ml	—	Bell *et al.* (1996)
American	0.437 ± 0.116 mg/ml	—	Santini *et al.* (2011)
	0.31–0.42 mg/ml	0.011–0.013 mg/ml	Fujioka and Shibamoto (2008)
ESPRESSO			
	0.5 < caffeine > 2.0 mg/ml	0.035–0.079 mg/ml	Niseteo *et al.* (2012)
	0.40–5.75 mg/ml	—	Ludwig *et al.* (2012)
	3.329 ± 0.476 mg/ml	—	Santini *et al.* (2011)
INSTANT			
	4.0 < caffeine > 5.0 mg/ml	0.236 mg/ml	Niseteo *et al.* (2012)
TURKISH			
	2.0 < caffeine > 3.0 mg/ml	0.315 mg/ml	Niseteo *et al.* (2012)
Boiled	0.28 < caffeine > 0.34 mg/ml	—	Bell *et al.* (1996)
	1.943 ± 0.354 mg/ml	—	Santini *et al.* (2011)
ITALIAN OR MOCCHA			
	5.399 ± 0.811 mg/ml	—	Santini *et al.* (2011)
NEAPOLITAN			
	1.888 ± 0.254 mg/ml	—	Santini *et al.* (2011)
BREWING PROCEDURE NOT INDICATED			
	0.174–0.762 mg/ml	0.008 mg/ml	Rodrigues *et al.* (2007)

Comparison of caffeine content of plain and decaffeinated coffee brews prepared by different brewing techniques. The displayed results are deduced from several studies and are provided as ranges of concentrations (mg/ml) for each category. The results are adapted from the original source for consistency.

espresso brews (determined with the ABTS and FRAP assays, respectively) when compared to plain espressos (Pellegrini *et al.*, 2003).

EFFECTS OF PREPARATION TECHNIQUES ON THE CONTENT OF TOCOPHEROLS

The main tocopherols identified in coffee brews are α- and β- isomers (Alves *et al.*, 2010; Kölling-Speer *et al.*, 2006). According to Alves *et al.* (2010) pure Arabica espressos contain 2.2 mg and 3.9 mg of α- and β-tocopherol, respectively, while Robusta espressos showed significantly lower amounts: 1.3 mg and 1.1 mg, correspondingly. When comparing the analyzed espresso coffee brews, no significant differences were found between the tocopherols content of caffeinated and decaffeinated blends. The mean total content of decaffeinated espresso coffee brews was 3.01 μg/30 ml of espresso coffee brews (range: 2.13–3.88 mg/30 ml of espresso coffee brews). Alves *et al.* (2010) also examined the influence of the extraction volume on the tocopherol content, since the espresso coffee length depends essentially on the consumer's preference. For Arabica coffees, no significant differences were found between the volumes analysed, while for the Robusta sample, significant differences were observed especially between 20-ml and 30-ml espresso coffee brews. Espressos prepared from servings contained ~5-fold higher contents of tocopherols than those prepared by the classic method. Compared with other coffee brews, an Arabica espresso cup contained an intermediate content of total tocopherols: 7.0 mg/30 ml, against 4.4 mg for filter (150 ml), 13.6 mg for press-pot (60 ml), 19.7 mg for Moka (60 ml), 25.4 mg for Turkish, and 40.1 mg for boiled (150 ml) coffees.

TABLE 10.4 Antioxidant Capacity of Regular and Decaffeinated Coffee Brews Assessed by ABTS and FRAP Assays

	Regular Brews		Decaffeinated Brews		
	ABTS	**FRAP**	**ABTS**	**FRAP**	**Reference**
FILTER					
	968–1695 μg Trolox/serving	2554–2686 μg Trolox/serving	1261–1337 μg Trolox/serving	2527–2801 μg Trolox/serving	Sánchez-González *et al.* (2005)
	—	39.6 –120.3 μmol Fe(II)/ml*	—	44.8–83.6 μmol Fe(II)/ml*	Moreira *et al.* (2005)
	8.82–12.29 μmol Trolox/ml	—	7.75–8.06 μmol Trolox/ml	—	Parras *et al.* (2007)
ESPRESSO					
	36.54 μmol Trolox/ml	129.38 μmol Fe(II)/ml	26.96 μmol Trolox/ml	93.01 μmol Fe(II)/ml	Pellegrini *et al.* (2003)
	636–756 μg Trolox/serving	1401–1554 μg Trolox/serving	826–829 μg Trolox/serving	1593–1776 μg Trolox/serving	Sánchez-González *et al.* (2005)
	15.23–39.89 μmol Trolox/ml	27.81–122.35 μmol Fe(II)/ml	24.64–24.85 μmol Trolox/ml	35.91–48.65 μmol Fe(II)/ml	Niseteo *et al.* (2012)
	6.97–10.51 μmol Trolox/ml	—	7.80–8.54 μmol Trolox/ml	—	Parras *et al.* (2007)
INSTANT					
	453 μg Trolox/serving	1012 μg Trolox/serving	454 μg Trolox/serving	1015 μg Trolox/serving	Sánchez-González *et al.* (2005)
	41.04–49.34 μmol Trolox/ml	235.90–44.82 μmol Fe(II)/ml	47.12 μmol Trolox/ml	235.46 μmol Fe(II)/ml	Niseteo *et al.* (2012)
	—	114.2–162.2 μmol Fe(II)/ml	—	105.8–114.0 μmol Fe(II)/ml	Moreira *et al.* (2005)
TURKISH					
	37.73–37.98 μmol Trolox/ml	82.76–87.40 μmol Fe(II)/ml	36.11 μmol Trolox/ml	77.72 μmol Fe(II)/ml	Niseteo *et al.* (2012)
ITALIAN OR MOCCHA					
	613–673 μg Trolox/serving	807–1310 μg Trolox/serving	546–599 μg Trolox/serving	561–1319 μg Trolox/serving	Sánchez-González *et al.* (2005)
	8.77–12.82 μmol Trolox /ml	—	9.68–10.85 μmol Trolox/ml	—	Parras *et al.* (2007)

**Coffee brew prepared by pouring with boiling water, stirring and filtering through filter paper.*

Comparison of antioxidant capacity of differently prepared plain and decaffeinated coffee brews, determined by using the ABTS [scavenging assay of 2,2-azino-bis(3-ethylbenzthiazoline-6-sulphonic acid) diammonium salt radical] and FRAP (ferric reducing antioxidant power) assays. The displayed results are deduced from several studies and represent ranges of molar concentrations (equivalents) for each assay. The results are adapted from the original source due to consistency.

EFFECTS OF PREPARATION TECHNIQUES ON THE CONTENT OF CAFESTOL

Values for cafestol content by brew mechanism from previous studies are often variable due to differing extraction parameters, grind sizes, coffee to water ratios, temperatures, and brewing techniques, e.g., coffee pads (Buchmann *et al.*, 2010; Kurzrock and Speer, 2001; Sehat *et al.*, 1993; Boekschoten *et al.*, 2006).

The cafestol content of a standard cup of coffee varied depending on the brewing technique, but is the highest in unfiltered preparation methods such as Scandinavian-type boiled coffee and Turkish coffee with up to 88.7 mg/l in some Turkish brews. Filtered coffees such as drip-filter and soluble coffee contained negligible levels of cafestol in the brew, as the paper filter in drip-filtered coffee retains the diterpenes and in soluble coffee the diterpenes are retained within the grounds during production (Gross *et al.*, 1997). Mišik *et al.* (2010) determined that the initially determined contents of kahweol and cafestol (90 mg/l and 182 mg/l, respectively) decreased after filtration of coffee brews to 0.2–0.3 mg/l. Zhang *et al.* (2012) evaluated the extraction yield of cafestol from roasted and ground

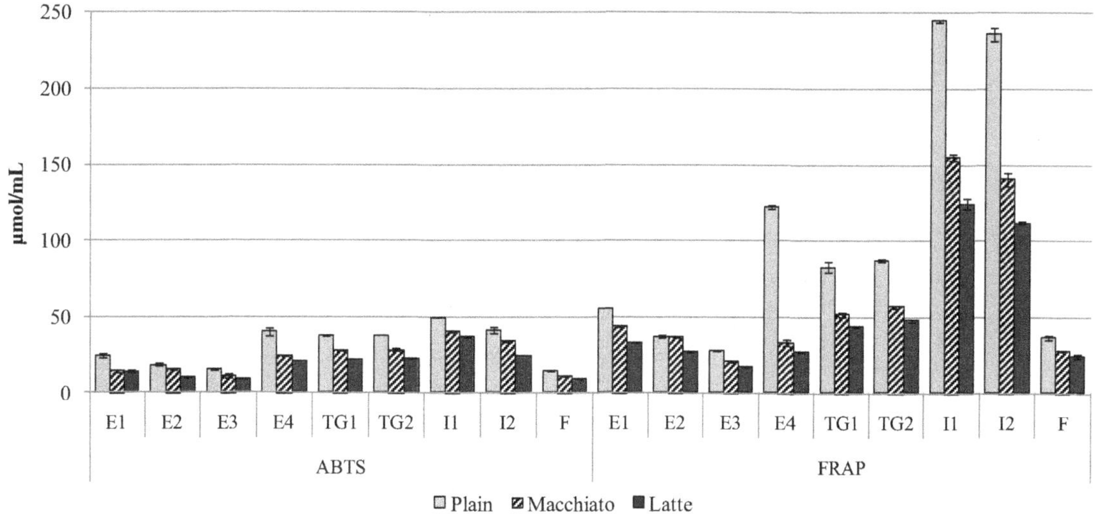

FIGURE 10.1 Antioxidant capacity, determined by ABTS [scavenging assay of 2,2-azino-bis(3-ethylbenzthiazoline-6-sulphonic acid) diammonium salt radical] and FRAP (ferric reducing antioxidant power) assays, of differently prepared coffee brews affected by the addition of milk (water, *macchiato*, *latte*). Comparison of antioxidant capacity of coffee brews prepared by different brewing techniques (E, espresso; TG, Turkish/Greek; I, instant; F, filter coffee) and with different amounts of milk (*macchiato* coffee brew: 30 ml of added milk; *latte*: 50 ml of added milk). The results are displayed as average values ± SD, expressed as μmol Trolox/ml for ABTS assay and μmol Fe (II)/ml for FRAP assay. *The results are excerpted from Niseteo* et al. *(2012) and adapted from the original source to provide a figurative display.*

coffee beans using brews prepared by four brewing techniques (boiled, Turkish, French press, and Mocha pot). The extraction yield of cafestol was shown to be dependent on the brewing technique and roasting time, with the lightest roasted coffee prepared by French press or boiled preparations having the highest cafestol extraction yield (6.5% and 5.84%), and dark-roasted Mocha and Turkish preparations having the lowest extraction yields, of 2.42% and 2.88%, respectively. Previously Kurzrock and Speer (2001) and Speer *et al.* (2000) summarized the work by Sehat *et al.* (1993) and suggested that in a Scandinavian-type brew up to 23% of the total diterpene esters are extracted from the coffee into the brew, whereas, for espresso and filter coffee brews extraction yields of 0.3% and 2.5%, respectively, were found. Sehat *et al.* (1993) demonstrated that for Scandinavian-style brews there was an impact of grind size on extraction yield, with very fine-ground coffee having a higher extraction yield when compared to coarse-grind size, supporting the conclusion that physical availability of cafestol within the roasted and ground coffee has a significant impact on the cafestol extraction yield.

EFFECTS OF MILK ADDITION ON THE ANTIOXIDANT CAPACITY OF COFFEE BREWS

In order to meliorate taste or abate caffeine effect, coffee is often consumed with milk. The preferred forms of coffee consumption are *macchiato* and *latte* brews prepared with the addition of different amounts of milk, which alters the bioactive composition and antioxidant capacity of coffee brew; an effect on the ability of polyphenols to interact with dietary proteins has been suggested (Prigent *et al.*, 2003). At acidic pH food matrices, these interactions would be reversible and involve hydrogen bonding between phenolic residues and polar groups of the protein, which would take place in hydrophobic pockets of the protein (Haslam *et al.*, 1992). As a consequence, these interactions could be a hindrance to polyphenols absorption and/or result in partial masking of their antioxidant power (Arts *et al.*, 2001), which would cause health-benefit impairment.

Concerning coffee and milk, Sánchez-González *et al.* (2005) found that *in vitro* antioxidant capacity decreased when milk was added to espresso coffee. Similarly, Niseteo *et al.* (2012) determined that the preparation of both *macchiato* and *latte* brews showed significantly lower total phenol content and antioxidant capacity when compared to plain brews (Figure 10.1). A similar trend was also observed for the total phenol content of decaffeinated coffee brews. This can be explained by the fact that up to one-third of the quantity of chlorogenic acids, the main antioxidant compounds, interact with milk proteins in coffee (Dupas *et al.*, 2006). However, according to Dupas *et al.* (2006), neither milk addition nor spray-drying of the prepared brew had a significant effect on its antioxidant power and total antioxidant capacity tests.

ANTIOXIDANT POTENTIAL OF INSTANT CAPPUCCINO BREWS

Besides coffee, instant cappuccino brews are also often consumed and can be prepared either by brewing with water or milk, depending on consumer preferences.

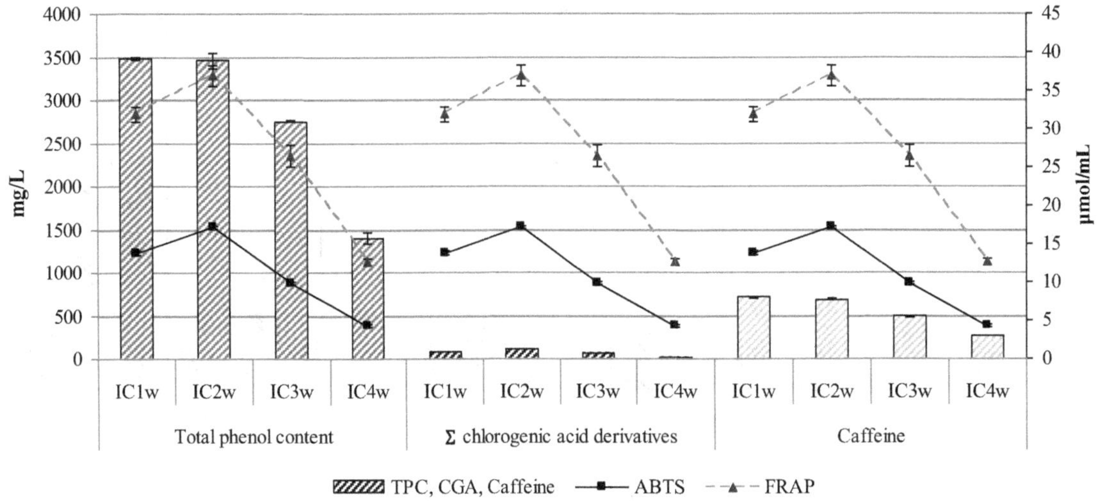

FIGURE 10.2 Total phenol content (mg/l GAE), chlorogenic acids (3-CQA, 4-CQA, and 5-CQA; mg/l) and caffeine contents (mg/l), as well as antioxidant capacity, determined by ABTS (µmol Trolox/ml) and FRAP (µmol Fe (II)/ml) assays, of instant cappuccino brews prepared with water. Comparison of the bioactive composition (total phenols, chlorogenic acids, and caffeine, displayed with bars) of different instant cappuccino brews prepared with water, with their corresponding antioxidant capacity (displayed with lines). The displayed results are shown as average values ± SD. GAE, gallic acid equivalents; 3-CQA, 3-caffeoylquinic acid; 4-CQA, 4-caffeoylquinic acid; 5-CQA, 5-caffeoylquinic acid; ABTS, scavenging assay of 2,2-azino-bis(3-ethylbenzthiazoline-6-sulphonic acid) diammonium salt radical; FRAP, ferric reducing antioxidant power assay. *The results are excerpted from Niseteo* et al. *(2012) and adapted from the original source to provide a figurative display.*

Niseteo *et al.* (2012) determined significant differences in the TPC of water-prepared instant cappuccino brews, with regards to the cappuccino composition (the content of instant coffee) (Figure 10.2). In contrast to the results obtained for coffee brews prepared with the addition of milk, instant cappuccino brews made with milk exhibited higher TPC than those made with water. This observation was, however, attributed to the presence of milk-derived compounds, which can directly interfere with accurate polyphenol determination by reducing analytical recovery, causing overestimation of milk-prepared cappuccino brews. This was also applicable to the antioxidant capacity of cappuccino brews. Additionally, it was observed that instant cappuccino brews provided higher caffeine content than decaffeinated coffee brews, almost comparable to regular espresso brews, and depended on the content of instant coffee in their composition. As can be seen in Figure 10.2, the ranking of antioxidant capacity of instant cappuccino brews corresponded to the TPC, CGA derivatives, and caffeine content. Therefore, it can be concluded that instant cappuccino brews also provide a substantial amount of polyphenolic antioxidants and caffeine, in the range of decaffeinated and filtered coffee.

References

Alves, R.C., Casal, S., Oliveira, M.B.P.P., 2010. Tocopherols in coffee brews: Influence of coffee species, roast degree and brewing procedure. J. Food Comp. Anal. 23, 802–808.

Arts, M.J.T.J., Haenen, G.R.M.M., Voss, H.P., Bast, A., 2001. Masking of antioxidant capacity by the interaction of flavonoids with protein. Food Chem. Toxicol. 39, 787–791.

Belkaid, A., Currie, J.-C., Desgagnés, J., Annabi, B., 2006. The chemopreventive properties of chlorogenic acid reveal a potent new role for the microsomal glucose-6-phosphate translocase in brain tumor progression. Cancer Cell Int. 6, 7–18.

Bell, L.N., Wetzel, C.R., Grand, A.N., 1996. Caffeine content in coffee as influenced by grinding and brewing techniques. Food Res. Int. 29, 785–789.

Blumberg, S., Frank, O., Hofmann, T., 2010. Quantitative studies on the influence of the bean roasting parameters and hot water percolation on the concentrations of bitter compounds in coffee brew. J. Agric. Food Chem. 58, 3720–3728.

Boekschoten, M.V., van Cruchten, S.T., Kosmeijer-Schuil, T.G., Katan, M.B., 2006. Negligible amounts of cholesterol-raising diterpenes in coffee made with coffee pads in comparison with unfiltered coffee. Ned. Tijdsch. Genees. 150, 2873–2875.

Buchmann, S., Zahm, A., Kölling-Speer, I., Speer, K., 2010. Lipids in coffee brews-impact of grind size, water temperature, and coffee/water ratio on cafestol and the carboxylic acid-5-hydroxytryptamides. In: 23rd International Conference on Coffee Science. ASIC, Bali, pp. 84–93.

Carlsen, M.H., Halvorsen, B.L., Holte, K., Bøhn, S.K., Dragland, S., Sampson, L., et al., 2010. The total antioxidant content of more than 3100 foods, beverages, spices, herbs and supplements used worldwide. Nutr. J. 9, 3–14.

Casal, S., Oliveira, M.B.P.P., Alves, M.R., Ferreira, M.A., 2000. Discriminate analysis of roasted coffee varieties for trigonelline, nicotinic acid, and caffeine content. J. Agric. Food Chem. 48, 3420–3424.

Daglia, M., Papetti, A., Gregotti, C., Berte, F., Gazzani, G., 2000. *In vitro* and *ex vivo* antihydroxyl radical activity of green and roasted coffee. J. Agric. Food Chem. 48, 1449–1454.

Del Castillo, M.D., Ames, J.M., Gordon, M.H., 2002. Effects of roasting on the antioxidant activity of coffee brews. J. Agric. Food Chem. 50, 3698–3703.

Dupas, C.J., Marsset-Baglieri, A.C., Ordonaud, C.S., Ducept, F.M.G., Maillard, M.-N., 2006. Coffee Antioxidant Properties: Effects of Milk Addition and Processing Conditions. J. Food Sci. 71, S253–S258.

Farah, A., Donangelo, C.M., 2006. Phenolic compounds in coffee. Braz. J. Plant Physiol. 18, 23–36.

Farah, A., 2009. Coffee as a functional beverage. Agro Food Ind. High Technol. 20 (Suppl.), 36–39.

Fisk, I.D., Massey, A.T., Hansen, N.A., 2011. Improvements in the preparation of beverage and liquid food products. UK patent number WO2011153065.

Fujioka, K., Shibamoto, T., 2008. Chlorogenic acid and caffeine contents in various commercial brewed coffees. Food Chem. 106, 217–221.

Gross, G., Jaccaud, E., Huggett, A.C., 1997. Analysis of the content of the diterpenes cafestol and kahweol in coffee brews. Food Chem. Toxicol. 35, 547–554.

Harland, B.F., 2000. Caffeine and nutrition. Nutrition 16, 522–526.

Haslam, E., Lilley, T.H., Warminski, E., Liao, H., Cai, Y., Martin, R., Gaffney, S.H., Goulding, P.N., Luck, G., 1992. Polyphenol complexation, a study in molecular recognition. In: Ho, C.T., Lee, C.Y., Huang, M.T. (Eds.), Phenolic compounds in foods and their effect on health I: Analysis, occurrence and chemistry, vol. 506. Amer. Chemical Society, Washington, D. C, pp. 8–50.

Hsu, J.L., Hung, W.-C., 2005. Packed coffee drink consumption and product attribute preferences of young adults in Taiwan. Food Qual. Prefer. 16, 361–367.

Kölling-Speer, I., Kurzrock, T., Speer, K., 2006. Tocopherols in green coffees harvested in different years. In: Proceedings of the 21st international scientific colloquium on coffee. ASIC, Montpellier.

Kroll, J., Rawel, H.M., Rohn, S., 2003. Reactions of plant phenolics with food proteins and enzymes under special consideration of covalent bounds. Food Sci. Technol. Res. 9, 205–218.

Kurzrock, T., Speer, K., 2001. Diterpenes and diterpene esters in coffee. Food Rev. Int. 17, 433–450.

Lau, O.W., Luk, S.F., Cheng, O.M., Chiu, T.P.Y., 1992. Background-correction methods for the determination of caffeine in beverages, coffee and tea by using second-derivative ultraviolet spectrophotometry. Analyst 117, 777–783.

Lee, H.S., O'Mahony, M., 2002. At what temperatures do consumers like to drink coffee? Mixing methods. J. Food Sci. 67, 2774–2777.

Lee, H.S., Carstens, E., O'Mahony, M., 2003. Drinking hot coffee: why doesn't it burn the mouth? J. Sens. Stud. 18, 19–32.

Lewis, R., 2004. Pursuing the perfect cup of coffee. Scientist 18, 56.

Lingle, T.R., 1996. The coffee brewing handbook. Specialty Coffee Association, Long Beach.

López-Galilea, I., de Peña, M.P., Cid, C., 2007. Correlation of selected constituents with the total antioxidant capacity of coffee beverages: Influence of the brewing procedure. J. Agric. Food Chem. 55, 6110–6117.

Ludwig, I.A., Sanchez, L., Caemmerer, B., Kroh, L.W., de Peña, M.P., Cid, C., 2012. Extraction of coffee antioxidants: Impact of brewing time and method. Food Res. Int. 48, 57–64.

Minamisawa, M., Yoshida, S., Takai, N., 2004. Determination of biologically active substances in roasted coffees using a diode-array HPLC system. Anal. Sci. 20, 325–328.

Mišik, M., Hoelzl, C., Wagner, K.H., Cavin, C., Moser, B., Kundi, M., Simic, T., Elbling, L., Kager, N., Ferk, F., Ehrlich, V., Nersesyan, A., Dušinská, M., Schilter, B., Knasmüller, S., 2010. Impact of paper filtered coffee on oxidative DNA-damage: results of a clinical trial. Mutat. Res. 692, 42–48.

Moreira, D.P., Monteiro, M.C., Ribeiro-Alves, M., Donangelo, C.M., Trugo, L.C., 2005. Contribution of Chlorogenic Acids to the Iron-Reducing Activity of Coffee Beverages. J. Agric. Food Chem. 53, 1399–1402.

Navarini, L., Ferrari, M., Suggi Liverani, F., Liggieri, L., Ravera, F., 2004. Dynamic tensiometric characterization of espresso coffee beverage. Food Hydrocol. 18, 387–393.

Nestle, 2007. Coffee: wherever, when ever, however. NutriPro Nestle Prof. Nutr. Mag. 1/07, 1–11.

Nicoli, M.C., Anese, M., Manzocco, L., Lerici, C.R., 1997a. Antioxidant properties of coffee brews in relation to the roasting degree. Lebensm. Wiss. Technol. 30, 292–297.

Nicoli, M.C., Anese, M., Parpinel, M., 1997b. Loss and formation of antioxidants during food processing and storage. Cancer Lett. 114, 71–74.

Niseteo, T., Komes, D., Belščak-Cvitanović, A., Horžić, D., Budeč, M., 2012. Bioactive composition and antioxidant potential of different commonly consumed coffee brews affected by their preparation technique and milk addition. Food Chem. 134, 1870–1877.

Nogueira, M., Trugo, L.C., 2003. Distribuição de isômeros de ácido clorogênico e teores de cafeína e trigonelina em cafés solúveis Brasileiros. Ciênc. Tecnol. Alim. 23, 296–299.

Nunes, F.M., Coimbra, M.A., 2007. Melanoidins from coffee infusions. Fractionation, chemical characterization, and effect of the degree of roast. J. Agric. Food Chem. 55, 3967–3977.

Oosterveld, A., Harmsen, J.S., Vorgen, A.G.J., Schols, H.A., 2003. Extraction and characterisation of polysaccharides from green and roasted *Coffea arabica* beans. Carbohyd. Polym. 52, 283–296.

Parras, P., Martínez-Tomé, M., Jiménez, A.M., Murcia, M.A., 2007. Antioxidant capacity of coffees of several origins brewed following three different procedures. Food Chem. 102, 582–592.

Pellegrini, N., Serafini, M., Colombi, B., Del Río, D., Salvatore, S., Bianchi, M., Brighenti, F., 2003. Total antioxidant capacity of plant foods, beverages and oils consumed in Italy assessed by three different *in vitro* assays. J. Nutr. 133, 2812–2819.

Pérez-Martínez, M., Caemmerer, B., De Peña, M.P., Cid, C., Kroh, L.W., 2010. Influence of brewing method and acidity regulators on the antioxidant capacity of coffee brews. J. Agric. Food Chem. 58, 2958–2965.

Petracco, M., 2005. Percolation. In: Illy, A., Viani, R. (Eds.), Espresso coffee: The chemistry of quality, vol. 2. Elsevier Academic Press London, London, pp. 259–290.

Petracco, M., 2001. Beverage preparation: brewing trends for the new millennium. In: Clarke, R.J., Vitzthum, O.G. (Eds.), Coffee recent developments. Blackwell Science, Oxford, p. 140.

Phan, T.T.D., Kuban, V., Kráčmar, S., 2012. Determination of caffeine contents of coffee brands in the vietnamese market. J. Microbiol. Biotech. Food Sci. 1, 995–1002.

Prigent, S.V.E., Gruppen, H., Visser, A.J.W.G., van Koningsveld, G.A., de Jong, G.A.H., Voragen, A.G.J., 2003. Effects of non-covalent Interactions with 5-O-caffeoylquinic acid (chlorogenic acid) on the heat denaturation and solubility of globular proteins. J. Agric. Food Chem. 51, 5088–5095.

Sacchetti, G., Di Mattia, C., Pittia, P., Mastrocola, D., 2009. Effect of roasting degree, equivalent thermal effect and coffee type on the radical scavenging activity of coffee brews and their phenolic fraction. J. Food Eng. 90, 74–80.

Sánchez-González, I., Jiménez-Escrig, A., Saura-Calixto, F., 2005. In vitro antioxidant activity of coffees brewed using different procedures (Italian, espresso and filter). Food Chem. 90, 133–139.

Santini, A., Ferracane, R., Mikusová, P., Eged, S., Srobárová, A., Meca, G., Mañes, J., Ritieni, A., 2011. Influence of different coffee drink preparations on ochratoxin A content and evaluation of the antioxidant activity and caffeine variations. Food Control 22, 1240–1245.

Sehat, N., Montag, A., Speer, K., 1993. Lipids in the coffee brew. In: 15th International Scientific Colloquium on Coffee. Montpellier, pp. 869–872.

Shi, X., Dalal, N.S., Jain, A.C., 1991. Antioxidant behaviour of caffeine: Efficient scavenging of hydroxyl radicals. Food Chem. Toxicol. 29, 1–6.

Speer, K., Hruschka, A., Kurzrock, T., Kölling-Speer, I., 2000. Diterpenes in coffee. In: Parliment, T.H., Ho, C.-T., Schieberle, P. (Eds.), Caffeinated Beverages—Health Benefits, Physiological Effects and Chemistry, vol. 754. ACS Symposium series, Washington, DC, pp. 241–251.

Speer, K., Kölling-Speer, I., 2006. The lipid fraction of the coffee bean. Braz. J. Plant Physiol. 18, 201–216.

Svilaas, A., Sakhi, A.K., Andersen, L.F., Svilaas, T., Stroem, E.C., Jacobs Jr., D.R., Ose, L., Blomhoff, R., 2004. Intakes of antioxidants in coffee, wine, and vegetables are correlated with plasma carotenoids in humans. J. Nutr. 134, 562–567.

Vignoli, J.A., Bassoli, D.G., Benassi, M.T., 2011. Antioxidant activity, polyphenols, caffeine and melanoidins in soluble coffee: the influence of processing conditions and raw material. Food Chem. 124, 863–868.

Zhang, C., Linforth, R., Fisk, I.D., 2012. Cafestol extraction yield from different coffee brew mechanisms. Food Res. Int. 49, 27–33.

CHAPTER

11

Applications of Enzymes in Processing Green Tea Beverages: Impact on Antioxidants

C.S. Murugesh, R. Subramanian

Academy of Scientific and Innovative Research, and Food Engineering Department, CSIR-Central Food Technological Research Institute, Mysore, India

CHAPTER POINTS

- Enzymes have been applied at various stages of green tea processing to improve its beverage quality.
- Application of commercial cell-wall-digesting enzymes improve the extraction efficiency including polyphenol extractability thereby increasing the antioxidant capacity of the extract.
- Tannase hydrolyzes the gallated polyphenolic compounds to improve the cold-water extractability/solubility. Ungallated catechins in green tea exhibit reduced protein–tannin interaction, leading to reduced cream formation. Thus, reduction in cream formation and improvement in clarity of the beverage can be achieved without any loss of catechins in the form of tea cream during storage.
- Tannase hydrolysis of monomeric catechins significantly enhances the antioxidant and metal ion chelation capacities of green tea extract through bioconversion of gallated catechins to their simpler forms accompanied by the release of gallic acid.
- When tannase treatment is followed by ethanolic extraction the concentration of monomeric catechins increases in the extract, thus facilitating the derivation of maximum physiological benefit from the beverages without adversely affecting their taste.

INTRODUCTION

Green tea is a widely consumed beverage in the world, so much so that the current annual world production of green tea of 1.3 million tons is estimated to double in the next 10 years (FAO, 2012). It is also projected that rate of growth of green tea (7.2%) will be four-times higher than that of black tea. The increase in popularity of green tea is due to its numerous physiological benefits. These physiological benefits are attributed to its rich content of polyphenols, especially the monomeric catechins (MC), epi-gallo catechin gallate (EGCG), epi-catechin gallate (ECG), epi-gallo catechin (EGC), and epicatechin (EC). These epicatechins can change to their epimers that are catechins, i.e., gallocatechin gallate (GCG), catechin gallate (CG), gallocatechin (GC), and catechin (C). Tea catechins have the most effective antioxidant activity compared to other tea polyphenols. They are efficient free-radical scavengers due to their one-electron reduction potential. The rate of reaction with free radicals and the stability of the resulting antioxidant radicals contribute to the reactivity of antioxidants (Ananingsih *et al.*, 2013). The order of scavenging activity of the catechins changes with respect to the type of free radical or reactive oxygen species. For instance the scavenging ability of tea catechins on superoxide anions ($O_2^{\bullet -}$), singlet oxygen ($^1O_2$), the free radicals generated from 2, 2P-azobis (2-amidinopropane) hydrochloride (AAPH) and 1, 1-diphenyl-2-picrylhydrazyl (DPPH) radicals has been reported (Guo *et al.*, 1999). The free-radical scavenging abilities of EGCG and GCG were higher than those of EGC, GC, EC, and C due to their gallate

Processing and Impact on Antioxidants in Beverages
http://dx.doi.org/10.1016/B978-0-12-404738-9.00011-8

group at the 3-position of the C ring. EGC and GC were stronger in their scavenging ability than EC and C because of a hydroxyl group at the 5′ position of B ring. GCG was more stable than EGCG because it has smaller steric hindrances. While the stability of GC and C was better than EGC and EC. Ananingsih *et al.* (2013) summarized the scavenging activity of tea catechins on various free radicals and reactive oxygen species (ROS) (Table 11.1).

Processing has a profound effect on the level of catechins as a result of epimerization and also due to degradation. It has been reported that even the brewing step could convert epicatechins to their corresponding epimers. These types of conversions continue to take place during the production of canned tea beverages and also during the subsequent storage period.

The role of enzymes in tea processing has been recognized for nearly four decades, along with its application to improve the quality of tea. The antioxidant capacity of green tea largely depends on the concentration and composition of the MC. Application of enzymes at various stages of processing will affect the concentration as well as composition of MC in the beverage, thus having a direct impact on the antioxidant capacity of the beverage. The major enzymatic methods proposed for processing black tea having a direct or indirect bearing on ready-to-drink (RTD) tea production, have been discussed along with their relative advantages and limitations in a recent review (Chandini *et al.*, 2013). The various major enzymatic methods proposed for processing green tea are discussed here, along with their intended benefits, limitations, and impact on the antioxidant capacity.

ENDO- AND EXOGENOUS ENZYMES

Endogenous Enzymes

Polyphenol oxidase (PPO) and peroxidase (POD) are the two natural endogenous enzymes present in fresh tea leaves responsible for fermentation (oxidation process) in the conversion of green tea to black tea. However, in the case of manufacture of green tea, it is necessary to inactivate these enzymes to prevent any enzymatic oxidation of catechins.

Exogenous Enzymes

Tannase (tannin acyl hydrolase EC3.1.1.20) is the most commonly employed exogenous enzyme for tea processing. This enzyme hydrolyzes esters of phenolic acids, including the gallated polyphenols found in tea. Filamentous fungi of the *Aspergillus* genus and bacteria of the *Bacillus* genus have been widely used for producing tannase (Bajpai and Patil, 1997). The tannase of some *Aspergillus* strains has a molecular weight around 150–350 kDa. Their activity and stability pH are 5.0–6.0 and 3.5–8.0, respectively, while optimal temperatures from 35°C to 40°C have been reported. Tannase is stable for several months at 30°C (Belmares *et al.*, 2004).

Various cell-wall-digesting enzymes have also been employed which react and modify plant cell wall biopolymers. Cell-wall-digesting enzyme is an enzyme which breaks down one or more tea cell-wall constituents to simpler materials and thus reduces the structural integrity or increases the permeability of the cell wall. Plant cell walls are composed primarily of cellulose,

TABLE 11.1 The Order of Scavenging Activities of Tea Catechins on Free Radicals and Reactive Oxygen Species

Free Radicals/Reactive Oxygen Species	The Order of Scavenging Activity	Reference
Singlet oxygen	EC, C > EGC, GC > EGCG, GCG	Guo *et al.* (1999)
	EGCG > ECG > EGC > EC > C	Mukai *et al.* (2005)
Hydroxyl radical	ECG > EGCG > EC > GC > EGC > C	Wiseman *et al.* (1997)
	ECG > EC > EGCG > EGC	Guo *et al.* (1996)
Lipid peroxyl radical	ECG = EGCG = EC = C > EGC	Salah *et al.* (1995)
$ABTS^{\bullet+}$ radical cation	ECG > EGCG > EGC > EC = C	Salah *et al.* (1995)
	ECG > EGCG > EGC > EC	Higdon and Frei (2003)
$DPPH^{\bullet}$ radical	EGCG = ECG > EGC > EC	Nanjo *et al.* (1996)
AAPH	EGCG > EGC > EC	Guo *et al.* (1999)

The scavenging ability of EGCG and GCG were generally higher than that of EGC, GC, EC, and C due to the specific position of the gallate group in their chemical structure. Similarly, EGC and GC were stronger than EC and C because of the specific position of their hydroxyl group. C, Catechin; EC, epi-catechin; ECG, epi-catechin gallate; EGC, epi-gallo catechin; EGCG, epi-gallo catechin gallate; GC, gallo catechin; GCG, gallo catechin gallate; AAPH-2, 2P-azobis (2-amidinopropane) hydrochloride; ABTS, 2,2-azino-bis-(3-ethyl-benzothiazoline-6-sulfonic acid); DPPH, diphenylpicrylhydrazyl.
Source: Ananingsih et al. *(2013).*

but contain lesser amounts of proteins, hemicellulose, pectins, and lipids. Accordingly, cell-wall digestive enzymes include cellulase and hemicellulase, pectinase, lysozyme, dextranase, lipases, and proteases like papain (Tsai, 1987).

Applications of Enzymes

Enzyme treatments have been given at three stages of green tea processing, that is, pre-treatment to green tea leaves, prior to/during extraction, and to the extract. These treatments aim at improving the soluble solids yield, cold water extractability/solubility, clarity, flavor and odor, as well as in reducing the tea cream formation and astringency/bitterness. Their impact on the antioxidant capacity has also been the subject of interest for many researchers. Figure 11.1 gives an overview of enzymatic treatments attempted in green tea processing.

EFFECTS OF PRODUCTION AND PROCESSING

Enzymatic Pre-treatment of Leaves

EGCG and ECG are the most abundant gallated catechins in fresh tea leaves and tannase treatment hydrolyzes EGCG to yield EGC and gallic acid (GA) and ECG to yield EC and GA by cleaving their ester bonds. Several studies have been carried out with tannase treatment on green tea leaves to simplify the mixture of catechins before the start of the fermentation in the conversion to black tea, aiming at improving the quality characteristics of black tea products. However, such treatment also offers benefits in the case of unfermented and semi-fermented tea products.

In one of the pioneering works, Sanderson *et al.* (1977) described the use of tannase for producing cold-water-extractable green, oolong, and black tea leaves. Accordingly, cold-water-extractable green tea leaves

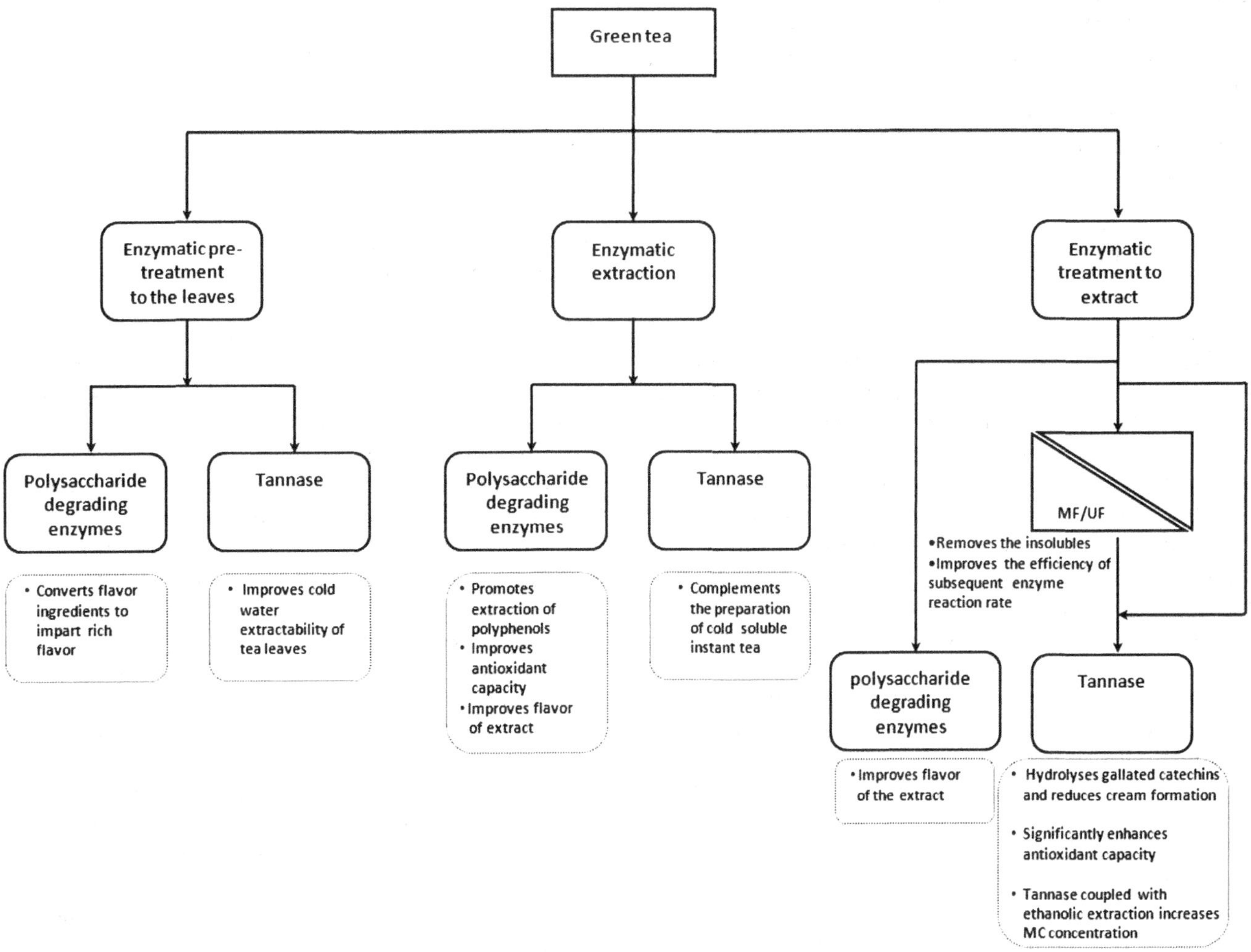

FIGURE 11.1 Overview of enzymatic treatments attempted in green tea processing. Enzymatic treatment offers specific advantages depending on the type of enzyme and the stage of application in the process. MC, monomeric catechins; MF, microfiltration; UF, ultrafiltration.

are prepared by first heat treating freshly harvested green tea leaves to inactivate tea leaf enzymes to prevent tea fermentation. Then the tea leaves are comminuted and treated with tannase to impart substantial cold-water extractability under anaerobic conditions and reduce the moisture content to below 5% to obtain fired green tea leaves. The product thus obtained has been claimed to possess good flavor and more extractable solids.

Yotsumoto (2009) gave a detailed account of the ability of various enzymes to produce a wide variety of typical flavors in tea extracts by enzymatic treatment to ground tea leaves prior to extraction. The method involves blending and reacting ground green tea leaves with enzyme solution to improve the quality by effectively converting its flavor ingredients to impart rich flavor, followed by extraction with warm or cold water. The invention has claimed a wide range of enzymes to choose from (polyphenol oxidase, β-glycosidase, lipase, chlorogenic acid esterase, nuclease, protease, lactase, invertase, pectinase, xylanase, and deaminase) for this application. The process has resulted in rich-flavored green tea extract having an excellent flavor. Further, it is envisaged that the flavor of tea extract could be 'designed' by combining the enzymatically treated leaves employing one or more enzymes. However, there are no reports on the impact of enzyme treatment on the antioxidant activity of the products.

Enzymatic Extraction

The extraction conditions employed in instant tea manufacturing aim for the maximum recovery of tea solids and are usually harsher compared to the brewing conditions used for tea preparations. Accordingly, the earlier studies on extractability (Blogg and Long, 1980; Long, 1977) were focused on the yield of black tea solids. Besides the extract yield, the quality of the soluble powder obtained is an equally important factor in deciding the quality of the final converted products. Considering the fact that not much effort has gone towards improving the extractability of polyphenols while improving the extract yield, research work was undertaken in the author's laboratory to study the influence of extraction conditions for maximal recovery of polyphenols and extractable solids yield from black tea without affecting the quality of black tea (Chandini *et al.*, 2011b). In a subsequent study, attempts were made using enzyme-assisted extraction to enhance the recovery of extractable solids and polyphenols and clarity, maintaining a good balance of tea quality, using pectinase and tannase (Chandini *et al.*, 2011a). More recently, Hong *et al.* (2013) reported a comparative study of improvement in the extraction efficiency of green tea polyphenols employing various enzymes.

Fresh tea leaf contains 30–40% of its dry weight as polyphenols (Arjuna, 2000), which are predominantly responsible for the excellent antioxidant capacity of green tea. These polyphenols are entrapped in the cell protected by the water-insoluble cell wall, composed of cellulose, hemicelluloses, pectin, and protein. Hong *et al.* (2013) reported the application of few commercial cell-wall-digesting enzymes for improving the extractability of polyphenols and the consequent change in the antioxidant capacity of the extract. The enzymes used were Celluclast (a cellulase preparation), Cytolase (containing β-glucopyranosidase and α-rhamnopyranosidase), Econase (a cellulase preparation), Pectinex (containing pectin transeliminase, polygalacturonase, and pectinesterase), Rapidase (containing pectinase, hemicellulase, and cellulase), Ultraflo (containing β-glucanase and xylanase), and Viscozyme (containing cellulase, β-glucanase, hemicellulase, and xylanase). The process involved incubating the selected enzyme with the extraction mixture of coarsely ground green tea leaves and water in the ratio 1:20 at 40°C for various pre-decided extraction periods.

The total polyphenol content and DPPH radical-scavenging activity of the extract obtained after treatment with various enzymes are depicted in Figure 11.2. Enzymatic extraction showed a significant increase (4–20%) in the total polyphenol content of the green tea extract as compared with non-enzymatic treatment. A similar trend was observed with the total flavonoid and catechin content of the extract. In particular, the polyphenol, flavonoid, and catechin contents of the tea extract obtained with Viscozyme were higher than that of the others.

Green tea extracts have higher antioxidant activity than black tea, and the total antioxidant potential correlates strongly with the total polyphenol content of tea (Langley-Evans, 2000). DPPH assay is widely used to determine the free-radical scavenging activity of various plants and pure compounds. All the seven commercial enzyme treatments of green tea produced significant ($p<0.05$) increases in DPPH radical-scavenging activity as compared with non-enzymatic treatment. As expected, the DPPH radical-scavenging capacity of the extract obtained using Viscozyme was found to be the highest owing to its greater extractability of polyphenols.

Viscozyme from *Aspergillus sp.* is a multi-enzyme complex containing a wide range of carbohydrate-hydrolyzing enzymes and was the most effective enzyme among the tested commercial enzymes. A time-course study on Viscozyme up to 8 h revealed that the total polyphenol content reached a maximum after 2 h of extraction and decreased thereafter. After 2 h, Viscozyme caused a high degree of tissue breakdown, releasing more non-phenolic components that contribute to the soluble solids of green tea extracts. The antioxidant activity measured as DPPH radical scavenging activity of the green tea extracts

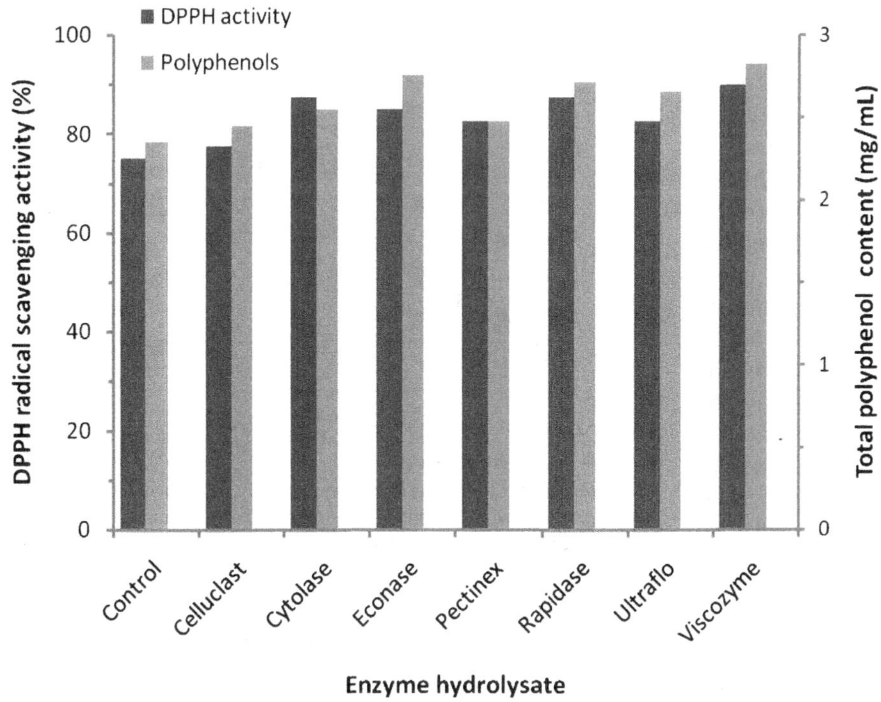

FIGURE 11.2 The total polyphenol content and DPPH radical scavenging activity of green tea extracts treated with various commercial enzymes. Enzymatic extraction employing cell-wall-digesting enzymes significantly improved the polyphenol content in green tea extract with a corresponding improvement in DPPH radical scavenging activity compared to control. Viscozyme was found to be more effective than the other six commercial enzymes used in the study. DPPH, diphenylpicrylhydrazyl. *Source: Hong* et al. *(2013).*

showed similar tendencies with respect to changes in the total polyphenol content. These authors claimed that Viscozyme is by far the superior enzyme for enzymatic extraction of green tea to improve the polyphenol content and thereby the antioxidant activity, and envisage its potential use in health foods and cosmetic products.

Attempts have also been made during extraction of green tea to improve its quality in terms of aroma, flavor, taste, and cold-water solubility as well as extraction yield with tannase and various cell-wall-digesting enzymes. These attempts have been primarily made by Japanese and Chinese researchers and are available as patent literature (Masakichi *et al.*, 1984; Yue *et al.*, 2008; Sakamaki *et al.*, 2006; Chen *et al.*, 2012). Although enzymatic extraction promotes extraction as well as improves the flavor of the extract, a relatively long time is required to exhibit the complete effects of enzymatic treatment. The prolonged treatment leads to quality deterioration during the extraction step since it is performed under a relatively large amount of water, as it is also intended to extract at the same time. Therefore, enzyme pre-treatment may be preferred over enzymatic extraction from the viewpoint of flavor improvement.

Enzymatic Treatment to Extract

Tannase is the most widely used enzyme at this stage of the process. Enzymatic treatment of extract was carried out mostly for improving the concentration of MC without affecting the taste of the beverage, often in combination with favorable chemical processes also aimed at the decreaming or clarification of the extract. As mentioned earlier, green tea polyphenols, especially MC, are responsible for the various health benefits associated with green tea. MC mainly include EC, EGC, ECG, and EGCG. The effects of these MC are believed to depend on their antioxidant capacity. In this context, it is pertinent to look at the impact of tannase treatment on the antioxidant activity of the treated green tea extract.

Effect of Tannase Treatment on Antioxidant Activity of Green Tea Infusion

The comparative antioxidant activity of tea catechins varied depending on the systems used for assessment (Lu and Chen, 2008). Furthermore, the differences in the chemical structure of the individual polyphenols contribute to the variations in their antioxidant activity and ability to chelate metals and to scavenge free radicals. Tea catechins are characterized by having a similar backbone (C6–C3–C6) with varying number and position of hydroxyl groups, with or without a galloyl group at the 3′-position. For instance, a hydroxyl group at the 5′-position may play an important role in contributing antioxidant activity as is evident from the observation that EGC is more effective than EC (Chen and Chan, 1996). Metal chelation is important for inhibiting metal-catalyzed lipid peroxidation. Among the green tea catechins, EGC was found to be most effective, followed by EGCG, EC, and ECG against lipid oxidation in hot canola oil (95°C) as measured by the oxygen consumption test (Chen and Chan, 1996). However, the actual composition of individual catechins and GA in tea extracts would play a great role in determining the beneficial effects of tea beverages.

Tannase catalyzes the hydrolysis of gallated compounds such as EGCG or ECG present in green tea. The GA released in this process is considered to be beneficial because of its significant antioxidant potency (Lekha and Lonsane, 1997). In addition, 1 mole of EGC and GA each is produced whenever 1 mole of EGCG is hydrolyzed. Accordingly, higher antioxidant activities are expected to be obtained, despite the fact that the relative activity of the individual catechins differs greatly, as the resultant concentrations of tea catechins in hydrolyzate increase. Lu and Chen (2008) investigated the alteration of the antioxidative activities of green tea catechins before and after tannase treatment by using assays for DPPH, superoxide anion, Trolox-equivalent antioxidant capacity (TEAC), hydrogen peroxide and metal ion chelation. Furthermore, the effect of such enzymatic treatment on the viability of a monocytic cell line was also evaluated to assess the oxidative deoxyribonucleic acid (DNA) damage induced by tea catechins in the presence of metal complexes such as Cu^{2+} and Fe^{2+}.

Hydrolysis reaction of the green tea extract was carried out by varying enzyme concentrations ranging from 0.5 to 3 mg/ml at pH6 and 35°C for 20 min. The gallated MC, EGCG and ECG were hydrolyzed by tannase to produce EGC and EC, respectively, along with a common product, GA. The relative concentrations of the MC changed with the degree of hydrolysis. At a tannase concentration of 3 mg/ml, EGC, EC and GA content increased by 2.5-, 5.8-, and 4.1-fold, respectively, accompanied by near-complete reduction in EGCG and ECG contents (100% enzyme hydrolysis).

DPPH assay based on its reaction with free-radical scavenging antioxidants showed a significant reduction in its concentration in both tannase-treated and untreated green tea samples. But the free-radical-scavenging capacity was greater for the treated sample than that of the untreated control (Figure 11.3). This increase in the scavenging capacity was due to the increase in the GA content in the treated sample. TEAC of tannase-treated green tea extract showed a 3.4-fold greater antioxidant activity compared to the control. The chelation ability of tannase-treated green tea extract for Fe^{+2}/Cu^{+2} ions was also superior compared to the untreated one. A similar trend was observed in the case of superoxide anion and hydrogen peroxide scavenging effects. The results validated that tannase hydrolysis of green tea catechins significantly enhanced the antioxidant and metal ion chelation capacities of green tea extract (Lu and Chen, 2008).

Battestin *et al.* (2008) conducted a similar study on enzymatic hydrolysis of green tea catechins using a tannase from *Paecilomyces variotii*, including a model experiment with a special focus on EGCG as substrate. The work established that EGCG could be entirely hydrolyzed using the tannase incubating the substrate mixture

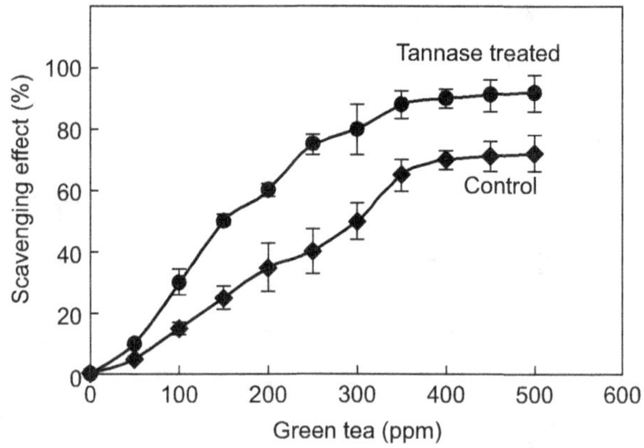

FIGURE 11.3 DPPH scavenging effect of green tea with or without tannase treatment. Tannase hydrolyzes EGCG and ECG present in the green tea extract to EGC and EC, respectively, along with the release of GA. The increase in GA in the treated sample was mainly responsible for its higher free radical scavenging capacity compared to the control sample. EC, epi-catechin; ECG, epi-catechin gallate; EGC, epi-gallo catechin; EGCG, epi-gallo catechin gallate; GA, gallic acid. *Source: Lu and Chen (2008).*

at pH6.0 for 30 min at 40°C. The antiradical properties and the reducing power of these samples were assessed using the DPPH and ferric reducing antioxidant power (FRAP) assays, respectively. The products of the enzyme reaction of EGCG, namely, EGC and GA, showed higher DPPH and FRAP activity than the EGCG. The results suggested that the tannase treatment improved the antioxidant activity of green tea extract.

A successive enzyme treatment approach was attempted by Hong *et al.* (2013) with Viscozyme to improve the extraction efficiency of polyphenols followed by tannase treatment to the extract to increase the radical-scavenging activities. As mentioned in the earlier section, Viscozyme-assisted extraction increased the polyphenol content by 20%. The subsequent tannase treatment reached equilibrium in ~20 min, and more than 90% of the EGCG and ECG was hydrolyzed to EGC or EC, respectively. There was a tendency for radical-scavenging activities to increase with increases in the EGC, EC, and GA contents of the green tea extracts. The radical-scavenging activities of successive treatments with Viscozyme and tannase were generally much higher than those of non-enzymatic and Viscozyme-alone treatment (Figure 11.4). The IC_{50} value of tannase treated extract with respect to DPPH and 2,2-azino-bis-(3-ethyl-benzothiazoline-6-sulfonic acid) (ABTS) radical-scavenging activity decreased by 46% and 58%, respectively, compared to its control. This evidenced the enhanced antioxidant activities of green tea after tannase treatment.

Increasing the Concentration of Monomeric Catechins

The MC present in green tea are believed to possess various health benefits, such as cholesterol-suppressing

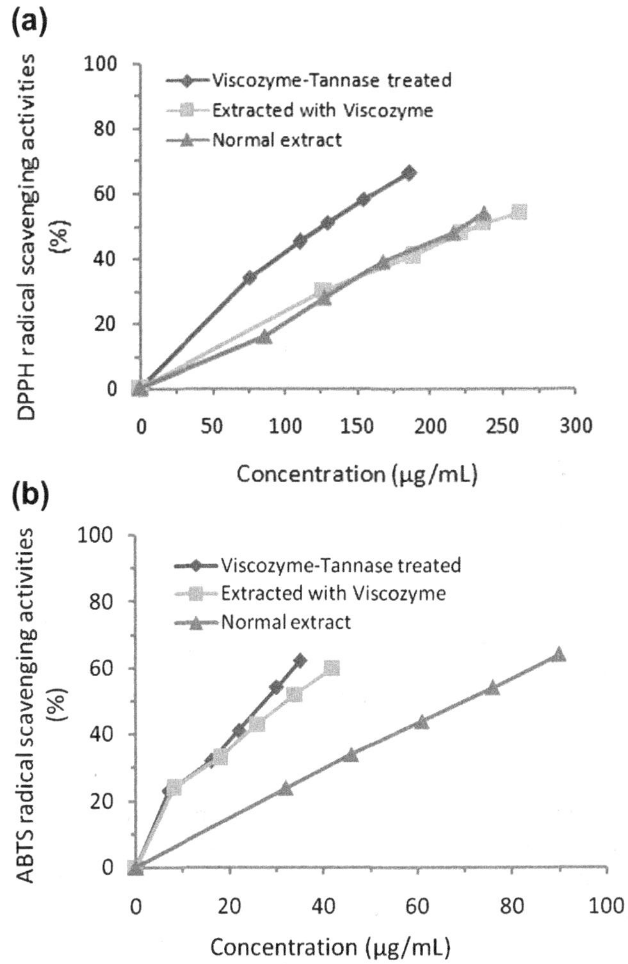

FIGURE 11.4 Half-maximum inhibitory concentration of green tea extracts on (a) DPPH and (b) ABTS hydroxyl radicals. Viscozyme treatment improved the extractability of polyphenols, thus showed (a) higher DPPH and (b) higher ABTS radical-scavenging activity compared to the normal extract. Successive tannase treatment to the Viscozyme-assisted extract hydrolyzed the gallated catechins further improving these activities. DPPH, diphenylpicrylhydrazyl; ABTS, 2,2-azino-bis-(3-ethylbenzothiazoline-6-sulfonic acid). *Source: Hong* et al. *(2013).*

effects, anti-cancer effects, and body-weight control. In order to derive all these physiological effects, an adult must drink four to five cups of green tea a day. This turned the research focus towards making a beverage with higher concentration of monomeric catechins. In one of the common approaches in this direction, a concentrate of green tea extract was incorporated in the beverage preparations. However, it ended up in proportionate increase in residual bitterness/astringency due to an increase in catechin gallates and miscellaneous tastes caused by other tea components derived from green tea, affecting the commercial value of the beverage. Monomeric catechins collectively encompass non-epicatechins (C, CG, GC, and GCG) as well as epicatechins (EC, ECG, EGC, and EGCG) including their gallates. It is well known that the gallate content in tea extract can be reduced by treating it with tannase (Masami *et al.*, 2004). Although the conventional tannase treatment reduces the bitterness/astringency, it does not reduce other tea components such as dietary fiber and caffeine having a bearing on the miscellaneous taste. Besides, protein being an allergic substance, there is a demand for reducing proteins derived from the enzyme or tea. Conversely, adsorption methods capable of reducing caffeine and foreign substances do not succeed in reducing bitterness/astringency.

With this backdrop, Abe *et al.* (2009) came up with a multi-step process to increase the concentration of the monomeric catechins in the tea extract that can improve the taste by reducing the bitterness/astringency and miscellaneous tastes, and also reduce the protein content. They reported that MC could be extracted efficiently from the tannase-treated tea extract by a mixed solution containing an organic solvent and water at a predetermined ratio. Among the polar solvents, ethanol was preferred from the viewpoint of food applications. Further, it was suggested that a gallate content of 5–48% in the MC could keep the bitterness/astringency at a favorable level. The invention also disclosed that a caffeine to MC ratio of 0–0.12 and protein plus dietary fiber to MC ratio of 0.1 or less could reduce the unwanted miscellaneous taste.

The multi-step process involves treating the green tea extract with an enzyme having a tannase activity, followed by extracting with aqueous ethanol and treating the extract with activated carbon. The concentration of MC when the green tea extract is treated with tannase is preferably from 0.5 to 16.5% of tea solids. The enzyme addition is preferably maintained between 5 and 150 U/g of the MC. From the viewpoint of suppressing bitterness/astringency, the gallate percentage in the MC in the green tea extract is preferably controlled to fall within 5–48% by controlling the terminal pH of the enzyme reaction preferably from 3.3 to 5.3. The hydrolysis reaction with the enzyme is terminated by thermal inactivation. The tannase-treated extract is concentrated or dried before ethanolic extraction. The precipitate formed with proteins and other impurities in the ethanolic extraction process along with activated clay is removed by known methods such as filtration. Then the extract is treated with activated carbon and subsequently ethanol content is reduced by distillation. When all the steps of the proposed process were followed, a very clear extract with a good color tone and superior overall score was obtained, reflecting the effectiveness of the process. The purified green tea extract thus obtained could be used in packaged beverages.

MC composition from green tea has found intense and wide applications in the preparation of health drinks. The Kao Corporation took a lead in this direction and filed several process patents based on tannase treatment with some process variations for producing green tea extract containing higher concentrations of MC (Fukuda *et al.*, 2006, 2007, 2009). All these methods proposed essentially involve obtaining a catechin composition with reduced gallates and caffeine contents from green tea extract.

Enzymatic Decreaming/Clarification of Green Tea Infusion

One of the most relevant problems encountered in the industrial production of natural, genuine and additive-free RTD cold tea is its instability due to development of tea haze and formation of tea cream. It gives discoloration and precipitation of complexed substances, affecting the visual appeal, flavor ,and color, besides reducing the shelf-stability of the beverage (Chandini *et al.*, 2011b). Polyphenols, caffeine, protein, carbohydrates, polysaccharides, free amino acids, flavones, catechins, and minerals were found to participate in the formation of green tea cream. However, only caffeine and polyphenol contents in the original green tea infusion were highly correlated to the amount of green tea cream. Based on the regression analysis, a relationship was proposed for the prediction of the amount of green tea cream from the contents of gallated catechins and caffeine present in the original tea infusion (Xu *et al.*, 2011).

Polyphenol complexation was suggested to be a major causative element in tea cream formation (Rutter and Stainsby, 1975). Caffeine–polyphenol and protein–polyphenol complexations were proposed to be predominant interactions in Paochung tea cream (Chao and Chiang, 1999a). Since gallated catechins have more active sites for hydrogen bonding, they have relatively greater tendency towards complex formation with protein as well as caffeine than the non-ester catechins (Chao and Chiang, 1999b). Caffeine does not play an essential part in the tea cream formation (Jobstl *et al.*, 2005) but stimulates cream formation in the presence of sufficient substances containing galloyl groups (Liang and Xu, 2003).

Tannase has been recommended for hydrolysis of tea cream to lower-molecular-weight compounds, reducing turbidity and increasing cold-water solubility (Sanderson and Coggon, 1974). This treatment of tea extract converts at least a portion of the insoluble solids of tea cream to a cold water-soluble form and is referred to as 'enzymatic solubilization of cream' and also as 'enzymatic clarification of extract.' Lu *et al.* (2009) carried out a detailed study on the effect of tannase treatment in the protein–polyphenol (especially the gallated epicatechins) aggregation. Tannase hydrolyzes both EGCG and ECG to EGC and EC, respectively, along with release of GA. Hydrolysis is expected to decrease the binding ability of gallated catechins with protein due to the cleavage of ester bond and thus avoid the aggregation of macromolecules and the formation of precipitates during storage of tea beverage. To elucidate the preferential tendency of the gallated catechins over ungallated catechins to interact with protein, bovine seum albumin (BSA) was used as a protein substrate in the experiments. The green tea with or without tannase treatment was mixed with dyed BSA to cause reaction and the precipitate obtained was quantified using spectrophotometric measurement. The binding rate increased with protein concentration indicating the potential of gallated catechins present in green tea (Figure 11.5). The binding ability of tannase-treated green tea was relatively much lower compared to the untreated one. This could be due to the near-complete hydrolysis of gallated catechins in the tannase-treated green tea that would otherwise offer more hydroxyl groups to form hydrogen bonds and finally precipitate as a complex. The cold-stored (4°C) tannase-treated green tea after 4 weeks showed less cream formation (0.2 g/100 g) in comparison with the untreated one (0.9 g/100 g) validating the above theory. Further, the content of both EGCG and ECG in green tea infusion (control) decreased significantly during the storage period while the content of EGC and EC for either the control or treated sample remained almost unchanged (Table 11.2). It could be inferred that tannase treatment reduces the concentration of the gallated catechins in the green tea, in turn reducing the protein–tannin interaction, leading to reduced cream formation. Besides, this process offers the additional benefit of preventing the loss of catechins in the form of tea cream as well as improving the antioxidant capacity of the beverage through bioconversion of gallated catechins to their simpler forms accompanied with release of GA.

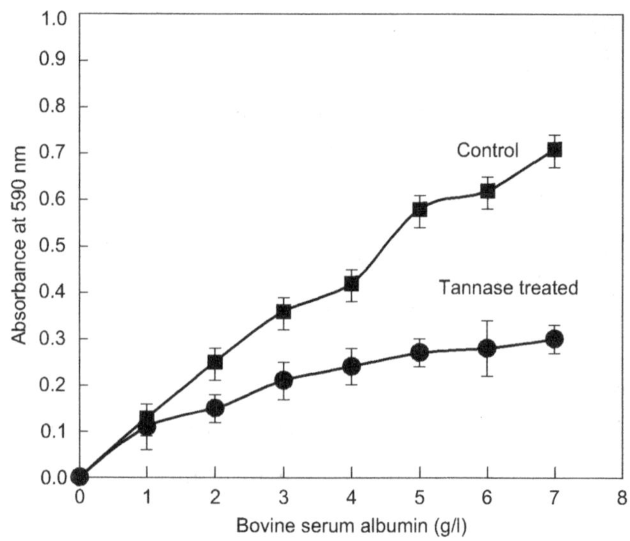

FIGURE 11.5 Effect of enzymatic treatment of green tea infusion on precipitation of model protein. Gallated catechins have a greater tendency than ungallated catechins to interact with protein. The untreated green tea extract contains more amounts of gallated catechins and increasingly bind with dyed bovine serum albumin as reflected in the increase in its absorbance compared to the tannase-hydrolyzed sample. *Source: Lu* et al. *(2009).*

TABLE 11.2 Effect of Enzymatic Hydrolysis on Individual Catechins in Green Tea Infusion During Storage

Green Tea	Storage Period (Weeks)	Tea Catechins (mg/g)			
		EGCG	EGC	ECG	EC
Control	0	25 ± 0.3^a	16 ± 0.5^a	3.9 ± 0.2^a	0.8 ± 0.1^a
	4	5.0 ± 0.3^b	15 ± 1.5^a	0.5 ± 0.1^b	0.7 ± 0.1^a
Tannase treated*	0	Trace	40 ± 2.1^b	Trace	4.8 ± 0.6^b
	4	Trace	39 ± 2.2^b	Trace	4.3 ± 0.7^b

Presence of galloyl groups in catechins facilitates the interaction between catechins and proteins leading to tea cream formation in tea infusions. Tannase hydrolyzes both EGCG and ECG preventing tea cream formation. The substantial reduction of EGCG and ECG in the cold-stored control sample indicates their greater participation in tea cream formation. EC, epi catechin; ECG, epi-catechin gallate; EGC, epi-gallo catechin; EGCG, epi-gallo catechin gallate.

a, bThe means (n = 3) with the same superscript in the same row are not significantly different at 5% levels with Duncan's multiple range tests. Data are expressed as mean ± standard deviation.

**Tannase at a concentration of 2.0 g/l was added to the green tea infusion (1 g/100 ml) and hydrolysis reactions were performed at 35°C for 20 min.*

Source: Lu et al. *(2009).*

Acknowledgment

We are grateful to L. Jaganmohan Rao for providing valuable suggestions.

References

Abe, T., Ueoka, H., Shibata, K., 2009. Process for producing purified green tea extract. US patent 0104337–A1.

Ananingsih, V.K., Sharma, A., Zhou, W., 2013. Green tea catechins during food processing and storage: A review on stability and detection. Food Res. Int. 50, 469–479.

Arjuna, T., 2000. Green Tea—Monograph. Altern. Med. Rev. 5, 372–375.

Bajpai, B., Patil, S., 1997. Tannin acyl hydrolase (EC 3.1.1.20) activity in some members of *Fungi imperfectii*. Enzyme. Microb. Technol. 20, 612–614.

Battestin, V., Macedo, G.A., Freitas De, V.A.P., 2008. Hydrolysis of epigallocatechin gallate using a tannase from *Paecilomyces variotii*. Food Chem. 108, 228–233.

Belmares, R., Contreras-Esquivel, J.C., Rodriguez-Herrera, R., Coronel, A.R., Aguilar, C.N., 2004. Microbial production of tannase: An enzyme with potential use in food industry. Food Sci. Technol. 37, 857–864.

Blogg, M.J., Long, V.D., 1980. Aqueous extraction of black leaf tea IV. Pilot-scale batch simulated continuous counter-current extraction. Int. J. Food Sci. Technol. 15, 369–382.

Chandini, S.K., Subramanian, R., Rao, L.J., 2013. Application of Enzymes in the Production of RTD Black Tea Beverages: A Review. Crit. Rev. Food Sci. Nutr. 53 (2), 180–197.

Chandini, S.K., Rao, L.J., Gowthaman, M.K., Haware, D.J., Subramanian, R., 2011a. Enzymatic treatment to improve the quality of black tea extracts. Food Chem. 127, 1039–1045.

Chandini, S.K., Rao, L.J., Subramanian, R., 2011b. Influence of extraction conditions on polyphenols content and cream constituents in black tea extracts. Int. J. Food Sci. Technol. 46, 879–886.

Chao, Y.C., Chiang, B.H., 1999a. Cream formation in a semifermented tea. J. Sci. Food Agric. 79, 1767–1774.

Chao, Y.C., Chiang, B.H., 1999b. The roles of catechins and caffeine in cream formation in a semi-fermented tea. J. Sci. Food Agric. 79 (12), 1687–1690.

Chen, F., Kawaguchi, R., Kino, H., Kato, S., Nagano, K., Murai, K., Fujita, R., 2012. Process for producing tea extract. WO patent 2012046349.

Chen, Z.Y., Chan, P.T., 1996. Antioxidative activity of green tea catechins in canola oil. Chem. Phys. Lipids. 82, 163–172.

FAO, 2012. Current situation and medium term outlook for tea. Committee on commodity problems. Sri Lanka, Colombo 30 Jan–1 Feb, 2012.

Fukuda, M., Itaya, E., Kusaka, R., Takahashi, H., 2009. Preparing non-polymer catechin containing refined green tea extract used in container-packed drink e.g. soft drink, by contacting tea extract processed using tannase and anion exchange resin, with activated carbon and acid clay. JP patent 225728 A.

Fukuda, M., Takahashi, H., Konishi, A., 2006. Manufacture of non-polymeric catechin composition useful as health drink, involves processing tannase processed water-soluble composition extracted from green tea leaf in synthetic adsorbent column, and eluting ethanol solution. JP patent 206482-A.

Fukuda, M., Takahashi, H., Konishi, A., 2007. Purified green tea extract for use drink contains mixture of aqueous ethanol solution containing non-polymer catechin and green tea extract, contacted with activated carbon/clay and/or acid clay, and processed with tannase. JP patent 338057.

Guo, Q., Zhao, B., Shen, S., Hou, J., Hu, J., Xin, W., 1999. ESR study on the structure antioxidant activity relationship of tea catechins and their epimers. Biochimica et Biophysica. Acta. 1427, 13–23.

Hong, Y.H., Jung, E.Y., Park, Y., Shin, K.S., Kim, T.Y., Yu, K.W., Chang, U.J., Suh, H.J., 2013. Enzymatic Improvement in the Polyphenol Extractability and Antioxidant Activity of Green Tea Extracts. Biosci. Biotechnol. Biochem. 77, 22–29.

Jobstl, E., Fairclough, J.P.A., Davies, A.P., Williamson, M.P., 2005. Creaming in Black Tea. J. Agric. Food Chem. 53, 7997–8002.

Langley-Evans, S.C., 2000. Antioxidant potential of green and black tea determined using the ferric reducing power (FRAP) assay. Int. J. Food Sci. Nutr. 51, 181–188.

Lekha, P.K., Lonsane, B.K., 1997. Production and application of tannin acyl hydrolase: state of the art. Adv. Appl. Microbiol. 44, 215–260.

Liang, Y., Xu, Y., 2003. Effect of extraction temperature on cream and extractability of black tea [*Camellia sinensis* (L.) O. Kuntze]. Int. J. Food Sci. Technol. 38, 37–45.

Long, V.D., 1977. Aqueous extraction of black tea-leaf solubility. J. Food Technol. 12, 459–472.

Lu, M.J., Chen, C., 2008. Enzymatic modification by tannase increases the antioxidant activity of green tea. Food Res. Int. 41, 130–137.

Lu, M.J., Chu, S.C., Yan, L., Chen, C., 2009. Effect of tannase treatment on protein–tannin aggregation and sensory attributes of green tea infusion. Food Sci. Technol. 42, 338–342.

Masakichi, A., Masao, T., Akira, S., Shigeru, F., Seij, H., 1984. Preparation of instant tea. JP patent 59–34849 A.

Masami, S., Hitoshi, T., Ogura, Y., 2004. Green tea extract useful as drink, is obtained by tannase processing of green tea extract containing non-polymer catechin, dissolving in aqueous solution, adjusting to specific pH and deactivating tannase. JP patent 2004321105A.

Rutter, P., Stainsby, G., 1975. The solubility of tea cream. J. Sci. Food Agric. 26, 455–463.

Sakamaki, S., Kawabata, Y., Kawaguchi, R., Komai, T., 2006. Antioxidant composition useful in foodstuffs contains tea extract containing epigallocatechin and gallic acid in preset ratio. JP patent 2006083352–A.

Sanderson, G.W., Coggon, P., 1974. Green tea conversion using tannase and natural tea enzymes. US patent 3812266.

Sanderson, G.W., Hoefler, A.C., Graham, N.H., Coggon, P., 1977. Cold water extractable tea leaf and process. US Patent 4051264.

Tsai, C.H., 1987. Enzymatic treatment of black tea. US Patent 4639375.

Xu, Y.Q., Chen, S.Q., Shen, D.Y., Yin, J.F., 2011. Effects of Chemical Components on the Amount of Green Tea Cream. Agric. Sci. China. 10 (6), 969–974.

Yotsumoto, Y., 2009. Method of enzymatically treating green tea leaves. US Patent 280215 A1.

Yue, P., Ouyang, X., Zhang, Y., 2008. Method for processing cold soluble type instant tea. CN patent 101189998-A.

CHAPTER

12

Antioxidant Capacity of Tea: Effect of Processing and Storage

Agnieszka Kosińska[*], [†], *Wilfried Andlauer*[†]

[*]Division of Food Sciences, Institute of Animal Reproduction and Food Research of the Polish Academy of Sciences, Olsztyn, Poland, [†]Institute of Life Technologies, University of Applied Sciences Valais, Sion, Switzerland

CHAPTER POINTS

- Processing of tea leaves results in profound differences in phenolic compound composition and antioxidant capacity of different types of tea.
- The more processed the tea, the lower its antioxidant capacity.
- Antioxidant activity of ready-to-drink tea beverages is much lower in comparison to tea-cup-prepared tea.
- Extended storage, being necessary to develop desirable taste and aroma of pu-erh and old oolong tea, can lead to loss of quality in white, yellow, green, and black tea.
- The lower the oxygen permeability of packaging material of ready-to-drink tea, the longer its storage stability.

INTRODUCTION

Tea plant is an evergreen shrub or small tree of the *Camelia* genus, native to China, with dark green shiny leaves and white flowers. *Camelia sinensis* var. *sinensis* and *Camelia sinensis* var. *assamica* are the two varieties most commonly used for the production of different types of tea. Freshly prepared infusion of dried leaves is a beverage consumed on a daily basis worldwide. Tea beverage is prepared at home by pouring boiling water over processed tea leaves. Tea is valued for its taste, aroma, health benefits, and cultural practices.

Fresh tea leaves are plucked manually or using a plucking machine. Most preferably, just two fresh leaves and the buds should be used, since it influences the quality and therefore the price of the tea. The younger the leaves used for the production are, the more expensive is the final product. Processing of tea leaves allows the production of various types of tea: white, yellow, oolong, green, black, and pu-erh, depending on the extent of the oxidation/fermentation process. Generally, leaves of the *sinensis* variety are used mainly for the production of green tea, while black and pu-erh tea are produced from *assamica* (Zhang *et al.*, 2012). It is estimated that about 78% of world production of tea accounts for black tea, 20% for green tea, and only 2% for other tea types. Black tea is consumed all over the world, whereas green and oolong tea, previously peculiar to Asia are also now gaining popularity in Europe and North America. White and pu-erh tea is mainly consumed in Asia. The increased popularity of green tea in Western countries results from a number of studies linking tea consumption with beneficial health effects. These beneficial effects are associated mainly with antioxidant properties of tea polyphenols. Moreover, antimutagenic, anticarcinogenic, hypocholesterolomic, antibacterial, and antiallergenic effects of tea have been reported (Cooper, 2011).

The composition of bioactive compounds, and therefore antioxidant capacity of tea, might be influenced by several parameters associated with growth conditions: genetic strain, climatic conditions, soil profile, growth altitude, horticultural practices, or plucking season. Interesting results on the effects of shade growth on tea phytochemical composition and antioxidant capacity were presented by Ku *et al.* (2010). Tencha, shade grown tea, showed generally lower antioxidant capacity (and epicatechin, epigallocatechin, and galloylquinic acid content) but higher content of amino acids than green tea.

Processing and Impact on Antioxidants in Beverages
http://dx.doi.org/10.1016/B978-0-12-404738-9.00012-X

The other factors that may profoundly influence profile of bioactives in tea are processing, storage, and home preparation practices. They are considered in this review.

ANTIOXIDANT COMPOUNDS OF TEA

Tea is well known for its antioxidant effect. Several compounds present in tea are responsible for this activity.

Flavan-3-ols

Green tea is one of the most abundant food sources of catechins, members of flavan-3-ols, a subclass of flavonoids. Flavan-3-ols have two chiral centers C2 and C3, which results in four isomers for each level of B ring hydroxylation. (+)-Catechins and (−)-epicatechin are widespread in nature whereas (−)-catechins and (+)-epicatechin are rare. Catechins are characterized by di-or tri-hydroxyl group substitution of the B ring and meta-5,7-dihydoxysubstitution of the A ring. The most abundant catechins in tea are (−)-epigallocatechin gallate (EGCG), (−)-epicatechin-3-gallate (ECG), (−)-epigallocatechin (EGC), (−)-epicatechin (EC), (+)-catechin (C), and (−)-gallocatechin gallate (GCG). Tea catechins can undergo reversible epimerization, mainly as an effect of high-temperature treatment. It was reported that radical-scavenging ability of EGCG was higher than that of other catechins due to the presence of a gallate moiety in the C ring (Xu *et al.*, 2004). The following order of antioxidant activity of tea catechins emerged from a number of studies: epigallocatechin ~ epigallocatechin gallate >> epicatechin gallate = epicatechin > catechins (Łuczaj and Skrzydlewska, 2005). The number of hydroxyl groups is the major factor contributing to antioxidant potential (Rice-Evans *et al.*, 1997).

Phenolic Acids

Gallic acid and its quinic acid ester, theogallin (Figure 12.1), are the most abundant simple polyphenols present in tea. During processing of black tea, the amount of gallic acid significantly increases due to oxidative degallation of phenolic esters during the fermentation. Recently, 15 different phenolic acid derivatives have been identified in black tea by Lin *et al.* (2008).

Flavonols and Flavones

Quercetin, myricetin, kaempferol and their mono-, di-, and tri-glycosides are the most abundant flavonols in tea. Recently, three flavonols, 19 *O*-glycosylated flavonols, 28 acylated glycosylated flavonols, and seven C-glycosylated flavonols have been identified in white, green, oolong, and black tea samples (Lin *et al.*, 2008).

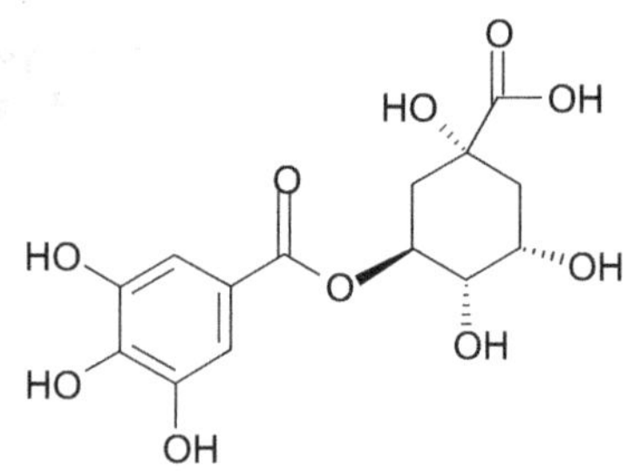

FIGURE 12.1 Chemical structure of theogallin. Theogallin and gallic acid are the most common simple phenolics in tea.

Theaflavins and Thearubigins

Theaflavins and thearubigins are characteristic products formed from catechins during enzymatic oxidation of tea during manufacturing. Theaflavins give an orange or orange–red color to tea and contribute to a mouthfeel sensation and an extent of cream formation. They are dimeric compounds that possess a benzotropolone skeleton that is formed from co-oxidation of selected pairs of catechins. The oxidation of the B ring of either (−)-epigallocatechin or (−)-epigallocatechin gallate is followed by loss of CO_2 and simultaneous fusion with the B ring of (−)-epicatechin or (−)-epicatechin gallate molecule (Figure 12.2). Four major theaflavins have been identified in black tea: theaflavin, theaflavin-3-monogallate, theaflavin-3′-monogallate, and theaflavin-3,3′-digallate. Additionally, their stereoisomers and derivatives can be present. Recently, the presence of theaflavin trigallate and tetragallate in black tea was reported (Chen *et al.*, 2012). The theaflavins can be further oxidized. They are probably also the precursors for the formation of polymeric thearubigins. However, the mechanism of reaction is not known hitherto. Thearubigins are red–brown or dark-brown pigments in black tea, their content accounts for up to 60% of dry weight of tea infusion. The structure of thearubigins is yet to be revealed. Presumably their molecular weight does not exceed 2100 Da (Haslam, 2003).

The antioxidant activity of theaflavins is comparable to that of catechins (Łuczaj and Skrzydlewska, 2005). Moreover, theaflavins were shown to be able to scavenge superoxide radicals more efficiently and 10-times faster than EGCG.

Theasinensins

Theasinensins are dimeric gallocatechins linked by C–C bonds between two B rings forming a biphenyl (Figure 12.3). They are present mainly in oolong tea. Theasinensin A is a product of oxidation of two EGCG molecules, B of EGCG and EGC, C of two EGC molecules, and F of EGCG and ECG. Additionally, theasinensins might be formed from gallocatechins in the gut during

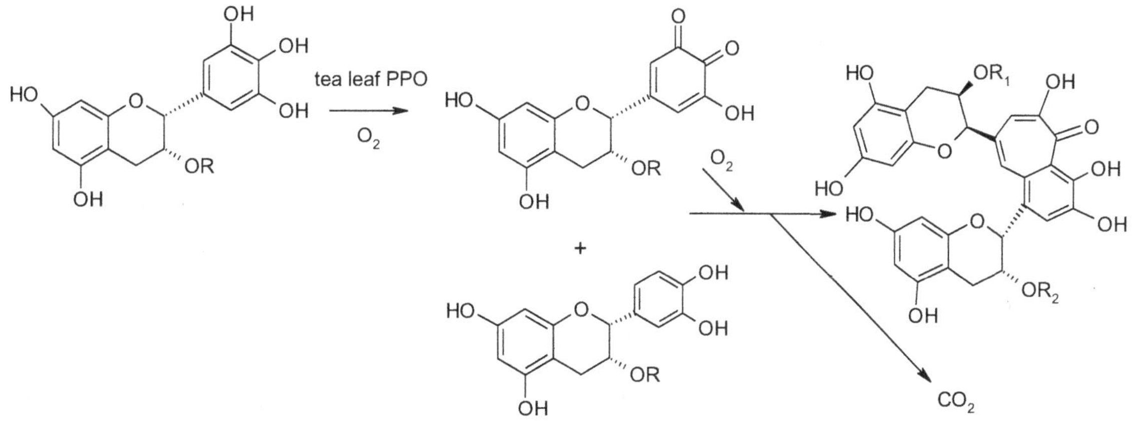

FIGURE 12.2 Mechanism of theaflavin formation. Theaflavins are formed as a result of co-oxidation of selected pairs of catechins catalyzed by polyphenol oxidase (PPO). R = H – EGC, R = galloyl group – EGCG, $R_1 = R_2 = H$ – theaflavin; R_1 = galloyl group, $R_2 = H$ and $R_1 = H$, R_2 = galloyl group – theaflavin gallate; $R_1 = R_2$ = galloyl group – theaflavin digallate.

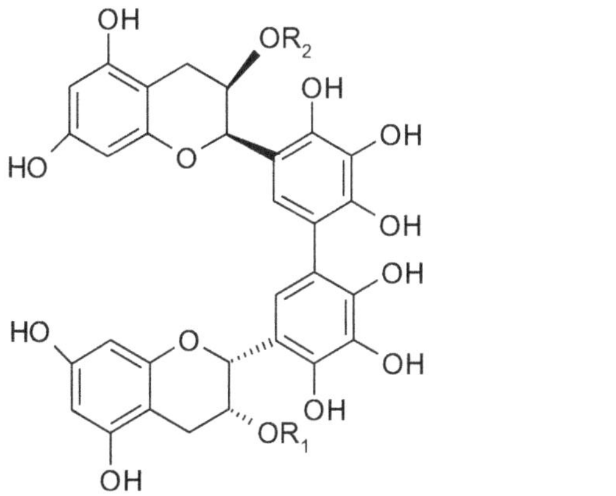

FIGURE 12.3 Chemical structure of theanisenisins: theanisenisin A: $R_1 = R_2$ = galloyl group; theanisenisin B: R_1 = galloyl group, R_2 = H and R_1 = H, R_2 = galloyl group; theanisenisin C: $R_1 = R_2 = H$.

digestion of tea (Neilson *et al.*, 2010). It was reported that radical-scavenging activity of theasinensin A and D is comparable with that of EGCG on a molar basis (Yoshino *et al.*, 1999). In the case of thearubigins, their contribution to the antioxidant capacity of black tea is assumed.

Theabrownins

Theabrownins are pigment compounds formed during oxidation of tea leaves. They dissolve easily in water and not in organic solvents like ethyl acetate. They are the main bioactive components in pu-erh tea, and are characterized by high molecular weight and complex structure. It is known that theabrownins are formed by oxidation of polyphenols. However, the mechanism of their formation is not clear. It was reported that the cellulases and pectinases produced by microorganisms, together with polyphenol oxidase, play an important role in theabrownin formation (Wang *et al.*, 2011).

EFFECT OF PRODUCTION AND PROCESSING

Different tea types are obtained as result of differentiated processing, which substantially influence tea antioxidant composition.

Phenolic Compounds in Fresh Tea Leaves

Fresh tea leaves and unfermented tea types contain mainly flavan-3-ols, *O*-glycosylated flavonols, *C*-glycosylflavones, proanthocyanidins, phenolic acids, and their derivatives, whereas fermented tea contains also theaflavins and thearubigins. Recently, 24 phenolic compounds have been identified in tea leaves of different varieties by Wu *et al.* (2012). Leaves of eight tea varieties that are used for production of green, oolong, and black teas, were harvested, steamed to preserve oxidation, dried, and subsequently analyzed to evaluate the content of individual phenolic compounds. A unique phenolic compounds profile of each variety was reported. The contents of total catechins in the tea varieties for green tea manufacturing were all lower than the other two groups. EGCG was the major catechin in all tea varieties, accounting for about half of the total catechins. In the cited study, varieties for oolong tea manufacturing had the lowest content of total flavone and flavonol glycosides, whereas the green tea varieties had the highest content of these compounds.

Manufacturing Process

On the basis of the production process, tea can be divided into non-fermented (white, yellow, and green tea), semi-fermented or partly fermented (oolong tea), fermented (black tea), and post-fermented (pu-erh).

There are several steps in the production process that, selectively combined, determine which kind of tea will be obtained. The flow chart of production processes for different types of tea is depicted in Figures 12.4 and 12.5.

Withering

After the plucking, tea leaves undergo wilting or withering. Usually it takes place in the sun or on a rack in a heated room. Withering aims to remove moisture and soften the leaves to prepare them for further rolling. It leads also to the development of aroma and partial oxidation due to a breakdown of cell walls caused by moisture loss. Duration of withering depends on type of tea: white tea leaves are withered for 4–5 h, whereas green, oolong and black tea are withered for at least twice as long (Cooper, 2011).

Steaming/Pan-Firing

Deactivation of enzymes, which initiates leaves oxidation, is a crucial process in the production of green tea. It can be achieved by quick application of high temperature either with steam, typical for Japanese-style green tea, or by dry heating called pan-firing, which is the traditional Chinese method. The temperature of pan-firing can reach 230°C, which is much higher than steaming temperature of 100°C (Wang *et al.*, 2000). Therefore, the method chosen for this step crucially affects the characteristics of the tea product obtained.

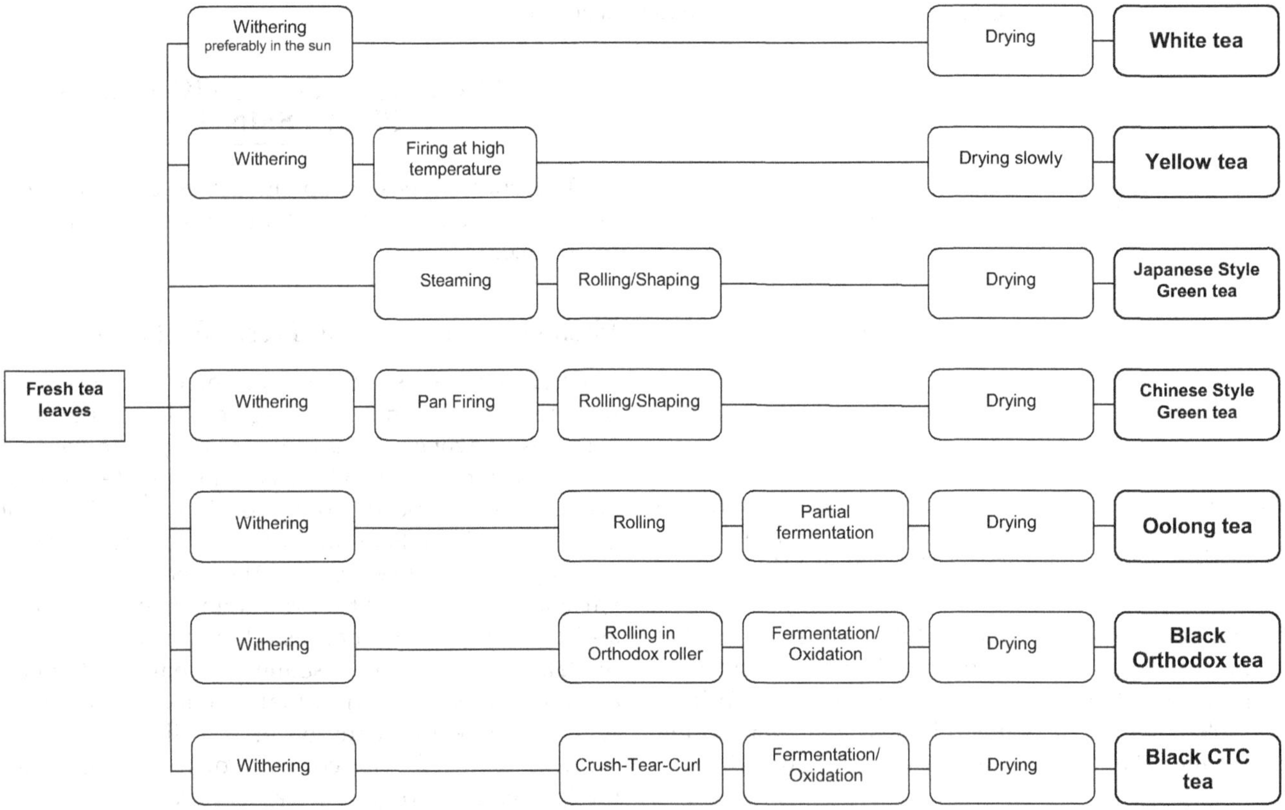

FIGURE 12.4 Flow chart of processing of different types of tea: white, yellow, green, oolong, black CTC, crush-tear-curl.

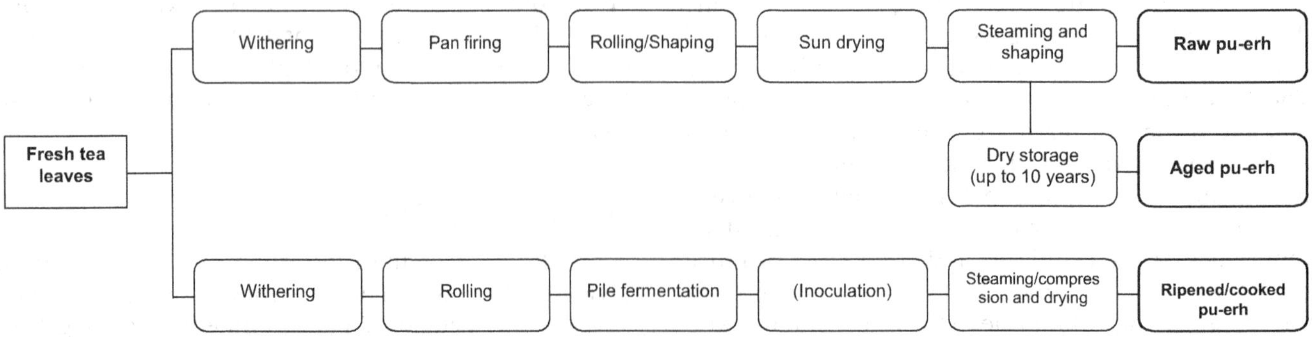

FIGURE 12.5 Flow chart of processing of different types of pu-erh tea.

Rolling

The rolling/twisting of tea leaves was originally done manually, whereas nowadays it is performed by machines. The aim of this process is to disrupt tea leaves in order to release tea oils and to enable the access of enzymes to polyphenols, which is essential for the oxidation process. There are two major types of processing: orthodox and cut-tear-curl (CTC). Orthodox rolling refers to hand processing or rolling with machines that imitate hand rolling. CTC machines enable more rapid and extensive leaf disruption, producing smaller particles and therefore greater surface areas for enzymatic oxidation. Tea obtained by this method is used mainly for commercial tea in the form of teabags.

Oxidation/Fermentation

Rolled leaves are placed on trays and left in a room with controlled temperature, humidity, and aeration for 2–4 h. The process is traditionally called fermentation; however, it is mainly enzymatic oxidation of polyphenols by polyphenol oxidases. The initial green color of leaves turns into light brown, and deep brown in the course of their oxidation, which indicates formation of oligomeric theaflavins and polymeric thearubigins.

Drying

The last step in tea manufacturing is drying of tea leaves to stop the fermentation and to reach the favorable moisture content to suppress microorganism growth. Thereafter, tea is sorted and packed.

Changes in Phenolic Composition Caused by Processing of Tea Leaves

The manufacturing process evokes profound changes in the phenolic compound profile, in the content of individual compounds as well as in the antioxidant capacity of final tea products. Therefore, depending on the type of tea, the final products contain highly distinct profiles of antioxidant compounds. The identification of 92 phenolic compounds in a total of 66 non-fermented and fermented teas collected from all around the world allowed the classification of all types of tea into five groups (Lin *et al.*, 2008). The first group, characterized by a high concentration of acylated flavonol glycosides, consisted of high-grade teas made of the younger buds and leaves harvested in the early leaf-growing stage. High content of catechins and glycosylated flavonols was typical for the group of common green teas. The profile of the teas belonging to third group (partially fermented oolong and pu-erh teas) was characterized by lower content of EGCG and presence of theaflavins, whereas in the teas of the fourth group (black) most of EGCG was oxidized to theaflavins. Only traces of EGCG were present in overfermented teas, i.e., pu-erh (fifth group).

A number of studies carried out have aimed to characterize of different types of commercially available teas in specific regions or around the world. An overview of these studies is presented in Table 12.1.

White Tea

White tea is made mainly from newly grown buds and young leaves with tiny, silvery hairs not exposed to sunlight to prevent chlorophyll production. Buds and leaves for white tea production are harvested only once a year in the early spring. Buds are plucked before they are open, then withered and air dried in the shade, under sunshine, or in a temperature-controlled room. White tea is the least processed type of tea. It is considered as a non-fermented type, however, a slight fermentation occurs since the processing lacks the step of enzyme deactivation. In line with this statement, the results presented by Zhang *et al.* (2011) revealed in white tea the presence of compounds characteristic for black tea mainly, and not found in any other tea type analyzed by the authors, i.e., aged pu-erh, ripened pu-erh, green, yellow, oolong (Table 12.1A). The compounds were identified as theaflavin-3-gallate, theaflavin-3′-gallate, and theaflavin-3,3′-digallate. Also the content of gallic acid in white tea is high and comparable to that of black and pu-erh teas. Another study showed that the total content of catechins in white tea was similar to that of green tea and higher than in green pu-erh tea (Zhao *et al.*, 2011). In all types of tea analyzed, EGCG was the predominant catechin, and its content was the highest in white tea, followed by green and green pu-erh. White tea was also the most abundant source of proanthocyanidins, phenolic acid derivatives and acylated glycosylated flavonols (Table 12.1B). Green tea showed significantly higher radical-scavenging activity compared to white tea (Unachukwu *et al.*, 2010). Total catechin content in both white tea and green tea was highly variable for different tea brands analyzed (Table 12.1C).

Yellow Tea

Yellow tea is the least produced and known type of tea. The production process involves mild pan-firing until 40% of dryness, and subsequently to 70% dryness. Still-hot tea leaves are packed in thick paper or cloth and stored in wooden or bamboo boxes for 7 days. The other possible way of obtaining yellow tea is to remove pan-fired tea leaves, cover them with cloth and allow resting for a few hours. The processing is repeated a few times until leaves are smothered and have gained appropriate taste and color. The final product is slightly oxidized and characterized by a milder taste. As can be seen in Table 12.1B, yellow tea has a phenolic composition like green tea,

TABLE 12.1 An Overview of the Studies Carried Out on the Phenolic Compounds Composition of Different Types of Tea

Compound	Green	Green/Aged Pu-erh	White	Yellow	Oolong	Black	Pu-erh
(A) Study characteristics 71 tea samples: 14 ripened pu-erh, 14 aged pu-erh, 17 green tea, 11 oolong tea, 7 black tea, 6 white tea, 2 yellow tea, water infusions were prepared; results were expressed as mg/g; average calculated for tea type (Zhang *et al.*, 2011)							
Gallic acid	1.21 ± 0.98	1.47 ± 1.27	4.62 ± 1.10	2.63	0.79 ± 0.69	3.91 ± 0.60	6.51 ± 2.98
Total catechins	66.38 ± 35.87	38.69 ± 16.51	70.73 ± 22.06	72.29	87.82 ± 35.13	7.79 ± 1.78	4.28 ± 1.01
(EGCG)	33.80 ± 19.20	9.43 ± 4.27	35.79 ± 12.38	35.58	30.81 ± 10.91	1.43 ± 0.43	0.08 ± 0.07
(B) Study characteristics 15 tea samples: 5 white, 5 green, and 5 green pu-erh, tea powder was extracted with 60% aqueous methanol, results expressed as mg/g; average calculated for tea type (Zhao *et al.*, 2011)							
Gallic acid	27.98 ± 9.08	46.56 ± 14.54	74.73 ± 31.81	—	—	—	—
Total catechins	213.74 ± 52.97	131.03 ± 58.91	200.94 ± 15.61	—	—	—	—
(EGCG)	144.22 ± 29.12	88.34 ± 37.54	169.77 ± 4.76	—	—	—	—
Total glycosylated flavonols	15.26 ± 3.98	11.75 ± 3.03	15.81 ± 3.67	—	—	—	—
Total acylated glycosylated flavonols	0.26 ± 0.11	0.36 ± 0.14	0.84 ± 0.71	—	—	—	—
Total proanthocyanidins	1.81 ± 0.95	0.87 ± 0.32	4.67 ± 2.72	—	—	—	—
(C) Study characteristics 26 tea samples: 8 types of white tea of two subtypes white peony and Yin Zhen silver needle; 18 samples of green tea representing 5 subtypes (dragonwell, gunpowder, jasmine pearl, sencha, and gykuro); results were expressed as mg/g; average calculated for tea type for the purpose of this review (Unachukwu *et al.*, 2010)							
Total catechins	109.71 ± 64.83	—	84.25 ± 117.89	—	—	—	—
(D) Study characteristics 8 tea samples: 4 green tea (Meifoo green, Shanghai green, Hangzhou Lung Ching and Jasmine tea), 2 oolong tea (Fujian and Jiangxi), 1 pu-erh, and 1 black tea; tea powder was extracted with 80% methanol; results were expressed as mg/g; average calculated for tea type for the purpose of this review (Zuo *et al.*, 2002)							
Gallic acid	1.02 ± 0.63	—	—	—	1.54 ± 0.18	2.06	5.53
Total catechins	110.29 ± 10.25	—	—	—	47.33 ± 8.73	15.31	12.78
EGCG	55.10 ± 5.02	—	—	—	25.20 ± 4.24	3.79	1.99
(E) Study characteristics Samples of Indonesian green and black tea. Water infusions were prepared by adding 18 ml of boiling water to 1 g of leaves and brewing for 3 min; the results were expressed as mg/l (Del Rio *et al.*, 2004)							
Gallic acid	6.0 ± 0.1	—	—	—	—	125 ± 7.5	—
Total catechins	4572	—	—	—	—	101	—

EGCG	1255 ± 63	—	—	—	—	19 ± 0.1	—
Total flavonols	778	—	—	—	—	570	—
Total theaflavins	nd	—	—	—	—	224	—
(F) Study characteristics 11 samples of green, oolong and black tea purchased in Hong Kong. The results were expressed as mg/g of dry leaves; average calculated for tea type for the purpose of this review (Chen *et al.*, 2001)							
Total catechins	111.2 ± 25.1	—	—	—	45.6 ± 3.9	3.8 ± 1.3	—
EGCG	68.74 ± 21.87	—	—	—	31.8 ± 3.5	2.0 ± 1.0	—
(G) Study characteristics 50 samples of black and green tea purchased on the German market, the results were expressed as mg/g; average calculated for tea type (Hilal and Engelhardt, 2007)							
Total catechins	15.1	—	—	—	—	14.2	—
Total theaflavins	nd	—	—	—	—	0.94	—
(H) Study characteristics 7 samples of black tea purchased on local market: 3 Chinese black tea (Keemun, Yingde and Yunnan black tea), 3 India black tea (Dajeeling, Assam and Nierjeeli) and 1 Sri Lanka black tea (Uva); were expressed as mg/g of dry leaves; average calculated for tea type for the purpose of this review (Wang *et al.*, 2010)							
Total catechins	—	—	—	—	—	19.24 ± 10.86	—
Total theaflavins	—	—	—	—	—	7.62 ± 3.77	—
(I) Study characteristics 6 samples of pu-erh tea purchased in the Yunnan region of China: 3 green pu-erh and 3 black pu-erh, results were expressed as % (w/w) (Wang *et al.*, 2009)							
Total polyphenols	—	21.97 ± 0.48	—	—	—	—	10.31 ± 016
Total theaflavins	—	0.07 ± 0.02	—	—	—	—	0.17 ± 0.01
Total thearubigins	—	4.31 ± 1.24	—	—	—	—	5.88 ± 0.88
Total theabrownins	—	2.85 ± 0.11	—	—	—	—	9.73 ± 0.11
Gallic acid	—	nd	—	—	—	—	0.62
EGCG	—	7.69	—	—	—	—	nd

EGCG, (−)-epigallocatechin gallate; nd, not detected.

without grassy taste of green tea (Zhang *et al.*, 2011). A recent study comparing different types of tea for their hydrogen-peroxide-scavenging activity showed the following order: yellow > green > black > white > oolong tea (Gorjanovic *et al.*, 2012). A similar order of activity was also reported for total reducing power and phenolic compounds content with the exception of oolong tea being more active than white. The ability to scavenge DPPH radicals decreased in the following order: green > black > yellow > oolong > white tea.

Green Tea

Amongst all types of tea, green tea is the richest in catechins (Table 12.1). Generally, phenolic compound composition of green tea resembles that of unprocessed tea leaves. Mainly, flavon-3-ols contribute to antioxidant capacity of green tea and its sensory properties. The content of individual phenolic compounds in green tea products varies widely. There are considerable differences in green tea processing between Chinese and Japanese producers. Chinese green tea is usually dry heated in order to deactivate oxidases, whereas in the case of Japanese green tea steaming is employed. Moreover, Japanese green tea is usually shade grown, and it contains a lower content of catechins and more amino acids compared to Chinese-style tea (Ku *et al.*, 2010).

Oolong Tea

Oolong tea is the most popular in Taiwan. It is referred to as semi-fermented tea and contains a mixture of non-oxidized monomeric polyphenols and higher-molecular-weight theaflavins. There are considerable differences in the reported content of catechins in oolong tea. On one study, it was reported to be comparable to that of green tea (Table 12.1B), whereas in another study it comprised only half of that noted for green tea (Table 12.1D and F). Oolong tea can be stored for several years; so-called old oolong tea has to be stored for at least 5 years. It is believed that the longer it is stored the better it tastes. When stored for long time, tea absorbs water and has to be dried again by heating at 120°C for 10h (Lee *et al.*, 2008). Considerable increase in gallic acid content during processing of old oolong tea was noted. At the same time 50% loss in EGCG was observed, which explains gallic acid content increase.

Black Tea

Black tea is produced in the process traditionally called fermentation, which is in fact an enzymatic oxidation evoked by polyphenol oxidases native to tea leaves. As an effect of the action of polyphenol oxidases, first quinones are formed from catechins, which further react to form dimers and oligomers.

There are two possible ways of processing in order to obtain black tea. In order to accelerate the oxidation process the leaf size is reduced by rupturing the withered tea leaves using 'orthodox rollers' or 'crush-tear-curl' (CTC) machines. It aims to disrupt the cells in order to release polyphenols oxidases and to enable them to react with catechins. Black tea produced using rollers is more suitable for large-leaf-type tea, whereas the CTC method leads to obtaining small particles suitable for teabags. CTC method creates a greater surface for enzymatic oxidation. Therefore, CTC tea has lower catechin content and antioxidant capacity compared to orthodox tea (Carloni *et al.*, 2013). Black tea was reported to contain around 20-times more gallic acid compared to green tea (Table 12.1E), whereas the other studies report only two-fold differences (Table 12.1D). During black tea manufacturing the most considerable changes occur in the content of catechins. A profound loss in catechin content is observed (Table 12.2B, D, F). At the same time new compounds are formed as a result of enzymatic oxidation: theaflavins and thearubigins.

Pu-erh Tea

Pu-erh tea is still the most popular in China. It is produced from the broad-leaf variety of tea plant *Camellia sinensis* var. *assamica* in the Yunnan Province of China. It is most often commercialized in the form of compressed brick, cake log, nest, or gourd shape (Ahmed *et al.*, 2010). After harvest, the leaves follow processing similar to that of green tea to obtain the raw material for pressed green pu-erh or raw pu-erh (Figure 12.5). The deactivation of enzymes by firing is not complete for pu-erh. Therefore oxidation can occur during long storage, which develops a smooth-tasting tea. The raw pu-erh is stored for at least 10 years in clay jars, bamboo wrapping, and underground pits at a temperature close to room temperature in order to obtain tea with optimal flavor. The oxidation in pu-erh tea occurs due to the enzymatic activity of microorganisms, like *Aspergillus* sp. The older the tea gets, the more prominent its color and aroma becomes. In order to imitate this long process of pu-erh tea aging, a microbial processing was developed that leads to the production of ripened/cooked pu-erh. It is manufactured by the piling of sun-dried leaves under controlled temperature and humidity, often inoculated by selected strains, and allowed to ripen for several weeks or months. The oxidized and fermented leaves are then steamed, compressed, and dried. In China, the tea obtained in this way is called black pu-erh.

Microorganisms oxidize phenolic compounds of tea and lead to considerable loss of catechins and formation of theaflavins, thearubigins, theabrownins, and gallic acid (Table 12.1I) (Wang *et al.*, 2011; Xie *et al.*, 2009). Recently, more and more studies have been carried out to reveal the detailed composition of pu-erh tea and its antioxidant activity. The effect of processing to obtain aged and black (ripened) pu-erh on catechins content and radical-scavenging capacity was studied by Ahmed *et al.* (2010). The results indicated higher total catechin content and free-radical-scavenging capacity of green pu-erh compared to aged green and black pu-erh tea. The results confirm the statement that a decrease in polyphenolic levels and antioxidant activity of pu-erh and other teas occur with an increase in processing and aging. However, the results reported by Zhang *et al.* (2012) showed an increase in radical scavenging activity of 1-week fermented pu-erh tea and total antioxidant and superoxide anion-scavenging activity of teas fermented for up to 3 weeks in comparison to unfermented sun-dried tea. The comparison of commercially available teas of different types (Table 12.1B) revealed the lower content of catechins, glycosylated flavonols, and procyanidins in green pu-erh tea than in green and white tea (Zhao *et al.*, 2011). The content of gallic acid in ripened pu-erh was the highest among all kinds of tea studied, i.e., aged pu-erh, green, oolong, black, white, and yellow (Table 12.1A) (Zhang *et al.*, 2011).

Generally, the real effect of processing can be analyzed only when the same raw material is subjected to different processing. Most of the literature available reports on the differences in phenolic compound profile between various types of tea obtained from the market worldwide, but history of production and storage of tea is not known. The studies exploring the effect of processing leading to obtaining different tea types are still scarce. However, two papers investigating this effect have been published lately (Carloni *et al.*, 2013; Kim *et al.*, 2011a). Carloni *et al.* studied the antioxidant activity of white, green, and black tea obtained from unified raw tea leaves processed differently in the same factory to obtain two black teas (orthodox and CTC), two green teas, and one white tea. The authors reported that the total phenolic content was higher in green tea, than in white and black tea. A similar trend was observed for the content of individual catechins, except for black orthodox tea for which total catechins as well as EGCG content were higher than for white tea. As it was expected, total theaflavins content was higher in black teas, and very low in the green tea. It is worth noting that white tea also showed a significant content of theaflavins, much higher than in green tea. Theaflavins are formed because during processing of white tea oxidase is not deactivated. Accordingly, the lowest antioxidant capacity was reported for black CTC tea whereas the highest was reported for green tea.

The second study carried out aimed to identify changes during processing of unfermented tea (green tea) through oolong (20–60% fermentation) to black tea (80% fermentation) (Kim *et al.*, 2011a). The largest changes were noticed in the content of four major tea catechins EGCG, EGC, EC, and ECG. Their content decreased by more than 50%, and even up to 91% for EGC, while free gallic acid content increased. During formation of theaflavins, galloyl groups are cleaved from gallated catechins, which results in increase in gallic acid content. At the same time total phenolic content decreased, whereas total theaflavins and thearubigins increased with increasing degree of fermentation. The content of total flavonol glycosides was strongly reduced (38%) during processing, and a clear tendency of decrease with increased degree of fermentation was noticed in the case of myricetin-3-glycoside, rutin, and quercetin-3-glucosyl-rhamnosyl-galactoside. Antioxidant activity of tea infusion decreased by 20% in the course of black tea processing. The authors suggested that the loss in antioxidant capacity noted was due to degradations of ascorbic acid, and flavonol glycosides.

Home Preparation of Tea Infusions

In recent years there has been more and more research on the effect of consumer preparation on composition and activity of tea infusions (Kyle *et al.*, 2007; Lakenbrink *et al.*, 2000; Molan *et al.*, 2009; Su *et al.*, 2006). The studies have examined the different culinary methods used in the domestic preparation of tea infusions, taking into account factors such as water temperature, infusion time, stirring and dosage form, i.e., loose-leaf tea *versus* tea bag.

The extraction temperature, extraction time, water quality and water-to-tea ratio, tea particle size, extraction pH, and the number of extractions are all important factors which directly affected the efficiency of the extraction of antioxidants (Vuong *et al.*, 2011).

Water Quality

Tap water, activated carbon-adsorbed water, deionized water, distilled water, reverse-osmosis water and ultra-pure water were compared for their impact on tea quality of green tea (Danrong *et al.*, 2009). Similar radical-scavenging activity was found for tea infusions prepared with tap water, activated carbon-adsorbed water, and deionized water. However, radical-scavenging activity in the case of distilled-water-, reverse-osmosis-water-, and ultra-pure-water-prepared tea was significantly higher. Most likely radical-scavenging activity was affected by the pH differences between the different kinds of water. The effect of pH in a wide range on the content of catechins in green tea infusion was studied, and much higher content of catechins at acidic conditions of brewing ($pH<6$) than under more neutral and basic conditions were noted. It was concluded that the effect is mainly due to different stability of catechins at different pH values, rather than to efficacy of extraction (Vuong *et al.*, 2011).

Temperature and Time

In the case of green tea it was reported that the higher temperature of brewing, the higher the reducing power of the infusion; however, no difference was noticed between extraction carried out for 10min and 30min (Molan *et al.*, 2009). Similarly, for black and oolong tea, the longer the time of brewing was, the higher the antioxidant activity as well as the total phenolic and catechin contents were (Kyle *et al.*, 2007). Oolong tea prepared at 95°C for 3min had the highest scores of aroma, bitterness, astringency, and sweetness, but the total polyphenol and radical-scavenging activity in tea solutions increased with higher temperatures and longer brewing times. The brewing conditions optimal for maximal TEAC values of green tea were reported to involve 80°C for 3–5min (Samaniego-Sanchez *et al.*, 2011). In terms of the maximal extraction of catechins per gram of green tea, the best extraction efficiency was achieved with water extraction at 80°C for 30min (Vuong *et al.*, 2011).

Tea Leaves/Tea Bags/Tea Powder

Antioxidant activity and catechin content of green tea depends on form of powder, loose leaves, or tea bags, as well as brewing conditions (Komes *et al.*, 2010). Powder and bagged green tea allowed the extraction of more phenolic compounds than leaves, regardless of the extraction temperature. Total flavan-3-ols amounted to around 1g/l, whereas for loose-leaf teas ranged from 0.35 to 0.9g/l for different tea brands. The highest concentration of flavon-3-ols was obtained for extractions performed at 80°C for 5min (powder), 15min (bagged), and 30min (loose leaf). Generally, higher antioxidant activity was noted for tea powder and tea bags than tea leaves. Antioxidant activity was positively influenced by increased extraction time and temperature. However, other studies showed that the extraction efficacy of phenolic compounds from white (Rusak *et al.*, 2008) and green tea (Samaniego-Sanchez *et al.*, 2011) was not significantly affected by form of loose leaf *versus* tea bag.

Serving Additives

Addition of milk can reduce antioxidant potential of black tea (Ryan and Petit, 2010). In the case of green tea the results are contradictory: The addition of milk to powdered Matcha green tea mildly decreased its antioxidant activity, whereas for two other types of tea (Gyokuro and Twinings) the opposite effect was observed (Komes *et al.*, 2010). The antioxidant activity of green tea was enhanced after addition of lemon juice, which might be explained by the synergistic effect of ascorbic acid on the polyphenols as well as increased stability of polyphenols due to lower pH.

Production of Ready-to-Drink Tea Beverages

Tea drinks are becoming more and more popular worldwide. Tea beverages are produced on the basis of tea extracts or tea infusions prepared as a part of beverage manufacturing process and available on the market bottled or canned (Cordero *et al.*, 2009). Generally, very low content of catechins was noted for most tea beverages studied in comparison to tea-cup- or tea-pot-prepared tea. Fourteen brands of canned and bottled green tea drinks were analyzed by Chen *et al.* and in 10 of them the total catechin content was lower than 3mg/100ml (Chen *et al.*, 2001). Only in three of all analyzed tea drinks produced in Japan was the content of catechins about 30mg/100ml. Moreover, the profile of phenolic compounds in tea beverage differs from that of original tea. For green tea beverages the content of GCG considerably exceeded that of EGCG. GCG is a product of epimerization of EGCG, which might occur due to temperature treatment. One of the steps during processing of tea beverages is sterilization, at a temperature of 120°C for several minutes. After heating for 20min, only 76% green tea catechins remained in water solution, and the stability was shown to be pH dependent (Chen *et al.*, 2001). Also heating for as little as 1min at 120°C caused loss of 15 or 36% of total catechins for steamed and roasted green tea, respectively (Wang *et al.*, 2000). Moreover, it was suggested that steamed green tea leaves are a better starting material than roasted tea for production of bottled or canned tea beverages since they contain more flavanols (Wang *et al.*, 2000). In this way, products rich in antioxidants and with good sensory qualities can be obtained.

Eighteen brands of canned and bottled drinks (10 green teas, two oolong teas, six black teas) available in Hong Kong were studied for their catechin profile and antioxidant activity. Epimers dominated in oolong and black tea beverages, whereas for green tea similar content of both forms of catechins was found (Xu *et al.*, 2004).

The content of catechins in fermented tea beverages of eight brands available on the Italian market was highly variable (Cordero *et al.*, 2009). The sample richest in catechins, (>200µg/ml) was produced from tea leaf infusion. It was suggested that products obtained by an industrial tea leaf infusion process are a better source of flavan-3-ols than those produced from reconstituted tea extracts. However, the impact of the quality of raw material, additives, thermal treatments, formulation, and packaging can not be neglected.

STORAGE OF TEA LEAVES AND TEA BEVERAGES

Tea leaves have a considerably long shelf-life due to their low moisture content. In the case of pu-erh tea and old oolong tea, long storage is even necessary for the development of desirable taste and aroma. However, for other tea types, storage for an extended period of time can lead to loss of quality of the product. It was reported that tea catechins are not stable during long-term storage. During 6 months' storage of green tea bags in their original containers at room temperature in the dark, the average content of EGCG decreased by one-third and ECG decreased by half. This suggests that ECG may be more susceptible to degradation than EGCG (Friedman *et al.*, 2009). Storage of black tea for up to 12 months can affect theaflavins and thearubigins content (Thomas *et al.*, 2008). The loss of theaflavins was greater for orthodox black tea (around 37%) than for CTC (22%), whereas for thearubigins the relation was opposite. It was indicated that γ-irradiation could reduce this deterioration to a great extent. The main factors influencing tea quality and catechin stability during storage are light, oxygen, and temperature. The optimal storage conditions for green tea powder to maintain catechin stability were indicated by Li *et al.* as ambient temperature and relative humidity below 43% or in a freezer at relative humidity below 60% (Li *et al.*, 2011).

In the case of ready-to-drink beverages, the great influence of packaging material on storage stability is evident. The shelf-life of the products is usually 12 months; nevertheless, product aging during that time was found to be extensive. Storage stability of green tea infusions depending on packaging material used (glass, PET, and retortable pouch) was evaluated by Kim *et al.* (2011b). Tea drinks were stored in the absence of light at a temperature of 3°C and catechins were found to be stable up to 6 weeks, and after that time the degradation appeared to a greatest extent in retortable pouch > PET > glass (Kim *et al.*, 2011b). The antioxidant capacity of green tea infusions decreased by 18, 14, and 30% in glass, PET, and retortable pouch, respectively, during the 12 weeks of storage. The differences were caused most probably by different oxygen permeability of packaging materials. When PET, polystyrene glass, and cans used as packaging for fermented tea drinks originating from the same batch were compared, the highest storage stability was found for drinks in cans (Cordero *et al.*, 2009). Nevertheless, also in cans there was a 50% loss of catechins. It was reported that cold storage at 4°C of tea beverages in PET bottles ensures a slower decrease in catechin content in white, black, and green teas (Nekvapil *et al.*, 2012).

References

Ahmed, S., Unachukwu, U., Stepp, J.R., Peters, C.M., Long, C.L., Kennelly, E., 2010. Pu-erh tea tasting in Yunnan, China: Correlation of drinkers' perceptions to phytochemistry. J. Ethnopharmacol. 132, 176–185.

Carloni, P., Tiano, L., Padella, L., Bacchetti, T., Customu, C., Kay, A., Damiani, E., 2013. Antioxidant activity of white, green and black tea obtained from the same cultivar. Food Res. Int. 53, 900–908.

Chen, H., Shurlknight, K., Leung, T., Sang, S., 2012. Structural identification of theaflavin trigallate and tetragallate from black tea using liquid chromatography/electrospray ionization tandem mass spectrometry. J. Agric. Food Chem. 60, 10850–10857.

Chen, Z.Y., Zhu, Q.Y., Tsang, D., Huang, Y., 2001. Degradation of green tea catechins in tea drinks. J. Agric. Food Chem. 49, 477–482.

Cooper, R., 2011. Green tea and theanine: health benefits. Int. J. Food Sci. Nutr. 63, 90–97.

Cordero, C., Canale, F., Del Rio, D., Bicchi, C., 2009. Identification, quantitation, and method validation for flavan-3-ols in fermented ready-to-drink teas from the Italian market using HPLC-UV/DAD and LC-MS/MS. J. Sep. Sci. 32, 3643–3651.

Danrong, Z., Yuqiong, C., Dejiang, N., 2009. Effect of water quality on the nutritional components and antioxidant activity of green tea extracts. Food Chem. 113, 110–114.

Del Rio, D., Stewart, A.J., Mullen, W., Burns, J., Lean, M.E.J., Brighenti, F., Crozier, A., 2004. HPLC-MS^n analysis of phenolic compounds and purine alkaloids in green and black tea. J. Agric. Food Chem. 52, 2807–2815.

Friedman, M., Levin, C.E., Lee, S.U., Kozukue, N., 2009. Stability of green tea catechins in commercial tea leaves during storage for 6 months. J. Food Sci. 74, H47–H51.

Gorjanovic, S., Komes, D., Pastor, F.T., Belscak-Cvitanovic, A., Pezo, L., Hecimovic, I., Suznjevic, D., 2012. Antioxidant capacity of teas and herbal infusions: Polarographic assessment. J. Agric. Food Chem. 60, 9573–9580.

Haslam, E., 2003. Thoughts on thearubigins. Phytochemistry 64, 61–73.

Hilal, Y., Engelhardt, U., 2007. Characterisation of white tea—Comparison to green and black tea. J. Consum. Prot. Food Saf. 2, 414–421.

Kim, Y., Goodner, K.L., Park, J.D., Choi, J., Talcott, S.T., 2011a. Changes in antioxidant phytochemicals and volatile composition of *Camellia sinensis* by oxidation during tea fermentation. Food Chem. 129, 1331–1342.

Kim, Y., Welt, B.A., Talcott, S.T., 2011b. The impact of packaging materials on the antioxidant phytochemical stability of aqueous infusions of green tea (*Camellia sinensis*) and yaupon holly (Ilex vomitoria) during cold storage. J. Agric. Food Chem. 59, 4676–4683.

Komes, D., Horzic, D., Belscak, A., Ganic, K.K., Vulic, I., 2010. Green tea preparation and its influence on the content of bioactive compounds. Food Res. Int. 43, 167–176.

Ku, K.M., Choi, J.N., Kim, J., Kim, J.K., Yoo, L.G., Lee, S.J., Hong, Y.S., Lee, C.H., 2010. Metabolomics analysis reveals the compositional differences of shade grown tea (*Camellia sinensis* L.). J. Agric. Food Chem. 58, 418–426.

Kyle, J.A.M., Morrice, P.C., McNeill, G., Duthie, G.G., 2007. Effects of infusion time and addition of milk on content and absorption of polyphenols from black tea. J. Agric. Food Chem. 55, 4889–4894.

Lakenbrink, C., Lapczynski, S., Maiwald, B., Engelhardt, U.H., 2000. Flavonoids and other polyphenols in consumer brews of tea and other caffeinated beverages. J. Agric. Food Chem. 48, 2848–2852.

Lee, V.S.Y., Dou, J.P., Chen, R.J.Y., Lin, R.S., Lee, M.R., Tzen, J.T.C., 2008. Massive accumulation of gallic acid and unique occurrence of myricetin, quercetin, and kaempferol in preparing old oolong tea. J. Agric. Food Chem. 56, 7950–7956.

Li, N., Taylor, L.S., Mauer, L.J., 2011. Degradation Kinetics of Catechins in Green Tea Powder: Effects of Temperature and Relative Humidity. J. Agric. Food Chem. 59, 6082–6090.

Lin, L.Z., Chen, P., Harnly, J.M., 2008. New phenolic components and chromatographic profiles of green and fermented teas. J. Agric. Food Chem. 56, 8130–8140.

Łuczaj, W., Skrzydlewska, E., 2005. Antioxidative properties of black tea. Preventive Medicine 40, 910–918.

Molan, A.L., De, S., Meagher, L., 2009. Antioxidant activity and polyphenol content of green tea flavan-3-ols and oligomeric proanthocyanidins. Int. J. Food Sci. Nutr 60, 497–506.

Neilson, A.P., Song, B.J., Sapper, T.N., Bomser, J.A., Ferruzzi, M.G., 2010. Tea catechin auto-oxidation dimers are accumulated and retained by Caco-2 human intestinal cells. Nutr. Res. 30, 327–340.

Nekvapil, T., Kopriva, V., Boudny, V., Hostovsky, M., Dvorak, P., Malota, L., 2012. Decrease in the antioxidant capacity in beverages containing tea extracts during storage. Sci. World J. 2012.

Rice-Evans, C.A., Miller, J., Paganga, G., 1997. Antioxidant properties of phenolic compounds. Trend Plant Sci. 2, 152–159.
Rusak, G., Komes, D., Likic, S., Horzic, D., Kovac, M., 2008. Phenolic content and antioxidative capacity of green and white tea extracts depending on extraction conditions and the solvent used. Food Chem. 110, 852–858.
Ryan, L., Petit, S., 2010. Addition of whole, semiskimmed, and skimmed bovine milk reduces the total antioxidant capacity of black tea. Nutr. Res. 30, 14–20.
Samaniego-Sanchez, C., Inurreta-Salinas, Y., Quesada-Granados, J.J., Blanca-Herrera, R., Villalon-Mir, M., de la Serrana, H.L.G., Lopez Martinez, M.C., 2011. The influence of domestic culinary processes on the Trolox Equivalent Antioxidant Capacity of green tea infusions. J. Food Compos. Anal. 24, 79–86.
Su, X.G., Duan, J., Jiang, Y.M., Shi, J., Kakuda, Y., 2006. Effects of soaking conditions on the antioxidant potentials of oolong tea. J. Food Compos. Anal. 19, 348–353.
Thomas, J., Senthilkumar, R.S., Kumar, R.R., Mandal, A.K.A., Muraleedharan, N., 2008. Induction of γ irradiation for decontamination and to increase the storage stability of black teas. Food Chem. 106, 180–184.
Unachukwu, U.J., Ahmed, S., Kavalier, A., Lyles, J.T., Kennelly, E.J., 2010. White and green teas (*Camellia sinensis*): Variation in phenolic, methylxanthine, and antioxidant profiles. J. Food Sci. 75, C541–C548.
Vuong, Q.V., Golding, J.B., Stathopoulos, C.E., Nguyen, M.H., Roach, P.D., 2011. Optimizing conditions for the extraction of catechins from green tea using hot water. J. Sep. Sci. 34, 3099–3106.
Vuong, Q.V., Golding, J.B., Stathopoulos, C.E., Roach, P.D., 2012. Effects of aqueous brewing solution pH on the extraction of the major green tea constituents. Food Res. Int. 53, 713–719.
Wang, D., Xiao, R., Hu, X., Xu, K., Hou, Y., Zhong, Y., Meng, J., Fan, B., Liu, L., 2009. Comparative safety evaluation of Chinese pu-erh green tea extract and pu-erh black tea extract in Wistar rats. J. Agric. Food Chem. 58, 1350–1358.
Wang, L.F., Kim, D.M., Lee, C.Y., 2000. Effects of heat processing and storage on flavanols and sensory qualities of green tea beverage. J. Agric. Food Chem. 48, 4227–4232.
Wang, Q.P., Peng, C.X., Gong, J.S., 2011. Effects of enzymatic action on the formation of theabrownin during solid state fermentation of Pu-erh tea. J. Sci. Food Agric. 91, 2412–2418.
Wang, Y.Y., Yang, X.R., Li, K.K., Li, C.R., Li, L.L., Li, J.X., Huang, H., He, Y., Ye, C., Song, X., 2010. Simultaneous determination of theanine, gallic acid, purine alkaloids, catechins, and theaflavins in black tea using HPLC. Int. J. Food Sci. Technol. 45, 1263–1269.
Wu, C., Xu, H., Héritier, J., Andlauer, W., 2012. Determination of catechins and flavonol glycosides in Chinese tea varieties. Food Chem. 132, 144–149.
Xie, G., Ye, M., Wang, Y., Ni, Y., Su, M., Huang, H., Qiu, M., Zhao, A., Zheng, X., Chen, T., Jia, W., 2009. Characterization of Pu-erh tea using chemical and metabolic profiling approaches. J. Agric. Food Chem. 57, 3046–3054.
Xu, J.Z., Yeung, S.Y.V., Chang, Q., Huang, Y., Chen, Z.Y., 2004. Comparison of antioxidant activity and bioavailability of tea epicatechins with their epimers. Br. J. Nutr. 91, 873–881.
Yoshino, K., Suzuki, M., Sasaki, K., Miyase, T., Sano, M., 1999. Formation of antioxidants from (−)-epigallocatechin gallate in mild alkaline fluids, such as authentic intestinal juice and mouse plasma. J. Nutr. Biochem. 10, 223–229.
Zhang, L.Z., Wang, D.L., Chen, W.X., Tan, X.D., Wang, P.C., 2012. Impact of fermentation degree on the antioxidant activity of pu-erh tea in vitro. J. Food Biochem. 36, 262–267.
Zhang, L., Li, N., Ma, Z.Z., Tu, P.F., 2011. Comparison of the chemical constituents of aged pu-erh tea, ripened pu-erh tea, and other teas using HPLC-DAD-ESI-MS^n. J. Agric. Food Chem. 59, 8754–8760.
Zhao, Y., Chen, P., Lin, L.Z., Harnly, J.M., Yu, L.L., Li, Z.W., 2011. Tentative identification, quantitation, and principal component analysis of green pu-erh, green, and white teas using UPLC/DAD/MS. Food Chem. 126, 1269–1277.
Zuo, Y.G., Chen, H., Deng, Y.W., 2002. Simultaneous determination of catechins, caffeine and gallic acids in green, Oolong, black and pu-erh teas using HPLC with a photodiode array detector. Talanta 57, 307–316.

CHAPTER

13

Antioxidant Quality of Tea (*Camellia sinensis*) as Affected by Environmental Factors

Lovedeep Kaur, Shiromani Jayasekera, Paul J. Moughan

Riddet Institute, Massey University, Palmerston North, New Zealand

CHAPTER POINTS

- The content of phenolic compounds and antioxidant properties of tea are influenced by a variety of environmental (production location, season, altitude, amount of sunlight) factors.
- Various husbandry (tea source, plucking standards and frequency, pruning frequency) factors also affect the antioxidant quality of tea.
- More *in vitro* and *in vivo* studies are required to explore the interactions between these factors and how such interactions affect the overall antioxidant quality of different teas.
- While both green and black teas are good sources of antioxidant polyphenols, more research is needed to establish the mechanisms associated with the antioxidant activity of black tea.

INTRODUCTION

Tea is the most commonly consumed beverage after water. Being an important dietary source of natural phenolic antioxidants, many beneficial health effects are purportedly associated with its regular consumption: reduced risk of cardiovascular disease, certain types of cancer, inflammatory bowel, liver, and neurodegenerative diseases, and diabetes, and even weight loss (Dufresne and Farnworth, 2001, Carloni *et al.*, 2012).

At present there are over 2000 varieties of tea and it is produced mainly in China, India, Kenya, Sri Lanka, Indonesia, and Turkey. All of the varieties of tea are produced from young and tender leaves of *Camellia sinensis* (L.) (family Theaceae), an evergreen shrub. Freshly plucked leaves are processed to give black (fermented), oolong (semi-fermented), and green (unfermented) teas (Carloni *et al.*, 2012), with black tea accounting for approximately 75% of global production. There are other tea products that are less commonly consumed or are limited to particular geographical locations or communities, such as pickled tea (made from tea leaves plucked in the rainy season and pickled in bamboo containers), brick tea (compressed black or green tea), and white tea (made from the delicate tea buds allowed to wither in natural sunlight before they are lightly processed to prevent further fermentation).

The content of phenolic compounds, antioxidant properties, and hence the quality of tea, are reportedly influenced by a variety of environmental (atmospheric temperature, rainfall, and the amount of sunlight) and husbandry (fertilization, soil, plucking standards and frequency) factors, including geographical location/season of tea picking, cultivar type, and tea manufacturing practices. All of these factors are reviewed and discussed in this chapter.

EFFECTS OF PRODUCTION AND PROCESSING

The fresh tea leaf is rich in the flavonoids (Table 13.1). Flavonoids are polyphenolic in nature and contain an aromatic ring bearing one or more hydroxyl groups. The most common flavonoid in tea is flavan-3-ol, which provides tea with its typical flavonoid pattern. The catechins are monomers and the principal catechins of young tea leaves are (−)-epigallocatechin gallate (EGCG), (−)-epigallocatechin (EGC), (−)-epicatechin gallate (ECG), (−)-gallocatechin (GC), (−)-epicatechin (EC), and (+)-catechin (C). Other polyphenols include flavonols (quercetin, kaempferol, myricetin), and their glycosides, and depsides such as chlorogenic acid. In addition, tea also contains amino acids (theanine,

Processing and Impact on Antioxidants in Beverages
http://dx.doi.org/10.1016/B978-0-12-404738-9.00013-1

TABLE 13.1 Chemical Composition of Young Shoots of Tea

Component	g/100 g (dry weight basis)
Flavanols	
(−)-Epigallocatechin 3-gallate (EGCG)	9–13
(−)-Epigallocatechin (EGC)	3–6
(−)-Epicatechin 3-gallate (ECG)	3–6
(−)-Epicatechin (EC)	1–3
Gallocatectin (GC)	1–2
(+)-Catechin (C)	1–2
Flavonols and their glycosides	3–4
Leucoanthocyanins	2–3
Phenolic acids: theogallin and others	4
Total polyphenols	27–40*
Caffeine	3–4
Amino acids:	
Theanine	2
Others	2
Carbohydrates	4
Organic acids	0.5
Volatile compounds	0.01
Polysaccharides	
Starch	2–5
Other	12
Protein	15
Ash	5
Cellulose	7
Lignin	6
Lipids	3

Modder and Amarakoon (2002).

gamma aminobutyric acid), carbohydrates, proteins, vitamins, minerals, trace elements, volatile compounds, carotenoids, and alkaloids, namely caffeine, theophylline, and theobromine (Kerio *et al.*, 2013). Caffeine is present at an average level of 3% along with very small amounts of methylxanthines, theobromine, and theophylline. Anthocyanidins are also found in the leaf. The amino acid theanine is unique to tea. Vitamin C is lost during the processing of the fresh leaf, but carotenoids and vitamin K are present in brewed tea. Tea also accumulates aluminium, potassium, fluoride, and manganese (Modder and Amarakoon 2002).

During black tea manufacture, the catechins are oxidized to theaflavins (TFs) and thearubigins (TRs) in the presence of polyphenoloxidase and oxygen. The TFs are catechin dimers (theaflavin, theaflavin 3-gallate, theaflavin 3′-gallate, theaflavin 3,3′-digallate) and TRs are catechin oligomers of unknown structure. These oxidized constituents are unique to black tea. TF formation requires the presence of both EC or ECG and EGC or EGCG (Wickremasinghe, 1978). In green tea manufacture, the first operation is the heat destruction of polyphenoloxidase by steaming (in Japan) or pan firing (in China). The primary difference between black and green tea is, therefore, that the former contains a mixture of native unoxidized polyphenols together with oxidized products such as TF and TR, whereas green tea contains unoxidized native polyphenols only. Oolong tea is a semi-fermented tea that is allowed to oxidize only partially to lock-in the rich flavor that is associated with the high quality of oolong tea. The polyphenol profile of oolong tea is similar to that of green or black tea, but it has a lower content of EGCG, EC, and ECG than that found in green tea because the fresh oolong tea leaves are subjected to a partial fermentation step before drying (Zuo *et al.*, 2002).

The concentrations of the phenolic compounds have long been used as a quality parameter of tea. The plain black teas are valued for their TF content, which is responsible for their taste, brightness, and which also contributes to their color, and for the TRs, which are responsible for thickness and color of both the liquors and infusion (Biswas *et al.*, 1973). Catechins and gallates are used as quality parameters for green tea. Bryce *et al.* (1970) and Coxon *et al.* (1970) found that flavonoids (C, EC, GC, ECG, GC, EGC, EGCG) and one phenolic acid (gallic acid) in the fresh leaf were responsible for the formation of TFs in the resultant black tea. The relative amounts of catechins and polyphenoloxidase in the tea leaf vary with the season, elevation, agricultural practices employed, maturity of the leaf, processing methods, manufacturing practices, and storage, and may affect tea quality. The strength of brewed tea and the infusion time may also affect the phenolic composition of brewed tea (Hilton, 1973).

Catechins and other polyphenols act as antioxidants *in vitro* by sequestering metal ions and by scavenging reactive oxygen and nitrogen species (Wiseman *et al.*, 1997). The antioxidant activity of teas generally decreases in the order: green > oolong > black (Roginsky *et al.*, 2003), although some studies have shown that black teas have greater activity than green teas (Venditti *et al.*, 2010), whereas others have reported the absence of any significant differences (Carloni *et al.*, 2012).

Ranking of total phenolic content (TPC), 2,2-diphenyl-1-picrylhydrazyl (DPPH) radical-scavenging activity (RSA), ferric-reducing power (FRAP), and ferrous-ion-chelating (FIC) ability has been reported to be: green > oolong ≈ black, green ≈ oolong > black, green > oolong > black, and black > green ≈ oolong, respectively (Chan *et al.*, 2010). Findings from the latter study deviated from that of Yokozawa *et al.* (1998), who reported that the ranking of RSA was green > black > oolong. Green teas have been reported to have significantly higher TPC, RSA, and FRAP, but poorer FIC ability, than black teas (Chan *et al.*, 2007). The FRAP values, and the TPC values have been reported to increase with increasing brewing temperature from 70°C to 100°C and with stirring at both temperatures for both unfermented (green) and fermented (black) tea leaves (Jayasekera *et al.*, 2011). For DPPH, values increased with increasing brewing temperature and stirring for unfermented leaves, but there was a statistically significant interaction effect for unfermented leaves (Table 13.2), such that the effect of temperature was dependent upon stirring.

Environmental Factors

Tea is often cultivated at elevations ranging from almost sea level to more than 2000 m above sea level. It has been shown that the quality of tea, especially that of black tea, is influenced by altitude and growing season (Owuor *et al.*, 1990, Jayasekera *et al.*, 2011), the genetic make-up of the plant, and the region of production and its climate (Owuor *et al.*, 2008). Studies in central Africa have shown that the concentration of flavonoids in fresh apical shoots of tea was highest during the cold season (Hilton and Palmer-Jones, 1973). In contrast, a lower level of catechin gallates was reported for Australian tea harvested in the cooler months compared to that in tea harvested in the warmer months (Caffin *et al.*, 2004). In the northern hemisphere, total flavanol content is greatest during the height of summer. Studies in Japan have shown that the EGC level is higher in spring while ECG and EGCG levels are higher in summer (Nakagawa and Torri, 1964).

Jayasekera *et al.* (2011) investigated the TPC as well as the FRAP and RSA of water extracts from fresh (green) and fermented (black) tea leaves grown in different regions of Sri Lanka. The selected regions included four highland plantations and two lower-altitude plantations. For high-grown teas, samples were also collected from the two harvesting seasons (dry and monsoon). Their results highlighted significant variations in antioxidant activity across the six plantations. FRAP and RSA for both the fermented and unfermented teas from the four highland plantations showed a statistically significant interaction between season and plantation (Table 13.3, Figure 13.1), such that there was an effect of season, but the nature of the effect was dependent on the plantation. A similar interaction between season and plantation was observed for total phenolics in unfermented teas from the four highland plantations. The variability of the total phenolics for fermented teas, however, was reported to be independent of seasonal variations. A significant correlation was observed between FRAP and total phenolics. The total catechin content and amounts of individual catechins, flavonols, and TFs of the above-mentioned samples were also determined (Jayasekera *et al.*, unpublished data), and a statistically significant effect of season was observed on all of the phenolic compounds for unfermented tea leaves, demonstrating a significant impact of the wet (monsoon) or dry climate on the polyphenolic profile of the unprocessed fresh tea leaves.

The altitude of tea planting and the season of tea picking have been reported to influence the composition and level of catechins in oolong teas during their growth (Chen *et al.*, 2010). The levels of EC, ECG, and EGCG were reported to be statistically significantly different in some Taiwan oolong teas grown in different seasons (spring and summer) (Wang *et al.*, 2006). Chen *et al.* (2010) demonstrated that the contents of EGCG, CG, and TC in the 'Huang Zhi Xiang' oolong teas grown at a high altitude were statistically significantly higher than those for plants grown at a low altitude. Furthermore, the contents of EGCG, CG, and TC in the autumn tea were statistically significantly higher than those from the spring tea grown at a low altitude. In addition, the contents of EGC, C, GC, and SC in the spring tea leaves were statistically significantly higher than those found in the autumn tea leaves grown at a low altitude, but they were not significantly different from those in the autumn tea leaves grown at a high altitude. Their work suggests that the production of the 'Huang Zhi Xiang' oolong tea grown at low altitudes should utilize the autumn tea leaves while the production of oolong tea grown at high altitudes can use both the spring and autumn tea leaves.

Black tea plain quality parameters (TFs, TRs, color, brightness, and sensory characteristics) have been reported to vary widely with geographical area of production (Owuor *et al.*, 1986). Owuor *et al.* (1990) showed that there was an increase in the quality of tea with an increase in altitude, which was explained by the decrease in growth rate with the increase in altitude. Hilton and Palmer-Jones (1973) showed that, in Central Africa, the quality of tea was inversely related to the growth rate of the tea bush. The impact of the plant's growth rate in Central Africa was also reflected in the tea's polyphenolic content. For the fast-growing tea, the polyphenolic content was lower and so was the TF content after fermentation. Similar observations have been made for tea produced in Himachal Pradesh in India (Gulati and Ravindranath, 1996). Sud and Baru (2000) also reported a lower level of TF and TR in black tea produced in the rainy season in Himachal Pradesh. The decreased TF

TABLE 13.2 Effect of Brewing Temperature and Stirring during Extraction on Ferric-Reducing Antioxidant Power (FRAP; mmol/l), 2,2-Diphenyl-1-Picrylhydrazyl Radical-Scavenging Activity (DPPH;%) and Total Phenolic Content [TPC; mg Gallic Acid Equivalent (GAE)/g Dry Leaves] for Unfermented and Fermented Tea Samples*

	Temperature (T)/Stirring (S)					Significance		
	70°C (unstirred)	70°C (stirred)	100°C (unstirred)	100°C (stirred)	SEM	T	S	T × S
FRAP								
Unfermented	9.2	9.9	101.1	10.6	0.073	$P<0.0001$	$P<0.0001$	NS
Fermented	8.0	8.8	9.0	9.9	0.142	$P<0.0001$	$P<0.0001$	NS
DPPH								
Unfermented	28.5	31.9	37.8	43.8	0.9	$P<0.0001$	$P=0.0003$	NS
Fermented	28.5	36.3	41.9	42.9	0.952	$P<0.0001$	$P=0.0015$	$P=0.0129$
TPC								
Unfermented	148.5	159.8	154.7	173.5	1.48	$P<0.0001$	$P<0.0001$	NS
Fermented	86.5	124.2	115.2	162.9	3.32	$P<0.0001$	$P<0.0001$	NS

**Values are the means of triplicate incubations for two separate runs (n = 6).*
Reproduced from Jayasekera et al. *(2011) © Elsevier (2011); reproduced with permission from Elsevier.*

TABLE 13.3 Effect of Plantation, Season and Plantation × Season on Mean (n = 6) Ferric-Reducing Antioxidant Power (FRAP; mmol/l), 2,2-Diphenyl-1-Picrylhydrazyl Radical-Scavenging Activity (DPPH;%) and Total Phenolic Content (TPC; mg gallic acid equivalent (GAE)/g dry leaves) for Unfermented and Fermented Teas*

	Plantation				Season			Significance		
	Pedro	Laxapana	Greatwestern	Sarnia	Dry	Monsoon	SEM	Plantation	Season	Interaction
UNFERMENTED										
FRAP	10.2	11.04	10.8	11.2	10.8	10.7	0.096	<0.001	NS	<0.001
DPPH	46.3	43.3	38.4	41.4	42.3	42.3	1.03	<0.05	NS	<0.001
TPC	170.72	173.66	169.92	179.48	172.8	174.16	2.24	<0.0001	0.008	<0.001
FERMENTED										
FRAP	10.4	10.8	9.9	9.7	10.9	9.5	0.125	<0.001	<0.0001	<0.0001
DPPH	44.8	41.1	33.5	44.6	44.1	44.1	1	<0.001	0.004	<0.05
TPC	171.9^{a}	169.2^{b}	157.5^{c}	153.6^{bc}	169.6^{a}	156.5^{b}	2.01	<0.0001	<0.0001	NS

**Means (± SE) of triplicate measurements for two separate runs (n = 6).*
Means sharing the same letter in a row within plantation and season effects were not significantly different at $P < 0.05$.
Reproduced from Jayasekera et al. *(2011) © Elsevier (2011); reproduced with permission from Elsevier.*

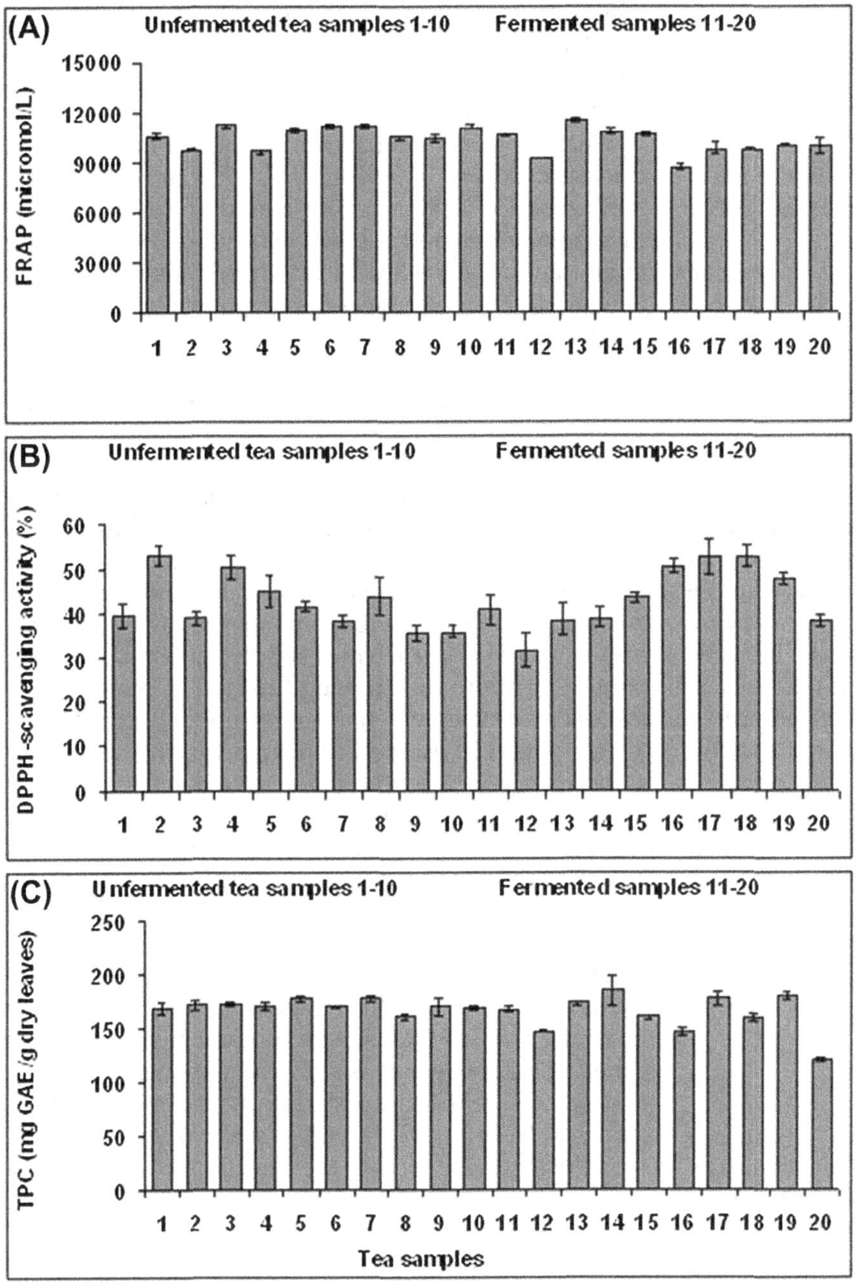

FIGURE 13.1 (A) Ferric-reducing antioxidant power (FRAP; μmol/l), (B) 2,2-diphenyl-1-picrylhydrazyl radical scavenging activity (DPPH; %) and (C) total phenolic content (TPC; mg gallic acid equivalent (GAE)/g dry leaves) for unfermented and fermented tea samples collected from the plantations listed in Table 13.3. Means (±SE) of triplicate measurements for two separate runs (n=6). *Reproduced from Jayasekera* et al. *(2011) © Elsevier (2011); reproduced with permission from Elsevier.*

and TR contents, however, were partly attributed to a low degree of withering and high chlorophyll content in the rainy season. They concluded that when more hot dry air was circulated through the withering troughs to increase the evaporation of leaf moisture, the quality of tea also increased.

Previous studies have assumed that large differences in climate are necessary for significant quality differences to be observed. It has been thought that a superior genotype selected in one location maintains its desirable attributes despite location. However, tea plants selected in one location and planted in other locations have usually not matched the performance at the site of selection (Wachira *et al.*, 2002). One reason for such differences has been altitude, which affects rates of growth, even when other agronomic/cultural practices are similar. From the results obtained in Kenya, where trials were conducted within a 10-km radius, making climatic variations minimal or nearly constant (Owuor *et al.*, 1990; Ng'etich *et al.*, 2001), it was assumed that yields and quality would change with the altitude in a predictable manner, though not necessarily at the same rates for different cultivars. Consequently, a genotype with desirable attributes would maintain these attributes relative to other genotypes, wherever they were grown within the same country. However, there are tea genotypes which are more resistant to yield variations with location of production (Wachira *et al.*, 2002). Owuor *et al.* (2010) assessed 20 commercial genotypes under identical management in three locations within Kenya and revealed significant differences in plain black tea quality parameters. Their results suggested that it is not possible to produce tea of the same quality even from the same cultivars when the production location is varied.

The production of black tea from the same vegetatively propagated cultivars in Kenya and Malawi has shown variations in both chemical composition and quality (Owuor *et al.*, 2008). Whereas it is possible to produce black teas with similar total TFs and individual TFs, brightness, and total color levels, black teas from Kenya generally have higher TRs, total volatile flavor compounds, and flavor index. However, the pattern in the changes in the individual TFs or theaflavins digallate equivalent followed that of total (flavognost) TFs, suggesting that the flavan-3-ols patterns in tea leaf might not have been affected by the geographical area of production. The variations in black tea quality between Malawi and Kenya were attributed to differences in environmental conditions, leading to different shoot growth rates and biochemical composition in the shoots (Owuor *et al.*, 2008).

Caffeine content has also been reported to be influenced by seasonal, genetic, agronomic and cultural factors (Cloughley, 1982). The highest levels of caffeine were reportedly produced during the peak harvesting season when shoot growth rate was most rapid. Teas produced in the off season contained about 50% less caffeine (Cloughley, 1982).

Husbandry and Varietal Factors

Different agronomic/cultural practices or cultivation methods have been reported to affect tea quality by altering the biosynthesis of flavonoids. Shade has been reported to be one such factor that effectively improves the quality of tea beverages by causing a reduction in the concentration of flavonoids, the main compounds that contribute to astringency, in the leaves (Wang *et al.*, 2012). Photoinhibition is easily generated in tea plants in high light and nutritional deficiency conditions (Mohotti and Lawlor 2002). Shade treatment can effectively relieve or eliminate the generation of photosynthetic inhibition, and improve the quality of tea beverages (Zhang *et al.*, 2004). The leaves of tea plants grown in the shade contain high amino acid but low catechin contents (Ku *et al.*, 2010). Wang *et al.* (2012) analyzed expression of the flavonoid pathway genes in relation to the qualitative and quantitative composition of flavonoids in the unfermented leaves of tea, with particular attention being paid to the differences between leaves exposed to direct sunlight and leaves grown under a shade net ($20 \pm 5\%$ light transmitted). They reported that shade had notable effects on flavonoid (including catechins, O-glycosylated flavonols) contents, suggesting activation of their biosynthesis by sunlight.

Plucking standards have an impact on the quality of black tea. Coarser plucking (three or four leaves and a bud) produces lower-quality teas than finer plucking (first two leaves and a bud), which has been reported to be due to the decline in catechin levels (Forrest and Bendall, 1969) and changes in polyphenol oxidase isoenzyme composition and activity (Thanaraj and Seshadri, 1990), leading to a general decline in the total theaflavin levels. The latter are also dependent on the tea fermentation time.

Tea bushes are pruned every few years to maintain them in a manageable condition. Pruning is followed by increased growth rate. Owuor and Langat (1988) showed that with a longer time gap after pruning, the quality of tea increased. It is known that the younger leaves and the bud are nurtured by the roots and the mature leaves in the upper half of the bush, which supply at least part of their photosynthate to the growing leaves. The more mature leaves in the lower part of the bush supply photosynthate to the roots. The excision of any part of the bush, such as pruning, can therefore influence metabolic functions of the plant and the quality of the tea (Wickremasinghe, 1978). Bokuchava and Skobeleva (1969) have reported that the levels of the two main flavonoids of tea (ECG and EGCG) fall sharply during the aging process of tea shoots, whereas the levels of another two catechins (EGC and EC) rise slightly (Bokuchava and Skobeleva 1969).

Also, the variety of the tea plant influences the quality of the tea. Tea plants were traditionally raised from genetically heterogeneous seed and referred to as 'seedling tea.' A vast majority of the tea plantations in the early days had 'seedling tea.' In the last six to seven decades the 'clonal tea,' which is the vegetatively propagated variety, became more popular and gradually replaced the seedling variety (Wickremasinghe, 1978).

In black tea production, the variety of tea preferred is *Camellia sinensis* var. assamica, which has relatively large leaves, whereas the preferred green tea is made from *Camellia sinensis* var. sinensis, where the leaves are very small. Chemically, the major difference is that the ratio of astringent compounds to nitrogenous compounds is higher in var. assamica than in var. sinensis (Haslam, 1989). Unoxidized polyphenols in green tea can produce a 'metallic taste' and, therefore, for high-quality green tea production, tea clones that produce low levels of polyphenols are required. These clones typically have very small, dark-green leaves. This property of being dark green is important because the desired dark appearance in tea is the result of the conversion of chlorophyll to phaeophytin during processing (Wickremasinghe, 1978). Aside from low polyphenol and high chlorophyll content, the other important property for green tea is a high level of amino acids. Therefore, the clones producing high levels of polyphenols and polyphenoloxidase, which are desirable for quality black tea, are not suited for quality green tea. A higher polyphenoloxidase in the leaf would require a longer period of steaming or pan firing to ensure enzyme deactivation, which would also lead to a loss of chlorophyll.

In efforts to enhance the health potency of tea, purple-colored-leaf tea cultivars have recently been developed in Kenya for the manufacture of a 'health tea product' (Kamunya *et al.*, 2009). Leaves from these cultivars are reported to contain anthocyanins and anthocyanidins, with the predominant anthocyanidin being malvidin. Kerio *et al.* (2013) reported that the tea products from the purple-colored-leaf tea cultivars had levels of TPC, total catechins and antioxidant activities (determined using DPPH) similar to those from the green-colored-leaf cultivars.

Manufacturing Practices

Fermentation of tea consists of the enzymatic oxidation by endogenous polyphenol oxidases and peroxidases of tea polyphenols, namely colorless flavanols that are partially converted into TFs and TRs, responsible for the characteristic aroma and color of black and oolong teas (Obanda *et al.*, 2004). This oxidation process is accelerated by rupturing the withered tea leaves using orthodox rollers or machines (CTC: crush-tear-curl), so that the oxidases released can more readily react with oxygen. In unfermented teas, fermentation of the withered leaves is prevented by inactivating the endogenous enzymes through steaming or heating using different methods (pan frying, roasting, baking) before the rolling and drying process. Instead, white tea is the least processed in that it goes only through natural (sun) withering and drying. Furthermore, for white teas only buds that are still covered with fine white hair, and one or two very young leaves, are used (Balentine, 1992, Carloni *et al.*, 2012).

The manufacturing conditions for tea vary from one factory to another. These variations become significant where tea is manufactured using the orthodox or traditional methods. This has made it difficult to establish the influence of the manufacturing process *per se* on the composition of the final product. Manufacturing practices have also been reported to influence the antioxidant properties of the teas. Carloni *et al.* (2013) explored what effect manufacturing has on the antioxidant properties of African teas coming from a single cultivar but processed differently to give a white, two black (orthodox and CTC), and two green (low-caffeine and non-decaffeinated) teas. Their results contribute to a better understanding of how the manufacturing process influences the antioxidant activity of tea when variables (geographical region, environmental conditions, cultivar type, plucking techniques) are controlled. Using the 2,2′-azino-bis(3-ethylbenzthiazoline-6-sulfonic acid) (ABTS), oxygen radical absorbance capacity (ORAC) and low-density lipoprotein (LDL) assays, the following 'antioxidant profile' was obtained: green ≥ low-caffeine green > white ≥ black orthodox > black CTC, with a statistically significant correlation between the ORAC and LDL assays, whereas TPC and flavonoids significantly correlated with ABTS. Metal-chelating activity, which was lowest in the green teas, did not correlate with antioxidant activity but was reported to be influenced by TFs content. Their results also showed that novel African green, white, and black orthodox teas, made from tea varieties typically used in black CTC tea production, may have potential health benefits comparable with commonly consumed teas.

Antioxidant Properties of Tea: *In Vitro* and *In Vivo* Studies

Polyphenols act as antioxidants by scavenging reactive oxygen and nitrogen species and chelating redox-active transition metal ions. They may also act as indirect antioxidants by: (1) inhibiting redox-sensitive transcription factors, nuclear factor κB and activator protein-1; (2) inhibiting pro-oxidant enzymes such as inducible nitric oxide synthase, lipoxygenase, cyclooxygenases, and xanthine oxidase; and (3) induction of phase II and antioxidant enzymes, such as glutathione S-transferases and superoxide dismutases (Frei and Higdon 2003).

Numerous *in vitro* studies have shown that tea catechin is a potent hydrophilic antioxidant that scavenges oxygen radicals (Rice-evans *et al.*, 1995; Salah *et al.*, 1995) and chelates metal ions. Among catechins, ECG has shown the strongest antioxidant activity *in vitro* (Katiyar *et al.*, 1994). TFs and TRs present in black tea have been reported to show antioxidant properties in the same way as green tea containing ECs (Yoshino *et al.*, 1994).

Antioxidative activity and the order of effectiveness of green tea polyphenols as radical scavengers is: ECG > EGCG > EGC > gallic acid > EC = catechins. The oxidation of low-density lipoprotein is inhibited by catechins, EC, ECG, and EGCT to a similar degree, but not as much as in the presence of EGC or gallic acid (Salah *et al.*, 1995). The relative antioxidative capacity of black tea polyphenols in both aqueous and lipophilic phases is: theaflavin digallate (TFDG) > theaflavin 3^1-gallate (TF3^1G) = theaflavin 3-gallate (TF3G) > TF. Furthermore, research has shown that the efficacy of antioxidant activity of a reconstituted green tea polyphenol mixture is lower than that of the extracted crude mixture itself, which indicates that other components in the mixture possess their own antioxidant activity or act synergistically with the catechins (Amarowicz and Shahidi, 1995).

Flavonoid compounds with similar chemical structures exhibit comparable trends in antioxidant activity (Fukumoto and Mazza, 2000). This activity usually increases with an increase in the number of hydroxyl groups and a decrease in glycosylation. Relative antioxidant effectiveness of tea polyphenols in an aqueous system has been found to be different from that tested in a lipid system (Aruoma, 1997), which indicates that the activity of flavonoids is influenced by environmental factors. Some flavonoids have shown synergism when they are tested in a mixture (Aruoma, 1997).

Green tea is known to have a high antioxidant potential because of its greater concentration of hydroxyl-rich polyphenols. TFs and TRs in black tea have also been shown to be a mixture of heterogeneous polyphenols (Yoshino *et al.*, 1994). Black tea has shown similar or higher antioxidative activity than green tea in the inhibition of lipid peroxidation in rat liver (Yoshino *et al.*, 1994). While both green and black teas are good sources of antioxidant polyphenols, more research is necessary to establish the mechanisms associated with antioxidant activity of black teas.

In contrast to *in vitro* studies, several *in vivo* studies have produced conflicting data in relation to the antioxidant activity of tea. Serafini *et al.* (1996) reported that in men a single dose equivalent to 300 ml (2 cups) of green tea increases resistance against radical initiator-induced peroxidation in plasma at 30–50 min after ingestion. Ishikawa *et al.* (1997) found that daily consumption of 750 ml (5 cups) of black tea for 4 weeks lowered the susceptibility of LDL to oxidative modification. However, Van het Hof *et al.* (1997) reported that daily consumption of 900 ml (6 cups) of green tea for 4 weeks had no effect on the resistance of LDL to oxidation. In this study, human plasma antioxidant activity was measured after overnight fasting (i.e., at least 12 h after they consumed tea). Therefore it may be that the effects of the catechins were not observed due to the catechins having been metabolized by the time the measurements were made.

In summary, the limited amount of research that has been undertaken has not generally revealed significant decreases in biomarkers for oxidative damage in humans. Lifestyle factors and genetic variability in the response of humans to oxidative stress may mask small changes in biomarkers induced by tea polyphenols.

Antioxidant capacity of the plasma after the consumption of tea remains as the main piece of indirect evidence for the absorption of tea polyphenols through the gut barrier (Serafini *et al.*, 1996, Van het Hof *et al.*, 1999) and their likely physiological effects.

The mechanisms related to TF and TR absorption however are far from clear.

CONCLUDING REMARKS

Tea is now viewed as a product containing physiologically active molecules with antioxidant properties. The experimental evidence generated to date suggests that consumption of tea or tea polyphenols results in modest transient increases in plasma antioxidant capacity in humans. However, the polyphenolic compounds present in tea have been reported to vary considerably with the source of tea, the production location, the season of tea production, and tea processing, ultimately affecting the antioxidant properties of tea. Based on the knowledge gained through this review, the authors feel that more research should be directed to explore the interactions between various environmental/husbandry factors and how these interactions affect the tea polyphenols and overall antioxidant quality (*in vitro* and human studies) of different teas.

References

Amarowicz, R., Shahidi, F., 1995. Antioxidant activity of green tea catechins in a β-carotene-linoleate model system. J. Food Lip. 2, 47–56.

Aruoma, O.I., 1997. Antioxidant and pro-oxidant actions of plant extracts and plant-derived compounds. Proc. Nutri. Soc. Aust. 21, 85–90.

Balentine, D., 1992. Manufacturing and chemistry of tea. In: Ho, C.-T., Osawa, T., Huang, M.-T., Rosen, R.T. (Eds.), Phenolic compounds in food and their effects on health. American Chemical Society, Washington, DC, pp. 102–117.

Biswas, A.K., Sarkar, A.R., Biswas, A.K., 1973. Biological and chemical factors affecting the valuations of North-East Indian plain teas. III. Statistical evaluation of the biochemical constituents and their effects on colour, brightness and strength of black teas. J. Sci. Food Agric. 24, 1457–1477.

Bokuchava, M.A., Skobeleva, N.I., 1969. The chemistry and biochemistry of tea and tea manufacture. Adv. Food Res. 17, 215–292.

Brown, S., Griffiths, L.A., 1981. New metabolites of the naturally occurring mutagen, quercetin, the pro-mutagen, rutin and taxifolin. Experimentia 39, 198–200.

Bryce, T., Collier, P.D., Fowlis, I., Thomas, P.E., Frost, D., Wilkins, C.K., 1970. The structures of the theaflavins of black tea. Tetrahedron Lett. 32, 2789–2792.

Carloni, P., Tiano, L., Padella, L., Bacchetti, T., Customu, C., Kay, A., Damiani, E., 2013. Antioxidant activity of white, green and black tea obtained from the same tea cultivar. Food Res. Int. 53, 900–908.

Chan, E.W.C., Lim, Y.Y., Chong, K.L., Tan, J.B.L., Wong, S.K., 2010. Antioxidant properties of tropical and temperate herbal teas. J. Food Com. Ana. 23, 185–189.

Chan, E.W.C., Lim, Y.Y., Chew, Y.L., 2007. Antioxidant activity of Camellia sinensis leaves and tea from a lowland plantation in Malaysia. Food Chem. 102, 1214–1222.

Chen, Y., Jiang, Y., Duan, J., Shi, J., Xue, S., Kakuda, 2010. Variation in catechin contents in relation to quality of 'Huang Zhi Xiang' Oolong tea (*Camellia sinensis*) at various growing altitudes and seasons. Food Chem. 119, 648–652.

Cloughley, J.B., 1982. Factors influencing the caffeine content of black tea: Part 1—The effect of field variables. Food Chem. 9, 269–276.

Caffin, N., D'Arcy, B., Yao, L., Rintoul, G., 2004. Developing and index of quality for Australian Tea. Rural Industries Res. Dev. Corporation Publication No. 04/033.

Coxon, D.T., Holmes, A., Ollis, W.D., Vora, V.C., 1970. The constitution and configuration of the theaflavin pigments of black tea. Tetrahedron Lett. 60, 5237–5240.

Dufresne, C.J., Farnworth, E.R., 2001. A review of latest research findings on the health promotion properties of tea. J. Nutr. Biochem. 12, 404–421.

Forrest, L.L., Bendall, D.S., 1969. The distribution of polyphenols in the tea plant (Camellia sinensis L.). Biochem. J. 113, 741–755.

Frei, B., Higdon, J., 2003. Antioxidant activity of tea polyphenols in vivo: evidence from animal studies. J. Nutr. 133, 3275–3284.

Fukumoto, L.R., Mazza, G., 2000. Assessing antioxidant and prooxidant activities of phenolic compounds. J. Agric. Food Chem. 48, 3597–3604.

Gulati, A., Ravindranath, S.D., 1996. Seasonal Variations in Quality of Kangra Tea (Camellia sinensis(L) O Kuntze) in Himachal Pradesh. J. Sci. Food Agric. 71, 231–236.

Haslam, E., 1989. Plant polyphenols: vegetable tannins revisited. Cambridge University Press, Cambridge, pp. 154–219.

Hilton, P.J., 1973. Tea. In: Snell, F.D., Ettre, L.S. (Eds.), Encyclopedia of Industrial Chemical Analysis, vol. 18. John Wiley and sons Inc., New York, pp. 455–518.

Hilton, P.J., Palmer-Jones, R., 1973. Relationship between the flavanol composition of fresh tea shoots and the theaflavin content of manufactured tea. J. Sci. Food Agric. 24, 813–818.

Ishikawa, T., Szukawa, M., Ito, T., Yoshida, H., Ayaori, M., Nishiwaki, M., Yonemura, A., Hara, Y., Nakamura, H., 1997. Flavonoid supplementation on the susceptibility of low-density lipoprotein to oxidative modification. Am. J. Clin. Nutr. 66, 261–266.

Jayasekera, S., Molan, A.L., Garg, M., Moughan, P.J., 2011. Variation in antioxidant potential and total polyphenol content of fresh and fully-fermented Sri Lankan tea. Food Chem. 125, 536–541.

Kamunya, S.M., Wachira, F.N., Pathak, R.S., Muoki, R.C., Wanyoko, J.K., Ronno, W.K., 2009. Quantitative genetic parameters in tea (*Camellia sinensis* (L.) O. Kuntze:): I. combining abilities for yield, drought tolerance and quality traits. Afr. J. Plant Sci. 3, 93–101.

Katiyar, S.K., Agarwal, R., Mukhtar, H., 1994. Inhibition of spontaneous and photoenhanced lipid peroxidation in mouse epidermal microsomes by epicatechin derivatives from green tea. Cancer Lett. 79, 61–66.

Kerio, L.C., Wachira, F.N., Wanyoko, J.K., Rotich, M.K., 2013. Total polyphenols, catechin profiles and antioxidant activity of tea products from purple leaf coloured tea cultivars. Food Chem. 136, 1405–1413.

Ku, K.M., Choi, J.N., Kim, J., Kim, J.K., Yoo, L.G., Lee, S.J., Hong, Y.S., Lee, C.H., 2010. Metabolomics analysis reveals the compositional differences of shade grown tea (*Camellia sinensis* L.). J. Agric. Food Chem. 58, 418–426.

Modder, W.W.D., Amarakoon, A.M.T., 2002. Tea and Health. The Tea Research Institute, Sri Lanka.

Mohotti, A.J., Lawlor, D.W., 2002. Diurnal variation of photosynthesis and photoinhibition in tea: effects of irradiance and nitrogen supply during growth in the field. J. Exp. Bot. 53, 313–322.

Nakagawa, M., Torri, H., 1964. Studies on the flavonols in tea. Part II. Variation in the flavonolic constituents during the development of tea leaves. Agric. Biol. Chem. 28, 497–504.

Ng'etich, W.K., Stephens, W., Othieno, C.O., 2001. Responses of tea to environment in Kenya. 3. Yield and yield distribution. Experim. Agric. 37, 361–372.

Obanda, M., Owuor, P.O., Mang'oka, R., 2004. Changes in thearubigen fractions and theaflavins levels due to variations in processing conditions and their effects on black tea liquor brightness and total colour. Food Chem. 85, 163–173.

Owuor, P.O., Langat, J.K.A., 1988. Changes in chemical composition of tea due to pruning. Trop. Sci. 28, 127–132.

Owuor, P.O., Obaga, S.O., Othieno, C.O., 1990. Effects of altitude on the chemical composition of black tea. J. Sci. Food Agric. 50, 9–17.

Owuor, P.O., Obanda, M., Nyirenda, H.E., Mandala, W.L., 2008. Influence of region of production on clonal black tea chemical characteristics. Food Chem. 108, 263–271.

Owuor, P.O., Tsushida, T., Horita, H., Murai, T., 1986. Comparison of the chemical composition of black teas from main tea producing parts of the world. Tea 7, 71–78.

Owuor, P.O., Wachira, F.N., Ng'etich, W.K., 2010. Influence of region of production on relative clonal plain tea quality parameters in Kenya. Food Chem. 119, 1168–1174.

Rice-evans, C.A., Miller, N.J., Bolwell, P.G., Bramley, P.M., Pridham, J.B., 1995. The relative antioxidant activities of plant-derived polyphenolic flavonoids. Free Rad. Res. 22, 375–383.

Roginsky, V., Barsukova, T., Hsu, C.F., Kilmartin, P.A., 2003. Chain-breaking antioxidant activity and cyclic voltammetry characterization of polyphenols in a range of green, oolong, and black teas. J. Agric. Food Chem. 51, 5798–5802.

Salah, N., Miller, N.J., Paganga, G., Tijburg, L., Bolwell, G.P., Rice-evans, C.A., 1995. Polyphenolic flavanols as scavengers of aqueous phase radicals and as chain-breaking antioxidants. Arch. Biochem. Biophy. 322, 339–346.

Serafini, M., Ghiselli, A., Ferro-Luzzi, A., 1996. In vivo antioxidant effect of green and black tea in man. Eur. J. Clin. Nutr. 50, 28–32.

Sud, R.G., Baru, A., 2000. Seasonal variations in theaflavins, thearubigins, total colour and brightness of Kangra orthodox tea (*Camellia sinensis* (L) O Kuntze) in Himachal Pradesh. J. Sci. Food Agric. 80, 1291–1299.

Thanaraj, S.N., Seshadri, S.R., 1990. Influence of polyphenol oxidase activity and polyphenol content of tea shoot on black tea. J. Sci. Food Agric. 51, 57–69.

Van het Hof, K.H., De Boer, H.S.M., Wiseman, S.A., Lien, N., Westsrtate, J.A., Tijburg, L.B.M., 1997. Consumption of green or black tea does not increase resistance of low density lipoprotein to oxidation in humans. Am J. Clin. Nutr. 66, 1125–1132.

Van het Hof, K.H., Wiseman, S.A., Yang, C.S., Tijburg, L.B., 1999. Plasma and lipoprotein levels of tea catechins following repeated tea consumption. Proc. Soc. Exp. Biol. Med. 220, 203–209.

Venditti, E., Bacchetti, T., Carloni, P., Greci, L., Damiani, E., 2010. Hot vs. cold water steeping of different teas: Do they affect antioxidant activity? Food Chem. 119, 1597–1604.

Wachira, F.N., Ng'etich, W., Omolo, J., Mamati, G., 2002. Genotype x environment interactions for tea yields. Euphtica 127, 289–296.

Wang, H.F., Tsai, Y.S., Lin, M.L., Ou, A.S.M., 2006. Effects of different variety, production area and season on the major chemical constituents in Taiwan GABA tea. Taiwanese J. Agric. Chem Food Sci. 44, 90–99.

Wang, Y., Gao, L., Shan, Y., Liu, Y., Tian, Y., Xia, T., 2012. Influence of shade on flavonoid biosynthesis in tea (*Camellia sinensis* (L.) O. Kuntze). Sci. Horticulturae 141, 7–16.

Wickremasinghe, R.L., 1978. Monographs on Tea Production in Sri Lankavol. 7. Facets of Tea research in Practice, TRI, Sri Lanka.

Wiseman, S.A., Balentine, D.A., Frei, B., 1997. Antioxidants in tea. Crit. Rev. Food Sci. Nutr. 37, 705–708.

Yokozawa, T., Dong, E., Nakagawa, T., Kashiwagi, H., Nakagawa, H., Takeuchi, S., Chung, H.Y., 1998. In vitro and in vivo studies on the radical scavenging activity of tea. J. Agric. Food Chem. 46, 2143–2150.

Yoshino, K., Hara, Y., Sano, M., Tomita, I., 1994. Antioxidative Effects of Black Tea Theaflavins and Thearubigin on Lipid Peroxidation of Rat Liver Homogenates Induced by tert-Butyl Hydroperoxide. Biol. Pharm. Bull. 17, 146–149.

Zhang, W.J., Liang, Y.R., Zhang, F.Z., Chen, C.S., Zhang, Y.G., Chen, R.B., Weng, B.Q., 2004. Effects on the yield and quality of oolong tea by covering with shading net. J. Tea Sci. 24, 276–282 (in Chinese).

Zuo, Y., Chen, H., Deng, Y., 2002. Simultaneous determination of catechins, caffeine and gallic acids in green, oolong, black and pu-erh teas using HPLC with a photodiode array detector. Talanta 57, 307–316.

CHAPTER

14

Antioxidants of Rooibos Beverages: Role of Plant Composition and Processing

Elizabeth Joubert, Dalene de Beer

Post-Harvest & Wine Technology Division, Agricultural Research Council, Infruitec-Nietvoorbij Institute, Stellenbosch, South Africa

CHAPTER POINTS

- Rooibos is popular as a herbal tea and source of food ingredient extract.
- Aspalathin, a novel dihydrochalcone C-glucoside, is the major flavonoid and antioxidant in the unprocessed plant material.
- Traditional processing of rooibos includes 'fermentation,' resulting in oxidative changes and development of red–brown leaf and infusion color.
- Processing of green rooibos aims to prevent oxidative changes and retention of green leaf color while having a yellow–orange infusion.
- Aspalathin is unstable under traditional processing (fermentation) conditions.
- Oxidation products of aspalathin include isoorientin, orientin, colorless dimers, yellow benzofurans, and brown polymers.

INTRODUCTION

Rooibos, produced from the endemic South African fynbos plant, *Aspalathus linearis* (Brum.f) Dahlg. (family Fabaceae; tribe Crotalarieae), has become popular on the global markets, experiencing commercial success as a caffeine-free herbal tea and antioxidant food ingredient (Joubert and de Beer, 2011). The potential of a flavone-C-glycoside-rich extract of fermented rooibos to stabilize acai fruit anthocyanins was recently demonstrated (Pacheco-Palencia and Talcott, 2010), indicating new applications. In 2010, the international market for rooibos was 5633 tons compared to 750 tons in 1993. Apart from South Africa, major markets are Germany and the Netherlands. With no commercial value at the beginning of the 20th century, its popularity has increased to such an extent that a recent report by the Swiss Business Hub South Africa (Anonymous, 2007) stated that "Rooibos appears to be headed towards becoming the second most commonly consumed beverage tea ingredient in the world after ordinary tea (*Camellia sinensis*)."

Originally different *Aspalathus* species and ecotypes were used to produce rooibos tea (Dahlgren, 1968, 1988), but today only the Rocklands type (so-called red type) of *A. linearis* subsp. *linearis* which turns red–brown when bruised, is of commercial importance. The Rocklands type is divided into the Cederberg type (wild-growing) and the selected, improved Nortier type. The latter has bright-green, needle-like leaves on straight, slender branches. The leaves of the Cederberg type are coarser and broader. Wild-harvesting is limited, as approximately 36,000 ha of the Nortier type are under cultivation, largely in the Cederberg area in the Western Cape Province of South Africa. Annual production is very dependent on rainfall, leading to a cycle of shortages and overproduction. Maximum production to date reached 18,000 tons in 2008, while exports peaked at 7176 tons in 2007 (Joubert and de Beer, 2011).

A common name used to describe rooibos is red bush (translation of rooibos). It refers to the traditional 'fermented' product with its red–brown leaf and infusion color. Another variation on the name, recently in use, is red rooibos. These descriptive names have become necessary to differentiate between the fermented and unfermented product. Since the early 2000s, a small demand exists for unfermented or green rooibos, a product with higher levels of antioxidants (Joubert and Schulz, 2006), but without the characteristic honey-like, woody and herbal–floral flavor notes and slightly sweet taste of the fermented product (Koch *et al.*, 2012). Green rooibos has a grassy, hay-like aroma, a green leaf color, and

Processing and Impact on Antioxidants in Beverages
http://dx.doi.org/10.1016/B978-0-12-404738-9.00014-3

yellow–orange infusion color. In many instances rooibos is sold on global markets as a flavored product (e.g., vanilla) or mixed with other herbs and/or dried fruits, but in South Africa the pure product is preferred, although more mixtures and flavored products are appearing on the market. Rooibos extract is used as 'instant tea' and an ingredient in food supplements, ready-to-drink flavored teas, jam, and food products such bread and yoghurt.

PHENOLIC COMPOSITION

Table 14.1 gives the structures of some of the novel and common phenolic compounds present in rooibos. Quantitative data on a number of rooibos phenolic compounds in the plant material, extracts, and infusions are summarized in Table 14.2. The plant is characterized by a high content of dihydrochalcones, in particular

TABLE 14.1 Structures of the Major Rooibos Phenolic Compounds

General Structure	Phenolic Compound	Substituents
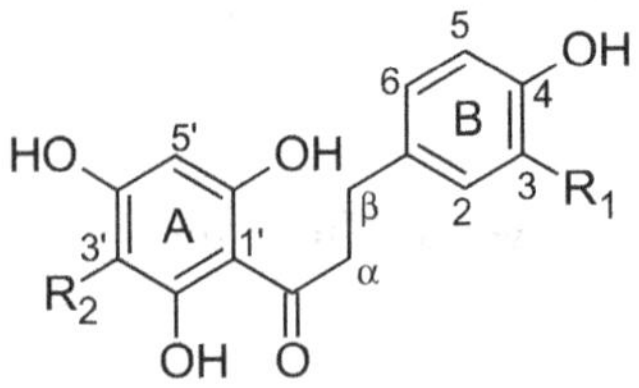	*Dihydrochalcones*	
	Aspalathin	R_1 = OH, R_2 = C-β-D-glucosyl
	Nothofagin	R_1 = H, R_2 = C-β-D-glucosyl
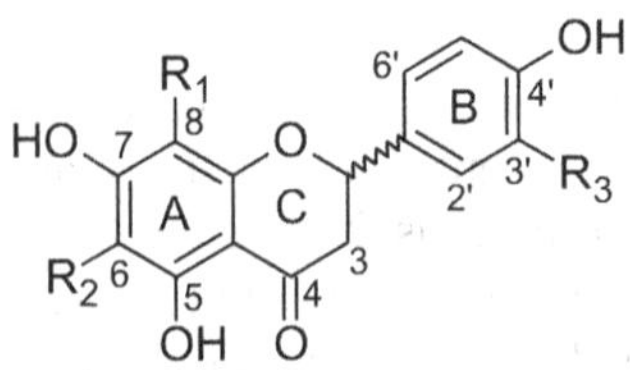	*Flavanones*[a]	
	(*R*)/(*S*)-eriodictyol-8-C-glucoside	R_1 = C-β-D-glucosyl, R_2 = H
	(R)/(*S*)-eriodictyol-6-C-glucoside	R_1 = H; R_2 = C-β-D-glucosyl
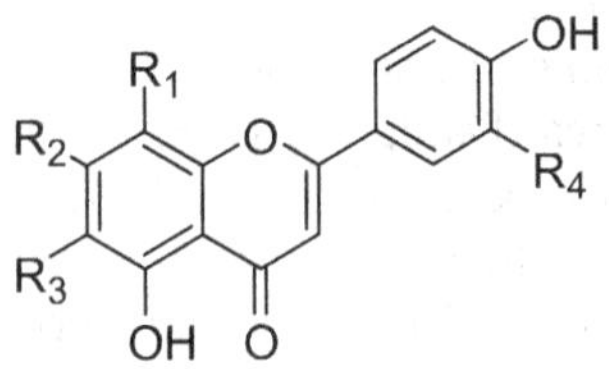	*Flavones*[a]	
	Orientin	R_1 = C-β-D-glucosyl, R_2 = R_4 = OH, R_3 = H
	Isoorientin	R_1 = H, R_2 = R_4 = OH, R_3 = C-β-D-glucosyl
	Vitexin	R_1 = C-β-D-glucosyl, R_2 = OH, R_3 = R_4 = H
	Isovitexin	R_1 = R_4 = H, R_2 = OH, R_3 = C-β-D-glucosyl
	Luteolin	R_1 = R_3 = H, R_2 = R_4 = OH
	Luteolin-7-*O*-glucoside	R_1 = R_3 = H, R_2 = *O*-β-D-glucosyl, R_4 = OH
	Chrysoeriol	R_1 = R_3 = H, R_2 = OH, R_4 = OCH_3
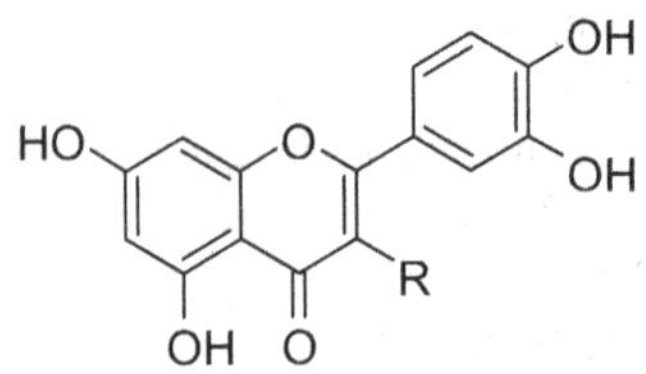	*Flavonols*[a]	
	Quercetin	R = OH
	Isoquercitrin	R = *O*-β-D-glucosyl
	Hyperoside	R = *O*-β-D-galactosyl
	Rutin	R = *O*-β-D-rutinosyl
	Quercetin-3-*O*-robinobioside	R = *O*-β-D-robinobiosyl
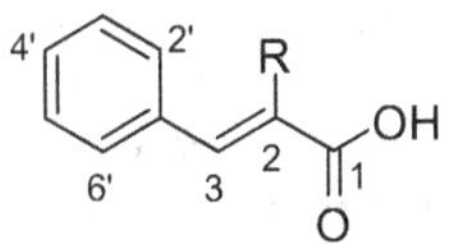	*Phenylpropanoid*	
	Enolic phenylpyruvic acid-2-*O*-glucoside (PPAG)	R = *O*-glucosyl
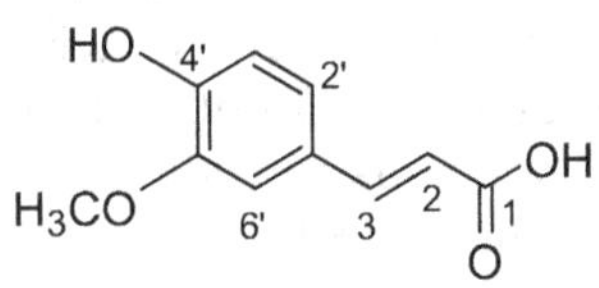	Hydroxycinnamic acid	
	Ferulic acid	

[a]*Numbering of the carbon skeleton for flavones and flavonols is the same as indicated for flavanones.*

TABLE 14.2 Phenolic Composition of Whole, Dried Rooibos Shoots[a], Dried Rooibos Leaves[a], Fermented Rooibos Product[a], Fermented Rooibos Infusions[b], and Fermented Rooibos Food Ingredient Extracts[c]

Compounds	Whole, Dried Shoots[d] (n = 20)	Ferm. Product[d] (n = 20)	Whole, Dried Shoots[e] (n = 97)	Ferm. Product[e] (n = 89)	Whole, Dried Shoots[f] (n = 3)	Ferm. Product[f] (n = 3)	Dried Leaves[g] (n = 54)	Ferm. Product[g] (n = 10)	Ferm. Infusion[h] (n = 114)	Ferm. Extract[e] (n = 9)	Ferm. Extract[i] (n = 74)
Aspalathin	4.9	0.1	6.6 (3.8–9.7)	0.3 (0.02–1.2)	2.56	0.42	9.7 (6.0–13.5)	0.10 (0.04–0.14)	5.8 (nd–3.3)	0.35	0.58 (0.16–1.52)
Nothofagin	—	—	0.7 (0.2–1.2)	0.1 (nd–0.4)	0.25	0.04	1.0 (0.5–1.8)	0.01 (0.01–0.02)	1.0 (nd–2.8)	0.11	0.069 (0.03–0.18)
Orientin	—	—	—	—	0.26	0.20	0.8 (0.5–1.2)	0.22 (0.19–0.24)	10.8 (10.3–14.3)	0.16	0.79 (0.44–0.90)
Isoorientin	—	—	—	—	0.45	0.33	1.5 (0.9–2.2)	0.34 (0.30–0.37)	15.0 (7.4–20.5)	0.05	0.83 (0.47–1.03)
Vitexin	—	—	—	—	0.04	0.04	—	—	2.3 (1.3–3.3)	0.24	—
Isovitexin	—	—	—	—	0.05	0.04	—	—	2.4 (1.4–3.3)	0.09	—
Luteolin	—	—	—	—	0.01	0.01	—	—	nd	traces	—
Luteolin-7-*O*-glc	—	—	—	—	0.02	0.02	—	—	nd	—	—
Chrysoeriol	—	—	—	—	traces	0.01	—	—	nd	traces	—
Quercetin	—	—	—	—	traces	0.01	—	—	nd	traces	—
Isoquercitrin	—	—	—	—	—	—	—	—	1.1 (nd–5.8)	—	—
Hyperoside	—	—	—	—	0.02	0.02	—	—	2.2 (nd–6.8)	—	—
Quercetin-3-*O*-rob	—	—	—	—	0.25[j]	0.17[j]	—	—	8.3 (0.9–18.4)	—	—
Rutin	—	—	—	—			—	—	1.7 (nd–5.7)	—	—
PPAG	—	—	—	—	—	—	0.1 (nd–0.8)	0.07 (0.03–0.11)	6.9 (2.7–14.8)	—	—
Ferulic acid	—	—	—	—	—	—	—	—	1.5 (nd–3.1)	—	—

[a]*g/100 g plant material*
[b]*mg/l infusion*
[c]*g/100 g dried extract*
[d]*Schulz* et al., *2003*
[e]*Joubert and Schulz, 2006*
[f]*Joubert and De Beer, 2011*
[g]*Joubert* et al., *2013*
[h]*Joubert* et al., *2012*
[i]*Joubert and De Beer, 2012*
[j]*sum of rutin and quercetin-3-O-robinobioside*
Ferm., fermented; glc, glucoside; PPAG, enolic phenylpyruvic acid-2-*O*-glucoside; rob, robinobioside.

aspalathin, a *C*-glucoside which is unique to *A. linearis*. The 3-deoxy analog of aspalathin, nothofagin, is not limited to rooibos, but its occurrence is not widespread in nature (Joubert and de Beer, 2011). Recently, the presence of a C5′-hexosyl derivative of aspalathin was demonstrated in an infusion of green rooibos (Beelders *et al.*, 2012b). Analysis of a large number of unprocessed (unfermented) plant material samples (n = 97), with most samples representing several shoots from an individual bush, showed large variations in aspalathin content, ranging from 3.84 to 9.66% (average 6.62%). Rooibos plants are cultivated from seedlings, resulting in large variations in phenolic composition between plants. The nothofagin content of the same sample set was substantially lower, ranging from 0.20 to 1.24% (average 0.67%) (Joubert and Schulz, 2006). High levels of aspalathin in individual samples were not necessarily accompanied by high levels of nothofagin. Values as high as 13.48 and 1.75% for aspalathin and nothofagin, respectively, have been found for leaves only (Joubert *et al.*, 2013).

Relatively high quantities of the flavone analogs of aspalathin, orientin, and isoorientin, are present in rooibos. The plant material normally contains more isoorientin than orientin. Green rooibos samples (n = 3) with an average aspalathin content of 2.56% contained these flavones at 0.26 and 0.45%, respectively (Joubert and de Beer, 2011). Much lower quantities were observed for vitexin and isovitexin, the flavone analogs of nothofagin.

Apart from interest in the major compounds of rooibos, the phenolic composition of the unprocessed plant material has not yet been studied extensively. Two-dimensional high-performance liquid chromatography (HPLC) has been applied for in-depth qualitative analysis of green rooibos (Beelders *et al.*, 2012a), confirming the presence of many compounds as yet unidentified. The focus to date has been on the identification of phenolic compounds in the traditional fermented and thus oxidized product and its infusions and extracts (Rabe *et al.*, 1994; Ferreira *et al.*, 1995; Kazuno *et al.*, 2005; Shimamura *et al.*, 2006; Beltrán-Debón *et al.*, 2011; Iswaldi *et al.*, 2011; Beelders *et al.*, 2012a,b), due to their commercial importance. Compounds identified include: the flavanones dihydro-orientin and dihydro-isoorientin (*R* and *S* configuration); flavones chrysoeriol, luteolin, luteolin-7-*O*-glucoside, carlinoside, neocarlinoside, isocarlinoside, and vicenin-2; flavonols quercetin and its *O*-linked glycosides, quercetin-3-*O*-robinobioside, hyperoside, isoquercitrin, and rutin; patuletin-7-*O*-glucoside; phenolic acids; lignans; and coumarins. The major phenolic acid is ferulic acid, while quercetin-3-*O*-robinobioside is the major quercetin glycoside and not rutin as previously believed (Beelders *et al.*, 2012b). Quercetin-3-*O*-robinobioside comprises ca. 0.9% of the soluble solids present in infusions of green and fermented rooibos, whilst their rutin contents were less than 0.5% (Beelders *et al.*, 2012b). It is noteworthy that the aglycones quercetin, luteolin, and chrysoeriol are either not detected in infusions or only present in trace amounts (Joubert and de Beer, 2012). Apart from ferulic acid, Rabe *et al.* (1994) demonstrated the presence of several benzoic and cinnamic acids in fermented rooibos, but none have been detected in infusions. A novel compound, 5,7-dihydroxy-6-*C*-β-D-glucopyranosyl chromone, is postulated to form from dihydro-isoorientin during fermentation (Ferreira *et al.*, 1995). Recently, Iswaldi *et al.* (2011) confirmed its presence in extracts of fermented rooibos. Another novel compound in rooibos is the enolic phenylpyruvic acid glucoside (Marais *et al.*, 1996; Joubert *et al.*, 2013).

Beelders *et al.* (2012b) studied the composition of hot-water infusions of both fermented and green rooibos, identifying 28 compounds of which all, except for the C5′-hexosyl derivative of aspalathin, have been previously identified in fermented rooibos. Aspalalinin, a cyclic dihydrochalcone, was not present in green rooibos. It is a novel compound and its structure suggests that it may be an oxidation product formed during fermentation. Dihydro-orientin and dihydro-isoorientin, oxidation products of aspalathin (Marais *et al.*, 2000; Krafczyk and Glomb, 2008), were also present in the infusion of green rooibos. It is not clear to what extent these compounds are present naturally in the fresh, unprocessed plant material, given the presence of relatively high quantities of their flavone analogs or whether conditions during processing of the plant material and/or preparation of an infusion are conducive to their formation. Rooibos is very susceptible to oxidation and it is very difficult to prevent oxidative changes during processing. The oxidation mechanism of aspalathin, as recently elucidated by Glomb and co-workers (Krafczyk and Glomb, 2008; Krafczyk *et al.*, 2009a; Heinrich *et al.*, 2012), will be discussed in the section *Processing of Fermented Rooibos and Oxidative Conversion of Aspalathin*, below.

Little structural information is available on the polymeric fraction of rooibos. A polymer isolated from unprocessed plant material was demonstrated to be an irregular procyanidin type heteropolymer, containing (+)-catechin and (−)-epicatechin chain extending units and (+)-catechin as terminal unit (Marais *et al.*, 1998). The dimer, procyanidin B3, and the trimer, bis-fisetinidol-(4β,6 : 4β,8)-catechin, are present in very low concentrations in fermented rooibos (Ferreira *et al.*, 1995).

ANTIOXIDANT ACTIVITY

Antioxidant Activity of Rooibos Flavonoids and Phenolic Acids

The antioxidant properties of aspalathin, as the major flavonoid in green rooibos and one of the major compounds in fermented rooibos, are of interest. It is a potent radical scavenger in spite of the lack of an extended conjugated system for electron delocalization and radical stabilization. Structural features of importance are the C2′- and C6′-hydroxyl groups, the keto-enol tautomerism of the carbonyl and α-methylene groups, and the catechol arrangement on the B-ring (Snijman *et al.*, 2009). Its potency has been established in relation to several other rooibos phenolic compounds (Table 14.3) and epigallocatechin gallate (EGCG), the well-known flavanol antioxidant of green tea (*Camellia sinensis*). These studies used different antioxidant assays, measuring the ability of the compounds to scavenge free radicals, inhibit lipid perioxidation, reduce Fe^{3+}, or chelate Fe^{2+}. Aspalathin is a good radical scavenger ($O_2^{\bullet -}$, DPPH$^{\bullet}$ (2,2-diphenyl-1-picrylhydrazyl), ABTS$^{\bullet +}$ [2,2′-azino-bis(3-ethylbenzthiazoline-6-sulfonic acid)], and Fremy radical) and of comparable activity to quercetin (von Gadow *et al.*, 1997a; Joubert *et al.*, 2004; Krafczyk *et al.*, 2009b; Snijman *et al.*, 2009). Although quercetin is only present in trace amounts in rooibos infusions (Table 14.2), and thus is not a major contributor to their antioxidant activity, it serves as a good reference compound because it is a common flavonoid and well-known antioxidant. Aspalathin also performed similarly to EGCG in the ABTS$^{\bullet +}$ scavenging assay (Snijman *et al.*, 2009). Conversely, aspalathin offered substantially less protection against microsomal Fe^{2+}-induced lipid peroxidation than both quercetin and EGCG. Quercetin also demonstrated better protection against lipid peroxidation than aspalathin in the β-carotene bleaching test (Table 14.3). This test makes use of coupled oxidation of β-carotene and linoleic acid so that a higher antioxidant activity coefficient (AAC) value is an indication of a slower degradation (bleaching) rate of β-carotene as a result of inhibiting oxidation of linoleic acid. Under the high-temperature oxidative conditions of the Rancimat test, aspalathin and quercetin (both 0.02% m/m of the lard) prolonged the induction time of lipid peroxidation to 2.55 h and 26.93 h, respectively (Table 14.3) compared to 0.63 h of the control (von Gadow *et al.*, 1997a). In this test, lipophilicity of the compounds is important, thus favoring quercetin.

Nothofagin without the catechol arrangement on the B-ring presents an interesting case. As one of the rooibos dihydrochalcones and a major compound of green rooibos, its relative antioxidant activity was compared to aspalathin in several assays (Table 14.3). In the ABTS$^{\bullet +}$ scavenging assay, carried out by two different groups (Krafczyk *et al.*, 2009b; Snijman *et al.*, 2009), nothofagin was less effective as a radical scavenger than aspalathin. This was expected due to the absence of the 3-OH group and thus the loss of the catechol arrangement, yet it remained one of the most potent rooibos antioxidants tested and outperformed the flavone analogs of aspalathin, orientin, and isoorientin. It was postulated that the most likely conformation of aspalathin and nothofagin could lead to hydrogen bonding of the C2′- and C4′-hydroxyl groups and, respectively, the C2″-hydroxyl group and the ethereal oxygen of the glucopyranosyl moiety with a subsequent effect on the bond dissociation energy of the phenolic groups. This may compensate to some extent for the enhanced susceptibility of oxygen radical formation at the catechol group of aspalathin (Snijman *et al.*, 2009). However, in the Fremy's salt assay, practically no scavenging of the Fremy radical was demonstrated for nothofagin (Krafczyk *et al.*, 2009b). Similarly, it performed very poorly as an inhibitor of Fe^{2+}-induced lipid peroxidation in rat liver microsomes (Snijman *et al.*, 2009), yet it was effective in scavenging radicals formed during oxidation of low-density lipoprotein (LDL) in the presence of Cu^{2+}, giving a shorter lag time than aspalathin, but a longer lag time than orientin and isoorientin (Krafczyk *et al.*, 2009b). Apart from the ability to scavenge peroxide radicals, the membrane structure of the microsomes introduces additional factors that play a role such as planarity and hydrophobicity of the molecule. Its positioning at the membrane interphase is also important. These factors will determine its partitioning into the lipid bilayer and access to the lipid core where the radicals are formed (Snijman *et al.*, 2009). It was postulated that aspalathin has a more 'open' structure than nothofagin on the basis of their C2-C1-Cβ-Cα dihedral angles of the most likely conformers, improving the accessibility of the catechol group to interact with the polar heads of the lipid bilayers and thus scavenging free radicals at the membrane interphase. Furthermore, secondary antioxidant activity, namely metal chelation, is important, since Fe^{2+}-induced lipid peroxidation was employed. Once again the absence of the catechol group as binding site for metals could be a contributing factor to the low activity of nothofagin.

The presence of flavanones and flavones in combination with the parent dihydrochalcone, aspalathin, gives insight into the structure–activity relationships and thus the changes in antioxidant activity observed with processing of rooibos as conversion of aspalathin to flavones via flavanones occurs (Krafczyk and Glomb, 2008). This conversion also takes place during heating and storage of ready-to-drink rooibos iced tea (Joubert *et al.*, 2009; de Beer *et al.*, 2012). Only one study to date investigated the antioxidant activity of the chiral flavanone isomers of isoorientin, (*S*)-eriodictyol-6-*C*-β-D-glucopyranoside and (*R*)-eriodictyol-6-*C*-β-D-glucopyranoside (Krafczyk *et al.*, 2009b). These enantiomers were less effective in scavenging ABTS$^{\bullet +}$ and the Fremy radical than their corresponding flavone. Futhermore, the chirality is important, since the *R*-enaniomer was more effective than the *S*-enantiomer. The position of the glucoside moiety of the flavones also affects antioxidant activity. The effect

TABLE 14.3 Antioxidant Activity of Phenolic Compounds from Rooibos and Selected Reference Compounds

	Free Radical Scavenging Assays						Lipid Peroxidation Assays			Metal Chelation
Compounds	ABTS (IC_{50}, μM)[a]	ABTS (TEAC)[b]	DPPH (% Inhibition)[c]	Superoxide (% Inhibition)[d]	Fremy's Salt (AC)[e]	β-Carotene Bleaching (AAC)[f]	Microsomes (IC_{50}, μM)[g]	LDL (Lag Time, h)[h]	Rancimat (Lag Time, h)[i]	Abs Shift (nm)[j]
Aspalathin	3.33	2.62	91.74 (87.62)	81.01	4.81	357.14	50.2	6.2	2.55	0
Nothofagin	4.04	2.06			0.02		1388	4.3		0
(*S*)-eriodictyol-6-*C*-β-D-glucopyranoside		0.88			2.67			2.1		
(*R*)-eriodictyol-6-*C*-β-D-glucopyranoside		1.04			3.13					
Orientin	11.43	1.47	– (88.65)	72.52	3.48		137.9	2.7		27
Isoorientin	11.25	1.54	– (82.18)	63.32	3		480.7	3.8		25
Vitexin	>2313	0.86	– (3.99)	10.15	1.57	306.28	>2323			0
Isovitexin	1224	0.81			1.58		1689			0
Luteolin	10.82		90.85 (88.01)	57.83		522.69	185.9			26
Chrysoeriol	21.54		– (2.02)	32.93			217.7			0
Rutin	10.47	1.2	91.18 (66.75)	68.16	3.7	80.7	240.1			23
Isoquercitrin	12.89	1.23	91.99 (86.59)	66.67	4.08	224.14	111.3	9.6	4.17	19
Hyperoside	8.55	1.33			4.21		283.2			5
Quercetin	3.6	2.7	93.27 (91.11)	81.45	4.48	455.26	17.5		26.93	46
(+)-Catechin	6.49		92.65 (71.11)			65.41	53.3		27.23	0
(–)-EGCG	3.46						22.3			0
Procyanidin B3			– (90.16)							
Caffeic acid			93.65 (–)			–66.67			18.85	
Ferulic acid						328.16			1.26	
p-coumaric acid			– (58.10)	5.31		135.26			1.08	
p-hydroxybenzoic acid			8.52* (–)			175.14			1.29	
Protocatechuic acid			79.8 (–)			52.63			15.5	
Syringic acid			85.80* (–)			202.88				
Vanillic acid			20.66* (–)			350.65				

[a]*Snijman* et al. *(2009)*
[b]*Krafczyk* et al. *(2009)*
[c]*% inhibition at 0.25 mol compound/mol DPPH, first value from Von Gadow* et al. *(1997a) and second value (in brackets) from Joubert* et al. *(2004)*
[d]*% inhibition at 12.5 μmol/mL, Joubert* et al. *(2004)*
[e]*mol Fremy's salt/mol compound, Krafczyk* et al. *(2009)*
[f]*antioxidant activity coefficient, Von Gadow* et al. *(1997)*
[g]*iron-induced microsomal lipid peroxidation, Snijman* et al. *(2009)*
[h]*copper-induced low-density lipoprotein oxidation at 1 nmol, Krafczyk et al. (2009)*
[i]*oxidation of lard at 0.02 μg compound/100 μg lard, Von Gadow* et al. *(1997)*
[j]*shift in band I absorption with Fe^{2+}, Snijman* et al. *(2009); EGCG, epigallocatechin gallate.*
AAC, antioxidant activity coefficient; Abs, Absorbance; ABTS, 2,2′-azino-bis-3-ethylbenzthiazoline-6-sulfonic acid; AC, antioxidant capacity; DPPH, 1,1-diphenyl-2-picrylhydrazyl; IC_{50}, concentration of sample needed to obtain 50% inhibition; LDL, low-density lipoprotein; TEAC, Trolox equivalent antioxidant capacity.

is, however, less pronounced for orientin and isoorientin, than for vitexin and isovitexin and depends on the assay conditions (Table 14.3). Due to the absence of the B-ring catechol group, the flavone analogs of nothofagin, vitexin, and isovitexin, were less effective than orientin and isoorientin. Luteolin, the aglycone of the latter two compounds, is more effective in scavenging $ABTS^{\bullet+}$ but less effective in scavenging $O_2^{\bullet-}$ than its glycosides. Its ability to inhibit lipid peroxidation in relation to orientin and isoorientin was only investigated in the Fe^{2+}-induced lipid peroxidation assay, showing it to be more effective than isoorientin but less effective than orientin (Table 14.3). Its importance as an antioxidant in rooibos beverages is limited due to being present in trace amounts only.

Quercetin and its glycosides offer another interesting comparison. Based on the data of Snijman *et al.* (2009) and other literature as discussed therein, the sugar moiety not only occupies the 3-OH of the flavonol, but the bulk of the sugar moiety and the spatial arrangement of its hydroxyl groups in relation to the 4′-OH deprotonation site on the flavonol could play a role in their ability to scavenge free radicals. Attachment of the sugar moiety to the 3-OH results in a twisted conformation and reduces coplanarity of the B- and C-rings and thus electron delocalization. As a radical scavenger, hyperoside, the 3-*O*-galactoside of quercetin, performed better than the 3-*O*-glucoside (isoquercitrin) and 3-*O*-rutinoside (rutin) (Table 14.3). The major quercetin glycoside, namely quercetin-3-*O*-robinobioside (Table 14.2), has not yet been tested.

The phenolic acids of rooibos are less effective than aspalathin as radical scavengers (Table 14.3). In the Rancimat assay, protocatechuic acid induced a longer lag time than aspalathin.

Antioxidant Activity of Rooibos Extracts and Infusions

The relationship between antioxidant activity and, respectively, total polyphenol content (TPC) and aspalathin content, has been investigated. Similar to many other plant extracts, very good correlation between TPC and $ABTS^{\bullet+}$ scavenging activity was obtained for water extracts from unprocessed plant material [$r=0.99$; Joubert *et al.* (2008b)]. The sample set ($n=73$) comprised plant material originating from different production years and differing in the age of the plantations, age of regrowth after harvesting, harvesting time, etc. The aspalathin content of the same extracts also correlated very well with their $ABTS^{\bullet+}$ scavenging activity ($r=0.96$; Joubert *et al.*, 2008b). This is in agreement with the results for commercial green rooibos ($n=15$, $r=0.90$), but contradictory to results obtained with unprocessed plant material ($n=20$; $r=0.39$). In the latter study by Schulz *et al.* (2003) small sample sets were used. The apparent contribution of aspalathin to total antioxidant activity of green rooibos was estimated at 22–57% (Schulz *et al.*, 2003), but these values do not take any synergistic effects into account. Based on their activity and quantity, nothofagin, orientin, and isoorientin could be expected to contribute substantially to radical scavenging activity of green rooibos infusions and extracts.

Interest in the antioxidant activity of rooibos has led to its inclusion in many studies (Table 14.4 and 14.5). Fermentation of rooibos reduces its antioxidant activity (Winterton, 1999; Standley *et al.*, 2001; Bramati *et al.*, 2003; Schulz *et al.*, 2003; Joubert *et al.*, 2004, 2005, 2008c), although one study showed no effect with fermentation (von Gadow *et al.*, 1997b). Pertinent information and results of some comparative studies with other teas, including green and black teas and/or herbal teas such as mate, chamomile, and peppermint, are summarized in Table 14.5. From these studies it can be concluded that both green and fermented rooibos have lower antioxidant activities than green tea. Only three studies comparing rooibos with other teas and herbal teas included green rooibos (von Gadow *et al.*, 1997b; Bramati *et al.*, 2003; Joubert *et al.*, 2008c). The relative order of activity depends on the assay; however, most of the studies used a very limited number of samples. In many cases only one sample of each herb or tea was used. This does not allow for natural variation in composition of plants.

A recent study investigated the variation in composition and antioxidant activity of hot-water extracts prepared from fermented rooibos (Joubert and de Beer, 2012). The samples ($n=74$) represented different production batches, mostly from different producers. The extraction process and conditions simulated those used by the South African industry when producing food ingredient extracts. The TPC of the extracts varied between 24.10 and 30.69% with aspalathin, nothofagin, orientin, and isoorientin comprising on average 2.22% of the extract. None of the individual compounds correlated well with the total antioxidant activity as measured by the $DPPH^{\bullet}$ scavenging and ORAC assays. The plant material was used unrefined, i.e., containing the dust (<40 mesh) and coarse (>10 mesh) fractions, which are removed when used for retail in loose or teabag packaging format. The coarse fraction mostly contains stems. Although weak ($r=-0.300$), a significant correlation ($P=0.009$) was observed between the coarse fraction and soluble solids content of the extracts, indicating that the stem content of the plant material used for extraction is important. An investigation of PPAG showed that its content is very low in the stems, i.e. 0.07% *versus* 0.34% in the leaves (Joubert *et al.*, 2013). Similar trends have been observed for other rooibos phenolic compounds, including aspalathin (unpublished results).

Fermented rooibos extract, when reconstituted to a 'cup-of-tea' strength (based on the average soluble solids content), would deliver the same amount of total

TABLE 14.4 Effects of Processing on Antioxidant Activity of Rooibos Aqueous Extracts

Samples	Assay	Results	Reference
Water extracts of green (n=5), semi-ferm. (n=7) and ferm. (n=7) rooibos	DPPH• scavenging (% inhibition)	Green≈ferm.	Von Gadow *et al.* (1997b)
	β-carotene bleaching (AAC)	Green≈ferm.; >semi-ferm.	
Water extract of ferm. rooibos extract (n=1) heated for varying times	β-carotene bleaching (AAC)	No changes	Von Gadow *et al.* (1997c)
	Rancimat lard oxidation (induction time)	Reduced activity after 30 min heating	
Water extracts of green and ferm. rooibos (n=1 each)	DPPH• scavenging (scavenging rate)	Green>ferm.	Winterton (1999)
	Sunflower oil-in-water emulsion oxidation (% inhibition; induction time)	Green>ferm.	
	Methyl linoleate micelles oxidation (% inhibition)	Green>ferm.	
Water extracts of green, ferm., sun-dried and steam-pasteurized rooibos (n=10 each)	DPPH• scavenging (EC_{50})	Green>steam-pasteurized>ferm.>sun-dried	Standley *et al.* (2001)
	$O_2^{\bullet-}$ scavenging (IC_{50})	Green>sun-dried≈ steam-pasteurized≈ferm.; sun-dried>ferm.	
Water extracts of green and ferm. rooibos (n=1 each)	$ABTS^{\bullet+}$ scavenging (TAC)	Green>ferm.	Bramati *et al.* (2003)
Water extracts of green and ferm. rooibos (n=20 each)	$ABTS^{\bullet+}$ scavenging (TAC)	Green>ferm.	Schulz *et al.* (2003)
Water extracts of green and ferm. rooibos (n=1 each)	DPPH• scavenging (EC_{50}; % inhibition)	Green>ferm.	Joubert *et al.* (2004)
	$O_2^{\bullet-}$ scavenging (IC_{50})	Green>ferm.	
Water extracts of green, ferm., sun-dried and steam-pasteurized rooibos (n=10 each)	Enhancement of •OH generation in a $FeCl_3$-EDTA/H_2O_2 system without ascorbate (TBARS)	Green>ferm., sun-dried, steam-pasteurized	Joubert *et al.* (2005)
Water extracts of green and ferm. rooibos (n=1 each)	Enhancement of •OH generation in a $FeCl_3$-EDTA/H_2O_2 system without ascorbate (TBARS)	Green > ferm.	
	Linoleic acid emulsion oxidation (% inhibition)	Green > ferm.	
Water extracts of green and ferm. rooibos (n=6 each)	$ABTS^{\bullet+}$ scavenging (TAC)	Green > ferm.	Joubert *et al.* (2008c)
	FRAP (TAC)	Green > ferm.	
	Microsomal lipid peroxidation (% inhibition)	Green > ferm.	

AAC, antioxidant activity coefficient; ABTS, 2,2′-azino-bis-3-ethylbenzthiazoline-6-sulfonic acid; DPPH, 1,1-diphenyl-2-picrylhydrazyl; EC_{50}, IC_{50}, concentration of sample needed to obtain 50% inhibition; EDTA, ethylene diamine tetraacetic acid; Ferm., fermented; FRAP, ferric-reducing antioxidant power; TAC, total antioxidant capacity; TBARS, thiobarbituric acid reactive substances.

polyphenols, aspalathin, and nothofagin, but slightly less orientin and isoorientin. The total antioxidant activity of reconstituted extracts was found to be slightly lower than that of the infusions as determined with the ORAC assay, while the opposite was observed for DPPH• scavenging assay (Joubert and de Beer, 2012). These findings are a further confirmation that some ready-to-drink rooibos iced tea manufacturers used inferior quality extracts or very little extract in their formulations (Joubert *et al.*, 2009).

EFFECTS OF PRODUCTION AND PROCESSING

Plant Production and Harvesting

The role of plant production and environmental factors on the phenolic composition of rooibos has not yet been studied. A preliminary study on the effect of harvest date showed that the aspalathin content of plant material harvested from the same plants over a

TABLE 14.5 Comparison of the Antioxidant Activity of Rooibos with that of Teas (*Camellia sinensis*) and Herbal Teas

Samples	Assay	Main Results	Reference
Water extracts of green (n = 5), semi-ferm. (n = 7) and ferm. (n = 7) rooibos, green tea (n = 5), oolong tea (n = 4), black tea (n = 7)	DPPH• scavenging (% inhibition)	Fermented rooibos ≈ black tea; < oolong tea, green tea	Von Gadow *et al.* (1997b)
	β-carotene bleaching (AAC)	Fermented rooibos ≈ black tea, oolong tea; < green tea	
Infusions of ferm. rooibos (n = 4), ferm. honeybush (n = 3), green tea (n = 5), oolong tea (n = 2), black tea (n = 8), various herbal teas (n = 1 each)	DPPH• scavenging (TAC)	Rooibos > honeybush, chamomile; < green tea, oolong tea, black tea, peppermint	du Toit *et al.* (2001)
Water extracts of ferm. rooibos, ferm. honeybush, Chinese green tea, jasmine green tea (n = 1 each)	Linoleic acid emulsion oxidation (% inhibition)	Rooibos > honeybush, Chinese green tea; ≈ jasmine green tea	Lindsey *et al.* (2002)
Infusions of ferm. rooibos, green tea, black tea, various herbal teas (n = 1 each)	H_2O_2 scavenging (TAC)	Rooibos < green tea, black tea, peppermint, others; > nettle	Paździoch-Czochra and Wideńska (2002)
Infusions of green and ferm. rooibos, green tea, black tea (n = 1 each)	$ABTS^{•+}$ scavenging (TAC)	Fermented rooibos < green tea, black tea	Bramati *et al.* (2003)
Infusions of ferm. rooibos, green tea, black tea, mate, burrito tea (n = 1 each)	$ABTS^{•+}$ scavenging (TAC)	Rooibos < other teas	Piccinelli *et al.* (2004)
	Chemiluminescence assay (TAC)	Rooibos ≈ burrito; < other teas	
Infusions of ferm. rooibos, ferm. honeybush, green tea, black tea, mate, various herbal teas (n = 1 each)	$ABTS^{•+}$ scavenging (TAC)	Rooibos ≈ honeybush; < green tea, black tea, mate; > chamomile, verbena	Ivanova *et al.* (2005)
Infusions of ferm. rooibos, ferm. honeybush, black tea, mate, various herbal teas (n = 1 each)	$ABTS^{•+}$ scavenging (TAC)	Rooibos ≈ honeybush; < black tea, mate	Kiselova *et al.* (2006)
Cold-water extracts of ferm. rooibos (n = 1), black tea (n = 1), *Athrixia phylicoides* (n = 2), *Athrixia elata* (n = 2)	$ABTS^{•+}$ scavenging (TAC)	Rooibos < black tea; ≈ *Athrixia elata*; > *Athrixia phylicoides*	McGaw *et al.* (2007)
Infusions of ferm. rooibos, green tea, black tea, red tea, white tea, peppermint, nettle, blended teas (n = 1 each)	$ABTS^{•+}$ scavenging (TAC)	Rooibos ≈ peppermint; > nettle; < other teas	Almajano *et al.* (2008)
	Sunflower oil-in-water emulsion oxidation (induction time)	Rooibos ≈ green tea; > black tea, red tea, white tea	
	Sunflower oil-in-water emulsion oxidation with BSA (induction time)	Rooibos > white tea; ≈ red tea; < green tea, black tea	
Water extracts of green and ferm. rooibos, green and ferm. honeybush (different species), green tea, oolong tea, black tea (n = 6 each)	$ABTS^{•+}$ scavenging (TAC)	Fermented rooibos > fermented honeybush; < green rooibos, green tea, oolong tea, black tea	Joubert *et al.* (2008c)
	FRAP (TAC)	Fermented rooibos > fermented honeybush; ≈ oolong tea, black tea; < green rooibos, green tea	
	Microsomal lipid peroxidation (% inhibition)	Fermented rooibos > fermented honeybush; < green rooibos, green tea, oolong tea, black tea	

AAC, antioxidant activity coefficient; ABTS, 2,2′-azino-bis-3-ethylbenzthiazoline-6-sulfonic acid; BSA, bovine serum albumin; ferm., fermented; DPPH, 1,1-diphenyl-2-picrylhydrazyl; FRAP, ferric reducing antioxidant power; IC_{50}, concentration of sample needed to obtain 50% inhibition; TAC, total antioxidant capacity.

one-year- period ranged from 2.33 to 3.76 g/100 g dry matter. Observations indicated that factors such as the presence of young growth, the age of the bush and the cultivation area could affect the quality of rooibos (Joubert and de Beer, 2011), which is not surprising given findings for other plants, including *Camellia sinensis*. The use of seedlings leads to large variation in phenolic composition, including aspalathin content. Although limited, harvesting of wild types has implications in terms of composition. The phenolic composition of wild rooibos showed qualitative and quantitative variation between populations and types, with some populations lacking aspalathin (van Heerden *et al.*, 2003). A wild type that was harvested and delivered to one of the processing yards contained a major compound, previously believed to be rutin, and little aspalathin (Joubert and Schulz, 2006). With a new HPLC method using a sub-2 micron stationary phase, Beelders *et al.* (2012b) could show that co-elution of rutin and quercetin-3-*O*-robinobioside occurred with previous HPLC methods used for analysis of rooibos. Another batch of wild-harvested rooibos contained more PPAG than aspalathin (Joubert *et al.*, 2013).

Harvesting takes place during the hot summer months and early autumn (January to April) by cutting all the shoots above the topping or previous harvest height. Young plants are topped after about 8 months to a height of 30 cm to stimulate branching. When harvested the active growth should not be more than 50 cm, otherwise the plant material will produce a weak infusion (Joubert *et al.*, 2008a).

Processing of Fermented Rooibos and Oxidative Conversion of Aspalathin

Processing of fermented rooibos basically entails the following steps: The tea shoots are cut into small pieces, with the length depending on whether the market requires long cut or fine cut. After cutting, the plant material is placed in fermentation heaps in the open air. Bruising and cell disruption caused by cutting initiates oxidation, an essential process of rooibos fermentation. The color on the outside of the heap rapidly changes to red–brown. Within 15 min of cutting of the plant material, its aspalatin content decreased by ca. 29%. After 210 min only ca. 32% of the aspalathin content remains (Joubert, 1996). The period between cutting and fermentation varies, depending, amongst others, on the capacity of the processing yard. In many instances the plant material is cut in the morning. Normally fermentation commences in the late afternoon when the plant material is bruised by rolling a heavy object over the heap, water is added and the heap is aerated to ensure proper oxidation. These steps accelerate the 'fermentation' process, initiated by cutting. Polyphenols released during bruising and wetting is absorbed by the shredded stems, coloring it red–brown to give a more uniformly colored product. The fermentation period can vary, depending on the plant material and ambient conditions, but it usually takes place for 12–14 h, whereafter the fermented plant material is spread out in a thin layer to dry in the sun (Joubert *et al.*, 2008a). Rooibos fermented for 14 h at 38°C and dried at 40°C for 3 h retains very little of the aspalathin present in the plant material at harvest (Joubert, 1996). The dried rooibos is sieved and steam-pasteurized before packaging. Steam-pasteurization was introduced during the 1980s for food safety reasons. It leads to changes in the flavor profile, a reduction in astringency, and a decrease in soluble solid, total polyphenol, and aspalathin contents of rooibos infusions (Koch *et al.*, 2013). Standley *et al.* (2001) found that the DPPH$^{\bullet}$ scavenging ability of steam-pasteurized rooibos was slightly higher than that of unpasteurized rooibos, while no effect could be demonstrated for $O_2^{\bullet-}$ scavenging ability. In the latter study, randomly selected samples (n = 10) were collected before and after pasteurization at a commercial processing yard so that the decrease in antioxidant activity is not necessarily the result of pasteurization, but could also be due to natural variation.

The chemical conversion of aspalathin has received attention due to its role in the formation of the red–brown color of fermented rooibos. The first study was undertaken by Koeppen and Roux (1965), who showed photochemical conversion of aspalathin to dihydro-isoorientin and dihydro-orientin, as well as the formation of unidentified polymeric substances in an ethanolic solution of aspalathin, exposed to sunlight and oxygen. Dihydro-isoorientin formed preferentially with traces of dihydro-orientin. Its concentration increased with time, attributed to the conversion of dihydro-isoorientin to dihydro-orientin. Under conditions mimicking fermentation, Marais *et al.* (2000) confirmed the oxidative cyclization of aspalathin to form the diastereomeric flavanone mixture, (*S*)- and (*R*)-eriodictyol-6-C-β-D-glucopyranoside (dihydro-isoorientin).

Recent studies by Glomb and co-workers (Krafczyk and Glomb, 2008; Krafczyk *et al.*, 2009a; Heinrich *et al.*, 2012) provide further insight into the oxidation of aspalathin and its role in color formation. Reaction products from the oxidation of aspalathin are depicted in Figure 14.1. Incubation of aspalathin at 37°C in a phosphate buffer (pH 7.4) led to its oxidation and intermediary products, i.e., the diastereomeric mixture of dihydro-isoorientin as major products and the diastereomeric mixture of dihydro-orientin as minor products in a stable ratio of ca. 10 : 1. Dihydro-isoorientin oxidizes to form isoorientin. Dihydro-orientin, however, does not oxidize directly to orientin, but its conversion is postulated to occur via the quinine methide to dihydro-isoorientin and isoorientin. The latter compound undergoes opening of its vinyl ester structure, followed by bond rotation and recyclization to form orientin irreversibly. Isoorientin and orientin concentrations peaked after 6 h in solution. No isoorientin was detectable after 10 h. The major conversion product(s) of isoorientin is, however, not orientin but unidentified brown material (Krafczyk and Glomb, 2008).

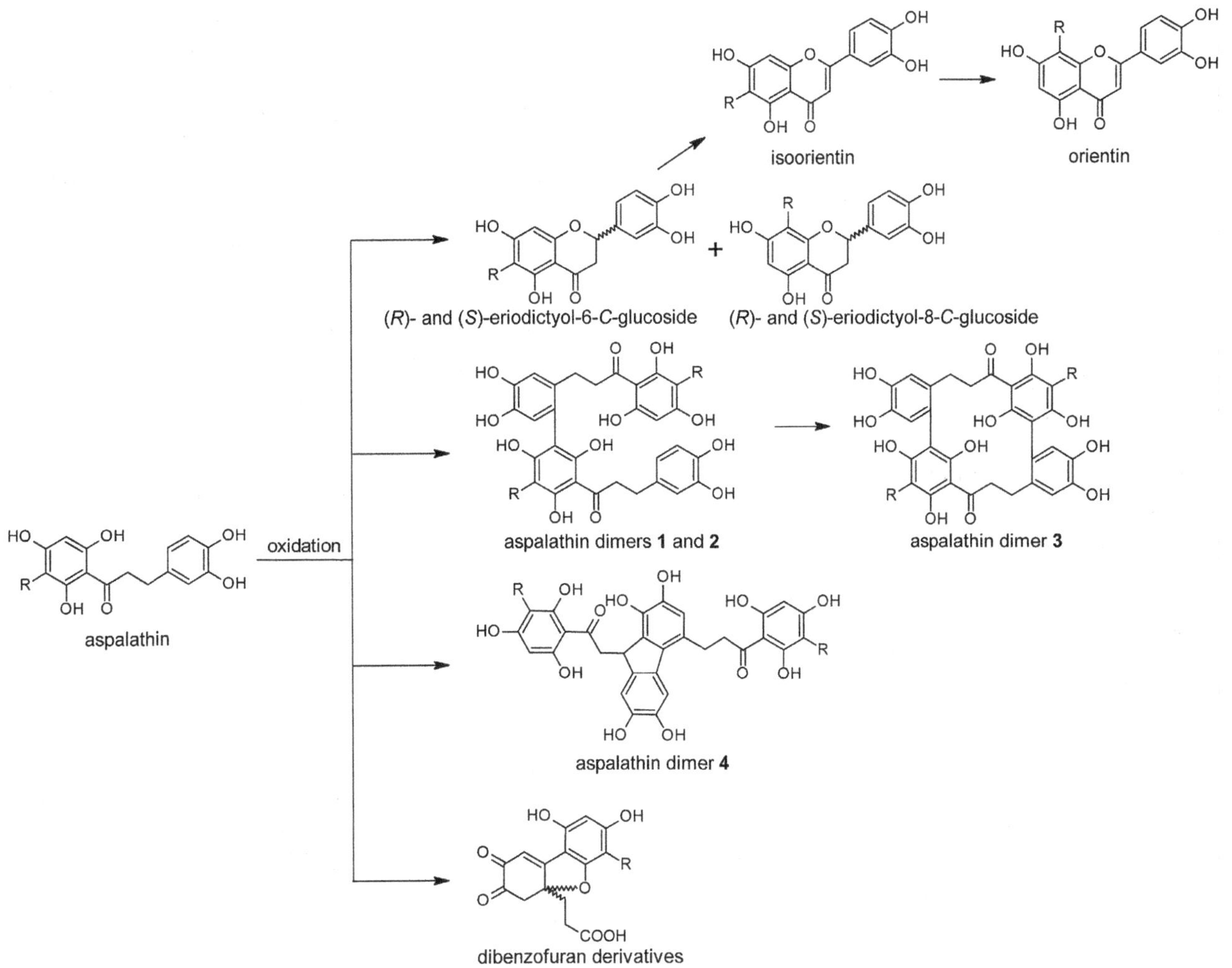

FIGURE 14.1 Reaction products from oxidation of aspalathin with R = glucosyl *(compiled from data in Krafczyk and Glomb (2009) and Heinrich et al. (2012)).*

Following fractionation of an HPLC eluate of an ethyl acetate fraction of fermented rooibos and incubation of the fractions to measure color formation (absorbance at 450 nm based on maximum absorbance for fermented rooibos), aspalathin was identified as the most important substrate for the formation of colored products (Krafczyk *et al.*, 2009a). Incubation of aspalathin under aerated and deaerated conditions confirmed that the browning reaction is oxygen-dependent. Other major findings were that nothofagin also oxidizes to form brown products, but at a much slower rate, and that the configuration of the C-ring in the case of the flavanones plays an important role. (*S*)-Eriodictyol-6-C-β-D-glucopyranoside showed results similar to aspalathin in terms of color formation. Aeration of aspalathin also led to the formation of colorless dimers (atropisomers), **1** and **2** (Figure 14.1). The *o*-quinone forms upon oxidation of aspalathin coupled to a second aspalathin through a nucleophilic attack by its A-ring. The concentration of these dimers peaked after 3 h in the ratio 47:53. In addition to these dimers, two other colorless dimers, **3** and **4**, were present in the reaction mixture (Heinrich *et al.*, 2012). Dimer **3** peaked after 6 h at a lower concentration than dimers **1** and **2**, postulated to be the precursors of dimer **3**. These dimers eventually completely degraded to brown end products with the characteristic absorbance maximum at 450 nm. Low yield of dimer **4** in combination with its reactivity precluded its isolation and structure elucidation, but it was postulated to form through an alternative pathway, starting with the nucleophilic attack of the *o*-quinone by the B-ring of aspalathin, finally leading to a five-membered ring system. Several other compounds formed in parallel in the reaction mixture. Two were identified as the colored dibenzofurans, (*S*)- and (*R*)-3-(7,9-dihydroxy-2,3-dioxo-6-β-D-glucopyranosyl-3,4-dihydrobenzo[*b*,*d*]furan-4*a*(2*H*)-yl) propionic acid with maximum absorbance at 430 nm. These compounds are reactive intermediates as they were degraded after 60 h, giving rise to high-molecular weight tannin-like structures.

Processing of Green Rooibos

Processing of green rooibos was instigated by the demand for a product with a higher antioxidant activity and/or aspalathin content than the traditional 'fermented' tea. Whereas fermentation is a key step in the development of the characteristic color, flavor, and taste of fermented rooibos, these reactions are detrimental to the character of green rooibos. To retain the green leaf color, oxidative changes should be kept to a minimum. High-end-quality products can be produced either by: rapid drying of shredded plant material without delay under vacuum; initial low-temperature drying followed by high-temperature drying of the shredded plant material; drying of whole shoots to a critical moisture content before shredding; or steaming of the fresh shoots to inactivate enzymes before shredding. Mostly, green rooibos is produced by spreading the shredded plant material in a thin layer in the sun for quick drying, as no additional equipment to that used for processing of fermented rooibos is required. However, this leads to loss of aspalathin content, and if not properly dried, to slow browning (Joubert and de Beer, 2011).

Extract Production

The effect of factors such as extraction temperature and time, tea-to-water ratio, and extraction process (stirred batch-wise and fixed-bed with flow through) on the recovery of soluble solids and their total flavonoid and total polyphenol contents has been investigated for fermented rooibos [references cited by Joubert and de Beer (2011)] at a period when antioxidant activity of rooibos extract was not yet of general interest. The increase in phenolic content of the soluble matter with increasing temperature, as well as the stability of its antioxidant activity during heat exposure (von Gadow *et al.*, 1997c), indicates that it would be beneficial to employ high extraction temperatures. Approximately 50% of the aspalathin in fermented rooibos is extracted within 5 min in water at 80°C, approaching steady state after 60 min (Jaganyi and Wheeler, 2003). No data on the extraction kinetics of other rooibos flavonoids are available.

Addressing the relatively poor solubility of fermented rooibos in hot water, enzyme cocktails of *Rhizopus oryzae* were employed to increase the extraction of soluble solids and polyphenols from rooibos (Pengilly *et al.*, 2008). Their effect differed, depending on the culture medium used. When cultured in yeast peptone-wheat straw medium the resulting enzyme cocktail of *R. oyzae* improved the soluble solids yield by 47%, while 99% and 95% of the TPC and antioxidant activity ($ABTS^{\bullet+}$ scavenging assay), respectively, were retained. When cultured in a potato dextrose medium, the resulting enzyme cocktail did not improve the extraction of soluble solids, but the TPC and antioxidant content of the soluble solids increased. Treatment of the plant material with a concentrated enzyme cocktail of *R. oryzae* (cultured in yeast peptone–wheat straw medium) at 40°C for 2 h, followed by extraction simulating the process used by industry, improved the soluble solids yield by 30% and the release of polyphenols by 31%.

Samples collected during commercial extract production showed that substantial losses of the major compounds, aspalathin, orientin, and isoorientin, as a result of different processing steps, i.e., microfiltration, reverse osmosis, vacuum concentration, and spray-drying, did not occur (Joubert *et al.*, 2009). Samples collected directly before and after spray-drying did not differ significantly. Although high inlet and outlet temperatures are employed during spray-drying of plant extracts, exposure to heat while still in a high-water-activity environment is very short due to the rapid evaporation rate when the concentrate is atomized.

Ready-to-Drink Rooibos

Experiments carried out on ready-to-drink (RTD) rooibos iced teas showed that the severity of the heat treatment is important for the stability of aspalathin, orientin and isoorientin (Joubert *et al.*, 2009, 2010). Pasteurization had little or no effect, while both normal temperature sterilization (121°C/15 min) and high-temperature sterilization (135°C/4 min) induced significant losses. Their effect on the retention of aspalathin, orientin and isoorientin depended on the formulation. The latter was varied in terms of type of rooibos extract (hot-water fermented rooibos extract and organic solvent-based aspalathin-enriched green rooibos extract), and the presence of common iced tea ingredients, ascorbic acid and/or citric acid. Addition of ascorbic acid and/or citric acid improved retention of these compounds. The greatest effect was observed for aspalathin, while orientin was least affected. The conversion of isoorientin to orientin could have contributed to this difference in stability of the flavones. Aspalathin was more stable in RTD formulations containing the aspalathin-enriched green rooibos extract, indicating that compositional differences introduced by the state of fermentation and solvent used for extraction could play a role. Encapsulation of an aspalathin-enriched green rooibos extract with ascorbic acid in a water-soluble nanomicelle-based carrier system did not offer additional stability.

Stability of aspalathin, orientin, and isoorientin during storage (12 weeks at 25°C in the dark) was also investigated for the same RTD formulations (de Beer *et al.*, 2012). Once again the extract matrix was shown to be important in terms of the stability of aspalathin with the best protection observed by the nanomicelles. Starting with ca. 16 mg aspalathin in a liter of fermented rooibos

RTD iced tea containing no ascorbic acid or citric acid, no aspalathin could be detected after 8 weeks. When adding citric acid alone or citric acid and ascorbic acid, losses of 39 and 25%, respectively, over the 12-week storage period were observed. A short-term storage experiment of 48h at different pH (3, 4, and 5) and temperature (5, 30, and 40°C) combinations showed that low temperature and low pH increased the retention of aspalathin. Orientin and isoorientin were less affected under similar conditions.

References

Almajano, M., Carbó, R., Jiménez, J., Gordon, M., 2008. Antioxidant and antimicrobial activities of tea infusions. Food Chem. 108, 55–63.

Anonymous, 2007. South Africa Sector Report Food and Beverages Industry. Swiss Business Hub South Africa, Pretoria, South Africa.

Beelders, T., Kalili, K.M., Joubert, E., de Beer, D., de Villiers, A.J., 2012a. Comprehensive two-dimensional liquid chromatographic analysis of rooibos (*Aspalathus linearis*) phenolics. J. Sep. Sci. 35, 1808–1820.

Beelders, T., Sigge, G.O., Joubert, E., de Beer, D., de Villiers, A., 2012b. Kinetic optimization of the reversed phase liquid chromatographic separation of rooibos tea (*Aspalathus linearis*) phenolics on conventional high performance liquid chromatographic instrumentation. J. Chromatogr. A. 1219, 128–139.

Beltrán-Debón, R., Rull, A., Rodríguez-Sanabria, F., Iswaldi, I., Herranz-López, M., Aragonès, G., Camps, J., Alonso-Villaverde, C., Menéndez, J.A., Micol, V., Segura-Carretero, A., Joven, J., 2011. Continuous administration of polyphenols from aqueous rooibos (*Aspalathus linearis*) extract ameliorates dietary-induced metabolic disturbances in hyperlipidemic mice. Phytomed. 18, 414–424.

Bramati, L., Aquilano, F., Pietta, P., 2003. Unfermented rooibos tea: quantitative characterization of flavonoids by HPLC-UV and determination of the total antioxidant activity. J. Agric. Food Chem. 51, 7472–7474.

Dahlgren, R., 1968. Revision of the genus *Aspalathus*. II. The species with ericoid and pinoid leaflets. 7. Subgenus Nortieria. With remarks on rooibos tea cultivation. Bot. Not. 121, 165–208.

Dahlgren, R., 1988. Crotalarieae (*Aspalathus*). Flora of Southern Africa 16, 84–90.

de Beer, D., Joubert, E., Viljoen, M., Manley, M., 2012. Enhancing aspalathin stability in rooibos (*Aspalathus linearis*) ready-to-drink iced teas during storage: the role of nano-emulsification and beverage ingredients, citric and ascorbic acids. J. Sci. Food Agric. 92, 274–282.

du Toit, R., Volsteedt, Y., Apostolides, Z., 2001. Comparison of the antioxidant content of fruits, vegetables and teas measured as vitamin C equivalents. Toxicol. 166, 63–69.

Ferreira, D., Marais, C., Steenkamp, J.A., Joubert, E., 1995. Rooibos tea as a likely health food supplement. In: Proceedings of the Recent Development of Technologies on Fundamental Foods for Health. Korean Society of Food Science and Technology, Seoul, Korea, pp. 73–88.

Heinrich, T., Willenberg, I., Glomb, M.A., 2012. Chemistry of color formation during rooibos fermentation. J. Agric. Food Chem. 60, 5221–5228.

Iswaldi, I., Arráez-Román, D., Rodríguez-Medina, I., Beltrán-Debón, R., Joven, J., Segura-Carretero, A., Fernández-Gutiérrez, A., 2011. Identification of phenolic compounds in aqueous and ethanolic rooibos extracts (*Aspalathus linearis*) by HPLC-ESI-MS (TOF/IT). Anal. Bioanal. Chem. 400, 3643–3654.

Ivanova, D., Gerova, D., Chervenkov, T., Yankova, T., 2005. Polyphenols and antioxidant capacity of Bulgarian medicinal plants. J. Ethnopharmacol. 96, 145–150.

Jaganyi, D., Wheeler, P.J., 2003. Rooibos tea: equilibrium and extraction kinetics of aspalathin. Food Chem. 83, 121–126.

Joubert, E., 1996. HPLC quantification of the dihydrochalcones, aspalathin and nothofagin in rooibos tea (*Aspalathus linearis*) as affected by processing. Food Chem. 55, 403–411.

Joubert, E., de Beer, D., 2011. Rooibos (*Aspalathus linearis*) beyond the farm gate: From herbal tea to potential phytopharmaceutical. S. Afr. J. Bot. 77, 869–886.

Joubert, E., de Beer, D., 2012. Phenolic content and antioxidant activity of rooibos food ingredient extracts. J. Food Comp. Anal. 27, 45–51.

Joubert, E., de Beer, D., Malherbe, C.J., Muller, N., Bonnet, S.L., van der Westhuizen, J.H., Ferreira, D., 2013. Occurrence and sensory perception of Z-2-(β-D-glucopyranosyloxy)-3-phenylpropenoic acid in rooibos (*Aspalathus linearis*). Food Chem. 136, 1078–1085.

Joubert, E., Gelderblom, W.C.A., Louw, A., de Beer, D., 2008a. South African herbal teas: *Aspalathus linearis, Cyclopia* spp. and *Athrixia phylicoides*—a review. J. Ethnopharmacol. 119, 376–412.

Joubert, E., Manley, M., Botha, M., 2008b. Evaluation of spectrophotometric methods for screening of green rooibos (*Aspalathus linearis*) and green honeybush (*Cyclopia genistoides*) extracts for high levels of bio-active compounds. Phytochem. Anal. 19, 169–178.

Joubert, E., Richards, E.S., van der Merwe, J.D., de Beer, D., Manley, M., Gelderblom, W.C.A., 2008c. Effect of species variation and processing on phenolic composition and in vitro antioxidant activity of aqueous extracts of *Cyclopia* spp. (honeybush tea). J. Agric. Food Chem. 56, 954–963.

Joubert, E., Schulz, H., 2006. Production and quality aspects of rooibos tea and related products. A review. J. Appl. Bot. Food Qual. 80, 138–144.

Joubert, E., Viljoen, M., de Beer, D., Malherbe, C.J., Brand, D.J., Manley, M., 2010. Use of green rooibos (*Aspalathus linearis*) extract and water-soluble nanomicelles of green rooibos extract encapsulated with ascorbic acid for enhanced aspalathin content in ready-to-drink iced teas. J. Agric. Food Chem. 58, 10965–10971.

Joubert, E., Viljoen, M., de Beer, D., Manley, M., 2009. Effect of heat on aspalathin, iso-orientin, and orientin contents and color of fermented rooibos (*Aspalathus linearis*) iced tea. J. Agric. Food Chem. 57, 4204–4211.

Joubert, E., Winterton, P., Britz, T.J., Ferreira, D., 2004. Superoxide anion and α,α-diphenyl-β-picrylhydrazyl radical scavenging capacity of rooibos (*Aspalathus linearis*) aqueous extracts, crude phenolic fractions, tannin and flavonoids. Food Res. Int. 37, 133–138.

Joubert, E., Winterton, P., Britz, T.J., Gelderblom, W.C.A., 2005. Antioxidant and pro-oxidant activities of aqueous extracts and crude polyphenolic fractions of rooibos (*Aspalathus linearis*). J. Agric. Food Chem. 53, 10260–10267.

Kazuno, S., Yanagida, M., Shindo, N., Murayama, K., 2005. Mass spectrometric identification and quantification of glycosyl flavonoids, including dihydrochalcones with neutral loss scan mode. Anal. Biochem. 347, 182–192.

Kiselova, Y., Ivanova, D., Chervenkov, T., Gerova, D., Galunska, B., 2006. Correlation between the in vitro antioxidant activity and polyphenol content of aqueous extracts from Bulgarian herbs. Phytother. Res. 965, 961–965.

Koch, I.S., Muller, M., Joubert, E., van der Rijst, M., Næs, T., 2012. Sensory characterization of rooibos tea and the development of a rooibos sensory wheel and lexicon. Food Res. Int. 46, 217–228.

Koch, I.S., Muller, N., de Beer, D., Næs, T., Joubert, E., 2013. Impact of steam pasteurization on the sensory profile and phenolic composition of rooibos (*Aspalathus linearis*) herbal tea infusions. Food Res. Int. 53, 704–712.

Koeppen, B.H., Roux, D.G., 1965. Aspalathin: a novel *C*-glycosylflavonoid from Aspalathus linearis. Tetrahedron Lett. 39, 3497–3503.

Krafczyk, N., Glomb, M.A., 2008. Characterization of phenolic compounds in rooibos tea. J. Agric. Food Chem. 56, 3368–3376.

Krafczyk, N., Heinrich, T., Porzel, A., Glomb, M.A., 2009a. Oxidation of the dihydrochalcone aspalathin leads to dimerization. J. Agric. Food Chem. 57, 6838–6843.

Krafczyk, N., Woyand, F., Glomb, M.A., 2009b. Structure–antioxidant relationship of flavonoids from fermented rooibos. Mol. Nutr. Food Res. 53, 635–642.

Lindsey, K.L., Motsei, M.L., Jäger, A.K., 2002. Screening of South African food plants for antioxidant activity. J. Food Sci. 67, 2129–2131.

Marais, C., Rensburg, W.J.V., Ferreira, D., Steenkamp, J.A., 2000. (*S*)- and (*R*)-eriodictyol-6-C-β-D-glucopyranoside, novel keys to the fermentation of rooibos (*Aspalathus linearis*). Phytochem. 55, 43–49.

Marais, C., Steenkamp, J.A., Ferreira, D., 1996. The occurrence of phenylpyruvic acid in woody plants: Biosynthetic significance. Tetrahedron Lett. 37, 5763–5764.

Marais, S.S., Marais, C., Steenkamp, J.A., Malan, E., Ferreira, D., 1998. Progress in the investigation of rooibos tea extractives. In: Gross, G.G., Hemingway, R.W., Yoshida, T. (Eds.), Abstracts of the 3rd Tannin Conference. Bend, Oregon, USA, pp. 129–130.

McGaw, L.J., Steenkamp, V., Eloff, J.N., 2007. Evaluation of *Athrixia* bush tea for cytotoxicity, antioxidant activity, caffeine content and presence of pyrrolizidine alkaloids. J. Ethnopharmacol. 110, 16–22.

Pacheco-Palencia, L.A., Talcott, S.T., 2010. Chemical stability of açai fruit (*Euterpe oleracea* Mart.) anthocyanins as influenced by naturally occurring and externally added polyphenolic cofactors in model systems. Food Chem. 118, 17–25.

Paździoch-Czochra, M., Wideńska, A., 2002. Spectrofluorimetric determination of hydrogen peroxide scavenging activity. Anal. Chim. Acta 452, 177–184.

Pengilly, M., Joubert, E., van Zyl, W.H., Botha, A., Bloom, M., 2008. Enhancement of rooibos (*Aspalathus linearis*) aqueous extract and antioxidant yield with fungal enzymes. J. Agric. Food Chem. 56, 4047–4053.

Piccinelli, A.L., De Simone, F., Passi, S., Rastrelli, L., 2004. Phenolic constituents and antioxidant activity of *Wendita calysina* leaves (Burrito), a folk Paraguayan tea. J. Agric. Food Chem. 52, 5863–5868.

Rabe, C., Steenkamp, J.A., Joubert, E., Burger, J.F.W., Ferreira, D., 1994. Phenolic metabolites from rooibos tea (*Aspalathus linearis*). Phytochem. 35, 1559–1565.

Schulz, H., Joubert, E., Schütze, W., 2003. Quantification of quality parameters for reliable evaluation of green rooibos (*Aspalathus linearis*). Eur. Food Res. Technol. 216, 539–543.

Shimamura, N., Miyase, T., Umehara, K., Warashina, T., Fujii, S., 2006. Phytoestrogens from *Aspalathus linearis*. Biol. Pharm. Bull. 29, 1271–1274.

Snijman, P.W., Joubert, E., Ferreira, D., Li, X.-C., Ding, Y., Green, I.R., Gelderblom, W.C.A., 2009. Antioxidant activity of the dihydrochalcones aspalathin and nothofagin and their corresponding flavones in relation to other rooibos (*Aspalathus linearis*) flavonoids, epigallocatechin gallate, and Trolox. J. Agric. Food Chem. 57, 6678–6684.

Standley, L., Winterton, P., Marnewick, J.L., Gelderblom, W.C.A., Joubert, E., Britz, T.J., 2001. Influence of processing stages on antimutagenic and antioxidant potentials of rooibos tea. J. Agric. Food Chem. 49, 114–117.

van Heerden, F., van Wyk, B.E., Viljoen, A.M., Steenkamp, P., 2003. Phenolic variation in wild populations of *Aspalathus linearis* (rooibos tea). Biochem. Syst. Ecol. 31, 885–895.

von Gadow, A., Joubert, E., Hansmann, C.F., 1997a. Comparison of the antioxidant activity of aspalathin with that of other plant phenols of rooibos tea (*Aspalathus linearis*), α-tocopherol, BHT, and BHA. J. Agric. Food Chem. 45, 632–638.

von Gadow, A., Joubert, E., Hansmann, C.F., 1997b. Comparison of the antioxidant activity of rooibos tea (*Aspalathus linearis*) with green, oolong and black tea. Food Chem. 60, 73–77.

von Gadow, A., Joubert, E., Hansmann, C.F., 1997c. Effect of extraction time and additional heating on the antioxidant activity of rooibos tea (*Aspalathus linearis*) extracts. J. Agric. Food Chem. 45, 1370–1374.

Winterton, P., 1999. Antioxidant activity of rooibos tea (*Aspalathus linearis*) in model lipid and radical generating systems. MSc. In: Food Science thesis. Stellenbosch University, Stellenbosch, South Africa.

CHAPTER

15

Antioxidant Activity of Maté Tea and Effects of Processing

Kleber Berté, Délia B. Rodriguez–Amaya†, Rosemary Hoffmann-Ribani*, Agenor Maccari, Junior***

*Graduate Program of Food Engineering – PPGEAL, Chemical Engineering Department, Federal University of Paraná (UFPR), Curitiba, PR, Brazil, †Faculty of Food Engineering, University of Campinas – UNICAMP, Department of Food Science, Campinas, SP, Brazil, **Chemical Engineering Department, Federal University of Paraná (UFPR), Curitiba, PR, Brazil

CHAPTER POINTS

- In South America, the leaves and stems of the yerba-maté (*Ilex paraguariensis*) are used in the preparation of several types of beverages appreciated for their peculiar bitter taste and functional properties (antioxidant, stimulant, diuretic, hypocholesterolemic, and hepatoprotective).
- Important compounds found in yerba-maté and maté products are caffeoyl derivatives (caffeic acid, mono- and dicaffeoylquinic acids), methylxanthines (caffeine and theobromine), and flavonoids (quercetin and kaempferol glycosides).
- Maté showed high antioxidant activity *in vitro* and *in vivo*, attributed to its high polyphenol content.
- Processing generally consists of blanching, drying, milling, sieving, and packaging. In Argentina and Uruguay, natural or forced aging is also carried out.
- Because the conditions for the processing of yerba-maté vary widely, depending on the manufacturer and the desired style and flavor of maté tea, somewhat diverging results are reported regarding the processing effects on bioactive compounds. However, in general, processing tends to reduce methylxanthines (e.g., caffeine) but increases the content of polyphenols (e.g., caffeoyl derivatives), resulting in higher antioxidant activity.

INTRODUCTION

Maté tea is a popular herbal beverage made from the leaves and stems of yerba-maté (*Ilex paraguariensis* Auguste Saint Hillaire), a plant that belongs to the family Aquifoliaceae, native to the southern part of South America. The yerba-maté trees occupy a 540,000-km^2 area located between Argentina, Brazil, and Paraguay. The perennial trees thrive at high subtropical and temperate altitudes (500–1500 m) with an annual average temperature of 15–21°C and moderate moisture. They grow in the wild to a height of 7–15 m, but in order to facilitate harvest in plantations, they are kept at 2–3 m in height by pruning. In South America, yerba-maté currently grows between 21° and 30° south latitude in the northeast of Argentina (Misiones, Corrientes, and Tucumã), southern Brazil (Paraná, Santa Catarina, Rio Grande do Sul, and Mato Grosso do Sul), and eastern Paraguay (Alto Paraná, Amambay, and San Pedro) (Valduga *et al.*, 2003; Maccari Junior, 2005). The main regions where yerba-maté is grown are shown in Figure 15.1. According to the Food and Agriculture Organization of the United Nations, the main maté producer is Brazil (443,635 tons in 2011), followed by Argentina (272,619 tons in 2011), and Paraguay (85,490 tons in 2011) (FAOSTAT, 2013), and most part of the produced yerba-maté is processed and sold as *chimarrão*.

The commercial product obtained by the processing of the aerial parts (leaves and stems) of *Ilex paraguariensis*, called maté tea, yerba-maté tea, or erva-maté tea, is used in the preparation of several types of beverages appreciated for their peculiar bitter taste, regionally known as *chimarrão* (dried green leaves brewed with hot water),

Processing and Impact on Antioxidants in Beverages
http://dx.doi.org/10.1016/B978-0-12-404738-9.00015-5

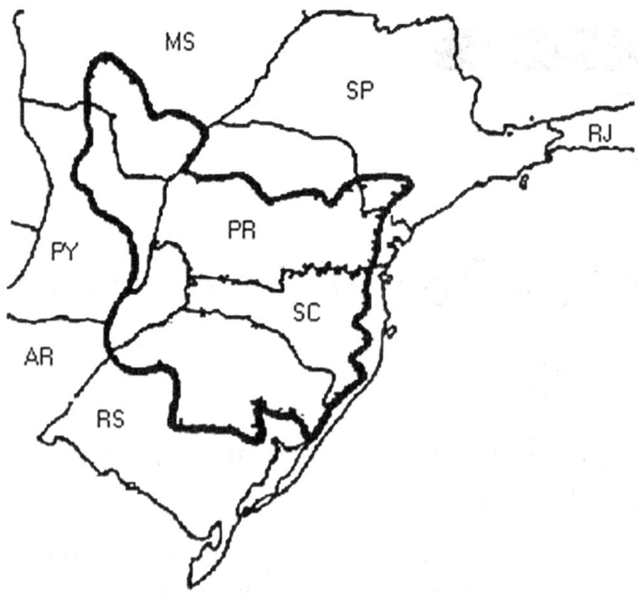

FIGURE 15.1 Map of South America showing regions producing yerba-maté *(Ilex paraguariensis)*. Argentina (AR); Brazil [Rio Grande do Sul (RS), Santa Catarina (SC), Paraná (PR), Mato Grosso do Sul (MS), São Paulo (SP) and Rio de Janeiro (RJ)]; Paraguay (PY). *Source: Rotta and Oliveira (2005).*

tererê (dried green leaves brewed with cold water), maté tea (tea-bag), roasted maté tea, aged maté tea, and instant teas (soluble roasted and unroasted). Considering consumers' increasing interest in health-promoting products, it would be advantageous for the maté industry if these herbal beverages contained high amounts of phenolic compounds, widely recognized as responsible for antioxidant properties, and caffeine, a stimulating compound (Maccari Junior, 2005; Bastos *et al.*, 2006a). The processing of yerba-maté fresh leaves into different products is presented schematically in Figure 15.2.

HARVESTING AND PROCESSING

The yerba-maté harvest occurs from about May until August in southern Brazil (Paraná, Santa Catarina, and Rio Grande do Sul). At harvest, the fresh green leaves and small stems are cut manually or mechanically from the trees, and then undergo several stages of processing before being packaged. The conditions for processing vary widely, depending on the manufacturer and the desired style and flavor of maté tea. Processing can vary in time and temperature of blanching and drying; however, the overall process is generally the same. It consists of blanching, drying, milling, sieving (classification), and packaging (Valduga *et al.*, 2003; Maccari Junior, 2005; Berté, 2011).

Industrial processing involves rapid transport of the harvested leaves and stems to a central processing plant. The leaves are first blanched (rapid drying) in a rotary dryer oven over direct fire at temperatures between 500°C and 700°C for 30 s to 4 min (Schmalko and Alzamora, 2001; Maccari Junior, 2005; Berté, 2011). Blanching can be briefly described as the process of heating vegetables to a temperature high enough to inactivate enzymes in the tissues. It stops enzyme action, sets the color, and shortens the drying and dehydration time. It is usually carried out in hot water or in steam. This technique is used by indigenous people to reduce or eliminate the bitterness of the vegetable and acid components that are common in leaves (Valduga *et al.*, 2003; Oboh, 2005).

During blanching, the moisture content of the leaves is reduced from about 70% to about 45–55%, so that the leaves become flaccid, a prerequisite for the next stage of processing (Schmalko and Alzamora, 2001; Maccari Junior, 2005). The blanched leaves are dried in mechanical dryers in a stream of hot air (80–110°C) for about 2–6 h, during which time the water content goes down to 5–10%. During this time the yerba-maté leaves lose about 40–45% of their weight (Valduga *et al.*, 2003; Maccari Junior, 2005). The dried yerba-maté leaves and stems are coarsely ground, separated from each other and sieved, producing the yerba-maté *cancheada* (large particles) to facilitate manipulation and transport. For the Brazilian market the *cancheada*, without an aging period, are further milled to smaller particles to obtain the maté tea for *chimarrão*, among other products (Figure 15.2).

For Argentinian and Uruguayan markets, the yerba-maté *cancheada* are left to age in storage warehouses for a period of time (usually between 36 and 48 weeks) necessary for the product to develop its optimal flavor, aroma, and color. The *cancheada* is stored under natural conditions of temperature and moisture for about 9–12 months for a natural aging process. In the forced aging process, the product is stored under controlled temperature (55–65°C), moisture (30–50%), and air circulation conditions for 30–60 days (Heck and Mejia, 2007; Isolabella *et al.*, 2010).

MATÉ TEA PRODUCTS

The bulk of harvested yerba-maté leaves and stems are used in the preparation of teas (known as maté tea), partial infusion drinks with hot water (*chimarrão*), and a total infusion cold drink (*tererê*) (Figure 15.2) (Filip *et al.*, 2000). Maté tea has a very important social role; the act of offering it and sharing it carries connotations similar to those of the tea ceremony for some oriental cultures (Bracesco *et al.*, 2011). In South America, about 30% of the population consumes more than 1 l of *chimarrão* and/or *tererê* per day, serving it as the main alternative to coffee and tea (*Camellia sinensis*) (Filip *et al.*, 2000; Turner *et al.*, 2011). The per capita consumption of maté herb in Brazil is estimated at 1.2 kg per year. Only the states of Rio Grande do Sul, Santa Catarina, and Parana

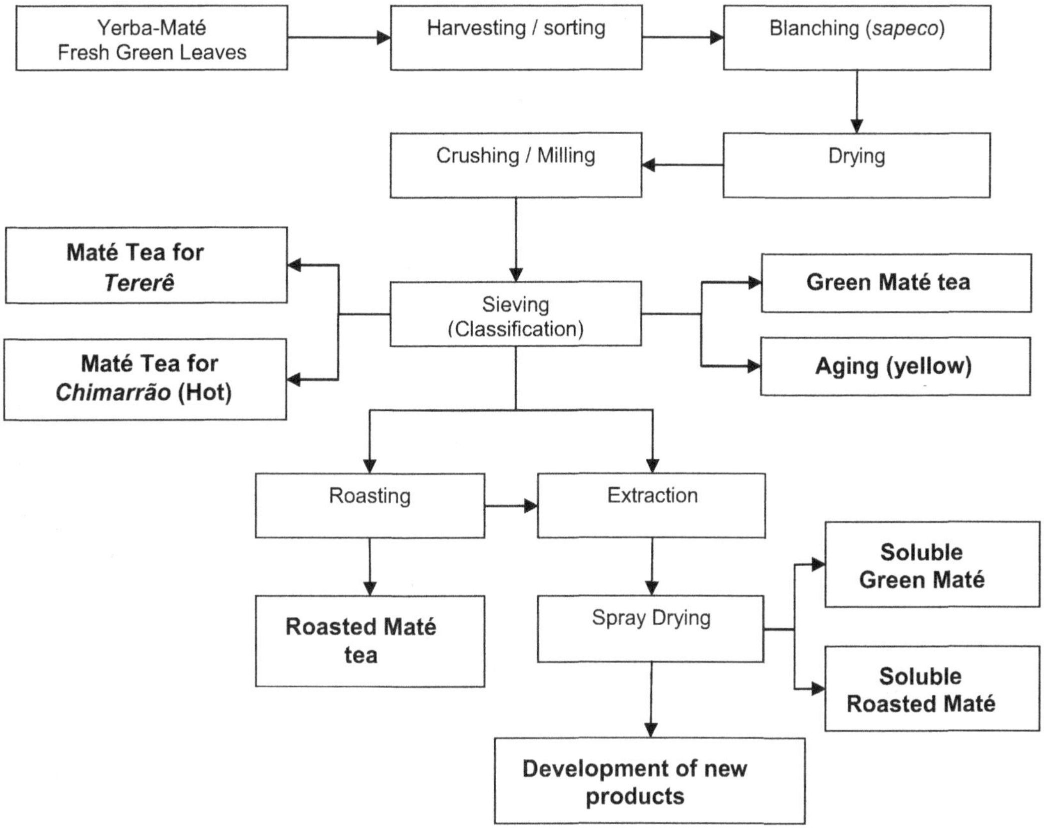

FIGURE 15.2 Generic flowsheet for the production of different products from yerba-maté. *Source: Berté (2011).*

have maté drinkers in their population (lowering the per capita intake), but up to 70% of the population in these three states drinks *chimarrão* daily. Argentina, the leading exporter, has a consumption of 5 kg/person per year. Uruguay has the highest per capita consumption, ranging from 6 to 8 kg/person per year (Heck and Mejia, 2007; Bracesco *et al.*, 2011).

Yerba-maté is listed by the European Council as a natural source of food flavoring. The category N2 for flavors indicates that yerba-maté can be added to foodstuffs in small quantities, with a possible limitation of an active principle in the final products. In South America, tea-like yerba-maté, commonly consumed as a beverage, is said to be less astringent than tea (*Camellia sinensis*). In the USA, yerba-maté is listed as GRAS (generally recognized as safe) (Newall *et al.*, 1996; Berté et al., 2011).

The unusual bitter taste of yerba-maté in the prepared form may cause aversion to yerba-maté-based beverages in some people, restricting its consumption to certain regions (Maccari Junior, 2005; Andrade *et al.*, 2012). However, there is an increasing number of yerba-maté products being developed, as well as growing interest in these products by countries whose populations do not traditionally consume maté tea (Bastos *et al.*, 2007; Berté *et al.*, 2011). There has been more interest in the development of new forms of yerba-maté products such as extracts, capsules, tablets, beers, soft drinks, candies, creams, and manufactured products in combination with other herbs as energy teas or as supplements. Products based on yerba-maté are currently exported to Asia, the United States, and Europe as a herbal drug (dry leaf) or as extracts to be employed in different functional foods, cosmetics, and phytopharmaceutical preparations (Heck and Mejia, 2007; Vieira *et al.*, 2008, Berté *et al.*, 2011). The different products of maté tea are presented in Table 15.1.

The polyphenol content of maté herbal tea is influenced by the varying processes used in yerba-maté manufacturing. For example, maté for *chimarrão* or *tererê* and green maté (tea-bag) herbal teas typically receive the least processing; therefore, their naturally occurring polyphenols are preserved with greater efficacy. Other types of maté tea (e.g., roasted) are subjected to more processing, reducing the polyphenol content of the final product. Tea extracts (instant or soluble), formulated for high polyphenol content, can contain the highest amounts (30%) of beneficial substances because they are concentrated formulations (Maccari Junior, 2005; Berté, 2011).

MATÉ TEA COMPOSITION

Maté tea is used as functional food, in popular medicine, and also included in commercial herbal preparations. Compositional studies have shown the presence of different chemical groups, such as phenolics, saponins,

TABLE 15.1 The Content of Total Phenolics and Caffeine in Maté Tea Prepared by Different Processing Techniques

Characteristics	Polyphenols	Caffeine	Reference
Yerba-maté (leaves) Leaves are dried in a forced draft oven	7.3–12.3 mg/g (5-CQA)	8.6–16.1 mg/g	Streit *et al.* (2007)
	79.1–95.9 mg/g (TP) 7.6–9.5 mg/g (5-CQA)	4.9–6.1 mg/g	Cardozo *et al.* (2007)
Chimarrão Beverage traditionally prepared in a container or calabash (cuia or porongo) (Figure 15.3) where maté fills two-thirds of its volume. The free volume is completed with hot water (70–80°C) forming a partial infusion. The resulting aqueous extract is sipped by the consumer through a metal straw called bombilla or bomba. The addition of water is repeated several times	92.1–100.3 mg/g (TP) 22.7–26.8 mg/g (5-CQA)	5.4–5.8 mg/g	Berté (2011)
	49.2–51.8 mg/g (TP)	7.9–8.3 mg/g	Meinhart *et al.* (2010)
Tererê Beverage prepared in a container by total infusion with cold water. Maté for *tererê* can be pure (containing only leaves and stems of the plant yerba-maté) or composed (added with other plant species) and may contain flavoring and/or sugar	41.8 mg/g (TP)	10.2 mg/g	Meinhart *et al.* (2010)
Green maté tea Green maté tea is obtained after blanching, drying, and coarsely crushing (milling) of leaves and stems of the yerba-maté. Green maté tea infusion is prepared with hot water, just off the boil, poured over 1.5–2 g (1 teaspoon) of maté and after 5 min passed through a tea strainer.	94.2 mg/g (TP)	—	Bastos *et al.* (2006b)
	77.3–81.2 mg/g	—	Bravo *et al.* (2007)
	107–133 mg/cup	19–33 mg/cup	Clifford and Ramirez-Martinez (1990)
Roasted maté tea Yerba-maté dry leaves are roasted (similar to processing of roasted coffee) at 160°C for approximately 12 min. Changes can occur in the antioxidant activity of green maté as a consequence of the roasting process. Roasted maté tea generally has a sweeter flavor and bitterless taste. Roasted maté tea infusion is prepared by boiling 1.5–2 g (1 teaspoon) of dried leaves and stems in 150 ml of water for 5 min	59.2 mg/g (TP)	—	Bastos *et al.* (2006b)
	16–41 mg/cup	12–20 mg/cup	Clifford and Ramirez-Martinez (1990)
Maté tea bags Maté tea leaves are packed in individual tea bags. The use of maté tea bags is easy and convenient, making them popular for many people (one tea bag per cup of hot water, brewed for 5 min)	10.9–20.2 mg/182 ml (5-CQA)	11.6–20.2 mg/182 ml	Bastos *et al.* (2005)
Instant or soluble maté tea Instant maté tea, very similar to instant coffee. Instant maté tea is a form of drink that is produced from yerba-maté by extracting (infusion) the soluble solids from processed leaves and concentrating the extract under low pressure, and drying the concentrate to a powder by spray-drying or freeze-drying. With its dried granulated form, it can be made into a convenient beverage with the addition of cold or hot water.	Instant green maté tea 178.3 mg/g (TP)	18.5–36.4 mg/g	Berté (2011)
	Yerba-maté spray-dried extract (soluble) 304.5 mg/g (TP)	6.5 mg/g	Andrade *et al.* (2012)
	Instant roasted maté tea 348.8 mg/g (TP) 32.2 mg/g (5-CQA)	5.8 mg/g	Arçari *et al.* (2009)
Yellow maté tea (aging or forded aging) Yerba-maté dry leaves are left to age in storage warehouses for a period of time necessary for the product to develop flavor, aroma, and color. Aged yerba-maté leaves are milled to smaller particles to obtain the yellow maté tea for *chimarrão*	86.4 mg/g (TP) 20.3 mg/g (5-CQA)	13.9 mg/g	Turner *et al.* (2011)
	18.5–48.1 mg/g (5-CQA)	—	Dutra *et al.* (2010)

5-CQA, 5-*O*-caffeoylquinic acid; TP, total phenolics; 1 cup = 150 ml.

alkaloids, and essential oil. The leaves also contain vitamins (A, C, B1, and B2), magnesium, calcium, iron, zinc, sodium, and potassium (Newall *et al.*, 1996; Alonso, 1998).

The phenolic fraction consists mainly of chlorogenic acid and its derivatives, ranging from 28% to 99%. Generically named chlorogenic acids (CGAs), these are esterified hydroxycinnamates. The major caffeoyl derivatives are

4,5-di-caffeoylquinic acid (4,5-diCQA), 5-CQA, 3-CQA, 4-CQA and 3,5-CQA (Clifford and Ramirez-Martinez, 1990; Filip *et al.*, 2001; Hoffmann-Ribani, 2006; Bastos *et al.*, 2007; Bravo *et al.*, 2007; Peres, 2007). Glycosides of quercetin and kaempferol were also found at low levels. Table 15.2 provides information about the composition of maté tea for *chimarrão* and instant maté tea (unroasted).

Using high-performance liquid chromatography–diode array detector–electrospray ionization/mass spectrometry (HPLC-DAD-ESI/MS) and analyzing *chimarrão* and *tererê* prepared from six commercial lots of maté from Rio Grande do Sul, Santa Catarina, and Parana, the methylxanthine caffeine varied from 124 to 268 μg/ml for *chimarrão* and from 114 to 246 μg/ml for *tererê*, while teobromine varied from 64 to 126 μg/ml for *chimarrão* and 54 to 114 μg/ml for *tererê* (Peres, 2007). The ranges for 3-CQA, 5-CQA, and 4-CQA were: 153–242, 183–263, and 123–188 μg/ml, respectively, for *chimarrão*; and 122–218, 164–209, and 102–169 μg/ml, respectively, for *tererê*. For 3,5-diCQA, 4,5-diCQA, and triCQA, the ranges were: 112–167, 238–289, and 125–154 μg/ml, respectively, for *chimarrão*; and 103–145, 206–265, and 102–143 μg/ml, respectively, for *tererê*. Much lower levels of quercetin-3-*O*-rhamnosylglucoside, quercetin-3-*O*-flucoside, and kaempferol-3-*O*-glucoside were also found.

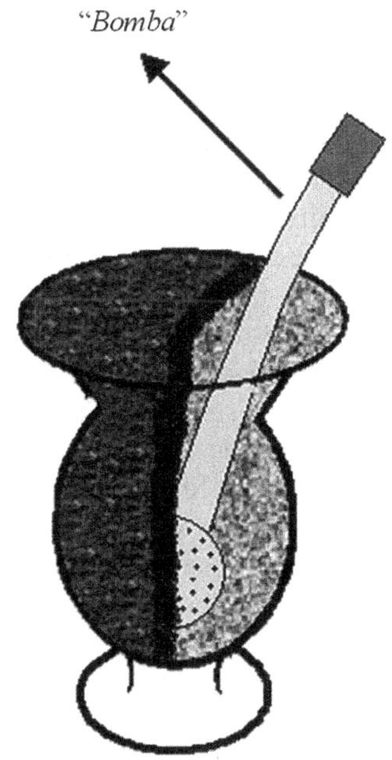

FIGURE 15.3 Container (*cuia*) with maté tea used to prepare the *chimarrão*. *Source: Meinhart* et al. *(2010).*

TABLE 15.2 Composition (%) of Maté Tea for *Chimarrão* and Instant Maté Tea (Unroasted)

Constituent	Maté tea (*chimarrão*)	Instant maté tea (unroasted)
Carbohydrates[a]	13.7–26.1	80.7–81.5
Protein	7.9–9.8	3.7–4.1
Lipids	3.8–4.2	0.7–0.9
Fiber alimentary	52.9–59.1	0.0
Moisture	3.6–8.0	3.9–4.8
Ash	5.1–5.7	9.5–10.1
Caffeine	0.6–1.0	1.8–3.6
Total phenolics	5.1–9.6	17.8–23.9

[a]*Calculated as the difference.*
Source: Berté (2011)

ANTIOXIDANT ACTIVITY

Maté showed *in vitro* and *in vivo* antioxidant activity, being able to scavenge free radicals and reactive oxygen species (ROS) (Carini *et al.*, 1998; Filip *et al.*, 2000; Chandra and Mejia, 2004; Bastos *et al.*, 2006b, 2007; Bravo *et al.*, 2007; Berté *et al.*, 2011). This antioxidant activity has been attributed to the presence of polyphenolic compounds. Bravo *et al.* (2007) found the antioxidant activity of maté, determined by the ferric ion-reducing antioxidant power (FRAP) and 2,2′-azinobis-3-ethylbenzothiazoline-6-sulphonic acid diammonium salt (ABTS) methods, to be slightly higher than that of wines, orange juice, and black tea, but lower than green tea.

An aqueous extract prepared from an infusion of *I. paraguariensis* inhibited the enzymatic and nonenzymatic lipid peroxidation in rat liver microsomes in a concentration-dependent manner (Schinella *et al.*, 2000). It also inhibited the hydrogen peroxide (H_2O_2)-induced peroxidation of red blood cell membranes and exhibited radical scavenging properties toward superoxide anion and 1,1-diphenyl-2-picrylhydrazyl (DPPH) radical.

Bracesco *et al.* (2003) analyzed the antioxidant properties of *I. paraguariensis* infusion using two experimental models: the induction of DNA double-strand breaks (DSB) by H_2O_2 and lethality in *Saccharomyces cerevisiae*, as well as peroxide and lipoxygenase-induced human low-density lipoprotein (LDL) oxidation. The maté infusion significantly decreased the dose dependent DSB number and the lethality induced by H_2O_2. It also inhibited peroxynitrite- and lipoxygenase-induced human LDL oxidation in a potent, dose-dependent fashion.

Herbal preparations, prepared the way they are usually drunk, and wines were compared using a protein nitration model and mammalian cell cytotoxity (Bixby *et al.*, 2005). Maté exhibited the highest inhibition of protein nitration and the highest promotion of cell survival, whereas green tea or red wines displayed significant but lesser effects at the same concentrations. Electron spin resonance experiments indicated that maté scavenged hydroxyl and superoxide radicals (Leonard *et al.*, 2010).

It also inhibited hydroxyl-radical-induced lipid peroxidation and DNA damage in cell culture systems.

The polyphenol concentration of maté tea showed a strong correlation with its total antioxidant capacity measured by the oxygen radical absorbance capacity (ORAC) assay (Chandra and Mejia, 2004). Maté tea showed a slightly higher polyphenol concentration, 7.73 mg chlorogenic acid/ml aqueous extract, than green tea (*Camellia sinensis*), 7.15 mg chlorogenic acid/ml aqueous extract. This corresponded to a higher antioxidant capacity for maté tea (90% inhibition of free radicals) than green tea (88% inhibition of free radicals) when the DPPH method was used (Bastos *et al.*, 2007).

The relation between polyphenol composition and antioxidant activity of a mixture of caffeic acid, chlorogenic acid, kaempferol, quercetin, and rutin was found to be satisfactorily predicted with a polynomial model (Valerga *et al.*, 2012). Quercetin was the highest contributor to the antioxidant activity, followed by kaempferol and caffeic; rutin and chlorogenic acid inputs were the lowest.

Based on the ABTS assay, the individual antioxidant activity of the phenolics of *chimarrão* was determined on-line, after separation and quantification of the phenolic compounds by HPLC (Peres, 2007). The antioxidant capacity was not proportional to the concentrations. 3-CQA, quercetin-3-*O*-glucoside, and quercetin-3-rhamnosylglucoside had the highest antioxidant activity, although the quercetin glycosides had concentrations less than 10-times that of 3-CQA. Tri-CQA, kaempherol-3-*O*-glucoside, and 4-*O*-CQA all had the second-highest antioxidant activity, although the CQAs had levels more than 10-times that of the kaempferol glycoside. The major phenolic compounds of the maté *chimarrão* infusion, 4,5-diCQA and 5-CQA, had lower activity. The quercetin and kaempferol glcycosides, therefore, apppeared to be much more potent antioxidants than the CQAs.

Methylxanthines in yerba-maté, such as caffeine, have a stimulatory activity on the central nervous system. In rat submandibulary glands, these compounds induce the secretion of antioxidant enzymes such as peroxidase (Anesini *et al.*, 2006). Moreover, a protective effect of caffeine on cells of the immune system has been demonstrated through its catalase-like (CAT) activity.

Affirming that caffeine in coffee could protect cells from death, Davicino *et al.* (2011) proposed that it might be more beneficial to drink normal coffee than decaffeinated coffee because caffeine as an antioxidant could maintain redox equilibrium in cells. The protective effect of an aqueous extract of *I. paraguariensis* (green leaves) against hemolysis of red blood cells induced by H_2O_2 was attributed to the presence of chlorogenic acid and caffeine, and to the antioxidant activity exerted by them in relation to the H_2O_2 scavenging effect (Peralta *et al.*, 2013).

Supplementation with maté tea acutely lowered lipid peroxidation in healthy young women, an effect that was maintained after prolonged administration (Matsumoto *et al.*, 2009). Increased total antioxidant status and level of antioxidant enzyme gene expression were also demonstrated after prolonged consumption.

EFFECTS OF PRODUCTION AND PROCESSING

Pre-harvest and post-harvest variation in the composition of *I. paraguariensis* leaves has been documented; further compositional changes occur during processing and storage. The amount of polyphenols in the leaves or in the beverages may vary considerably due to numerous factors, such as agronomic procedures, processing technology, and brewing methodology. It is well known that the parameters involved in the extraction procedure (such as particle size of leaves and brewing conditions) greatly influence the solubility of bioactive compounds. Growing conditions and genetic characteristics also highly influence the content of such substances in plants (Astill *et al.*, 2001). Some of the different results for phenolic composition of yerba-maté leaves and beverages are shown in Table 15.3.

The difference in the yerba-maté chemical composition might be due to climate/season, soil characteristics, plant variety, leaf age, type of processing, and part of the plant used (Esmelindro *et al.*, 2004; Maccari Junior, 2005). Overall, the leaves from trees grown in plantations had the highest levels of nearly all the polyphenols. Several phenolic compounds are produced by plants as a response to environmental stimuli, generally protecting them from environmental factors, such as stress, pests, and the sun (Meyer *et al.*, 2006). Leaves of *I. paraguariensis* exposed to the sun exhibited higher levels of caffeoyl derivatives, caffeine, theobromine, and rutin, as compared with those from plants grown in a protected environment (under the shaded forest canopy) (Dartora *et al.*, 2011). Lower rainfall, temperature, and drying conditions had varying effects on the phenolics, but on average, sun-exposed plantation-grown maté teas had a greater concentration of phenolics than shaded forest-grown maté teas (Heck *et al.*, 2008).

During the industrial processing stages (i.e., harvesting, blanching, drying, milling, roasting, and aging) some changes in the profile and concentration of bioactive compounds of yerba-maté may occur, which can modify its antioxidant activity (Bastos *et al.*, 2006a; Isolabella *et al.*, 2010; Turner *et al.*, 2011). Bastos *et al.* (2006a) reported that the soluble solid content of the extract prepared with dried yerba-maté leaves was approximately four-fold higher than that of the extract prepared with fresh leaves. The extraction of caffeine and 5-CQA was found to be more effective when dried leaves are used. This was attributed to cell disruption due to mechanical impact and heat during processing.

TABLE 15.3 Yerba-Maté Polyphenol Content Determined in Several Research Studies (1990–2012)

Sample/extraction	Analytical methodology	Results	Reference
Methanolic extract from 5 samples of roasted maté tea and green maté tea.	Reversed-phase HPLC, detection at 313 nm	Chlorogenic acids content varied from 16 to 41 mg/cup from the brownish samples and from 107 to 133 mg/cup for the greenish samples	Clifford and Ramirez-Martinez (1990)
Water infusion from 18 samples of maté tea for *chimarrão* from south Brazil (1.5 g/30 ml)	RP-HPLC, Detection at 313 nm (5-CQA and caffeic acid) Folin–Ciocalteau method (total phenolics)	Mean values for chlorogenic acids and caffeic acid were 0.48 mg/ml and 0.34 mg/ml, respectively. Total phenolics varied from 0.78 to 1.6 mg/ml	Mazzafera (1997)
Water infusion from dried leaves at 30% (p/v)	Caffeoyl derivatives were determined by spectrophotometry (330 nm) with 5-CQA as standard	Phenolic content was 10.7% (w/w) for *Ilex paraguariensis* and varied from 0.96 to 6.8% (p/p) for the other *Ilex* species	Filip *et al.* (2000)
Aqueous extract from yerba-maté 5% (w/v) fresh leaves from Argentine.	EP-HPLC, detection at 325 nm for caffeic acid derivatives; rutin detection at 255 nm	Total phenolic content for yerba-maté was 9.6% (w/w). Caffeic acid content was 0.23 mg/g and rutin was 0.60 mg/g	Filip *et al.* (2001)
Water infusions from maté-tea (1 bag/cup) and *chimarrão*-maté (3 g/60 ml) prepared as *chimarrão* (hot water) or *tererê* (cold water).	RP-HPLC, detection at 323 nm	5-CQA content varied from 0.06 to 0.13 mg/ml for the maté-tea; from 0.43 to 0.46 mg/ml for the *chimarrão* beverage and from 0.27 to 0.38 mg/ml for the *tererê* beverage	Bastos *et al.* (2005)
Ethanolic extract from maté-tea for *chimarrão* of different regions of Brazil	RP-HPLC, detection at 325 nm for caffeic acid derivatives; rutin detection at 370 nm	5-CQA content varied from 13.1 to 24.7 mg/g and rutin concentration varied from 2.4 to 8.0 mg/g	Hoffmann-Ribani (2006)
Aqueous extract from 4 samples of yerba-maté harvested in 2 different regions of south Brazil, divided into native and reforested according to the cultural practice used	RP-HPLC, detection at 274 nm for caffeic acid and 5-CQA	Caffeic acid content varied from 0.24 to 0.29 mg/g and 5-CQA varied from 10.0 to 12.7 mg/g for the yerba-maté native; caffeic acid content varied from 0.04 to 0.37 mg/g and 5-CQA varied from 7.3 to 9.0 mg/g for the samples yerba-maté reforested	Streit *et al.* (2007)
Ethanolic extract from 6 samples of yerba-maté *cancheada* that had undergone natural aging and forced aging	RP-HPLC, detection at 325 nm for caffeic acid derivatives; rutin detection at 370 nm	Caffeic acid was 0.18 mg/g, 5-CQA was 34.9 mg/g and rutin 7.1 mg/g from the yerba-maté natural aging. For yerba-maté forced aging showed slightly higher phenolics; caffeic acid content was 0.23 mg/g, 5-CQA 36.1 mg/g, and rutin 7.2 mg/g	Dutra *et al.* (2010)
Ethanolic extract from yerba-maté dehydrated leaves and maté tea soluble from Brazil (Paraná) were analyzed	RP-HPLC, detection at 325 nm for caffeic acid derivatives; rutin detection at 370 nm; Folin–Ciocalteu colorimetric method detection at 765 nm for total phenolics	Yerba-maté dehydrated leaves: caffeic acid 0.72 mg/g; 5-CQA 24.8 mg/g; rutin 3.1 mg/g; and total phenolics 96.2 mg/g; Soluble green maté tea: caffeic acid 1.5 mg/g; 5-CQA 91.4 mg/g; rutin 3.1 mg/g and total phenolics 178.3 mg/g.	Berté *et al.* (2011)
Instant maté tea was obtained by spray drying of the yerba-maté aqueous extract and dissolved in water	RP-HPLC, detection at 280 nm for chlorogenic acid; Folin–Ciocalteu colorimetric method detection at 765 nm for total phenols.	Total phenol content for yerba-maté spray-dried extract (soluble) was 304.5 mg/g and chlorogenic acid content was 39.8 mg/g.	Andrade *et al.* (2012)

HPLC, high-performance liquid chromatography; RP-HPLC, Reverse phase HPLC; 5-CQA, 3-*O*-caffeoylquinic acid.

The influence of industrial processing in the physicochemical parameters and in the content of caffeine has been studied by several authors. Esmelindro *et al.* (2002) reported that industrial processing reduced the amount of caffeine significantly, especially during blanching (*sapeco*) and drying, due to the high temperatures used. The most considerable loss of caffeine (roughly 20%) occurred during the drying process (Schmalko and Alzamora, 2001). The results obtained by Isolabella *et al.* (2010) agree with those of the previous studies, however, the loss of caffeine was found to be lower (8.7%). These quantitative differences could be attributed to the different processing conditions employed, mainly the temperature and time span (Esmelindro *et al.*, 2002).

According to Valerga *et al.* (2012), the chlorogenic acid content increased with blanching (*sapeco*) from 0.08 to 6.9 mg/g dried leaves. A subsequent pre-drying treatment duplicated the amount present in the blanched samples. No changes were observed in the dried samples, however, forced aging had a very strong effect, reducing CGA concentration by 92%. Industrial processing increased the total chlorogenic derivative content (calculated as the sum of the individual levels) from 1.6 to 62.5 mg/g dried leaves. Blanching and pre-drying did not affect the caffeic acid level, but drying and especially forced aging caused a considerable increment. Isolabella *et al.* (2010) also reported that samples from blanched, dried, and aged yerba-maté leaves had higher concentrations of caffeoyl derivatives than those obtained from freshly harvested leaves; however, the results of Valerga *et al.* (2012) showed a much stronger impact of the industrial process. The latter authors attributed these differences to variations in the sample handling procedure between the two studies.

Processed yerba-maté leaves for *chimarrão* showed a decrease in the concentration of xanthines when compared with yerba-maté green leaves (unprocessed). Conversely, the leaves subjected to blanching and drying (*chimarrão* type) contained more phenolic compounds and consequently a more intense antioxidant activity (Dartora *et al.*, 2011).

Turner *et al.* (2011) reported that the leaves of yerba-maté subjected to industrial processing, involving blanching (250–550°C/2–4 min), drying (hot air at 90–110°C for 3–6 h until a 3% of moisture) and forced aging (product was placed in a chamber with temperature of 60 ± 5°C and humidity of 40 ± 5°C for 30–60 days), displayed higher total polyphenol contents (86.4–88.5 mg/g) than yerba-maté green leaves (75.9 mg/g). The caffeic acid content was similar in all samples, varying from 0.33 to 0.35% of dried plant material. Chlorogenic acid content was slightly lower in yerba-maté green leaves (1.84%) when compared with the samples subjected to blanching, drying, and aging, no significant differences ($p > 0.05$) being found among samples from these three stages (2.03–2.06%). The major differences occurred in the caffeine and rutin contents. Samples prepared with green leaves had lower levels of caffeine (0.91%) and rutin (0.98%) than the samples subjected to industrial processing. In these three stages, the caffeine content varied from 1.36 to 1.49% and the rutin content varied from 1.25 to 1.35%.

An increase in the content of methylxanthines was observed in blanching (*sapeco*) when compared to the yerba-maté green leaves, followed by a decrease during the drying process. The same behavior was observed for the caffeoyl derivatives (i.e., chlorogenic acid and the isomers), the content of which was found to be significantly higher in the blanched leaves, as compared to the yerba-maté green leaves. Caffeic acid was the only compound that had a constant level throughout the process. Turner *et al.* (2011) and Isolabella *et al.* (2010) suggested that both processes would be contributing to the increase in the levels of the major bioactive compounds found in the samples subjected to heat during the manufacturing process.

Samples obtained after blanching, drying and aging were all found by Isolabella *et al.* (2010) to have higher contents of biologically active principles (caffeoyl derivatives, methylxanthines, and flavonoids) when compared with green leaves. There were no differences between the natural and forced aging processes.

Industrial processing (drying) increased the antioxidant activity of maté tea, evaluated by both the DPPH and the beta-carotene-linoleic acid assays (Dartora *et al.*, 2011). Bastos *et al.* (2006b) obtained equivalent antioxidant activity, measured by the ferric thiocyanate method, for green maté and maté tea infusion, indicating that the roasting stage did not affect the antioxidant property. Conversely, in the study of Turner *et al.* (2011), the total antioxidant activity of extracts of *I. paraguariensis*, measured by both the DPPH and the ferric thiocyanate assays, decreased during the industrial process.

References

Alonso, J.R., 1998. Tratado de fitomedicina. Editora Isis, Buenos Aires, Argentina 992–995.

Andrade, F., Albuquerque, C.A.C., Maraschin, M., Silva, E.L., 2012. Safety assessment of yerba maté (*Ilex paraguariensis*) dried extract: Results of acute and 90 days subchronic toxicity studies in rats and rabbits. Food Chem. Toxicol. 50, 328–334.

Anesini, C., Turner, S., Manuele, M.G., Ferraro, G., Filip, R., 2006. Pharmacological activity of caffeine isolated from Ilex paraguariensis on peroxidase secretion in rat submandibulary glands. Pharmacology-Online 3, 372–375.

Arçari, D.P., Bartchewsky, W., Santos, T.W., Oliveira, K.A., Funck, A., Pedrazzoli, J., Souza, M.F.F., Saad, M.J., Bastos, D.H.M., Gambero, A., Carvalho, P.O., Ribeiro, M.L., 2009. Antiobesity effects of yerba maté extract (*Ilex paraguariensis*) in high-fat diet-induced obese mice. Obesity 17, 2127–2133.

Astill, C., Birch, M.R., Dacombe, C., Humphrey, P.G., Martin, P.T., 2001. Factors affecting the caffeine and polyphenol content of black and green tea infusions. J. Agric. Food Chem. 49, 5340–5347.

Bastos, D.H.M., Fornari, A.C., Queiroz, Y.S., Manolio, R., Torres, E.F.S., 2005. The chlorogenic acid and caffeine content of yerba maté (*Ilex paraguariensis*) beverages. Acta. Farm. Bonaerense 24, 91–95.

Bastos, D.H.M., Fornari, A.C., Queiroz, Y.S., Torres, E.A.F.S., 2006a. Bioactive compounds content of chimarrão infusions related to the moisture of yerba maté (*Ilex paraguariensis*) leaves. Braz. Arch. Biol. Technol. 49, 399–404.

Bastos, D.H.M., Ishimoto, E.Y., Marques, M.O.M., Ferri, A.F., Torres, E.A.F.S., 2006b. Essential oil and antioxidant activity of green maté and maté tea (*Ilex paraguariensis*) infusions. J. Food Comp. Anal. 19, 538–543.

Bastos, D.H.M., Saldanha, L.A., Catharino, R.R., Sawaya, A.C.H.F., Cunha, I.B.S., Carvalho, P.O., Eberlin, M.N., 2007. Phenolic antioxidants identified by ESI-MS from yerba-maté (*Ilex paraguariensis*) and green tea (*Camelia sinensis*) extracts. Molecules 12, 423–432.

Berté, K.A.S., Beux, M.R., Spada, P.K.W.D.S., Salvador, M., Hoffmann-Ribani, R., 2011. Chemical composition and antioxidant activity of yerba-maté (*Ilex paraguariensis* A.St.-Hil., Aquifoliaceae) extract as obtained by spray drying. J. Agric. Food Chem. 59, 5523–5527.

Berté, K.A.S., 2011. Tecnologia da erva-maté solúvel. Doctoral Thesis in Food Technology. Universidade Federal do Paraná, Curitiba, PR, Brasil 160.

Bixby, M., Spieler, L., Menini, T., Gugliucci, A., 2005. *Ilex paraguariensis* extracts are potent inhibitos of nitrosative stress: A comparative study with green tea and wines using a protein nitration model and mammalian cell cytotoxity. Life Sci. 77, 345–358.

Bracesco, N., Dell, M., Rocha, A., Behtash, S., Menini, T., Gugliucci, A., Nunes, E., 2003. Antioxidant activity of a botanical extract preparation of *Ilex paraguariensis*: Prevention of DNA double-strand breaks in *Saccharomyces cerevisiae* and human low-density lipoprotein oxidation. J. Altern. Complement. Med. 9, 370–387.

Bracesco, N., Sanchez, A.G., Contreras, V., Menini, T., Gugliucci, A., 2011. Recent advances on *Ilex paraguarensis* research: Mini review. J. Ethnopharmacol. 136, 378–384.

Bravo, L., Goya, L., Lecumberri, E., 2007. LC/MS characterization of phenolic constituents of maté (*Ilex paraguariensis*, St. Hil.) and its antioxidant activity compared to commonly consumed beverages. Food Res. Int. 40, 393–405.

Cardozo, E.L., Ferrarese Filho, O., Cardozo Filho, L., Ferrarese, M.L.L., Donaduzzi, C.M., Sturion, J.A., 2007. Methylxanthines and phenolic compounds in maté (Ilex paraguariensis St. Hil.) progenies grown in Brazil. J. Food Comp. Anal. 20, 553–558.

Carini, M., Facino, R.M., Aldini, G., Calloni, M., Colombo, L., 1998. Characterization of phenolic antioxidants from maté (*Ilex paraguayensis*) by liquid chromatography/mass spectrometry and liquid chromatography/tandem mass spectrometry. Rapid Commun. Mass Spectrom 12, 1813–1819.

Chandra, S., Mejia, E.G., 2004. Polyphenolic compounds, antioxidant capacity, and quinone reductase activity of an aqueous extract of *Ardisia compressa* in comparison to maté (*Ilex paraguariensis*) and green (*Camellia sinensis*) teas. J. Agric. Food Chem. 52, 3583–3589.

Clifford, M.N., Ramirez-Martinez, J.R., 1990. Chlorogenic acids and purine alkaloids contents of maté (*Ilex paraguariensis*) leaf and beverage. Food Chem. 35, 13–21.

Dartora, N., Souza, L.M., Santana-Filho, A.P., Iacomini, M., Valduga, A.T., Gorin, P.A.J., Sassaki, G.L., 2011. UPLC-PDA-MS evaluation of bioactive compounds from leaves of *Ilex paraguariensis* with different growth conditions, treatments and ageing. Food Chem. 129, 1453–1461.

Davicino, R., Alonso, R., Anesini, C., 2011. Comparison between normal coffee and decaffeinated coffee effects on lymphocytes and macrophages: role of the antioxidant activity of caffeine. J. Food Biochem. 35, 877–897.

Dutra, F.L.G., Hoffmann-Ribani, R., Ribani, M., 2010. Determinação de compostos fenólicos por cromatografia líquida de alta eficiência isocrática durante estacionamento da erva-maté. Quim. Nova 33, 119–123.

Esmelindro, M.C., Toniazzo, G., Waczuk, A., Dariva, C., Oliveira, D., 2002. Effects of industrial processing steps on the physico-chemical characteristics of maté tea leaves. Ciênc. Tecnol. Aliment 22, 193–204.

Esmelindro, A.A., Girardi, J.S., Mossi, A., Jacques, R.A., Dariva, C., 2004. Influence of agronomic variables on the composition of maté tea leaves (*Ilex paraguariensis*) extracts obtained from CO_2 extraction at 30°C and 175bar. J. Agric. Food Chem. 52, 1990–1995.

FAOSTAT. Food and Agricultural Organization (FAO) Statistics Division 2013. Available from: http://faostat.fao.org/site/339/default.aspx. (accessed 07.02.13).

Filip, R., Lopez, P., Gilbert, G., Coussio, J., Ferraro, G., 2001. Phenolic compounds in seven South American *Ilex* species. Fitoterapia 72, 774–778.

Filip, R., Lotito, S.B., Ferraro, G., Fraga, C.G., 2000. Antioxidant activity of Ilex paraguariensis and related species. Nutr. Res. 20, 1437–1446.

Heck, C.I., Mejia, E.G., 2007. Yerba maté tea (*Ilex paraguariensis*): A comprehensive review on chemistry, health implications, and technological considerations. J. Food Sci. 72, R138–R151.

Heck, C.I., Schmalko, M., Mejia, E.G., 2008. Effect of growing and drying conditions on the phenolic composition of maté teas (Ilex paraguariensis). J. Agric. Food Chem. 56, 8394–8403.

Hoffmann-Ribani, R., 2006. Compostos fenólicos em erva-maté e frutas. Doctoral Thesis in Food Science. University of Campinas, Campinas, SP, Brazil p. 137.

Isolabella, S., Cogoi, L., López, P., Anesini, C., Ferraro, G., Filip, R., 2010. Study of the bioactive compounds variation during yerba maté (*Ilex paraguariensis*) processing. Food Chem. 122, 695–699.

Leonard, S.S., Hogans, V.J., Coppes-Petricorena, Z., Peer, C.J., Vining, T.A., Fleming, D.W., Harris, G.K., 2010. Analysis of free-radical scavenging of yerba maté (*Ilex paraguariensis*) using electron spin resonance and radical-induced DNA damage. J. Food Sci. 75, C14–C20.

Maccari Junior, A., 2005. Análise do pré-processamento da erva-maté para chimarrão. Doctoral Thesis in Agricultural Engineering. University of Campinas, Campinas, SP, Brazil p. 215.

Matsumoto, R.L.T., Bastos, D.H.M., Mendonça, S., Nunes, V.S., Bartchewsky Jr., W., Ribeiro, M.L., Carvalho, P.O., 2009. Effects of maté tea (*Ilex paraguariensis*) ingestion on mRNA expression of antioxidant enzymes, lipid peroxidation, and total antioxidant status in healthy young women. J. Agric. Food Chem. 57, 1775–1780.

Mazzafera, P., 1997. Maté drinking: caffeine and phenolic acid intake. Food Chem. 60, 67–71.

Meinhart, A.D., Bizzotto, C.S., Ballus, C.A., Rybka, A.C.P., Sobrinho, M.R., Cerro-Quintana, R.S., Teixeira-Filho, J., Godoy, H.T., 2010. Methylxanthines and phenolics content extracted during the consumption of maté (Ilex paraguariensis St. Hil) beverages. J. Agric. Food Chem. 58, 2188–2193.

Meyer, S., Cerovic, Z.G., Goulas, Y., Montpied, P., Demontes-Mainard, S., Bidel, L.P.R., Moya, I., Dreyer, E., 2006. Relationships between optically assessed polyphenols and chlorophyll contents, and leaf mass per area ratio in woody plants: A signature of the carbon–nitrogen balance within leaves? Plant Cell Environ. 29, 1338–1348.

Newall, C.A., Anderson, L.A., Phillipson, J.D., 1996. Herbal medicines. The Pharmaceutical Press, London, U.K. 189–190.

Oboh, G., 2005. Effect of blanching on the antioxidant properties of some tropical green leafy vegetables. Food Sci. Technol 38, 513–517.

Peralta, I.N., Cogoi, L., Filip, R., Anesini, C., 2013. Prevention of hydrogen peroxide-induced red blood cells lysis by *Ilex paraguariensis* aqueous extract: Participation of phenolic and xanthine compounds. Phytother. Res. 27, 192–198.

Peres, R.G., 2007. Aplicações de CE-DAD e HPLC-DAD-ESI/MS na determinação de compostos fenólicos, metilxantinas e ácidos orgânicos em bebidas. Doctoral Thesis in Food Science. University of Campinas, Campinas, SP, Brazil 179.

Rotta, E., Oliveira, Y.M.M., 2005. Cultivo da erva-maté. Embrapa florestas, n.1, Colombo.

Schinella, G.R., Troiani, G., Davila, V., Buschiazzo, P.M., Tournier, H.A., 2000. Antioxidant effects of an aqueous extract of Ilex paraguariensis. Biochem. Biophys. Res. Commun. 269, 357–360.

Schmalko, M.E., Alzamora, S.M., 2001. Colour, chlorophyll, caffeine, and water content variation during yerba maté processing. Dry Technol. 19, 599–610.

Streit, N.M., Hecktheuer, L.H.R., Canto, M.W., Mallmann, C.A., Streck, L., Parodi, T.V., Canterle, L.P., 2007. Relation among taste-related compounds (phenolics and caffeine) and sensory profile of erva-maté (*Ilex paraguariensis*). Food Chem. 102, 560–564.

Turner, S., Cogoi, L., Isolabella, S., Filip, R., Anesini, C., 2011. Evaluation of the antioxidant activity and polyphenols content of *Ilex paraguariensis* (maté) during industrialization. Adv. J. Food Sci. Technol. 3, 23–30.

Valduga, A.T., Finzer, J.R.D., Mosele, S.H., 2003. Processamento de erva-maté. Erechim: Edifapes, 183.

Valerga, J., Reta, M., Lanari, M.C., 2012. Polyphenol input to the antioxidant activity of yerba maté (*Ilex paraguariensis*) extracts. LWT – Food Sci. Technol 45, 28–35.

Vieira, M.A., Rovaris, A.A., Maraschin, M., Simas, K.N., Pagliosa, C.M., Podesta, R., Amboni, R.D.M.C., Barreto, P.L.M., Amante, E.R., 2008. Chemical characterization of candy made of erva-maté (*Ilex paraguariensis* A. St. Hil.) residue. J. Agric. Food Chem. 56, 4637–4642.

CHAPTER

16

Antioxidants in Goji Berry Juice (*Lycium barbarum*) and Effects of Processing Steps

Harunobu Amagase

FreeLife International, Phoenix, AZ, USA

CHAPTER POINTS

- *Lycium barbarum* is an interesting herb and food, and has wide variety of biological effects shown in various human clinical, *in vivo* animal and *in vitro* studies.
- Antioxidant ingredients in *Lycium barbarum* fruit are located in both juice and other parts, such as skin and seeds.
- Concentration of an active ingredient, *Lycium barbarum* polysaccharide (LBP), is within a generally accepted level of specification range and is relatively stable throughout 2–3 years' shelf-life.
- LBP-standardized *Lycium barbarum* juice has been shown to significantly stimulate antioxidant capacities in human plasma indicated by increased superoxide dismutase, glutathione peroxidase and reduced malondialdehyde levels, which are more reliable markers of antioxidant efficacies *in vivo* than is the *in vitro* oxygen radical absorbance capacity (ORAC) test
- *Lycium barbarum* offers skin protection against immune suppression and oxidative stress by sunlight simulated ultraviolet (SSUV) radiation.

INTRODUCTION

Lycium barbarum L. (*L. barbarum*) is a Solanaceous defoliated shrub that grows in China, Tibet and other parts of Asia. Its fruits are 1–2 cm-long, bright orange–red ellipsoid berries (Figure 16.1). The ripe fruit has been used in Asian countries as a traditional herbal medicine and functional food (Amagase and Farnsworth, 2011). Concentrated extracts and infusions prepared from the berries have a history of use as ingredients in various soft or alcoholic drinks that have been marketed (UK Food Standard Agency, 2007; Amagase and Farnsworth, 2011) for their benefits to anti-aging, vision, kidney and liver functions. In support of these traditional properties, recent studies indicate that extracts from *L. barbarum* fruit and one of its active compounds, *Lycium barbarum* polysaccharides (LBP), possess a range of biological activities including antioxidant properties, effects on aging, neuroprotection, anti-fatigue/endurance, increased metabolism, glucose control in diabetics, glaucoma, immunomodulation, anti-tumor activity and cytoprotection (Amagase and Farnsworth, 2011). We have conducted various clinical and basic studies to examine the traditional effects of *L. barbarum* fruits provided in the form of a juice that is standardized for LBP. These were provided daily and changes in subjective ratings of feelings of general well-being, fatigue, stress, neurological/psychological traits, gastrointestinal and musculoskeletal complaints, cardiovascular effects (blood pressure and pulse rate), visual acuity (Amagase and Nance, 2008; Hsu *et al.*, 2012), plasma levels of antioxidant factors (Amagase *et al.* 2009a), immune factors and any side effects (Amagase *et al.* 2009b), waist circumference reduction and post-prandial energy expenditure stimulation (Amagase and Nance, 2011), anti-stress and anti-fatigue effects through glucocorticoid control (Amagase and Nance 2012) were determined at the end of the treatment period. *L. barbarum* fruit have recently been marketed as foods and dietary supplements in many countries including North America (McLaughlin, 2006; Sohn, 2008; Karp, 2009), Caribbean countries, EU (UK Food Standard Agency, 2007; Dutch authorities, 2004), Australia, New Zealand, and Southeast Asia in various retail outlets, including major supermarkets and direct-marketing channels.

Processing and Impact on Antioxidants in Beverages
http://dx.doi.org/10.1016/B978-0-12-404738-9.00016-7

FIGURE 16.1 Picture of *Lycium barbarum* fruit taken in Ningxia province in People's Republic of China. *Shared by FreeLife International, Inc.*

Oxygen radical absorbance capacity (ORAC) is a method of measuring antioxidant capacities in biological samples *in vitro* (Cao *et al.*, 1993; Ou *et al.*, 2001). A wide variety of foods have been tested using this method, with certain spices, berries, and legumes rated highly in extensive tables once published by the United States Department of Agriculture (USDA), but these were withdrawn in 2012 as biologically invalid (USDA, 2012), as no physiological proof *in vivo* exists that the free-radical theory is valid. Consequently, the ORAC method, derived only in *in vitro* experiments, cannot be interpreted as relevant to human diets or biology. Thus, we have examined *in vivo* effects of LBP-standardized *L. barbarum* fruit juice in a randomized, double-blind, placebo-controlled human clinical study by measuring serum levels of antioxidant markers, such as superoxide dismutase (SOD), glutathione peroxidase (GSH-Px) and malondialdehyde (MDA). These are the reliable surrogate markers of antioxidant effects *in vivo*. This chapter summarizes current knowledge and information of *L. barbarum* for its antioxidant capacities, including our clinical and basic studies and the active ingredients. And we have tested stability of LBP in the LBP-standardized *L. barbarum* juice in both accelerated and real-time conditions for 12 or 24 months, respectively, from a quality control standpoint based on the internationally harmonized (ICH) guideline to confirm its stability during shelf-life.

CHEMICAL CONSTITUENTS

Polysaccharides

According to the Chinese understanding of *Lycium* extracts and products, the content of a group of water-soluble glycoconjugates, (*Lycium barbarum* polysaccharides; LBP) is important in the efficacy of *L. barbarum*. Many plant- and fungal-derived bioactive polysaccharides with a broad range of immunomodulatory activities are found in traditional Chinese medicine and, therefore, a high content of polysaccharides with proven pharmacological activities is considered to be an indicator of the medicinal status of a natural product (Chang, 2002; Wong *et al.*, 1994).

Among the chemical constituents of *L. barbarum* fruit, the most well-researched components are LBP, which are estimated to comprise 5–8% of the dried fruits (Wang *et al.*, 1991). However, quantitative data in the literature diverge considerably. The LBP group has a molecular weight range of 8–290 kDa, and several LBP have been isolated and purified from aqueous *L. barbarum* extracts by methods such as diethylaminoethyl cellulose (DEAE) ion-exchange cellulose, gel-permeation chromatography and high-performance liquid chromatography (HPLC) (Luo *et al.*, 2000; Peng *et al.*, 2001a,b,c; Peng and Tian, 2001; Tian and Wang, 2006; Ouyang *et al.*, 2007; Yin and Dang, 2008). LBP consists of a complex mixture of highly branched and only partly characterized polysaccharides and proteoglycans. Their structural composition has been studied by sodium dodecylsulfate polyacrylamide gel electrohoresis (SDS-PAGE), gas chromatography, amino acid automatic analysis, partial acid hydrolysis, periodate oxidation and nucleic magnetic resonance (NMR) spectrum regio-selective enzymatic, alkaline and acidic hydrolyses and spectroscopic methods involving gas-chromatographic-mass-spectrometry (GCMS), high resolution electrospray ionization mass spectrometry (HRESIMS) and 1D and 2D NMR (Gao *et al.*, 2008); they have been found to be complex glycopeptides consisting of acidic heteropolysaccharides and polypeptides or proteins. Although they differ somewhat in composition, the LBP contain six monosaccharides (arabinose, glucose, galactose, mannose, rhamnose and xylose) (Peng and Tian, 2001; Huang *et al.*, 1998a, 1999a; Peng *et al.*, 2001b, 2001c; Gan *et al.*, 2004; Zhao *et al.*, 1996, 1997; Chen *et al.*, 2008; Duan *et al.*, 2001), mainly containing xylose and glucose with smaller amounts of arabinose, rhamnose, mannose and galactose. LBP also contain galacturonic acid and 18 amino acids. They share a glycan-O-Ser glycopeptide structure (Tian and Wang, 2006). The main chains of the glycan backbones of LBP have been found to be either alpha-(1→6)-D-glucans or alpha-(1→4)-D polygalacturonans (Duan *et al.*, 2001). LBP have been also analyzed by preparative high-performance size exclusion chromatography (HPSEC). After protein hydrolysis, two major fractions were obtained with a molecular weight of 79,250 and 24,470 amu, respectively (Wang *et al.*, 2009a). The monosaccharidic composition of the whole polysaccharidic fraction has been determined by gas chromatography after hydrolysis. There are however some discrepancies with regard to the presence of fucose and mannose as well as the content of glucose. Thus the composition was found to be Rha, Ara, Xyl, Man, Glc, and Gal (molar ratio of 0.3:2.7:0.3:0.2:2.7:0.9) (Wang *et al.*, 2009a) or Rha, Ara, Xyl, Fuc, Glc, Gal (1:2.1:1.1:2.3:3.6:10.1) (Li *et al.*, 2007b), respectively. An overview

TABLE 16.1 Polysaccharide (LBP) Constituents Purified from *Lycium barbarum*

LBP/ Glycoconjugate	Molecular Weight	Carbohydrate Content	Monosaccharides (Molar Ratio or %)	Reference
LbGp2	68,200	90.7	Ara, Gal (4:5)	Peng and Tian (2001)
LbGp3	92,500	93.6	Ara, Gal (1:1)	Huang *et al.* (1998a, 1999a)
LbGp4	214,800	85.6	Ara, Gal, Rha, Glc (1.5:2.5:0.43:0.23)	Huang *et al.* (1998a, 1999a)
LbGp5	23,700	8.6	Rha, Ara, Xyl, Gal, Man, Glc (0.33:0.52:0.42:0.94:0.85:1)	Huang *et al.* (1998a)
LbGp5B	23,700		Rha, Ara, Glc, Gal (0.1:1:1.2:0.3), Galu (0.9%)	Peng *et al.* (2001b)
LBP3p	157,000	92.4	Gal, Glc, Rha, Ara, Man, Xyl (1:2.12:1.25:1.10:1.95:1.76)	Gan *et al.* (2004)
$LBPC_2$	12,000	92.8	Xyl, Rha, Man (8.8:2.3:1)	Zhao *et al.* (1996, 1997)
$LBPC_4$	10,000	95	Glc	Zhao *et al.* (1996, 1997)
$LBPA_1$	18,000		Heteroglycan	Zhao *et al.* (1997)
$LBPA_3$	66,000		Heteroglycan	Zhao *et al.* (1997)
LBP1a-1	11,500		Glc	Duan *et al.* (2001)
LBP1a-2	9400		Glc	Duan *et al.* (2001)
LBP3a-1	10,300		GalA	Duan *et al.* (2001)
LBP3a-2	8200		GalA	Duan *et al.* (2001)
LBPF1	ca150,000	48.2*		Chen *et al.* (2008)
LBPF2	ca150,000	30.5*		Chen *et al.* (2008)
LBPF3	ca150,000	34.5*		Chen *et al.* (2008)
LBPF4	ca150,000	20.3*		Chen *et al.* (2008)
LBPF5	290,000	23.5*		Chen *et al.* (2008)

**The carbohydrate content has been determined by the phenol-sulfuric acid assay. The content is much higher when it is put in relation to protein content.*

Polysaccharide (LBP) constituents purified from *Lycium barbarum*. Cited from Potterat (2010), permission from *Planta Medica* under the license number 3061401086576.

of the fractions that have been purified in homogenous forms are shown in Table 16.1 (Poterrat, 2010).

Effects of Production Processing and Stability

LBP is relatively stable during the production process and product shelf-life. ORAC is generally used for *in vitro* evaluation of the material itself. As ORAC can neither predict nor indicate antioxidant activity *in vivo*, we have instead examined more reliable surrogate markers of antioxidant blood markers in humans. The material used was a commercially available LBP-standardized product *L. barbarum* juice, GoChi®, a liquid dietary supplement containing reconstituted juice from fresh whole *Lycium barbarum* fruit. GoChi has been shown to stimulate antioxidant capacities as indicated by increased serum levels of SOD, and GSH-Px, and by reduced level of MDA (Figure 16.2).

GoChi is standardized to contain, per daily 120-ml serving, a content of LBP equivalent to that found in at least 150 g of fresh fruit, the amount customarily consumed in Traditional Chinese Medicine (Zhu, 1998; Dharmananda, 2007). The types of analyses applied to the assay and identification of LBP are not standardized worldwide. A general method has been applied based on identification by Fourier transform near-infrared spectroscopy, the gravimetric assay of the isolated polysaccharide fraction, and a determination of sugars after hydrolysis using gas chromatography (German Foodstuffs and Commodities Act, 1986) and/or HPLC with the glucan analysis reported (AOAC International). The amount of LBP in GoChi is 2023 mg/daily serving (120 ml) in an average of eight lots and based on previous report (Amagase *et al.*, 2009a). The value of LBP concentration was maintained within 20% of initial level at the end of the 12-month accelerated shelf-life study (Table 16.2). This is typical of stability range allowances imposed by regulatory authorities in many countries. Various pharmacopoeias (USP, EP, BP, and JP) recognize such ranges for herbal products and their marker compounds. Therefore, LBP is a relatively stable and useful marker compound in *L. barbarum* fruit juice.

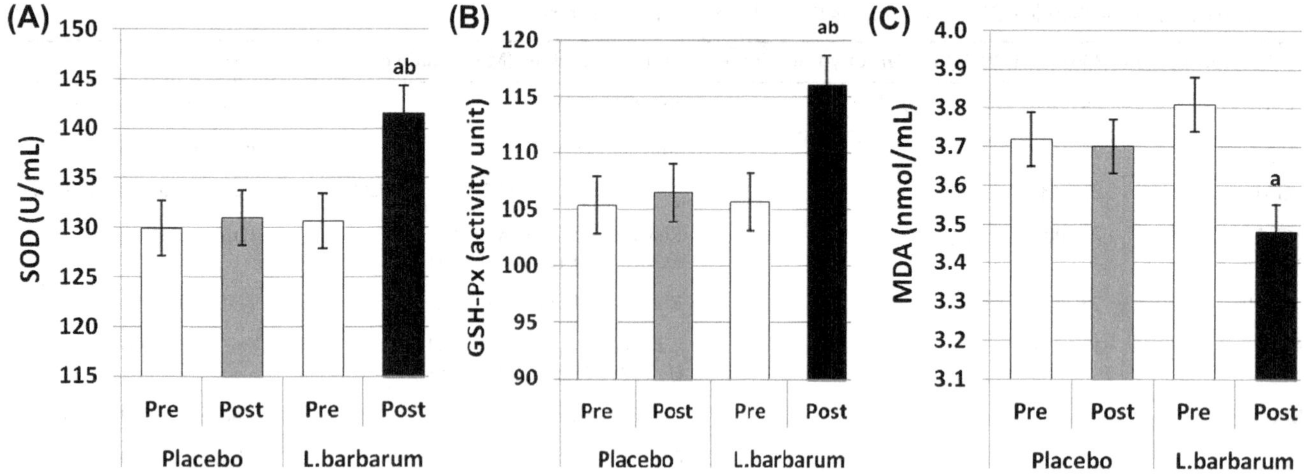

FIGURE 16.2 *In vivo* antioxidant effects of a standardized *Lycium barbarum* fruit juice in human serum as indicated by (A) superoxide dismutase (SOD), (B) glutathione peroxidase (GSH-Px), and (C) malondialdehyde (MDA) levels obtained in a randomized, double-blind, placebo-controlled clinical study (30-day intervention trial). Serum SOD, GSH-Px, and MDA levels were measured individually in enzyme immunoassay (EIA) by commercially available test kits (Northwest Lifescience Specialties, LLC., WA). Each value indicates the mean ± SEM. T-tests were used for self-paired data. Group t-tests were used for mean comparisons, with homogeneity of variance between the *Lycium barbarum* group and placebo group as a prerequisite. Otherwise, t-tests were performed after homogeneity of variance was achieved through variable conversion. Rank-sum tests were used if the variance still lacked homogeneity. 'Pre' and 'Post' in the figures indicate pre-intervention and post-intervention, respectively. Differences were considered significant at $P < 0.05$. (a) Significant difference compared to pre-intervention ($P < 0.05$), and (b) Significant difference compared to placebo ($P < 0.05$). *The figures were created based on the data published in* Nutrition Research *29 (2009) 19–25 and permitted from this journal under the license number 3037861357030.*

TABLE 16.2 Stability Test Result (Original Data)

	Months after the Stability Study Commencement		
LBP level (mg/120 ml)	**Initial Value**	**12 Months**	**24 Months**
Accelerated condition (40°C, 75% humidity)	2023.0*	1782.4	—
Real-time condition (30°C, 65% humidity)		—	1618.4

**Average value from eight lots.*

A standardized *Lycium barbarum* fruit juice has been tested in two different conditions adapted to the ICH guideline for its shelf-life stability measured by its active ingredient, *Lycium barbarum* polysaccharides (LBP). Each value indicates LBP level (mg in 120 ml).

Flavonoids from *L. barbarum* were also found to be antioxidants. Mitochondrial lipid peroxidation, measured as MDA, was significantly inhibited by total flavonoids from *L. barbarum* in a dose-dependent manner, and the fluidity of the mitochondrial membrane was also protected. The shape of red blood cells (RBC) was protected (Huang *et al.* 1999b). Scavenging effects were also found by electron spin resonance (ESR) spin-trapping technique. Flavonoids scavenged free radicals in xanthine/xanthine oxidase (Xan/XO) system, and in Fenton reaction in a concentration-dependent manner. It inhibited the heat output from phorbol 12-myristate 13-acetate (PMA)-stimulated polymorphonuclear leukocyte and inhibited the heat output from L1210 cells (Huang *et al.*, 1998b). To date, no newer studies on flavonoids from *L. barbarum* have been seen since these publications.

Total antioxidant capacity assay showed that *L. barbarum* extracts/fractions possessed antioxidant activity. Water and methanolic fruit extracts and crude polysaccharide extracts exhibited stronger antioxidant activity than purified polysaccharide fractions because crude extracts were identified to be rich in antioxidants (Luo *et al.*, 2004). However, since ORAC is not purported to represent actual antioxidant capacity in the body, ORAC comparison *in vitro* does not necessarily translate to real antioxidant activities *in vivo*.

Antioxidant Effects

Several clinical studies, various pre-clinical *in vivo* animal and *in vitro* cell culture studies have shown the efficacy of *L. barbarum* and LBP as antioxidants in protecting against various peroxidation-related conditions, as summarized in Table 16.3 (Amagase *et al.*, 2009a; Gong *et al.*, 2005; Huang *et al.*, 1998b, 1999b, 2001, 2003; Li *et al.*, 2000, 2002, 2007a; Luo *et al.*, 2006; Ni *et al.*, 2004; Reeve *et al.*, 2010; Ren *et al.*, 1995; Sui *et al.*, 1996; Wang *et al.*, 2002; Wu *et al.*, 2004; Zhang, 1993; Zhang *et al.*, 1997; Zhao *et al.*, 2005).

TABLE 16.3 Antioxidant Effects of *Lycium barbarum* and LBP in Various *In Vivo* and *In Vitro* Experimental Oxidation Systems Reported in the Literature Listed in PubMed (Searched and Summarized by the Author)

Experimental System	Tested Material	Antioxidant Markers	Reference
CLINICAL STUDY			
A randomized clinical trial (healthy older adults)	LBP-standardized *L. barbarum* fruit juice	Serum superoxide dismutase (SOD), glutathione peroxidase (GSH-Px), and malondialdehyde (MDA) levels compared to pre-intervention and placebo	Amagase *et al.* (2009a)
Non-insulin dependent diabetes mellitus (NIDDM) patients	LBP	Red blood cell fragility and abnormality	Li *et al.* (2000)
ANIMAL MODEL			
Mice model	*L. barbarum* LBP	The activities of SOD, catalase (CAT) and total antioxidative capacity (TAOC), MDA level, creatine kinase activities, endogenous lipid peroxidation, and the oxidative stress induced by exhaustive exercise	Li *et al.* (2002), Wu *et al.* (2010), Niu *et al.* (2008), Li *et al.* (2007a), Lin *et al.* (2009), Sui *et al.* (1996), Zhu (1998)
Doxorubicin-treated rats *In vitro* cytotoxic study	*L. barbarum*	The loss of myofibrils, heart function as evidenced from lower mortality, normalization of antioxidative activity and serum aspartate aminotransferase and creatine kinase. Arrhythmias and conduction abnormalities Antitumor activity of doxorubicin was not compromised by *L. barbarum*.	Xin *et al.* (2007)
Hairless mice	Goji juice in the drinking water (1–10%)	Skin protection against immune suppression and oxidative stress by solar simulated UV (SSUV) radiation Systemic immunosuppression measured as the contact hypersensitivity reaction (CHS) Endogenous cutaneous antioxidant HO-1 Actinic skin damage leading to cancers	Reeve *et al.* (2010)
Testicular tissue damage induced by 43°C heat exposure in male rats	LBP	SOD activity, and raised sexual hormone levels Copulatory performance and reproductive function of hemicastrated male rats	Luo *et al.* (2006)
Irradiation or chemotherapy-induced myelosuppressive mice	LBP	Peripheral blood recovery Peripheral blood mononuclear cells (PBMCs) to produce granulocyte colony-stimulating factor (G-CSF)	Gong *et al.* (2005)
Hydrogen peroxide (H_2O_2) in rats	*L. barbarum*, LBP Betaine	Lipid peroxidation of RBC membrane	Ren *et al.* (1995)
Ischemia/reperfusion-induced liver injury	*L. barbarum*	Hydroxyl radical release Nitric oxide (NO) release	Wang *et al.* (2009b)
TISSUE CULTURE			
Rat liver homogenate	*L. barbarum* extract	MDA formation Superoxide anion scavenging Anti-superoxide formation activities	Wu *et al.* (2004)
In vitro	LBP	Effective level as 0.25–1.0 mg/ml	Ni et al. (2004)
Brain tissue and tissue mitochondria	*L. barbarum*	Hydroxyl and superoxide radicals $FeCl_2$/ascorbic acid-induced dysfunction Reactive oxygen species Lipid peroxidation Activities of complex I, complex II, and glutamate cysteine ligase	Feng et al. (2010)
Oxidative stress and oxidative damage	LBP	Free radicals DNA damage	Huang et al. (2003)

(Continued)

TABLE 16.3 Antioxidant Effects of *Lycium barbarum* and LBP in Various *In Vivo* and *In Vitro* Experimental Oxidation Systems Reported in the Literature Listed in PubMed (Searched and Summarized by the Author)—cont'd

Experimental System	Tested Material	Antioxidant Markers	Reference
Murine seminiferous epithelium	*L. barbarum*	Time- and hyperthermia-induced structural damage	Wang et al. (2002)
Membrane	LBP	Passive electrical membrane parameters of the injured membrane	Zhang et al. (1997), Zhang (1993)
Whole human skin system	LBP	Selective matrix metalloproteinase Viability of the skin Skin expansion under mechanical stress collagen type I, the matrix metalloproteinase-1 substrate	Zhao et al. (2005)
Irradiation- or chemotherapy-induced cultured human PBMCs	LBP	G-CSF production	Gong et al. (2005)
In vitro	LBP and their glycans, LbGp5, glycoconjugate	Low-density lipoprotein (LDL) peroxidation measured by the oxidative production of thiobarbituric acid-reactive substances (TBARS)	Huang et al. (2001)

Clinical Studies

A randomized clinical trial showed intrinsic antioxidant effects of orally consumed *L. barbarum*, administered as a LBP-standardized *L. barbarum* fruit juice, in healthy older adults (Amagase *et al.*, 2009a). As shown in Figure 16.2, serum SOD and GSH-Px levels in the *L. barbarum* group were significantly increased over the starting point, and MDA was decreased significantly. SOD and GSH-Px in the *L. barbarum* group were also significantly improved over placebo. No significant differences were detected in the placebo group. As most of the antioxidant studies with other food materials or supplements have been done under disease conditions (Wang *et al.*, 1998; Bose and Agrawal, 2007; Micke *et al.*, 2001) or deficient conditions (Machefer *et al.*, 2007), it is meaningful that *L. barbarum* intake produced a nearly 10% increase in serum antioxidant capacities in human subjects under normal conditions. LBP also improved red blood cell fragility and abnormality in non-insulin dependent diabetes mellitus (NIDDM) patients so that they either approached or were restored to normal states, and thus may be useful as a supplementary treatment for NIDDM (Li *et al.*, 2000).

Animal Experiments

In addition to increased SOD activity and decreased MDA levels shown in the clinical studies, catalase (CAT) and total antioxidative capacity (TAOC), and creatine kinase activities have been improved by *L. barbarum* treatment in murine models (Li *et al.*, 2002, 2007a; Lin *et al.*, 2009; Niu *et al.*, 2008; Sui *et al.*, 1996; Wu *et al.*, 2010; Zhu, 1998). Pretreatment with *L. barbarum* significantly prevented doxorubicin- and solar simulated UV (SSUV) radiation-induced oxidation in murines (Xin *et al.*, 2007; Reeve *et al.*, 2010).

Tissue culture

Antioxidant effects of *L. barbarum* extract against various oxygen radical species *in vitro* also has been reported in many papers (Feng *et al.*, 2010; Gong *et al.*, 2005; Huang *et al.*, 2001, 2003; Ni *et al.*, 2004; Wang *et al.*, 2002; Wu *et al.*, 2004; Zhang *et al.*, 1997; Zhang, 1993; Zhao *et al.*, 2005). IC_{50} values of *L. barbarum* extract to inhibit MDA formation, scavenge superoxide anion and form anti-superoxide activities were reported as 0.77–2.55 μg/ml (Wu *et al.*, 2004), and the effective level of LBP *in vitro* was reported as 0.25–1.0 mg/ml (Ni *et al.*, 2004).

Anti-Inflammation Through Antioxidation

The immunostimulatory action of *L. barbarum*, as well as its anti-inflammatory effect in response to a septicemia endotoxin challenge with regards to tumor necrosis factor (TNF) and interleukin-6 (IL-6) was seen in rats (Nance *et al.*, 2009). *L. barbarum* produced a significant increase in basal plasma and splenic levels of IL-6 and plasma levels of IL-1β. Endotoxin increased levels of all cytokines and corticosterone; however, the increase in splenic levels of IL-6 and TNF-α was attenuated in the *L. barbarum* group. Consumption of *L. barbarum* produced a significant reduction in body weight gain and an increase in plasma and splenic levels of IL-6 and plasma levels of IL-1β. In contrast to the immunostimulatory effects of *L. barbarum* treatment alone, *L. barbarum* attenuated endotoxin-induced IL-6 and TNF-α levels in the spleen. *L. barbarum* had no effect on endotoxin-induced increases in interferon gamma (IFN-γ) and corticosterone levels. Thus, *L. barbarum* juice can have both an immunostimulatory and an anti-inflammatory action on basal and stimulated cytokine production. Since the anti-cancer effect of LBP was proposed to be via an immunostimulatory action (Cao *et al.*, 1994), the immunostimulatory effects seen in this study are reasonable and consistent with previous observations.

L. barbarum has photoprotective effects, and might thereby protect humans against actinic skin damage leading to cancers. It was found that mice drinking 1–10% *L. barbarum* were dose-dependently protected from SSUV radiation-induced systemic immunosuppression. It was confirmed that the protection was an innate property of the goji berry, because it was not observed with any excipients of *L. barbarum* fruit juice or vitamin C, which is found at high concentration in *L. barbarum* berry. It was observed that 5% *L. barbarum* juice induced the expression of haem oxygenase-1 (HO-1) mRNA in the mouse skin. Furthermore, inhibition of HO enzyme activity by injecting mice with the substrate antagonist abrogated the photoimmune protection. *L. barbarum* offers skin protection against immune suppression and oxidative stress by SSUV radiation (Reeve *et al.*, 2010).

References

Amagase, H., Farnsworth, N.R., 2011. A Review of botanical characteristics, phytochemistry, clinical relevance in efficacy and safety of *Lycium barbarum* fruit (goji). Food Res. Int. 44, 1702–1717.

Amagase, H., Nance, D.M., 2008. A Randomized, Double-blind, Placebo-controlled, clinical study of the general effects of a standardized *Lycium barbarum* (Goji) Juice, GoChi™. J. Alter. Complemen. Med. 14, 403–412.

Amagase, H., Nance, D.M., 2011. Lycium barbarum *increases caloric expenditure and decreases waist circumference in healthy overweight men and women: Pilot Study*. J. Am. Coll. Nutr. 30 (5), 304–309.

Amagase, H., Nance, D.M., 2012. *Lycium barbarum* Fruit (Goji) Attenuates the adrenal steroid response to an exercise challenge and the feeling of tiredness: A randomized, double-blind, placebo-controlled human clinical study. J. Food Res. 1 (2), 3–12 2012.

Amagase, H., Sun, B., Borek, C., 2009a. *Lycium barbarum* (goji) juice shows significant *in vivo* antioxidant effects in human serum in a randomized, double-blind, placebo-controlled clinical study. Nutr. Res. 29, 19–25.

Amagase, H., Sun, B., Nance, D.M., 2009b. Immunomodulatory effects of a standardized *Lycium barbarum* fruit juice in Chinese older healthy human subjects. J. Med. Foods 12 (5), 1159–1165.

AOAC International. Beta-D-Glucan in Barley and Oats. Method number 995.16 In AOAC International Official Methods of Analysis. http://www.eoma.aoac.org/methods/info.asp?ID=21738; accessed on 01.15.08.

Bose, K.S.C., Agrawal, B.K., 2007. Effect of lycopene from tomatoes (cooked) on plasma antioxidant enzymes, lipid peroxidation rate and lipid profile in grade-I hypertension. Ann. Nutr. Metab. 51, 477–481.

Cao, G., Alessio, H., Cutler, R., 1993. Oxygen-radical absorbance capacity assay for antioxidants. Free Radic. Biol. Med. 14 (3), 303–311. http://dx.doi.org/10.1016/0891-5849(93)90027-R PMID 8458588.

Cao, G.W., Yang, W.G., Du, P., 1994. Observation of the effects of LAK/IL-2 therapy combining with *Lycium barbarum* polysaccharides in the treatment of 75 cancer patients. Zhonghua Zhong Liu Za Zhi. 16 (6), 428–431.

Chang, R., 2002. Bioactive polysaccharides from traditional Chinese medicine herbs as anticancer adjuvants. J. Altern. Complement. Med. 8 (5), 559–565.

Chen, Z., Tan, B.K.H., Chan, S.H., 2008. Activation of T lymphocytes by polysaccharide–protein complex from *Lycium barbarum* L. Int. Immunopharmacol. 8, 1663–1671.

Dharmananda, S., 2007. *LYCIUM* FRUIT, Food and Medicine, Institute for Traditional Medicine. Portland, Oregon, August http://www.itmonline.org/arts/lycium.htm.

Duan, C.L., Qiao, S.Y., Wang, N.L., Zhao, Y.M., Qi, C.H., Yao, X.S., 2001. Studies on the active polysaccharides from *Lycium barbarum* L. Yao Xue Xue Bao. 36 (3), 196–199.

Dutch Authorities, 2004. Recommendation by Centre for Substances and Integrated Risk Assessment of National Institute of Public Health and Environment in The Netherlands dated October 6. http://www.rivm.nl/en/.

Feng, Z., Jia, H., Li, X., Bai, Z., Liu, Z., Sun, L., Zhu, Z., Bucheli, P., Ballèvre, O., Wang, J., Liu, J., 2010. A milk-based wolfberry preparation prevents prenatal stress-induced cognitive impairment of offspring rats, and inhibits oxidative damage and mitochondrial dysfunction in vitro. Neurochem. Res. May 35 (5), 702–711.

Gan, L., Zhang, S.H., Yang, X.L., Xu, H.B., 2004. Immunomodulation and antitumor activity by a polysaccharide–protein complex from Lycium barbarum. Int. Immunopharmacol. 1, 563–569.

Gao, Z., Ali, Z., Khan, I.A., 2008. Glycerogalactolipids from the fruit of *Lycium barbarum*. *Phytochemistry*. Nov. 69 (16), 2856–2861.

German Foodstuffs and Commodities Act, November 1986. Food Analysis: Detection of natural thickeners in foodstuffs; In Official Collection of Test Procedures under Section 35 of the German Foodstuffs and Commodities Act (Lebensmittel- und Bedarfsgegenstandsgesetz [LMBG]).

Gong, H., Shen, P., Jin, L., Xing, C., Tang, F., 2005. Therapeutic effects of *Lycium barbarum* polysaccharide (LBP) on irradiation or chemotherapy-induced myelosuppressive mice. Cancer Biother. Radiopharmacy 20 (2), 155–162.

Hsu, C.H.P., Nance, D.M., Amagase, H., 2012. A Meta-Analysis of Clinical Improvements of General Well-being by a Standardized Lycium barbarum. J. Med. Foods Nov. 15 (11), 1006–1014.

Huang, L., Lin, Y., Tian, G., Ji, G., 1998a. Isolation, purification and physicochemical properties of immunoactive glycoconjugates from fruit of *Lycium barbarum* L. Yaoxue Xuebao 33, 512–516.

Huang, L., Tian, G.Y., Ji, G.Z., 1999a. Structure elucidation of glycan of glycoconjugate LbGp3 isolated from the fruit of *Lycium barbarum* L. J. Asian Nat. Prod. Res. 1, 259–267.

Huang, L.J., Tian, G.Y., Wang, Z.F., Dong, J.B., Wu, M.P., 2001. Studies on the glycoconjugates and glycans from *Lycium barbarum* L in inhibiting low density lipoprotein (LDL) peroxidation. Yao Xue Xue Bao. 36 (2), 108–111.

Huang, X., Yang, M., Wu, X., Yan, J., 2003. Study on protective action of *Lycium barbarum* polysaccharides on DNA imparments of testicle cells in mice. Wei Sheng Yan Jiu. 32 (6), 599–601.

Huang, Y., Tan, A., Shen, Y., Lu, J., 1998b. Scavenging effect of total flavonoids of *Lycium barbarum* L on active oxygen radicals and inhibitory effects on heat output from L1210 cells. Wei Sheng Yan Jiu. 27 (2), 109–111 115.

Huang, Y., Lu, J., Shen, Y., Lu, J., 1999b. The protective effects of total flavonoids from *Lycium Barbarum* L. on lipid peroxidation of liver mitochondria and red blood cell in rats. Wei Sheng Yan Jiu. 28 (2), 115–116.

Karp, D., 2009. Goji taunts North American farmers—The berry, cultivated in China and said to convey health benefits, is eagerly sought in the West. Growers here are trying but are discovering they have much to learn. August 05 , Los. Angels Times. http://articles.latimes.com/2009/aug/05/food/fo-goji5.

Li, G., Yang, J., Ren, B., Wang, Z., 2002. Effect of *Lycium barbarum* L on defending free radicals of mice caused by hypoxia. Wei Sheng Yan Jiu. 31 (1), 30–31.

Li, W., Wang, L., Deng, X., Jiang, L., Zhang, C., Zhang, C., 2000. Study of the fragility and abnormality rate of red blood cells in patients with type-2 diabetes and the effects of *Lycium barbarum* polysaccharides. Hebei J. Traditional Chinese Med. August 22 (8), 585–586.

Li, X.M., Ma, Y.L., Liu, X.J., 2007a. Effect of the *Lycium barbarum* polysaccharides on age-related oxidative stress in aged mice. J. Ethnopharmacol. 111 (3), 504–511.

Li, X., Zhou, A.G., Li, X.M., 2007b. Inhibition of *Lycium barbarum* polysaccarides and Ganoderma lucidum polysaccharides against oxidative injury induced by γ-irradiation in rat liver mitochondria. Carbohydr. Polymers 69, 172–178.

Lin, C.L., Wang, C.C., Chang, S.C., Inbaraj, B.S., Chen, B.H., 2009. Antioxidative activity of polysaccharide fractions isolated from *Lycium barbarum* Linnaeus. Int. J. Biol. Macromol. Aug. 1 45 (2), 146–151.

Luo, Q., Yan, J., Zhang, S., 2000. Isolation and purification of *Lycium barbarum* polysaccharides and its antifatigue effect. Wei Sheng Yan Jiu. 29 (2), 115–117.

Luo, Q., Cai, Y., Yan, J., Sun, M., Corke, H., 2004. Hypoglycemic and hypolipidemic effects and antioxidant activity of fruit extracts from Lycium barbarum. Life Sci. 76 (2), 137–149.

Luo, Q., Li, Z., Huang, X., Yan, J., Zhang, S., Cai, Y.Z., 2006. *Lycium barbarum* polysaccharides: Protective effects against heat-induced damage of rat testes and H_2O_2-induced DNA damage in mouse testicular cells and beneficial effect on sexual behavior and reproductive function of hemicastrated rats. Life Sci. 79 (7), 613–621.

Machefer, G., Groussard, C., Vincent, S., Zouhal, H., Faure, H., Cillard, J., Radak, Z., Gratas-Delamarche, A., 2007. Multivitamin-mineral supplementation prevents lipid peroxidation during "The Marathon des Sables". J. Am. Coll. Nutr. 26, 111–120.

McLaughlin, L., 2006. Food: A Taste of the Future. TIME magazine Sunday, Jul. 16. 9171, 1214958,00.html http://www.time.com/time/magazine/article/0.

Micke, P., Beeh, K.M., Schlaak, J.F., Buhl, R., 2001. Oral supplementation with whey proteins increases plasma glutathione levels of HIV-infected patients. Eur. J. Clin. Invest. 31, 171–178.

Nance, D.M., Amagase, H., Luczy-Bachman, G., 2009. Effects of *Lycium barbarum* on basal and lps-induced cytokine production. Abstract 94 Brain Behavior Immun. 23 (Suppl. 2), S51 PNIRS 2009, Psycho-NeuroImmunology Research Society Annual Meeting 2009.

Ni, H., Qing, D., Kaisa, S., Lu, J., 2004. The study on the effect of LBP on cleaning hydroxygen free radical by EPR technique. Zhong Yao Cai. 27 (8), 599–600.

Niu, A.J., Wu, J.M., Yu, D.H., Wang, R., 2008. Protective effect of Lycium barbarum polysaccharides on oxidative damage in skeletal muscle of exhaustive exercise rats. Int. J. Biol. Macromol. Jun. 1 42 (5), 447–449.

Ou, B., Hampsch-Woodill, M., Prior, R., 2001. Development and validation of an improved oxygen radical absorbance capacity assay using fluorescein as the fluorescent probe. J. Agri. Food Chem. 49 (10), 4619–4626. http://dx.doi.org/10.1021/jf010586o PMID 11599998.

Ouyang, H.X., Li, Y.Q., Xiao, Q.W., 2007. Simultaneous determination of monosaccharides and oligosaccharides in *Lycium barbarum* L. by high performance liquid chromatography. Sichuan Da Xue Xue Bao Yi Xue Ban. Nov. 38 (6), 1040–1042.

Peng, X., Tian, G., 2001. Structural characterization of the glycan part of glycoconjugate LbGp2 from *Lycium barbarum* L. Carbohyd. Res. 331 (1), 95–99.

Peng, X.M., Wang, Z.F., Tian, G.Y., 2001a. Physico-chemical properties and activity of glycoconjugate LbGp2 from *Lycium barbarum* L. Yao Xue Xue Bao. 36 (8), 599–602.

Peng, X., Qi, C., Tian, G., Zhang, X.X., 2001b. Physico-chemical properties and bioactivities of a glycoconjugate LbGp5B from *Lycium barbarum* L. Chin. J. Chem. 19, 842–846.

Peng, X., Huang, J., Qi, C., Zhang, Y.X., Tian, G.Y., 2001c. Studies on chemistry and immuno-modulating mechanism of a glycoconjugate from *Lycium barbarum* L. Chin. J. Chem. 19, 1190–1197.

Potterat, O., 2010. Goji (*Lycium barbarum* and *L. chinense*): Phytochemistry, pharmacology and safety in the perspective of traditional uses and recent popularity. Planta Medica. 76 (1), 7–19.

Reeve, V.E., Allanson, M., Arun, S.J., Domanski, D., Painter, N., 2010. Mice drinking goji berry juice (*Lycium barbarum*) are protected from UV radiation-induced skin damage via antioxidant pathways. Photochemistry and Photobiological Science 9, 601–607.

Ren, B., Ma, Y., Shen, Y., Gao, B., 1995. Protective action of *Lycium barbarum* L. (LbL) and betaine on lipid peroxidation of erythrocyte membrane induced by H_2O_2. Zhongguo Zhong Yao Za Zhi. 20 (5), 303–304.

Sui, D.Y., Yu, X., Lv, Z., Gao, P., Piao, Y., Song, G., 1996. The effects of the crude polysaccharides of Gou Qi Zi, Wu Wei Zi, and Huang Jin on immunity and lipid peroxidation. J. Bethune Med. Univ. 22 (6), 606–607.

Sohn, E., 2008. Superfruits, super powers? Fruits from faraway lands have been showing up in a growing number of products lately. Los Angels Times. March 10. 0, 1331890,full.story http://www.latimes.com/features/health/la-hew-superfruit.

Tian, M., Wang, M., 2006. Studies on extraction, isolation and composition of *Lycium barbarum* polysaccharides. Zhongguo Zhong Yao Za Zhi. 31 (19), 1603–1607.

UK Food Standard Agency., 2007. GOJI BERRIES , http://www.food.gov.uk/news/newsarchive/2007/jun/goji.

http://www.food.gov.uk/multimedia/pdfs/gojiberriesrep.pdf; http://www.food.gov.uk/multimedia/pdfs/gojiberriesannx.pdf; Monday 18 June.

http://www.food.gov.uk/multimedia/pdfs/gojiberriesannx.pdf.

United States Department of Agriculture, Agricultural Research Service, 2012. Withdrawn: Oxygen Radical Absorbance Capacity (ORAC) of Selected Foods. 16 May Release 2 (2010). Retrieved 13.06.12. http://www.ars.usda.gov/services/docs.htm?docid=15866.

Wang, C.C., Chang, S.C., Chen, B.H., 2009a. Chromatographic determination of polysaccharides in *Lycium barbarum* Linneaus. Food Chem. 116, 595–603.

Wang, N.T., Lin, H.I., Yeh, D.Y., Chou, T.Y., Chen, C.F., Leu, F.C., Wang, D., Hu, R.T., 2009b. Effects of the antioxidants *Lycium barbarum* and ascorbic acid on reperfusion liver injury in rats. Transplant Proceedings Dec. 41 (10), 4110–4113.

Wang, Q., Chen, S., Zhang, Z., 1991. Determination of polysaccharide contents in Fructus. Lycii. Chin. Tradit. Herb. Drugs 22 (2), 67–68.

Wang, X.L., Adachi, T., Sim, A.S., Wilcken, E.L., 1998. Plasma Extracellular Superoxide Dismutase Levels in an Australian Population With Coronary Artery Disease. Arthriosclerosis Thromb. Vasc. Biol. 18, 1915–1921.

Wang, Y., Zhao, H., Sheng, X., Gambino, P.E., Costello, B., Bojanowski, K., 2002. Protective effect of Fructus *Lycii* polysaccharides against time and hyperthermia-induced damage in cultured seminiferous epithelium. J. Ethnopharmacol. 82 (2–3), 169–175.

Wong, C.K., Leung, K.N., Fung, K.P., Choy, Y.M., 1994. Immunomodulatory and anti-tumour polysaccharides from medicinal plants. J. Int. Med. Res. 22 (6), 299–312.

Wu, S.J., Ng, L.T., Lin, C.C., 2004. Antioxidant activities of some common ingredients of traditional chinese medicine, *Angelica sinensis, Lycium barbarum* and *Poria cocos*. Phytother. Res. 18 (12), 1008–1012.

Wu, H.T., He, X.J., Hong, Y.K., Ma, T., Xu, Y.P., Li, H.H., 2010. Chemical characterization of *lycium barbarum* polysaccharides and its inhibition against liver oxidative injury of high-fat mice. Int. J. Biol. Macromol. Jun. 1 46 (5), 540–543.

Xin, Y.F., Zhou, G.L., Deng, Z.Y., Chen, Y.X., Wu, Y.G., Xu, P.S., Xuan, Y.X., 2007. Protective effect of *Lycium barbarum* on doxorubicin-induced cardiotoxicity. Phytotherapy Res. Nov. 21 (11), 1020–1024.

Yin, G., Dang, Y., 2008. Optimization of extraction technology of the *Lycium barbarum* polysaccharides by Box–Behnken statistical design. Carbohydr. Polymers 74, 603–610.

Zhang, X., 1993. Experimental research on the role of Lycium barbarum polysaccharide in anti-peroxidation. Zhongguo Zhong Yao Za Zhi. 18 (2), 110–112 128.

Zhang, B., Zhang, X., Li, W., 1997. The injury of Xenopus laevis oocytes membrane and its acetylcholine receptor by free radical and the protection of *Lycium barbarum* polysaccharide. Zhongguo Ying Yong Sheng Li Xue Za Zhi. 13 (4), 322–325.

Zhao, C.J., He, Y.Q., Li, R.Z., Cui, G.H., 1996. Chemistry and pharmacological activity of petidoglycan from Lycium barbarum. Chin. Chem. Lett. 11, 1009–1010.

Zhao, C., Li, R., He, Y., Chui, G., 1997. Studies on chemistry of Gouqi polysaccharides. Yie Daxue Xuebao. 29, 231–232 240 (CAN 129: 120129).

Zhao, H., Alexeev, A., Chang, E., Greenburg, G., Bojanowski, K., 2005. *Lycium barbarum* glycoconjugates: effect on human skin and cultured dermal fibroblasts. Phytomed. 12 (1–2), 131–137.

Zhu, Y.P., 1998. Gou Qi Zi.. In: Chinese Materia Medica Chemistry, Pharmacology and Applications. Harwood Academic Publishers, Amsterdam, Netherlands, pp. 642–646.

CHAPTER

17

Açaí (*Euterpe oleracea* Mart.) Liquefied Pulp for Drinking and their Antioxidant Capacities During Processing

Liuping Miao, Tong Wu

Shanghai Institute of Pharmaceutical Industry, State Key Laboratory of New Drug & Pharmaceutical Process, Shanghai, China

CHAPTER POINTS

- Açaí pulp has potent antioxidant activities due to its high antioxidant content.
- Freeze-drying is a well-known method of processing açaí, and it can preserve the health benefits of açaí, but its efficiency is low and the cost is high.
- Preserve spray drying açaí with maltodextrin 10DE as a carrier agent in a relatively high-temperature and lower-humidity environment can maintain its antioxidant capacities.
- Sugar and salt all had a negative effect on the anthocyanin stability, and therefore the pure açaí pulp has relatively high antioxidant capacities.
- Anthocyanin is sensitive to temperature: the degradation rate of anthocyanin increased with the higher temperature, so it is better to store açaí in low-temperature conditions.

INTRODUCTION

Numerous epidemiological studies have demonstrated that a high intake of fruits and vegetables is associated with a lower risk of many chronic diseases, and this has been widely explained by the high antioxidants in plant foods (Van Duyn and Pivonka, 2000; Boeing *et al.*, 2012). Açaí, fruits of *Euterpe oleraceae* M., has drawn special attention due to its extremely high antioxidant capacity (Schauss *et al.*, 2006a). It is distributed in northern regions of South America and is especially abundant in flood plains of the Amazon River. It is commonly used to make beverages, and used as folk medicine by local people (Lee and Balick, 2008; Matheus *et al.*, 2006).

Açaí fruit tastes like a blend of berry and chocolate. It is round, reddish-purple, and 1–2 cm in diameter, with the seeds constituting about 80% of the fruit. However, consumers who live outside Brazil cannot eat this fresh fruit because it perishes quickly after picking. In order to satisfy consumers from outside Brazil, açaí has been made into a variety of products, including juices, powders, tablets, and capsules. According one report, a gram of raw açaí berry off the tree may have an oxygen radical absorption capacity (ORAC) over 1200, but by the time it arrives on a consumer's kitchen counter in the US that number may be under 100! (What is an ORAC Value?). Therefore, it is necessary to investigate the factors that influence the antioxidant properties of açaí. This chapter describes the liquefied açaí pulp for drinking and its antioxidant activities, as well as how these antioxidants are altered as a consequence of production and processing.

AÇAÍ PULPS AND THEIR ANTIOXIDANT CAPACITIES

Studies have found that healthy volunteers who have been drinking açaí juice and pulp have had an increase in plasma antioxidant capacity (Mertens-Talcott *et al.*, 2008). Antioxidant capacity of açaí has been demonstrated using different antioxidant assays including the ORAC assay, the peroxynitrite averting capacity, and hydroxyl

Processing and Impact on Antioxidants in Beverages
http://dx.doi.org/10.1016/B978-0-12-404738-9.00017-9

TABLE 17.1 Anthocyanin and Polyphenolic Content (mg/l Fresh Pulp) of Açaí

Anthocyanins and Polyphenolics	Content (mg/l Fresh Pulp)
Cyanidin 3-glucoside	1040 ± 58.2
Pelargonidin 3-glucoside	74.4 ± 2.90
Ferulic acid	212 ± 5.29
(−)-Epicatechin	129 ± 3.28
p-hydroxybenzoic acid	80.5 ± 2.00
Gallic acid	64.5 ± 1.64
Protocatechuic acid	64.4 ± 1.64
(+)-Catechin	60.8 ± 0.98
Ellagic acid	55.4 ± 1.39
Vanillic acid	33.2 ± 1.39
p-coumaric acid	17.1 ± 1.23
Gallic acid derivative-1	47.3 ± 1.40
Gallic acid derivative-2	18.4 ± 0.89
Gallic acid derivative-3	17.3 ± 1.25
Gallic acid derivative-4	13.3 ± 0.96
Gallic acid derivative-5	3.9 ± 0.18
Ellagic acid derivative	19.5 ± 0.40

Source: Del Pozo-Insfran et al. *(2004).*

radical averting capacity. Açaí was shown to inhibit the generation of reactive oxygen species (ROS) (Schauss *et al.*, 2006a). The pulp contains low levels of sugar and carbohydrates, but high amounts of unsaturated fats, phytosterols and fiber, which can improve antioxidant status (De Souza *et al.*, 2010). Açaí pulp was found to have a relatively high antioxidant content (48.6 μmol Trolox equivalents (TE)/ml) (Del Pozo-Insfran *et al.*, 2004) with respect to other dark-colored berries and fruits such as highbush blueberries (4.6–31.1 μmol TE/g) (Ehlenfeldt and Prior, 2001), strawberries (18.3–22.9 μmol TE/g) (Kalt *et al.*, 1999), raspberries (19.2–22.6 μmol TE/g) (Kalt *et al.*, 1999), blackberries (13.7–25.1 μmol TE/g) (Wang and Lin, 2000), cranberries (8.20–145 μmol TE/g) (Wang and Stretch, 2001) and muscadine grape juice (18.2–26.7 μmol TE/g) (Talcott *et al.*, 2003). The anthocyanins and other polyphenolic compounds in açaí have been determined (Table 17.1, Figure 17.1). In addition to the whole açaí pulp, the antioxidant capacities of different fractions of açaí pulp have also been investigated. Findings suggested that the whole pulp, C_{18}-retained phenolics (phenolic acids and anthocyanins), and the anthocyanins alone had appreciably higher contributions to the total antioxidant content (Figure 17.2) (Del Pozo-Insfran *et al.*, 2004). The antioxidant activity of açaí is associated with its high anthocyanin and polyphenolic contents (Coïsson *et al.*, 2005; Schauss *et al.*, 2006b). Anthocyanins and proanthocyanidins are in general considered as major antioxidants in berries and fruits; however, their concentrations are relatively low in açaí (Schauss *et al.*, 2006a). The contributions of the anthocyanins to the overall antioxidant capacities of açaí were estimated to be only approximately 10% (Lichtenthäler *et al.*, 2005). The flavonoids were found to be another group of major polyphenols in açaí (Schauss *et al.*, 2006; Gallori *et al.*, 2004). (2S, 3S)-dihyrokaempferol 3-O-β-D-glucoside, (2R,3R)-dihydrokaempferol 3-O-β-D-glucoside, isovitexin, velutin, and 5,4′-dihydroxy-7,3′,5′-trimethoxyflavone are found in açaí pulp, and these five compounds have antioxidant activities through ORAC assay (Kang *et al.*, 2011). Many other studies have demonstrated that flavonoids have strong antioxidant activities (Beara *et al.*, 2009; Leong *et al.*, 2010; Li *et al.*, 2009). Further study is needed to clarify which compounds are the predominant antioxidants in açaí.

EFFECTS OF PRODUCTION AND PROCESSING

It is well-known that production and processing changes the physical and chemical properties of foods; thus, it may affect the antioxidant activity of açaí berries (Madiwate *et al.*, 2011; Price *et al.*, 1997; Dewanto *et al.*, 2002). The açaí berry is a highly perishable fruit, and it quickly loses its nutritional properties after being picked. Therefore, in order to preserve its nutritional value, the berries must be properly stored and processed. Reports indicate that nearly half of the anthocyanins degrade within 2 days after the fruit is picked (Murdock and Schauss, 2006). At the same time, anthocyanins are very unstable to processing and storage, since they are sensitive to factors like temperature, light, pH, oxygen, etc. (Ersus and Yurdagel, 2007; Amr and Al-Tamimi, 2007; Kirca *et al.*, 2007). Anthocyanin degradation usually follows a first-order kinetics, i.e., anthocyanin content exponentially decreases with time (Dyrby *et al.*, 2001; Kirca and Cemeroglu, 2003). Thus the food industry has to look for novel and inexpensive processes to increase the açaí shelf-life, improve anthocyanin stability, and preserve antioxidant activity.

Drying

Fresh açaí berries are not available around the world because they are perishable, but there is a high demand for açaí berries among people who wish to receive the health benefits of these fruits. In order to preserve the açaí berries, freeze-drying and spray drying are used to reduce the water content in açaí so that biological growth and chemical reactions are inhibited, and, thus, the health benefits of açaí are preserved.

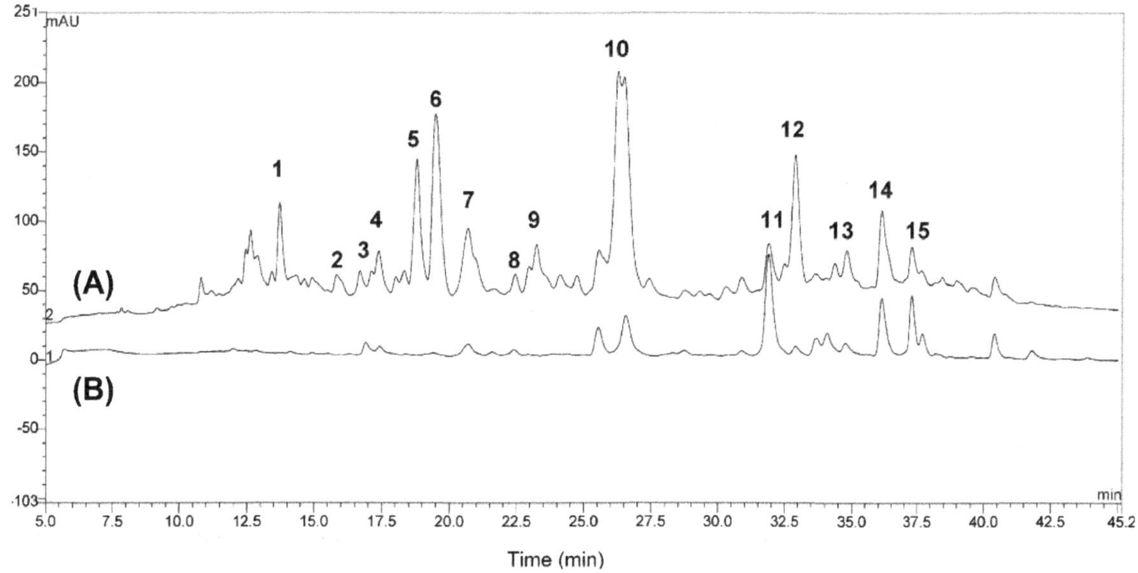

FIGURE 17.1 High-performance liquid chromatography (HPLC) chromatogram of (A) phenolic acids monitored at 280nm and (B) flavonoids monitored at 360nm present in açaí (*Euterpe oleracea* Mart.). Peak assignments: 1, gallic acid; 2, *p*-coumaric acid; 3, protocatechuic acid; 4, (+)-catechin; 5, *p*-hydroxybenzoic acid; 6, vanillic acid; 7, gallic acid derivative-2; 8, gallic acid derivative-5; 9, gallic acid derivative-3; 10, gallic acid derivative-1; 11, ferulic acid; 12, (−)-epicatechin; 13, gallic acid derivative-4; 14, ellagic acid; and 15, ellagic acid derivative. *Source: Del Pozo-Insfran* et al. *(2004).*

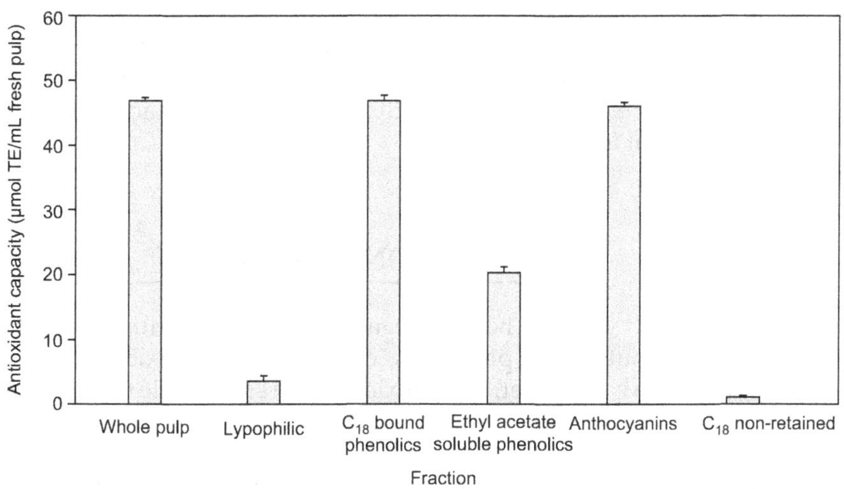

FIGURE 17.2 Antioxidant capacity of different phytochemical fractions (whole pulp, liphophilic extract, C_{18}-bound polyphenolics, ethyl-acetate-soluble phenolics, anthocyanins, and C_{18} non-retained) of acüai (*Euterpe oleracea* Mart.). Bars represent standard error of the mean ($n=6$). Antioxidant capacity quantified using Trolox equivalents (TE). *Source: Del Pozo-Insfran* et al. *(2004).*

Freeze-Drying

Freeze-drying is a process that removes the moisture from a substance while preserving most or all of its desired properties. The use of frozen pulp is necessary because of the short shelf-life of açaí, and due to the distance between the harvest and the processing sites. It is a popular way to preserve açaí since the antioxidants, vitamins, as well as other health and nutritional benefits of the fresh fruits, are preserved. Schauss *et al.* (2006a) have measured the antioxidant value of freeze-dried açaí pulp, and found that it exhibits extremely higher antioxidant value by ORAC (1027 μmol TE/g), higher than that of any fruit or vegetable that have ever been analyzed and reported (Wu *et al.*, 2004). However, freeze-drying still has several disadvantages, such as high costs and low efficiency, and also, it cannot eliminate bacteria.

Spray Drying

Spray drying is an alternative process for producing açaí with high anthocyanin content. This technique has been widely used in the microencapsulation of food ingredients susceptible to deterioration by external agents and consists of entrapping an active agent in a polymeric matrix in order to protect it from adverse conditions (Re, 1998). Tonon *et al.* (2010) evaluated the anthocyanin stability and antioxidant activity of açaí produced by spray drying using four different types of

TABLE 17.2 Bulk Density, Absolute Density and Porosity of Powders Produced with Different Carrier Agents

Carrier agent	Bulk Density (g/cm^3)	Absolute Density (g/cm^3)	Porosity (%)	Anthocyanin Content (mg/100 g Juice Dried Matter)	Antioxidant Activity (mol TE/g Juice Dried matter)
Maltodextrin 10DE	0.390 ± 0.015^a	1.531 ± 0.004^a	74.50 ± 1.01^a	3436.85 ± 79.18^a	1165.84 ± 35.29^a
Maltodextrin 20DE	0.370 ± 0.016^a	1.511 ± 0.004^b	75.49 ± 1.07^a	3402.30 ± 167.33^a	1101.73 ± 36.11^a
Gum Arabic	0.377 ± 0.011^a	1.491 ± 0.008^c	74.70 ± 0.72^a	3415.96 ± 68.08^a	1118.37 ± 48.02^a
Tapioca starch	0.480 ± 0.005^b	1.514 ± 0.001^b	68.33 ± 0.30^b	3247.15 ± 69.70^b	1010.87 ± 41.97^b

[a–c]*Means with different superscript letters in the same column indicate significant differences ($p \leq 0.05$).*
Antioxidant activity was determined by the DPPH free radical, according to the scavenging method.
Source: Tonon et al. *(2010).*

carrier agents (maltodextrin 10DE , maltodextrin 20DE, gum Arabic, and tapioca starch), stored at different temperatures (25°C and 35°C) and water activities (0.328 and 0.529) throughout 120 days. It was found that anthocyanin and antioxidant activity degraded differently in the spray-dried process based on the different carrier agents. Water activity negatively influenced anthocyanin stability and antioxidant activity, while temperature negatively affected anthocyanin stability, but positively affected the antioxidant activity. The results are shown in Tables 17.2 and 17.3, and Figures 17.3, and 17.4. In conclusion, it is better to use maltodextrin 10DE as a carrier agent, and after spray drying, storing the powders at a relatively high temperature and lower humidity environment can result in a nutritionally rich product with good anthocyanin retention and antioxidant activity preservation storage.

Production

Açaí anthocyanins in food products and beverages may be destroyed through the presence of oxygen and light, as well as the medium composition, such as the presence of co-pigments and sugars (Tonon *et al.*, 2010; de Rosso and Mercadante, 2007a,b). De Rosso and Mercadante (2007a,b) studied the influence of added sugars, salts, and fluorescent light on the stability of açaí anthocyanin. They observed that the addition of sugars and salts, as well as exposure to light all had negative effects on the stability of açaí anthocyanins (Table 17.4, Figure 17.5). Since anthocyanin is related to the antioxidant activity of açaí, it is necessary to avoid adding sugars and salt to açaí products, and to avoid exposing them to the sunlight. Açaí pulp has been widely used in beverages, ice cream, jelly, and shakes, so it is not surprising to see low-antioxidant activity of these food products. Therefore, it is better for consumers to select sealed products without sugars and salts.

Storage

Thermal degradation of dietary anthocyanins has been reported in numerous studies, including studies on red cabbage (Dyrby *et al.*, 2001), black carrot, blood orange (Kirca and Cemeroglu, 2003; Kirca *et al.*, 2006, 2007), strawberry (Garzon and Wrolstad, 2002) and blackberry (Wang and Xu, 2007). The knowledge of degradation kinetic parameters of anthocyanins is essential to predict qualitative changes during storage and during heat-treatment processes such as pasteurization (Zozio *et al.*, 2011). Zozio *et al.* (2011) tested anthocyanin stability of açaí juice stored at 4, 20, 30 or 50°C for 60 days. They found that there was no degradation at 4°C, but the degradation started when the storage temperature reached 20°C. The degradation rate of anthocyanin increased with the increasing temperature. Therefore, the recommendation is to store the açaí juice or other açaí products at low temperatures.

CONCLUSION

The açaí berry is regarded as a 'super fruit' due to its nutritional properties, especially its antioxidant capacity. However, its antioxidants are susceptible to temperature, humidity, light, etc. It is necessary to adopt proper methods to dry, process, and store açaí and açaí products in order to preserve the antioxidant activity. Ingredients such as sugars and salts both have a negative influence on the stability of anthocyanins and antioxidant activity of açaí, hence, the açaí products with less additives are the better choice.

Although açaí is popular throughout Brazil and abroad, it should be noted that açaí has been reported to be linked with the outbreak of Chagas' disease (Pereira *et al.*, 2009; Nóbrega *et al.*, 2009). Meanwhile, there is not enough information so far to identify the actual or major antioxidants, though anthocyanins, proanthocyanidins, and flavonoids are widely considered. It is possible that other yet-to-be-identified compounds may also contribute to the antioxidant activity of açaí. Further study is warranted to resolve these questions in order to scientifically promote açaí berries.

TABLE 17.3 Kinetic Parameters of Anthocyanin Degradation in the Powdered Açaí Juice Produced by Spray Drying using Maltodextrin 10DE, Maltodextrin 20DE, Gum Arabic and Tapioca Starch as Carrier Agents

Storage Conditions						
Temperature (°C)	**Water Activity (a_w)**	**Time Storage**	**k ($days^{-1}$)**	**$t_{½}$(days)**	**R^2**	**Actual $t_{½}$ (days)**
MALTODEXTRIN 10DE						
25	0.328	$t<t_2$	0.0014	495.11	0.987	1248.29
		$t>t_2$	0.0005	1386.29	0.970	
	0.529	$t<t_2$	0.0026	266.60	0.985	962.58
		$t>t_2$	0.0006	1155.25	0.996	
35	0.328	$t<t_2$	0.0021	330.07	0.985	880.35
		$t>t_2$	0.0007	990.21	0.996	
	0.529	$t<t_2$	0.0057	121.60	0.987	411.55
		$t>t_2$	0.0010	693.15	0.987	
MALTODEXTRIN 20DE						
25	0.328	$t<t_2$	0.0032	216.61	0.991	909.58
		$t>t_2$	0.0006	1155.25	0.970	
	0.529	$t<t_2$	0.0045	154.03	0.971	677.64
		$t>t_2$	0.0007	990.21	0.943	
35	0.328	$t<t_2$	0.0039	177.73	0.995	516.15
		$t>t_2$	0.0011	693.15	0.946	
	0.529	$t<t_2$	0.0084	82.52	0.978	260.55
		$t>t_2$	0.0010	693.15	0.953	
GUM ARABIC						
25	0.328	$t<t_2$	0.0029	239.02	0.999	978.75
		$t>t_2$	0.0006	1155.25	0.943	
	0.529	$t<t_2$	0.0049	141.46	0.950	826.25
		$t>t_2$	0.0006	1155.25	0.976	
35	0.328	$t<t_2$	0.0041	169.06	0.985	767.07
		$t>t_2$	0.0007	990.21	0.986	
	0.529	$t<t_2$	0.0091	76.17	0.976	328.55
		$t>t_2$	0.0010	693.15	0.994	
TAPIOCA STARCH						
25	0.328	$t<t_2$	0.0015	462.10	0.963	813.81
		$t>t_2$	0.0008	866.43	0.988	
	0.529	$t<t_2$	0.0030	231.05	0.996	695.18
		$t>t_2$	0.0008	866.43	0.962	
35	0.328	$t<t_2$	0.0028	247.55	0.985	655.27
		$t>t_2$	0.0009	770.16	0.997	
	0.529	$t<tt_2$	0.0057	121.60	0.959	248.14
		$t>t_2$	0.0018	385.08	0.993	

Source: Tonon et al. *(2010)*

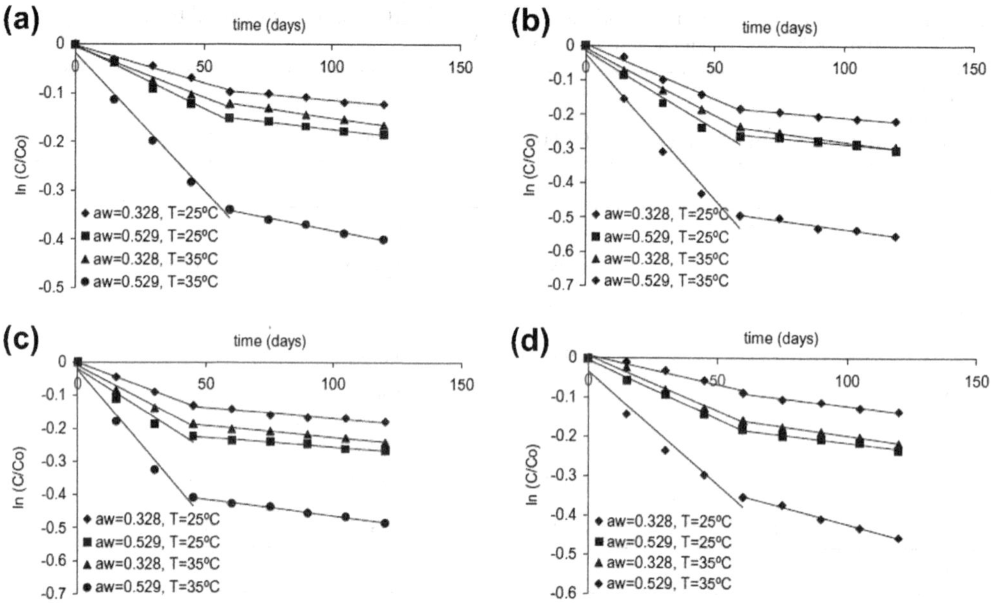

FIGURE 17.3 Anthocyanin degradation kinetics of particles produced with: (a) maltodextrin 10DE, (b) maltodextrin 20DE, (c) gum Arabic, and (d) tapioca starch. *Source: Tonon* et al. *(2010).*

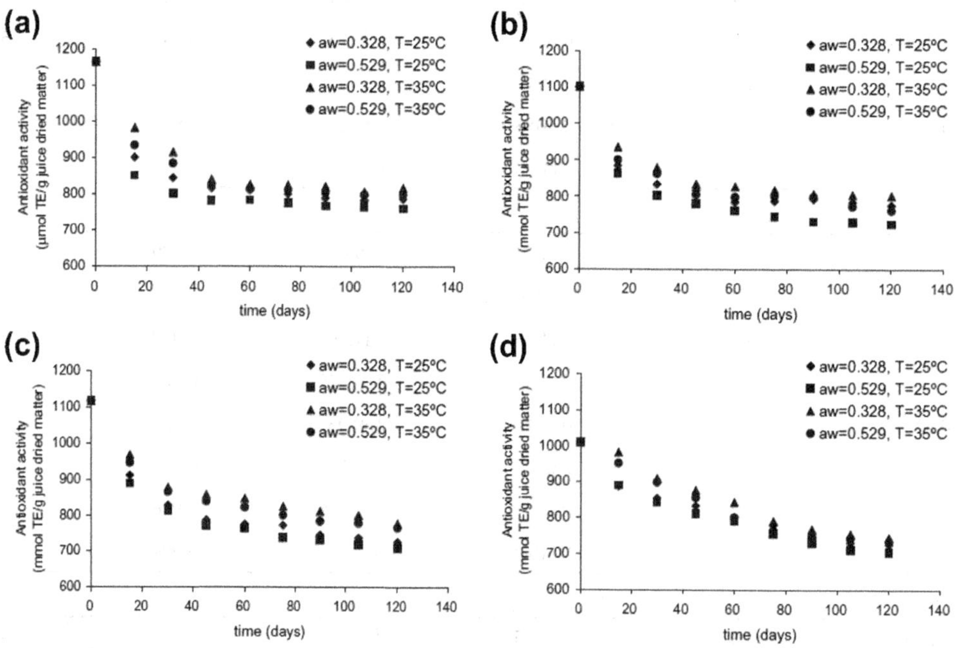

FIGURE 17.4 Antioxidant activity of particles produced with: (a) maltodextrin 10DE, (b) maltodextrin 20DE, (c) gum Arabic, and (d) tapioca starch. *Source: Tonon* et al. *(2010).*

TABLE 17.4 Rate Constant Values (k_{obs}) and Half-Life Times ($t_{½}$) for Anthocyanin Degradation in Açaí in Citrate-Phosphate at pH 2.5 and at 20°C

Systems	k_{obs} (h^{-1})[a]	$t_{½}$(h)[a]	r
Açaí-CE air/light	7.64×10^{-4}	909.0	0.987
Açaí-CE air/dark	1.07×10^{-4}	6456.2	0.926
Açaí-CE N_2/light	5.43×10^{-4}	1275.6	0.993
Açaí-CE N_2/dark	8.41×10^{-5}	8235.0	0.914
Açaí-CE + AA (276 mg/100 ml) air/light	1.65×10^{-2}	42.0	0.997
Açaí-CE + AA (276 mg/100 ml) air/dark	1.25×10^{-2}	55.2	0.994
Açaí-CE + AA (276 mg/100 ml) N_2/light	1.04×10^{-2}	66.6	0.971
Açaí-CE + AA (276 mg/100 ml) N_2/dark	9.18×10^{-2}	75.5	0.984
Açaí-CE + AA (138 mg/100 ml) air/light	1.46×10^{-2}	47.6	0.999
Açaí-CE + AA (1386 mg/100 ml) air/dark	1.18×10^{-2}	58.4	0.981
Açaí-CE + AA (30 mg/100 ml) air/light	1.03×10^{-2}	66.9	0.991
Açaí-CE + AA (30 mg/100 ml) air/dark	8.87×10^{-3}	78.1	0.989

[a]*Average of two experiments.*
AA, ascorbic acid; açaí-CE, crude extract (CE) of anthocyanins from açaí.
Source: de Rosso and Mercadante (2007b)

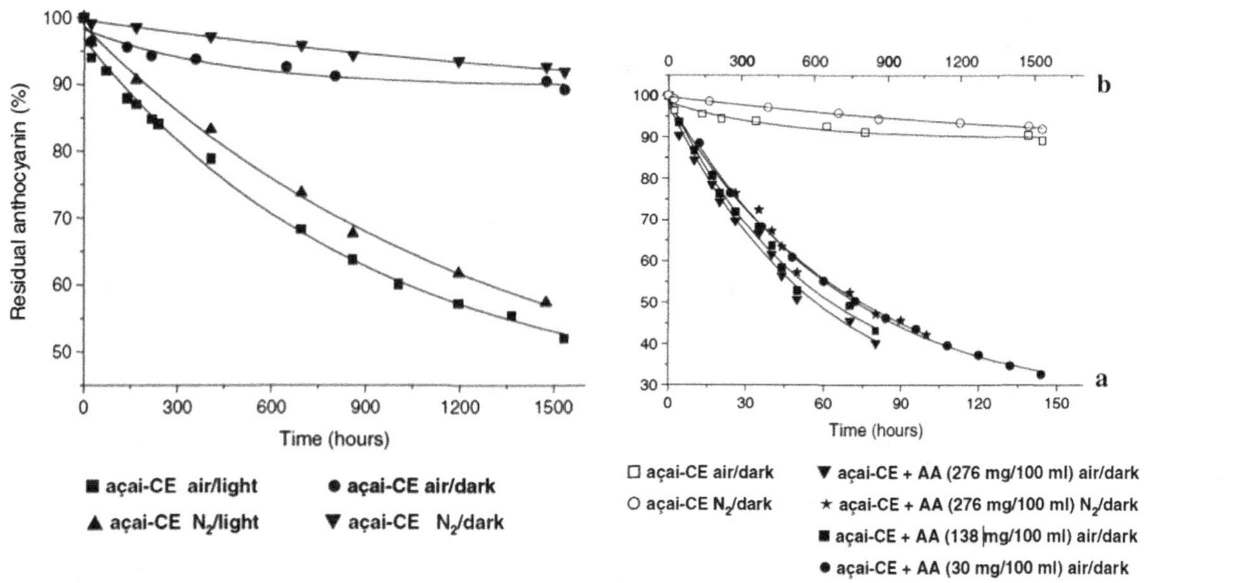

FIGURE 17.5 Degradation curves of anthocyanin extracts from açaí in the presence/absence of light/air/AA, in citrate-phosphate buffer at pH 2.5. Scale a: açaí-CE+AA, Scale b: açaí-CE. AA, ascorbic acid; açaí-CE, crude extract (CE) of anthocyanins from açaí. *Source: de Rosso and Mercadante (2007b).*

References

Amr, A., Al-Tamimi, E., 2007. Stability of the crude extracts of Ranunculus asiaticus anthocyanins and their use as food colourants. Int. J. Food Sci. Tech. 42, 985–991.

Beara, I.N., Lesjak, M.M., Jovin, E.D., Balog, K.J., Anackov, G.T., Orcić, D.Z., Mimica-Dukić, N.M., 2009. Plantain (Plantago L.) species as novel sources of flavonoids antioxidants. J. Agric. Food Chem. 57, 9268–9273.

Boeing, H., Bechthold, A., Bub, A., Ellinger, S., Haller, D., Kroke, A., Leschik-Bonnet, E., Müller, M.J., Oberritter, H., Schulze, M., Stehle, P., Watzl, B., 2012. Critical review: vegetables and fruit in the prevention of chronic diseases. Eur. J. Nutr. 51, 637–663.

Coïsson, J.D., Travaglia, F., Piana, G., Capasso, M., Arlorio, M., 2005. Euterpe oleracea juice as a functional pigment for yogurt. Food Res. Int. 38, 893–897.

De Rosso, V.V., Mercadante, A.Z., 2007a. Evaluation of colour and stability of anthocyanins from tropical fruits in an isotonic soft drink system. Innovative Food Sci. Emerg. Technol. 8, 347–352.

De Rosso, V.V., Mercadante, A.Z., 2007b. The high ascorbic acid content is the main cause of the low stability of anthocyanin extracts from acerola. Food Chem. 103, 935–943.

De Souza, M.O., Silva, M., Silva, M.E., Oliveira Rde, P., Pedrosa, M.L., 2010. Diet supplementation with açaí (Euterpe oleracea Mart.) pulp improves biomarkers of oxidative stress and the serum lipid profile in rats. Nutrition 26, 804–810.

Del Pozo-Insfran, D., Brenes, C.H., Talcott, S.T., 2004. Phytochemical Composition and Pigment Stability of Açaí (Euterpe oleracea Mart.). J. Agric. Food Chem. 52, 1539–1545.

Dewanto, V., Wu, X., Adom, K.K., Liu, R.H., 2002. Thermal processing enhances the nutritional value of tomatoes by increasing total antioxidant activity. J. Agric. Food Chem. 50, 3010–3014.

Dyrby, M., Westergaard, N., Stapelfeldt, H., 2001. Light and heat sensitivity of red cabbage extract in soft drink model systems. Food Chem. 72, 431–437.

Ehlenfeldt, M.K., amd Prior, R.L., 2001. Oxygen radical absorbance capacity (ORAC) and phenolic and anthocyanin concentrations in fruit and leaf tissues of highbush blueberry. J. Agric. Food Chem. 49, 2222–2227.

Ersus, S., Yurdagel, U., 2007. Microencapsulation of anthocyanin pigments of black carrot (Daucuscarota L.) by spray dryer. J. Food Engin. 80, 805–812.

Gallori, S., Bilia, A.R., Bergonzi, M.C., Barbosa, W.L.R., Vincieri, F.F., 2004. Polyphenolic constituents of fruit pulp of Euterpe oleracea Mart. (açaí palm). Chromatographia 59, 739–743.

Garzon, G.A., Wrolstad, R.E., 2002. Comparison of the stability of perlargonidin-based anthocyanins in strawberry juice and concentrate. J. Food Sci. 67, 1288–1299.

Kalt, W., Forney, C.F., Martin, A., Prior, R.L., 1999. Antioxidant capacity, vitamin C, henolics, and anthocyanins after fresh storage of small fruits. J. Agric. Food Chem. 47, 4638–4644.

Kang, J., Xie, C., Li, Z., Nagarajan, S., Schauss, A.G., Wu, T., Wu, X., 2011. Flavonoids from açaí (Euterpe oleracea Mart.) pulp and their antioxidant and anti-inflammatory activities. Food Chem. 128, 152–157.

Kirca, A., Cemeroglu, B., 2003. Degradation kinetics of anthocyanins in blood orange juice and concentrate. Food Chem. 81, 583–587.

Kirca, A., Ozkan, M., Cemeroglu, B., 2007. Effects of temperature, solid content and pH on the stability of black carrot anthocyanins. Food Chem. 101, 212–218.

Kirca, A., Ozkan, M., Cemeroglu, B., 2006. Stability of black carrot anthocyanins in various fruit juices and nectars. Food Chem. 97, 598–605.

Lee, R., Balick, M.J., 2008. Palms, people, and health. Explore 4, 59–62.

Leong, C.A.N., Kinjo, Y., Tako, M., Iwasaki, H., Oku, H., Tamaki, H., 2010. Flavonoid glycosides in the shoot system of Okinawa Taumu (Colocasia esculenta S.). Food Chem. 119, 630–635.

Li, C., Du, H., Wang, L., Shu, Q., Zheng, Y., Xu, Y., Zhang, J., Zhang, J., Yang, R., Ge, Y., 2009. Flavonoid composition and antioxidant activity of tree peony (paeonia section moutan) yellow flowers. J. Agric. Food Chem. 57, 8496–8503.

Lichtenthäler, R., Rodrigues, R.B.,S., Maia, J.G., Papagiannopoulos, M., Fabricius, H., Marx, F., 2005. Total oxidant scavenging capacities of Euterpe oleracea Mart. (Açaí) fruits. Int. J. Food Sci. Nutr. 56, 53–64.

Madiwate, G.P., Reddivari, L., Holm, D.G., Vanamala, J., 2011. Storage elevates phenolic content and antioxidant activity but suppresses antiproliferative and proapoptotic properties of colored-flesh potatoes against human colon cancer cell lines. J. Agric. Food Chem. 59, 8155–8166.

Matheus, M.E., Bessa de Oliveria Fermandes, S., Silveira, C.S., Rodrigues, V.P., Menezes, F., Fernandes, P.D., 2006. Inhibitory effects of Euterpe oleracea Mart. On nitricoxide production and iNOS expression. J. Ethnopharmacol. 107, 291–296.

Mertens-Talcott, S.U., Rios, J., Jilma-Stohlawetz, P., Pacheco-Palencia, L.A., Meibohm, B., Talcott, S.T., Derendorf, H., 2008. Pharmacokinetics of anthocyanins and antioxidant effects after the consumptionof anthocyanin-rich açaí juice and pulp (Euterpe oleracea Mart.) in human healthy volunteers. J. Agric. Food Chem. 56, 7796–7802.

Murdock, K.A., Schauss, A.G., 2006. Jucara and acai fruit-based dietary supplements [P]. EP1622632 A2.

Nóbrega, A.A., Garcia, M.H., Tatto, E., Obara, M.T., Costa, E., Sobel, J., Araujo, W.N., 2009. Oral transmission of Chagas disease by consumption of açaí palm fruit, Brazil. Emerg. Infect. Dis. 15, 653–655.

Pereira, K.S., Schmidt, F.L., Guaraldo, A.M., Franco, R.M., Dias, V.L., Passos, L.A., 2009. Chagas' disease as a foodborne illness. J. Food Prot. 72, 441–446.

Price, K.R., Bacon, J.R., Rhodes, M.J.C., 1997. Effect of storage and domestic processing on the content and composition of flavonol glucosides in onion (Allium cepa). J. Agric. Food Chem. 45, 938–942.

Re, M.I., 1998. Microencapsulation by spray drying. Drying Technol. 16 (6), 1195–1236.

Schauss, A.G., Wu, X., Prior, R.L., Ou, B., Huang, D., Owens, J., Agarwal, A., Jensen, G.S., Hart, A.N., Shanbrom, E., 2006a. Antioxidant capacity ond other bioactivities of the freeze-dried amazonian palm berry, Euterpe oleracea Mart. (Açaí). J. Agric. Food Chem. 54, 8604–8610.

Schauss, A.G., Wu, X., Prior, R.L., Ou, B., Patel, D., Huang, D., Kababick, J.P., 2006b. Phytochemical and Nutrient Composition of the Freeze-Dried Amazonian Palm Berry, Euterpe oleracea Mart. (Açaí). J. Agric. Food Chem. 54, 8598–8603.

Talcott, S.T., Brenens, C.H., Pries, D.M., Del Pozo-Insfran, D., 2003. Phytochemical stability and color retention of copigmented and processed muscadine grape juice. J. Agric. Food Chem. 51, 957–963.

Tonon, R.V., Brabet, C., Hubinger, M.D., 2010. Anthocyanin stability and antioxidant activity of spray-dried açai (Euterpe oleracea Mart.) juice produced with different carrier agents. Food Res. Int. 43, 907–914.

Van Duyn, M.A., Pivonka, E., 2000. Overview of the health benefits of fruit and vegetable comsumption for the dietetics professional: Selected literature. J. Am. Diet. Assoc. 100, 1511–1521.

Wang, S.Y., Lin, H.S., 2000. Antioxidant activity in fruits and leaves of blackberry, raspberry, and strawberry varies with cultivar and developmental stage. J. Agric. Food Chem. 48, 140–146.

Wang, S.Y., Stretch, A.W., 2001. Antioxidant capacity in cranberry is influenced by cultivar and storage temperature. J. Agric. Food Chem. 49, 969–974.

Wang, W.D., Xu, S.Y., 2007. Degradation kinetics of anthocyanins in blackberry juice and concentrate. J. Food Eng. 82, 271–275.

Wu, X., Beecher, G.R., Holden, J.M., Haytowitz, D.B., Gebhardt, S.E., Prior, R.L., 2004. Lipophilic and hydrophilic antioxidant capacities of common foods in the United States. J. Agric. Food Chem. 52, 4026–4037.

What is an ORAC Value? (http://www.acaiberryproducts.org/ORAC_Value.html). accessed 12.20.12.

Zozio, S., Pallet, D., Dornier, M., 2011. Evaluation of anthocyanin stability during storage of a coloured drink made from extracts of the Andean blackberry (Rubus glaucus Benth.), açai (Euterpe oleracea Mart.) and black carrot (Daucus carota L.). Fruits 66, 203–215.

CHAPTER

18

The Impact of Processing and Storage on the (Poly)Phenolic Fraction of Pomegranate (*Punica granatum* L.) Juices

Pedro Mena, Nuria Martí†, Cristina García-Viguera**

*Phytochemistry Laboratory, Department of Food Science and Technology, CEBAS-CSIC, Espinardo, Murcia, Spain,
†Instituto de Biología Molecular y Celular, Universidad Miguel Hernández, Elche, Alicante, Spain

CHAPTER POINTS

- The industrial transformation of pomegranate (*Punica granatum* L.) fruits into juice allows the use of large quantities of second-quality and over-ripe fruits.
- Pomegranate juice is a valuable source of (poly) phenolic compounds, such as anthocyanins and ellagitannins, with proven health-promoting features.
- The (poly)phenolic fraction of pomegranate juice depends on factors such as genomic differences, environmental conditions, fruit quality, processing, and storage conditions.
- Solving color alterations of pomegranate juice by preserving their phytochemical composition along processing and storage is a major challenge for the juice industry.
- The extraction process determines the phenolic composition of the raw pomegranate juice and affects the evolution of its phytochemical characteristics during processing and storage.
- Clarification limits the post-bottling haze formation by removing the phenolic substances and, at the same time, protecting (poly)phenolic fraction throughout storage.
- The concentration step significantly alters the anthocyanin content and color parameters of pomegranate juices.
- Industrially realistic thermal pasteurization of pomegranate juice reduces at the minimum the detrimental effects on the (poly)phenolic fraction reported for batch-pasteurized juices.
- Appropriate cultivar selection, processing, and storage conditions may favor an appealing color with reduced browning and phytochemical degradation over storage.
- Emerging approaches for pomegranate juice processing should be compared with realistic processes to verify their promising potential in solving industrialization problems and preserving the (poly)phenolic fraction of pomegranate juices.

INTRODUCTION

In recent years, pomegranate (*Punica granatum* L.) (Figure 18.1) has been revealed to be a promising source of bioactive phytochemicals with a wide range of biological properties related to protection against oxidative stress, which is related to pathologies such as cardiovascular or neurodegenerative diseases and cancer. This myriad of health-promotion features, mainly attributed to its (poly) phenolic constituents, has prompted the substantive rise in the popularity and worldwide marketing of the fruit (Mena *et al.*, 2011b). Nevertheless, owing to the unique pomegranate fruit morphology, fresh consumption may be inconvenient. Therefore, the industrial transformation of the fruit

Processing and Impact on Antioxidants in Beverages
http://dx.doi.org/10.1016/B978-0-12-404738-9.00018-0

FIGURE 18.1 Pomegrante (*Punica granatum* L.) fruit. Pomegranate juice is a valuable source of (poly)phenolic compounds, such as anthocyanins and ellagitannins, with proven health-promoting features. *Photo by Cristina García-Viguera.*

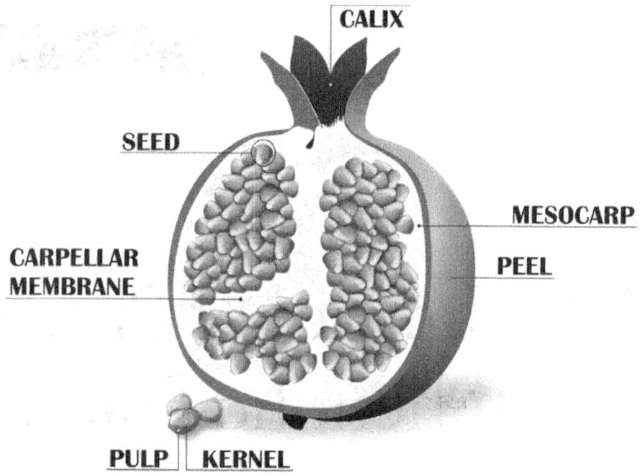

FIGURE 18.2 Cross-sectional illustration of a pomegranate fruit. *Picture by Raúl Domínguez-Perles, previously unpublished.*

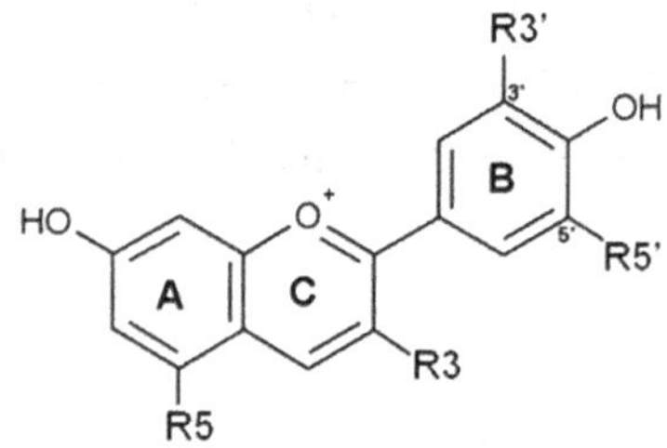

Anthocyanin	R3'	R5'	R3	R5
Pelargonidin 3 glucoside	H	H	Glc	OH
Pelargonidin 3,5 diglucoside	H	H	Glc	Glc
Cyanidin 3 glucoside	OH	H	Glc	OH
Cyanidin 3,5 diglucoside	OH	H	Glc	Glc
Delphinidin 3 glucoside	OH	OH	Glc	OH
Delphinidin 3,5 diglucoside	OH	OH	Glc	Glc

FIGURE 18.3 The chemical structure of the anthocyanins found in pomegranate juice. Glc, glucose. *Structures by Pedro Mena Parreño, previously unpublished.*

into easy-to-consume products, mainly pomegranate juice, may boost its sale and consumption. Pomegranate processing for juice production allows the use of large quantities of second-quality and over-ripe pomegranate fruits not suitable for commercialization, despite the optimal conditions of the edible part (i.e., with low development of husk color, non-homogenous ripening, showing scratches, and sunburn) (Melgarejo and Salazar, 2003). Furthermore, the processing of these underused fruits wasted as by-products generates more profits and minimizes production losses along with the subsequent environmental benefits. Other alternatives for the development of pomegranate processed products by the food industry are wines, jellies, jams, dried and fresh arils, and also nutraceutical ingredients such as seed oil and polyphenol extracts.

The pomegranate fruit consists of a rounded berry, with thick leathery skin or rind (exocarp), typically yellow, pink, or red (Figure 18.2). The inside the fruit is white, spongy, bitter-tasting tissue (albedo and carpellar membranes—mesocarp) that creates spaces filled with crunchy seeds. Each seed (aril) has a surrounding juice-laden pulp ranging in color from white to deep red or purple with a woody inner part (kernel) (Stover and Mercure, 2007). Juice is obtained from the juicy pulp, which comprises the 24–49% of the total fruit weight (Mena *et al.*, 2011a). Aril juice contains 85% water, 10–12% total sugars, 0.2–2% organic acids, pectin, fatty acids, amino acids, and antioxidant phenolics (Fadavi *et al.*, 2005). The total phenolic content in pomegranate juice varies from 0.2 to 2.0% depending on cultivar and processing (Gil *et al.*, 2000; Tzulker *et al.*, 2007).

(POLY)PHENOLIC ANTIOXIDANTS OF POMEGRANATE JUICE

(Poly)phenolic compounds have been reported to be the major antioxidants of pomegranate and to be capable of preventing diseases and promoting health features (Mena *et al.*, 2011b). Among different kinds of phenolic compounds, pomegranate contains phenolic acids, anthocyanins, other non-colored flavonoids, ellagitannins, and gallotannins (Mena *et al.*, 2012).

Anthocyanins are a group of natural pigments responsible for the red–blue color of many fruits, including pomegranate (Figure 18.3). Pomegranate presents an anthocyanin profile characterized by six anthocyanins: cyanidin 3,5-di- and 3-*O*-glucoside, delphinidin 3,5-di- and 3-*O*-glucoside, pelargonidin 3,5-di- and 3-*O*-glucoside; and recently, three new cyanidin derivatives have been described in pomegranate juice: pentoside, pentoside-hexoside, and rutinoside (Fischer *et al.*, 2011a). It is important to note that the amount of these colored flavonoids is largely affected by the cultivar group; for instance, 'Wonderful' variety is related to cyanidin 3,5-diglucoside instead of cyanidin 3-glucoside, the main anthocyanin for Spanish 'Mollar de Elche' cultivars (Gil *et al.*, 2000; Pérez-Vicente *et al.*, 2004). Anthocyanins

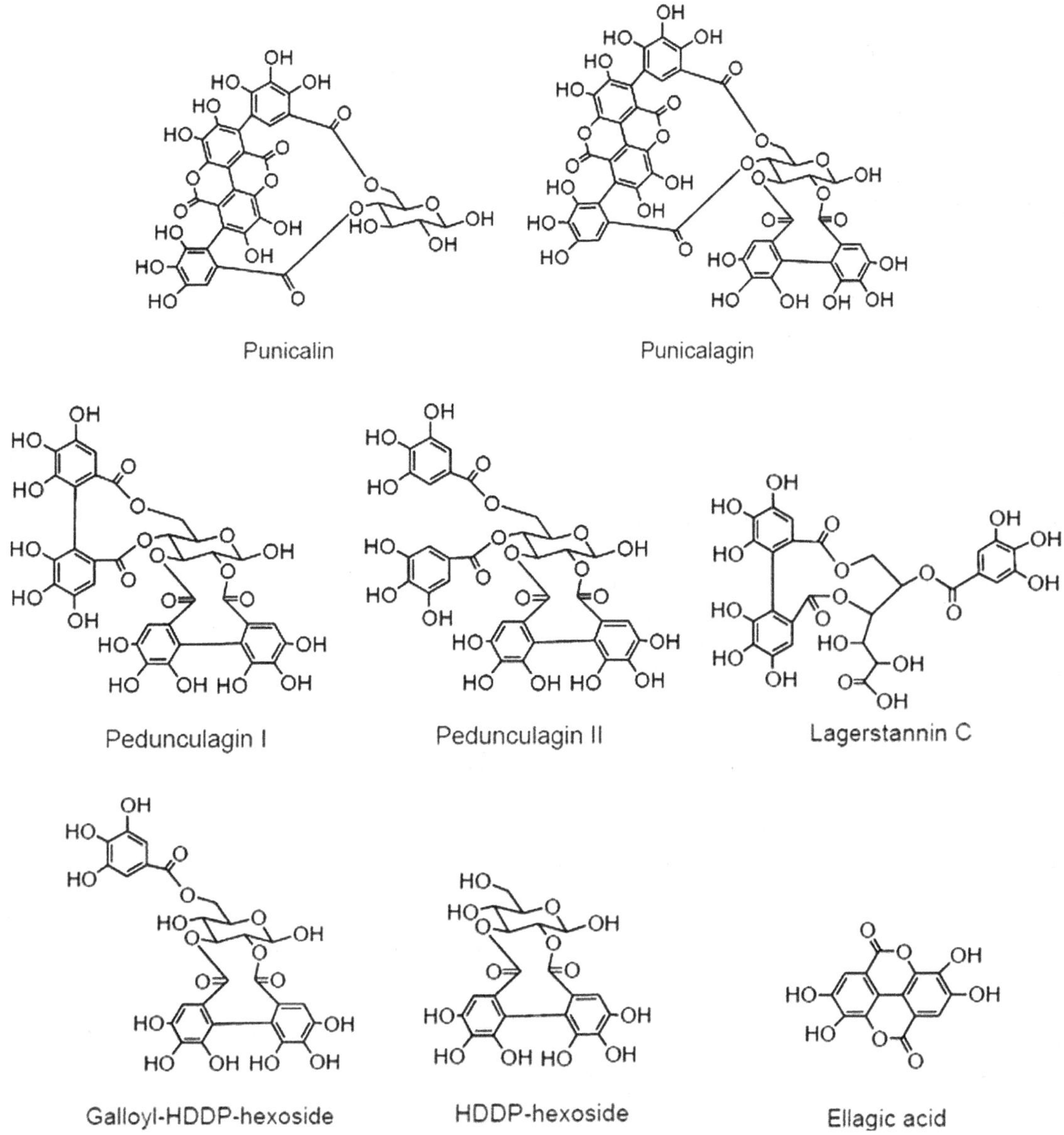

FIGURE 18.4 Some ellagitannins identified in pomegranate juices. *Structures by Alan Crozier.*

are prone to degradation during processing and storage. Causes behind these drops could be found in: (1) polymerization reactions involving the condensation of anthocyanins; (2) enzymatic browning if enzymes have not been inactivated as result of pasteurization; and (3) by direct oxidation of anthocyanin with oxygen (Brownmiller *et al.*, 2008).

Ellagitannins are polymeric structures including, but not limited to, different numbers of galloyl and hexahydroxydiphenoyl (HHDP) units esterified with glucose (Figure 18.4). A broad array of ellagitannin structures have been found in pomegranate juice, hydrolysable tannins being the main class of identified (poly)phenols in pomegranate juice (Borges *et al.*, 2010; Fischer *et al.*, 2011b; Mena *et al.*, 2012). Ellagitannins are extensively found in pomegranate husk, mainly punicalagins (Gil *et al.*, 2000), and are usually extracted into juice during processing. Thus, the extraction process determines the amounts achievable in juice, displaying important differences between whole-fruit juices and juices made with arils (Gil *et al.*, 2000). These compounds, when exposed to pH variations, are hydrolyzed and the HHDP spontaneously rearranges into the water-insoluble ellagic acid. Likewise, this hydrolysis also renders punicalin, a gallagyl residue bounded to glucose (Clifford and Scalbert, 2000).

Different flavonoids belonging to five subclasses of non-colored flavonoids (flavan-3-ols, flavonols, flavanones, dihydrochalcones, and flavones) have been identified recently (Mena *et al.*, 2012). Catequin and gallocatechin have also been identified as the major flavan-3-ols (De Pascual-Teresa *et al.*, 2000). Likewise, prodelphinidins, a kind of condensed tannins derived from the polymerization of gallocatechin, have been found in pomegranate peel (Plumb *et al.*, 2002). In addition, 35 flavanol–anthocyanin adducts have been detected in pressure-extracted pomegranate juice (Sentandreu *et al.*, 2010). Other phytochemicals such as phenolic acids and lignans have also been described in this juice (Bonzanini *et al.*, 2009; Mena *et al.*, 2012).

EFFECTS OF PRODUCTION AND PROCESSING

Factors Affecting the (Poly)Phenolic Composition of Pomegranate Juice

The content of bioactive compounds in pomegranate juice depends on various factors such as genomic differences, environmental conditions, fruit quality, processing, packaging, and storage conditions (Fischer *et al.*, 2011b; Gil *et al.*, 1995; Pérez-Vicente *et al.*, 2004; Rodríguez *et al.*, 2012; Tzulker *et al.*, 2007; Zaouay *et al.*, 2012).

Fruit condition, determined by maturity and sanitary status, has a crucial role in the final quality of pomegranate juice as the starting point of the production process. Fruit quality is affected, to a certain extent, by the environmental conditions, cultural practices, harvest time, and postharvest procedures (Gil *et al.*, 1995; Rodríguez *et al.*, 2012; Tzulker *et al.*, 2007; Zaouay *et al.*, 2012). As previously mentioned, second-quality and over-ripe pomegranate fruits are suitable for juice production. Nonetheless, cracked fruits should be discarded since they represent a potential source of microbiological contamination and, hence, severe pasteurization treatments could be required to reduce the microbial load in pomegranate juice.

The quality and sensory properties of pomegranate juices during processing and storage are also conditioned by genotype, due to their poly(phenolic) composition. For example, beverages derived from the Spanish 'Mollar de Elche' variety change into undesirable brown colors, whereas others, produced with more tannic and acid varieties, present an unacceptable taste (Mena *et al.*, 2011a). Diversity among phytochemical profiles of pomegranate cultivars has been widely reported (Alighourchi *et al.*, 2008b; Ardekani *et al.*, 2011; Cristofori *et al.*, 2011; Hassan *et al.*, 2012; Mena *et al.*, 2011a; Ozgen *et al.*, 2008; Poyrazoǧ;lu *et al.*, 2002; Tzulker *et al.*, 2007; Zaouay *et al.*, 2012). From these studies, the importance of the cultivar in the phenolic profile of juices can be deduced, as significant differences among the levels of anthocyanin and hydrolyzable tannins have been pointed out (Table 18.1). Furthermore, diversity can be employed by the juice-processing industry to select those cultivars and properties suitable for the development of pomegranate juices with attractive organoleptical properties and rich in antioxidant compounds.

TABLE 18.1 Phytochemical Composition (mg/l) of Pomegranate Juices from Different Accessions

Accession	Total Anthocyanins	Total Ellagic Acid	Punicalagin Isomers
M.63	303.9	3.6	7.5
M.55	263.5	5.4	4.9
M.49	208.3	6.9	8.9
M.29	263.9	6.4	7.9
M.Leon.1	148.8	14.0	15.2
M.Leon.2	196.8	14.8	20.1
V.46i	34.2	11.6	29.3
V.111	39.3	10.5	33.9
VSN	108.4	11.6	28.1
W.3	717.4	152.8	2.0
W.2	637.5	25.1	44.1
W.6	655.8	119.3	1.4
W.5	287.2	19.5	3.1
W.7	539.5	29.3	3.1
WSN	1074.6	4.9	9.5

Source: Mena et al. *(2011a).*

Processing techniques, packaging, and storage conditions are among the parameters influencing, to a great extent, the (poly)phenolic fraction of pomegranate juices and are detailed in the next sections.

Influence of Processing on Pomegranate Juice Phytochemicals

Juice processing displays some disadvantages since changes occurring may alter the phytochemical composition and quality parameters (color, aroma, etc.) (Fischer *et al.*, 2011b; Maskan, 2006; Turfan *et al.*, 2011; Yildiz *et al.*, 2009). Therefore, solving color alterations of pomegranate juice by preserving their phytochemical composition during processing is a major challenge for the beverage industry.

According to the *Association of the Industry of Juices and Nectars of the European Union (AIJN) Codes & Reference Guidelines* (AIJN, 2008), pomegranate juice is defined as an unfermented juice obtained from mature and sound fruit of the species *Punica granatum* by mechanical processes and is treated by physical means. Pomegranate juice can be marketed as fruit juice, fruit juice from concentrate, concentrated fruit juice, dehydrated/powdered fruit juice, and fruit nectar. In addition, pomegranate juice can be cloudy or clear. All these categories and possibilities result in a broad array of juices with their own particular characteristics and processing. A detailed

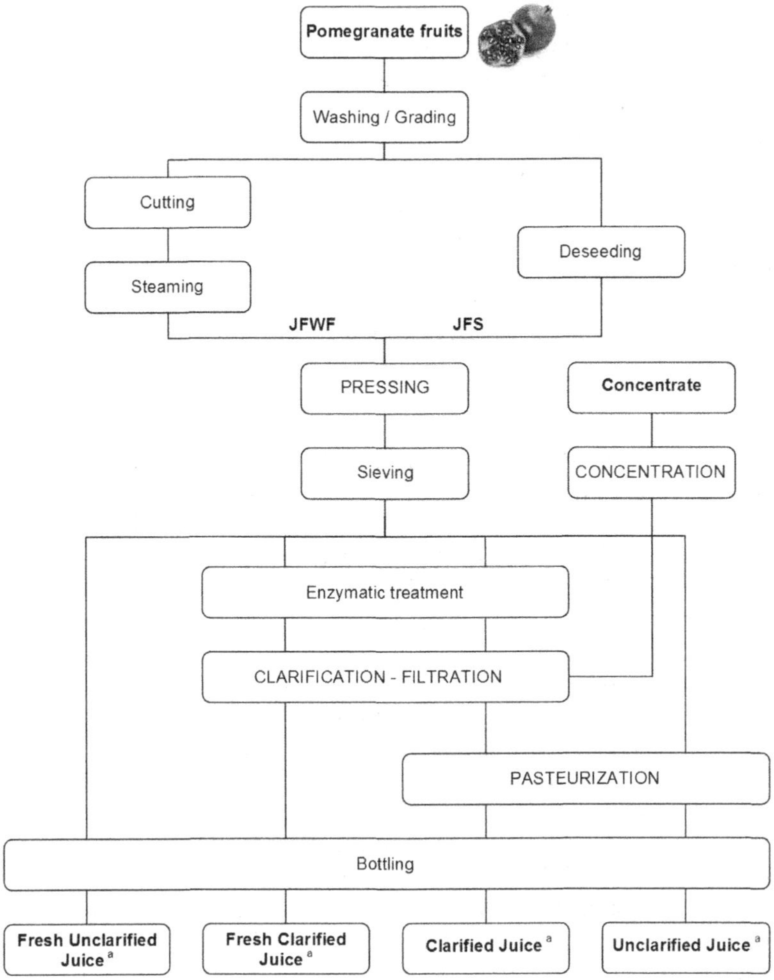

FIGURE 18.5 Flow diagram of processing operations for pomegranate juice production. JFWF, juice from whole fruit; JFS, juice from only sacs (arils). [a]All juices can be produced as both JFWF and JFS. *Scheme by Pedro Mena Parreño, previously unpublished.*

flow diagram of pomegranate juice production is shown in Figure 18.5.

Preliminary Steps

Weighing, material selection, grading, and washing are the front-end operations. Some authors have recommended steaming fruits before pressing to inactivate native enzymes that may cause polyphenol degradation (Fischer *et al.*, 2011b).

Juice extraction is usually performed by pressing either whole fruits or arils. Whole-fruit pressing can be accompanied by cutting the pomegranate fruit into pieces whereas juice obtained from only arils entails a previous step of deseeding. The phenolic composition and its related quality parameters are modified depending on the juice extraction process (Fischer *et al.*, 2011b; Turfan *et al.*, 2011; Türkyilmaz *et al.*, 2013; Tzulker *et al.*, 2007). Owing to the different content in (poly)phenolics of the distinct pomegranate parts (seed, rind, and membranes), which is higher in the rind and membranes (Tzulker *et al.*, 2007), a different content of substances are expected to pass into the juice during the pressing step. However, the migration of phenolics can confer bitter and astringent taste to the pomegranate juice if presented to some degree (Gil *et al.*, 2000; Tzulker *et al.*, 2007). Consequently, controlling the quantities of albedo and carpellar membranes in the juice is mandatory as it can affect the sensory quality (Vázquez-Araújo *et al.*, 2011), and at the same time that could make the clarification process difficult (Vardin and Fenercioğlu, 2003).

Juice pressing performed by using the whole fruit increases the levels of total phenolics (about four-fold) and antioxidant capacity [ferric reducing antioxidant power (FRAP) method] with respect to juices made only from arils (Tzulker *et al.*, 2007; Yildiz *et al.*, 2009). Increasing pressing pressure also augments the content in phenolic compounds and antioxidant activity [2,2′-azino-bis(3-ethylbenzothiazoline-6-sulphonic acid ($ABTS^+$) method] of the juices (Fischer *et al.*, 2011b; Vardin and Fenercioğlu, 2003). None the less, Türkyilmaz *et al.* (2013) did not note significant differences among the total phenol content of pomegranate juices produced with different pressing programs, although they realized that the extraction of phenolics into the juice depended on the pressure applied (increasing pressing pressure by 1.2bar resulted in an 18% increase in total phenolics). Furthermore, increasing applied pressure could modify the non-colored (poly)phenolic fraction by increasing ellagitannins instead of gallotannins and phenolic acids (Fischer *et al.*, 2011b).

Content of total monomeric anthocyanins of juices obtained from arils and from whole fruits was 382mg and 291mg/l, respectively (Turfan *et al.*, 2011). Increased pressing pressure may result in drops (Türkyilmaz *et al.*, 2013) or increases (Fischer *et al.*, 2011b) in the anthocyanin content, which may be related to the peel pigmentation of different cultivars. In addition, the anthocyanin profile may vary by augmenting the presence of delphinidin glucosides as the pressure increases (Fischer *et al.*, 2011b). Conversely, both polymeric pigments and condensed tannins could be higher in juice samples obtained from whole fruits than in those obtained only from arils (Turfan *et al.*, 2011). This fact could be related to the higher content of condensed tannins present in the peel, which are extracted as the pressing pressure is increased (Türkyilmaz *et al.*, 2013). As a consequence of these phenomena, color quality of juices obtained from whole pomegranates may be lower than for those obtained from arils, with enhanced color characteristics (Yildiz *et al.*, 2009).

Clarification and Filtration

As mentioned before, pomegranate juice is marketed as both cloudy and clear. Nonetheless, turbid appearance makes it undesirable to consumers and poses one of the greatest hindrances to its marketability (Mirsaeedghazi *et al.*, 2010b). Clarification meets the need for juice producers of limiting post-bottling haze formation and color change during storage. (Poly)phenolic compounds of pomegranate juice contribute to haze formation through mechanisms involving prior polymerization or condensation leading to the formation of protein–phenolic complexes. These complexes increase the turbidity of fruit juices, collecting at the bottom of the bottle when stored, in such a way that clarification is required to obtain a stable bright, clear pomegranate juice (Alper *et al.*, 2005; Vardin and Fenercioğlu, 2003). Moreover, the taste of the product is improved by means of clarification as tannins conferring bitter taste are removed (Vardin and Fenercioğlu, 2003). Hence, the main purpose of this processing step is to reduce the amount of (poly)phenolic substances.

Clarification can be produced by different but complementary techniques, namely conventional fining and membrane filtration. Concerning this, Alper *et al.* (2005) assayed diverse methods and noted that conventional fining together with polyvinylpolypyrrolidone (PVPP) was more effective than both conventional fining and ultrafiltration to remove phenolic compounds. Total phenolic reductions were 21% for conventional fining, 32% for conventional fining together with PVPP, and only 6% for ultrafiltration. Decreases in anthocyanin and ellagic acid content as well as color variations followed a similar trend to that reported for total phenolics (Alper *et al.*, 2005). On the contrary, Fischer *et al.* (2011b) found a higher loss of anthocyanins when microfiltration was applied (44%) instead of conventional fining (6%). However, phenolic compounds were removed to a large extent by fining with agents (54%) than by microfiltration (26%). Contradictory data could come from different concentrations of fining agents used as well as differences between membrane filtration systems.

CONVENTIONAL FINING

Conventional fining of pomegranate juice is performed by addition of clarification agents used to enhance settling of formed flocs. Settled solids and other suspended matter are removed afterwards by filtration with kieselguhr or active charcoal (Alper *et al.*, 2005; Oziyci *et al.*, 2013). The common fining agents used for clarification of fruit juices include bentonite, gelatine, PVPP, and kieselsol, or a combination of these. Moreover, natural sedimentation is also suitable for pomegranate juice clarification although it is less effective than when assisted with agents (Vardin and Fenercioğlu, 2003).

Whole-fruit-pressed pomegranate juice has been clarified with gelatine and PVPP at different concentrations and compared to natural sedimentation (Vardin and Fenercioğlu, 2003). The most effective fining on the reduction of the (poly)phenolic compounds was achieved after the application of 1g/l gelatine. In fact, total phenolics decreased 0.7, 7, and 18% for natural sedimentation, PVPP, and gelatine, respectively, at the end of 16h. In contrast to this, anthocyanins decreased by 0.3, 6, and 16% for natural sedimentation, gelatine, and PVPP, respectively. Therefore, gelatine but not PVPP allowed a satisfactory fining, favoring the preservation of the color density (Vardin and Fenercioğlu, 2003). Conversely, a progressive reduction of phenolics was recorded with increasing time and agent concentration following first-order reaction kinetics (Vardin and Fenercioğlu, 2003). Turfan *et al.* (2011) showed reductions in total anthocyanins between juice from sacs (4%) and juice from whole fruit (19%) after clarification with gelatine. In addition, they pinpointed the stability of each single anthocyanin as a result of the clarification, and the stability of cyanidin-3,5-diglucoside unlike delphinidin-3-glucoside was highlighted (Turfan *et al.*, 2011).

MEMBRANE FILTRATION

Membrane technology for clarification of pomegranate juice has been applied mainly by using ultrafiltration and microfiltration. Clarification using membranes decreases phenolics levels owing to their inability to pass through the membranes as they are blocked in a cake layer. This fact prevents permeate flux in a process known as membrane fouling (Mirsaeedghazi *et al.*, 2009).

Both microfiltration (MF) (pore sizes ranging from 0.1 to 10 μm) and ultrafiltration (UF) (pore sizes ranging from 0.01 to 0.1 μm) are pressure-driven processes applied to clarify pomegranate juice. Concerning this, Mirsaeedghazi *et al.* (2012) compared MF with UF and MF resulted in a greater permeate efficiency, entailing a higher yield, whereas turbidity of the juices was similar. Total phenolic content, anthocyanins, and antioxidant activities were also similar when using both membrane processes. Nonetheless, membrane fouling was greater in the UF process than in MF and did not provide any advantage in relation to the levels of valuable attributes such as anthocyanin content and antioxidant capacity. Complementary further UF of a previously microfiltered juice was also ineffective (Mirsaeedghazi *et al.*, 2012). Even though enzymatic pre-treatment of juice with laccase has been pointed out to be effective for clarification (Neifar *et al.*, 2011), it could provoke a loss of the natural color of the pomegranate juice (Alper and Acar, 2004). Hence, taking into consideration these drawbacks and the higher cost of the UF process, MF seems to be a more convenient process for clarification of pomegranate juice.

The effect of the membrane pore size and type of membrane on antioxidants of pomegranate juice was evaluated (Mirsaeedghazi *et al.*, 2010a). Drops in the total phenolic content, anthocyanins, and antioxidant activity occurred at the same rate irrespective of the membrane tested, thus falls in the bioactive composition of filtered pomegranate juices are not affected by the membrane selected (Mirsaeedghazi *et al.*, 2010a).

Concentration of Pomegranate Juice

Pomegranate juice is concentrated to ensure long-term storage and easier transportation. Concentration of fruit juices requires partial removal of water without changes in solids composition, preserving the original composition apart from water. This major unit operation is of critical importance as many reactions affecting the color and phytochemical composition of pomegranate juice take place during thermal processing. Among them, the most common are anthocyanin degradation and browning reactions (i.e., Maillard reaction, enzymatic browning, and oxidation of ascorbic acid) (Maskan, 2006).

Concentration can be performed by evaporation methods or membrane processing. Different evaporation processes such as microwave heating, rotary vacuum evaporator, and heating at atmospheric pressure have been used to produce concentrated pomegranate juice (Altan and Maskan, 2005). It has been proved that these three evaporation methods decrease the color parameters significantly (CIE*L*: 55%, CIE*a*: 27%, CIE*b*: 22%, total color differences: 16.2) although there are not significant differences between them (Maskan, 2006). Pomegranate juice can turn brown due to fading of the red color due to heat-destroyed pigments and polymerization of anthocyanins with other phenolics (García-Viguera *et al.*, 1999). In addition, color degradation increases with soluble solids content since sugar and its degradation products may accelerate anthocyanin breakdown and non-enzymatic browning during thermal processing (Maskan, 2006; Suh *et al.*, 2003). Yousefi *et al.* (2012) also concentrated pomegranate juice by conventional heating and microwave energy at different operational pressures (12, 38.5, and 100 kPa), concluding that microwave heating decreases required times to effective concentration and reduces the degradation of color, anthocyanins, and antioxidant activity. Moreover, when compounded further by the use of lower process pressures, the degradation rates were reduced. This fact gained relevance in the case of the antioxidant activity measured by the 2,2-diphenyl-1-picrylhydrazyl (DPPH$^\bullet$) method, where increasing the pressure from 12 to 100 kPa caused an increase in the degradation rate by about 40% and 60% when conventional methods and microwave heating were applied, respectively (Yousefi *et al.*, 2012).

For water removal during the concentration step, processes without heating can also be used. Membrane concentration processes like osmotic distillation might overcome limitations linked to conventional evaporation and protect thermo-sensitive compounds of pomegranate juice (Cassano *et al.*, 2011). Cassano *et al.* (2011) noted an efficient preservation of the antioxidant capacity of pomegranate juice made from arils during the concentration step. Reduction in anthocyanins of ultrafiltered and subsequently concentrated juice was about 14%, which represents a very low drop when compared with conventional or microwave heating. Unfortunately, membrane technology is far from being used as a standard concentration method of pomegranate juices owing to inherent problems, including higher costs and required prior UF (Yousefi *et al.*, 2012).

Although dehydration cannot not be considered a classical process for concentrate production, it also makes it possible to extend the shelf-life of juices by removing water. Dehydration by spray drying is used extensively for a broad array of foodstuffs in dry particulate form as powders and agglomerates (Sagar and Suresh Kumar, 2010). Spray-dried pomegranate juice powders can be reconstituted instantly or served as an anthocyanin-rich additive. Yousefi *et al.* (2011) assayed the role of different carriers (maltodextrin, gum Arabic, and waxy starch) in the drying of pomegranate juice. The best color properties and anthocyanin levels of the reconstituted juice were achieved by using maltodextrin as carrier, although gum Arabic was the carrier leading to better physical properties. Thus, the use in combination of these two carriers might offer the best solution (Yousefi *et al.*, 2011). Nonetheless, it should be taken into account that pomegranate juice powder exhibited only

58% of its original antioxidant capacity by applying optimal conditions for spray drying (Horuz *et al.*, 2012).

Pasteurization

Pasteurization treatments preserve pomegranate juice and extend its shelf-life. Heating inactivates spoilage microorganisms and inhibits native enzymes that cause brown color formation by oxidizing polyphenols like anthocyanins (Skrede *et al.*, 2000). In contrast, thermal processing entails color alteration and browning, which may decline marketable quality attributes of some slightly colored pomegranate varieties (Mena *et al.*, 2013). Thus, color variations linked to anthocyanins degradation, provoked at the pasteurization step and occurring during storage, should be avoided or limited by using accurate and protective pasteurization methods.

Different methods have been developed for juice quality preservation. Non-thermal processes are gaining strength as emerging alternatives for juice quality preservation (Ferrari *et al.*, 2010; Yildiz *et al.*, 2009). However, conventional thermal pasteurization is still regarded as the most suitable approach for being inexpensive, efficient, and environmentally friendly (Silva and Gibbs, 2010).

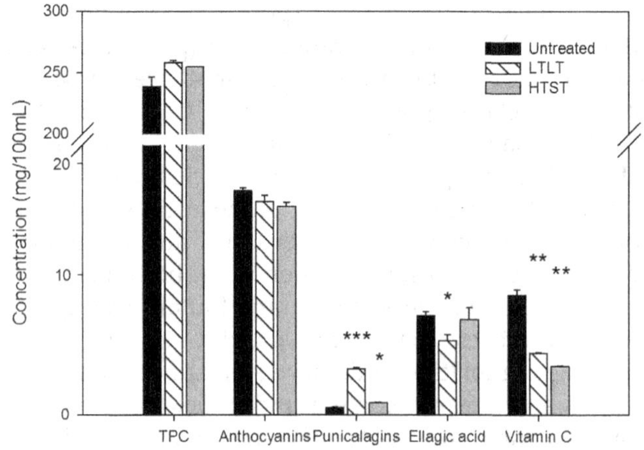

FIGURE 18.6 Influence of LTLT- and HTST-pasteurization in selected parameters of fresh 'Mollar de Elche' juice. Bars with different numbers of asterisks indicate significant differences (at $^{*}p<0.05$, $^{**}p<0.01$, and $^{***}p<0.001$) between the kind of pasteurization treatment and the untreated juice. *Modified from Mena* et al. *(2013).*

THERMAL TREATMENTS

Use of thermal pasteurization is widespread and typical commercial heat treatments to achieve effective stability vary from 60 to 95°C for 5–60 s. The process is accomplished in plate- or tubular-type heat exchangers that allow the reduction of phytochemical degradation by using short processing times. However, as far as we know, the scientific literature reporting the role of heating on pomegranate juice (poly)phenolics has been limited to batch systems using long pasteurization times (over 5 min) (Alighourchi *et al.*, 2008b; Alper *et al.*, 2005; Fischer *et al.*, 2013; Fischer *et al.*, 2011b; Pérez-Vicente *et al.*, 2004; Turfan *et al.*, 2011). These works have provided valuable information on phytochemical behavior even though they are far from being realistic for the juice processing industry. Nevertheless, Mena *et al.* (2013) succeeded in pasteurizing pomegranate juice with times no longer than 30 s by using a continuous and highly efficient process, more realistically reflecting industrial conditions. In this work, the authors assessed the effects of two different thermal treatments [high temperature–short time (HTST), 90°C for 5 s; and low temperature–long time (LTLT), 65°C for 30 s] on the phytochemical composition and color of different pomegranate juices, and observed the lack of reduction in anthocyanins as a result of the thermal pasteurization by both treatments (Figure 18.6). They found that anthocyanin content decreased after pasteurization by 8–13% depending on the juice extraction process (Turfan *et al.*, 2011). No significant changes were observed for conventionally clarified juices although non-clarified and ultrafiltered ones decreased anthocyanin values by 13 and 20%, respectively (Alper *et al.*, 2005). In addition, the total amount of anthocyanins slightly increased after pasteurization of some cultivars while it decreased for others (Alighourchi *et al.*, 2008b).

HTST and LTLT treatments slightly influenced the color of pasteurized pomegranate juices (Mena *et al.*, 2013). Pasteurization only increased the *Hue* angle values among color parameters, which indicated their yellowing. HTST treatment displayed higher increases in *Hue* values than LTLT treatment (Mena *et al.*, 2013). Batch pasteurization increased the color density of pomegranate juices produced by the different extraction and clarification methods (Alper *et al.*, 2005; Turfan *et al.*, 2011). Non-filtered juices slightly increased CIEa^* and *Chroma* values after pasteurization whereas microfiltered ones decreased their values (Fischer *et al.*, 2011b). Polymeric color and browning degree were also increased for all juices to a different degree attending to their previous operational units (Alper *et al.*, 2005).

The role of thermal pasteurization in the hydrolyzable tannins of pomegranate juice has been scarcely studied. In relation to this, Mena *et al.* (2013) have demonstrated that punicalagins increase after thermal treatment, while other compounds, such as punicalagin-like and ellagic acid (both free and glucoside forms), decrease. Overall, other authors have described that thermal pasteurization does not modify ellagitannins appreciably apart from punicalagin isomers, pointing to depolymerization of higher-molecular-weight ellagitannins (Hager *et al.*, 2010).

Total phenolic content of pomegranate juices is depleted between 1 and 14% depending on the previous processing steps (Alper *et al.*, 2005; Fischer *et al.*, 2011b; Pérez-Vicente *et al.*, 2004). Conversely, two different industrially realistic pasteurization methods have not shown variations in the phenolic content of pomegranate juice (Mena *et al.*, 2013). Regarding antioxidant capacity, heat treatments (HTST and LTLT) reduce the values by ~10% and ~20% for $ABTS^+$ and DPPH• methods, respectively. Antioxidant capacity could be altered irrespective of the pasteurization treatment assayed, indicating that the mere thermal processing affected the antioxidant capacity (Mena *et al.*, 2013).

Ohmic heating has also been proposed as an alternative fast-heating method for pomegranate juice. Ohmic heating is a thermal–electrical process based on the passage of electrical current through a food product that serves as an electrical resistance. Pomegranate juice was heated both ohmically and conventionally by matching the same thermal history for the two treatments (Yildiz *et al.*, 2009). Ohmic heating does not modify the color and phenolic content, in contrast to conventional heating. Therefore, ohmic heating could be considered as an alternative to conventional thermal pasteurization.

NON-THERMAL TREATMENTS

Non-thermal technologies for preservation of juices may be an alternative to thermal pasteurization, which has adverse effects (Pala and Toklucu, 2011). Among the different approaches developed recently, UV-C irradiation, gamma irradiation, high hydrostatic pressure (HHP), and pulsed electric fields (PEF) are those currently evaluated for pomegranate juice preservation.

UV-C irradiation has a germicidal effect on microorganisms such as bacteria, yeast, mold, and viruses, by preventing both DNA transcription and translation (Koutchma, 2009). Pomegranate juice was subjected to UV-C irradiation and variations in major quality characteristics were compared with heat-treated (90°C for 2 min) samples (Pala and Toklucu, 2011). In this work, total anthocyanin content did not vary significantly after UV-C treatment and losses were of up to 8% for different UV-C treatments compared with 24% for heat treatment. This fact caused higher polymeric color for heat-treated samples than for UV-C-treated ones (Pala and Toklucu, 2011). However, there were no significant changes in total phenolic content and antioxidant capacity of juice after UV-C and thermal treatments (Pala and Toklucu, 2011).

Ionizing radiations can also improve the safety of food. Gamma irradiation was tested on pomegranate juice at different dosages (0–10 kGy) but its use was not recommended due to the adverse effect on the stability of anthocyanins at all effective doses for microbial inactivation (Alighourchi *et al.*, 2008a).

Ferrari *et al.* (2010) assessed the potential application of HHP for the preservation of pomegranate juice. The main factor affecting the final quality of pressure-treated juices was the operating temperature. High-pressure treatments improved the redness of the juice while at the same time reducing the concentration of anthocyanins with the intensity of the treatment in terms of pressure level and processing time (Ferrari *et al.*, 2010). Varela-Santos *et al.* (2012) also assayed the role of HHP and its effects on the phenolic compounds of pomegranate juice but by using shorter processing times (ranging from 30 to 150 s) under refrigeration at 4°C. These processing conditions, sufficient to be considered effective with regard to microbial reduction, were softer than those tested by Ferrari *et al.* (2010). Regardless of this, reddish color of juice was reduced at processing, revealing a noticeable effect of pressure on color characteristics. Both phenolic content and antioxidant activity increased slightly (3–12%) for treated samples (Varela-Santos *et al.*, 2012). Unfortunately, the influence of HHP under refrigeration on the individual (poly)phenolics of pomegranate juice was not taken into account.

PEF have also been applied for the sanitation of pomegranate juice. Cui *et al.* (2007) subjected pomegranate juice to this non-thermal pasteurization and noted that no significant changes in the color and phenolic content of the juices were observed after PEF treatment. However, they pointed out that the role of electric-field intensity and treatment time are factors that should be borne in mind.

Despite the promising potential of all these emerging approaches, comparison with realistic industrial thermal pasteurization is required to verify their potential in solving problems associated with heating and preserving the (poly)phenolic fraction of pomegranate juices. Furthermore, studies should be conducted to analyze different factors affecting juice quality but also the complete (poly)phenolic profile of pomegranate juice. Economical criteria for selection of the most suitable method should not be avoided.

Stability of (Poly)Phenolics of Pomegranate Juice during Storage

Solving color alterations of pomegranate juice and other red juices during storage is another challenge for the juice-processing industry. Storage conditions are among the main factors affecting the final quality of juices and, thus, efforts should be made to understand how storage alters the (poly)phenolic fraction and antioxidant capacity of pomegranate juices. With this aim, Pérez-Vicente *et al.* (2004) studied the influence of the packaging material [transparent and green glass, and paperboard carton with polyethylene layers (Minibrik-200)] on the phytochemical quality of pomegranate juice during storage (for 160 days, at 24/18°C, and 16-h photoperiod). The degradation rate of anthocyanins was clearly influenced by the packaging material, being higher for samples stored in Minibrik-200 (95% loss), than for those stored in glass bottles (77–78% loss). The type of glass did not modify the pigment degradation and, hence, these authors concluded that oxygen but not light was the main factor affecting anthocyanin stability (Pérez-Vicente *et al.*, 2004). However, Fischer *et al.* (2011b) pointed out that light exposure can boost anthocyanin degradation and, thus, controlled illumination should be taken into consideration. Increasing storage temperature also affects the stability of anthocyanins (Alighourchi and Barzegar, 2009; Martí *et al.*, 2002). Oxygen permeability was also the cause behind the higher color alteration [increases in lightness (CIE*L*) and *Hue* angle, and decreases in redness (CIEa^*)] suffered by samples stored in Minibrik-200 than by those stored in glass bottles (irrespective of the type of glass) (Pérez-Vicente *et al.*, 2004). Similar results have been reported but they also emphasize the role of light in prompting color variations (Fischer *et al.*, 2011b). A general decrease (over 20%) in the total phenolic content was detected after 160 days of storage without noting differences in the packaging materials (Pérez-Vicente *et al.*, 2004). In addition, light exposure and storage temperature did not influence the degradation rates of total phenolic content and antioxidant capacity [Trolox equivalent antioxidant capacity (TEAC) and FRAP methods] (Fischer *et al.*, 2011b).

Nevertheless, previous processing steps may determine the evolution of (poly)phenolic compounds during storage. Oziyci *et al.* (2013) investigated in depth the effect of the extraction method, clarification, and filtration on color parameters during storage (6 months, at 5°C, in darkness). Minor color differences were recorded for juices extracted from arils, filtered by kieselguhr, and cold clarified, whereas the same processing but hot clarified yielded the higher color variation (Oziyci *et al.*, 2013). Concerning solely clarification, total phenolic content remained almost constant for clarified juices after 6 months' storage under different conditions whilst the content for the unclarified juice decreased significantly, particularly under light exposure (Fischer *et al.*, 2011b).

The impact of heat treatments and storage on pomegranate juice was evaluated by using comprehensive principal component analysis (PCA) (Mena *et al.*, 2013). The degradation of juices during storage corresponded to falls in anthocyanins, punicalagin-like compound, and vitamin C. Remarkable color variations were attributed to drops in color purity along with increases in lightness and browning. In addition, color variations as well as drops in the phenolic content of monovarietal juices made from Spanish cultivar 'Mollar de Elche' were eased by the combination of two widely distinct varietal juices (75% 'Mollar de Elche' + 25% 'Wonderful') (Mena *et al.*, 2013). HTST treatment favored the retention of the initial characteristics of the blended juices to a greater extent than LTLT. Interestingly, although browning was smoothed for blended juices, anthocyanin degradation rates were higher for these juices than for monovarietal juices. This fact was related to the superior pigment content of combined juices contrary to monovarietal ones (Mena *et al.*, 2013). In addition, higher retention of anthocyanins in pasteurized pomegranate juices was undertaken by diglucosides whereas monoglucosides were the most unstable ones (Alighourchi *et al.*, 2008b; Mena *et al.*, 2013; Pérez-Vicente *et al.*, 2004). Regarding pomegranate ellagitannins, the kind of heat treatment affected the trends of individual compounds during storage, provoking in LTLT treated juice higher changes in ellagitannins than in HTST (Mena *et al.*, 2013).

Conversely, the extent of industrially realistic pasteurization treatments did not determine the maintenance of the antioxidant activity ($ABTS^+$ and $DPPH^\bullet$ methods) during storage (Mena *et al.*, 2013).

References

AIJN (Association of the Industry of Juices and Nectars from Fruits and Vegetables of the European Union), 2008. Reference Guideline for Pomegranate Juice. Code of Practice for Evaluation of Fruit and Vegetable Juices 6.21.

Alighourchi, H., Barzegar, M., Abbasi, S., 2008a. Effect of gamma irradiation on the stability of anthocyanins and shelf-life of various pomegranate juices. Food Chem. 110, 1036–1040.

Alighourchi, H., Barzegar, M., Abbasi, S., 2008b. Anthocyanins characterization of 15 Iranian pomegranate (*Punica granatum* L.) varieties and their variation after cold storage and pasteurization. Eur. Food Res. Technol. 227, 881–887.

Alighourchi, H., Barzegar, M., 2009. Some physicochemical characteristics and degradation kinetic of anthocyanin of reconstituted pomegranate juice during storage. J. Food Eng. 90, 179–185.

Alper, N., Acar, J., 2004. Removal of phenolic compounds in pomegranate juices using ultrafiltration and laccase-ultrafiltration combinations. Nahrung Food 48, 184–187.

Alper, N., Bahçeci, K.S., Acar, J., 2005. Influence of processing and pasteurization on color values and total phenolic compounds of pomegranate juice. J. Food Process. Preserv. 29, 357–368.

Altan, A., Maskan, M., 2005. Rheological behavior of pomegranate (*Punica granatum* L.) juice and concentrate. J. Texture Stud. 36, 68–77.

Ardekani, M.R.S., Hajimahmoodi, M., Oveisi, M.R., Sadeghi, N., Jannat, B., Ranjbar, A.M., Gholam, N., Moridi, T., 2011. Comparative antioxidant activity and total flavonoid content of Persian pomegranate (*Punica granatum* L.) cultivars. Iran. J. Pharm. Res. 10, 519–524.

Bonzanini, F., Bruni, R., Palla, G., Serlataite, N., Caligiani, A., 2009. Identification and distribution of lignans in Punica granatum L. fruit endocarp, pulp, seeds, wood knots and commercial juices by GC-MS. Food Chem. 117, 745–749.

Borges, G., Mullen, W., Crozier, A., 2010. Comparison of the polyphenolic composition and antioxidant activity of European commercial fruit juices. Food Funct. 1, 73–83.

Brownmiller, C., Howard, L.R., Prior, R.L., 2008. Processing and storage effects on monomeric anthocyanins, percent polymeric color, and antioxidant capacity of processed blueberry products. J. Food Sci. 73, H72–H79.

Cassano, A., Conidi, C., Drioli, E., 2011. Clarification and concentration of pomegranate juice (*Punica granatum* L.) using membrane processes. J. Food Eng. 107, 366–373.

Clifford, M.N., Scalbert, A., 2000. Ellagitannins – Nature, occurrence and dietary burden. J. Sci. Food Agric. 80, 1118–1125.

Cristofori, V., Caruso, D., Latini, G., Dell'Agli, M., Cammilli, C., Rugini, E., Bignami, C., Muleo, R., 2011. Fruit quality of Italian pomegranate (*Punica granatum* L.) autochthonous varieties. Eur. Food Res. Technol. 232, 397–403.

Cui, X., Yang, R., Zhao, W., Sun, Z., Yu, Q., 2007. Pasteurization of pomegranate juice using high-intensity pulsed electric fields. Nongye Gongcheng Xuebao/Trans. Chin. Soc. Agri. Eng. 23, 252–256.

De Pascual-Teresa, S., Santos-Buelga, C., Rivas-Gonzalo, J.G., 2000. Quantitative analysis of flavan-3-ols in Spanish foodstuffs and beverages. J. Agric. Food Chem. 48, 5331–5337.

Fadavi, A., Barzegar, M., Azizi, M.H., Bayat, M., 2005. Note. physicochemical composition of ten pomegranate cultivars (*Punica granatum* L.) grown in Iran. Food Sci. Technol. Int. 11, 113–119.

Ferrari, G., Maresca, P., Ciccarone, R., 2010. The application of high hydrostatic pressure for the stabilization of functional foods: Pomegranate juice. J. Food Eng. 100, 245–253.

Fischer, U.A., Carle, R., Kammerer, D.R., 2011a. Identification and quantification of phenolic compounds from pomegranate (*Punica granatum* L.) peel, mesocarp, aril and differently produced juices by HPLC-DAD-ESI/MS^n. Food Chem. 127, 807–821.

Fischer, U.A., Dettmann, J.S., Carle, R., Kammerer, D.R., 2011b. Impact of processing and storage on the phenolic profiles and contents of pomegranate (*Punica granatum* L.) juices. Eur. Food Res. Technol. 233, 797–816.

Fischer, U.A., Carle, R., Kammerer, D.R., 2013. Thermal stability of anthocyanins and colourless phenolics in pomegranate (*Punica granatum* L.) juices and model solutions. Food Chem. 138, 1800–1809.

García-Viguera, C., Zafrilla, P., Romero, F., Abellán, P., Artés, F., Tomás-Barberán, F.A., 1999. Color stability of strawberry jam as affected by cultivar and storage temperature. J. Food Sci. 64, 243–247.

Gil, M.I., Cherif, J., Ayed, N., Artés, F., Tomás-Barberán, F.A., 1995. Influence of cultivar, maturity stage and geographical location on the juice pigmentation of Tunisian pomegranates. Zeitschrift für Lebensmittel-Untersuchung und -Forschung 201, 361–364.

Gil, M.I., Tomás-Barberán, F.A., Hess-Pierce, B., Holcroft, D.M., Kader, A.A., 2000. Antioxidant activity of pomegranate juice and its relationship with phenolic composition and processing. J. Agric. Food Chem. 48, 4581–4589.

Hager, T.J., Howard, L.R., Prior, R.L., 2010. Processing and storage effects on the ellagitannin composition of processed blackberry products. J. Agric. Food Chem. 58, 11749–11754.

Hassan, N.A., El-Halwagi, A.A., Sayed, H.A., 2012. Phytochemicals, antioxidant and chemical properties of 32 pomegranate accessions growing in Egypt. World Appl. Sci. J. 16, 1065–1073.

Horuz, E., Altan, A., Maskan, M., 2012. Spray Drying and Process Optimization of Unclarified Pomegranate (*Punica granatum*) Juice. Drying Technol. 30, 787–798.

Koutchma, T., 2009. Advances in ultraviolet light technology for non-thermal processing of liquid foods. Food Bioprocess Technol. 2, 138–155.

Martí, N., Pérez-Vicente, A., García-Viguera, C., 2002. Influence of storage temperature and ascorbic acid addition on pomegranate juice. J. Sci. Food Agric. 82, 217–221.

Maskan, M., 2006. Production of pomegranate (*Punica granatum* L.) juice concentrate by various heating methods: Colour degradation and kinetics. J. Food Eng. 72, 218–224.

Melgarejo, P. and Salazar, D. M. (2003). El granado. In Tratado de fruticultura para zonas áridas y semiáridas (vol. II): Algarrobo, granado y jinjolero, pp. 167–365. A. Madrid Vicente - Mundi-Prensa, Madrid.

Mena, P., García-Viguera, C., Navarro-Rico, J., Moreno, D.A., Bartual, J., Saura, D., Martí, N., 2011a. Phytochemical characterisation for industrial use of pomegranate (*Punica granatum* L.) cultivars grown in Spain. J. Sci. Food Agric. 91, 1893–1906.

Mena, P., Gironés-Vilaplana, A., Moreno, D.A., García-Viguera, C., 2011b. Pomegranate fruit for health promotion: myths and realities. In: Jiménez, A., García-Viguera, C. (Eds.), Antioxidant Properties of Crops III, 5. Functional Plant Science and Biotechnology, pp. 33–42.

Mena, P., Calani, L., Dall'Asta, C., Galaverna, G., García-Viguera, C., Bruni, R., Crozier, A., Del Rio, D., 2012. Rapid and comprehensive evaluation of (Poly)phenolic compounds in pomegranate (*Punica granatum* L.) Juice by UHPLC-MS^n. Molecules 17, 14821–14840.

Mena, P., Martí, N., Saura, D., Valero, M., García-Viguera, C., 2013. Combinatory Effect of Thermal Treatment and Blending on the Quality of Pomegranate Juices. Food Bioprocess Technol. 6, 3186–3199. http://dx.doi.org/10.1007/s11947-012-0961-z.

Mirsaeedghazi, H., Emam-Djomeh, Z., Mousavi, S.M., Aroujalian, A., Navidbakhsh, M., 2009. Changes in blocking mechanisms during membrane processing of pomegranate juice. Int. J. Food Sci. Technol. 44, 2135–2141.

Mirsaeedghazi, H., Emam-Djomeh, Z., Mousavi, S.M., Ahmadkhaniha, R., Shafiee, A., 2010a. Effect of membrane clarification on the physicochemical properties of pomegranate juice. Int. J. Food Sci. Technol. 45, 1457–1463.

Mirsaeedghazi, H., Emam-Djomeh, Z., Mousavi, S.M., Aroujalian, A., Navidbakhsh, M., 2010b. Clarification of pomegranate juice by microfiltration with PVDF membranes. Desalination 264, 243–248.

Mirsaeedghazi, H., Mousavi, S.M., Emam-Djomeh, Z., Rezaei, K., Aroujalian, A., Navidbakhsh, M., 2012. Comparison between ultrafiltration and microfiltration in the clarification of pomegranate juice. J. Food Process Eng. 35, 424–436.

Neifar, M., Ellouze-Ghorbel, R., Kamoun, A., Baklouti, S., Mokni, A., Jaouani, A., Ellouze-Chaabouni, S., 2011. Effective clarification of pomegranate juice using laccase treatment optimized by response surface methodology followed by ultrafiltration. J. Food Process Eng. 34, 1199–1219.

Ozgen, M., Durgaç, C., Serçe, S., Kaya, C., 2008. Chemical and antioxidant properties of pomegranate cultivars grown in the Mediterranean region of Turkey. Food Chem. 111, 703–706.

Oziyci, H.R., Karhan, M., Tetik, N., Turhan, I., 2013. Effects of processing method and storage temperature on clear pomegranate juice turbidity and color. J. Food Process Preserv. 37, 899–906.

Pala, T.U., Toklucu, A.K., 2011. Effect of UV-C light on anthocyanin content and other quality parameters of pomegranate juice. J. Food Compos. Anal. 24, 790–795.

Pérez-Vicente, A., Serrano, P., Abellán, P., García-Viguera, C., 2004. Influence of packaging material on pomegranate juice colour and bioactive compounds, during storage. J. Sci. Food Agric. 84, 639–644.

Plumb, G.W., De Pascual-Teresa, S., Santos-Buelga, C., Rivas-Gonzalo, J.C., Williamson, G., 2002. Antioxidant properties of gallocatechin and prodelphinidins from pomegranate peel. Redox Rep. 7, 41–46.

Poyrazoğlu, E., Gökmen, V., Artik, N., 2002. Organic acids and phenolic compounds in pomegranates (*Punica granatum* L.) grown in Turkey. J. Food Compos. Anal. 15, 567–575.

Rodríguez, P., Mellisho, C.D., Conejero, W., Cruz, Z.N., Ortuño, M.F., Galindo, A., Torrecillas, A., 2012. Plant water relations of leaves of pomegranate trees under different irrigation conditions. Environ. Exp. Bot. 77, 19–24.

Sagar, V.R., Suresh Kumar, P., 2010. Recent advances in drying and dehydration of fruits and vegetables: a review. J. Food Sci. Technol. 47, 15–26.

Sentandreu, E., Navarro, J.L., Sendra, J.M., 2010. LC-DAD-ESI/MS^n determination of direct condensation flavanol–anthocyanin adducts in pressure extracted Pomegranate (*Punica granatum* L.) juice. J. Agric. Food Chem. 58, 10560–10567.

Silva, F.V.M., Gibbs, P.A., 2010. Non-proteolytic *Clostridium botulinum* spores in low-acid cold-distributed foods and design of pasteurization processes. Trends Food Sci. Technol. 21, 95–105.

Skrede, G., Wrolstad, R.E., Durst, R.W., 2000. Changes in anthocyanins and polyphenolics during juice processing of highbush blueberries (*Vaccinium corymbosum* L.). J. Food Sci. 65, 357–364.

Stover, E., Mercure, E.W., 2007. The pomegranate: A new look at the fruit of paradise. HortScience 42, 1088–1092.

Suh, H.J., Noh, D.O., Kang, C.S., Kim, J.M., Lee, S.W., 2003. Thermal kinetics of color degradation of mulberry fruit extract. Nahrung Food 47, 132–135.

Turfan, O., Türkyilmaz, M., Yemi, O., Özkan, M., 2011. Anthocyanin and colour changes during processing of pomegranate (*Punica granatum* L.; Cv. Hicaznar) juice from sacs and whole fruit. Food Chem. 129, 1644–1651.

Türkyilmaz, M., Taği, S., Dereli, U., Özkan, M., 2013. Effects of various pressing programs and yields on the antioxidant activity, antimicrobial activity, phenolic content and colour of pomegranate juices. Food Chem. 138, 1810–1818.

Tzulker, R., Glazer, I., Bar-Ilan, I., Holland, D., Aviram, M., Amir, R., 2007. Antioxidant activity, polyphenol content, and related compounds in different fruit juices and homogenates prepared from 29 different pomegranate accessions. J. Agric. Food Chem. 55, 9559–9570.

Vardin, H., Fenercioğlu, H., 2003. Study on the development of pomegranate juice processing technology: Clarification of pomegranate juice. Nahrung Food 47, 300–303.

Varela-Santos, E., Ochoa-Martinez, A., Tabilo-Munizaga, G., Reyes, J.E., Pérez-Won, M., Briones-Labarca, V., Morales-Castro, J., 2012. Effect of high hydrostatic pressure (HHP) processing on physicochemical properties, bioactive compounds and shelf-life of pomegranate juice. Innovative Food Sci. Emerging. Technol. 13, 13–22.

Vázquez-Araújo, L., Chambers, E., Adhikari, K., Carbonell-Barrachina, A.A., 2011. Physico-chemical and sensory properties of pomegranate juices with pomegranate albedo and carpellar membranes homogenate. LWT Food Sci. Technol. 44, 2119–2125.

Yildiz, H., Bozkurt, H., Icier, F., 2009. Ohmic and conventional heating of pomegranate juice: Effects on rheology, color, and total phenolics. Food Sci. Technol. Int. 15, 503–512.

Yousefi, S., Emam-Djomeh, Z., Mousavi, S.M., 2011. Effect of carrier type and spray drying on the physicochemical properties of powdered and reconstituted pomegranate juice (*Punica granatum* L.). J. Food Sci. Technol. 48, 677–684.

Yousefi, S., Emam-Djomeh, Z., Mousavi, S.M.A., Askari, G.R., 2012. Comparing the Effects of Microwave and Conventional Heating Methods on the Evaporation Rate and Quality Attributes of Pomegranate (*Punica granatum* L.) Juice Concentrate. Food Bioprocess Technol. 5, 1328–1339.

Zaouay, F., Mena, P., Garcia-Viguera, C., Mars, M., 2012. Antioxidant activity and physico-chemical properties of Tunisian grown pomegranate (*Punica granatum* L.) cultivars. Ind. Crops Prod. 40, 81–89.

CHAPTER

19

Influence of High-Pressure and Ultra-High-Pressure Homogenization on Antioxidants in Fruit Juice

*Ángela Suárez Jacobo**, *Jordi Saldo*†, *Ramón Gervilla*†

†Departamento de Ciencia Animal y de los Alimentos, Universitat Autònoma de Barcelona, Edificio V. Campus de la UAB, Cerdanyola del Vallès, Spain, *CIATEJ, Unidad Noreste, Apodaca, Nuevo León, México

CHAPTER POINTS

- Ultra-high-pressure homogenization is a technology used to preserve fruit juices with a very limited heat effect.
- In some products, such as apple juice, it is possible to obtain a shelf-stable product.
- Fruit juices processed by ultra-high-pressure homogenization retain most of the vitamin C, polyphenols, and carotenoids present in the original juice.
- To ensure the protection of the health-related compounds it is important to prevent exposure to high temperatures or to oxygen.
- Antioxidant activity of ultra-high-pressure homogenization processed juices is well preserved.

HIGH-PRESSURE HOMOGENIZATION

In 1900, Auguste Gaulin invented homogenization and presented it at the World Fair in Paris with great success (French Patent no. 295.596). Since then, homogenization has been introduced into the food industry. Homogenization is the breaking down and mixing of the components of an emulsion or dispersion. A major use of homogenizers is to break down and disperse milk fat into the bulk of skim milk, which delays creaming of milk fat globules. Homogenizers are also used to process other emulsion such as silicone oil and to process dispersions such as pigments, antacids, and various paper coatings (Pandolfe, 1983). High-pressure homogenization (HPH), as it was called, was first used in the biochemical industry (Middelberg, 1995), where the main objective was the recovery of intracellular metabolites (such as protein and enzymes) released due to cell disruption (cell death) by using pressures in the range of 1–60 MPa. The term 'high-pressure' had not been precisely defined previously. In the early days, 34 MPa was considered as a high pressure for homogenization.

Between the 1980s and 1990s, a new technology was introduced for the production of fine emulsions (US patent no. 4,533,254). A new generation of homogenizers, referred to as high-pressure homogenizers (HPH), was developed. This new technology (HPH) has a different reaction chamber geometry that can withstand pressures 10- to 15-times higher than the classical homogenizers (up to 500 MPa) (Diels and Michiels, 2006; Donsì *et al.*, 2009b). HPH itself can be sub-divided in the following the way

1. Low-/medium-pressure homogenization: 5–35 MPa, typical of dairy-industry applications.
2. Medium-pressure homogenization: 35–70 MPa, various applications.
3. High-pressure homogenization (HPH): 70–200 MPa, typical for cell rupture, oil in water emulsions, liposomes.
4. Ultra-high-pressure homogenization (UHPH): 200–400 MPa, typical for novel food products.

The development of equipment operating at increasingly higher pressures has created possibilities for new applications and product enhancements not available with

Processing and Impact on Antioxidants in Beverages
http://dx.doi.org/10.1016/B978-0-12-404738-9.00019-2

lower-pressure operations (Diels and Michiels, 2006). One of the most recent applications is the microbial load reduction of food systems. Up until now, ultra-high-pressure homogenization as a unit process has been successfully applied in the preservation of food and other products and has been extensively reviewed by Dumay *et al.* (2013).

HIGH- AND ULTRA-HIGH-PRESSURE HOMOGENIZATION EQUIPMENT

Inspired by the capacity of HPH to disrupt cells, several research groups have developed an interest in homogenization equipment that operates at increasingly higher pressures. Different types of equipment in this category now exist, both prototypes and at an industrial level. Examples of these are: APV-Gaulin (USA), APV-Rannie (Denmark), and Panda (Gea Niro Soavi S.P.A., Parma, Italy), equipment capable of achieving pressures up to 150 MPa. Microfluidics (Microfluidics™ International Corporation, Newton, MA, USA), Nanojet (Nanojet™, Haskel, USA), Emulsiflex (Avestin®, Canada), and Stansted (Stansted Fluid Power Ltd, Essex, UK) are designed to work at pressures higher than 200 MPa (UHPH equipments).

A homogenizer consists of a high-pressure generator, typically a positive displacement pump, coupled with a pressure intensifier that is used to force the fluid through the homogenizing valve where the work is done (Middelberg, 1995; Floury *et al.*, 2004). In any type of homogenizer valve, the processed liquid passes under high pressure through a convergent section called the 'homogenization gap' and then expands (Floury *et al.*, 2004). The homogenization valve design is one of the major parameters that influences the efficiency of the process. Thus, it is understandable that the main differences between homogenizers from different manufacturers are in the homogenizing valves. In classic homogenization, the fluid is forced under pressure through a small orifice between the valve and the valve seat (Figure 19.1A). The fluid leaves the gap in the form of a radial jet that collides against an impact ring (Middelberg, 1995). The fluid exits at low velocity and at essentially atmospheric pressure.

A common small-scale (55 l/h) machine is the APV-Gaulin 15M homogenizer, which has a single acting pump capable of operating at 75 MPa. For large-scale processing, homogenizers with multiple pistons providing an almost constant pressure in excess of 100 MPa are available (HPH). The studies with this equipment have been focused on the reduction of fat globules and cell disruption to release intracellular products (Middelberg, 1995; Pandolfe, 1999), although citrus juice pasteurization has also been studied (Clark *et al.*, 1993).

The Stansted valve, however, works differently from the valve used in APV-Gaulin homogenizers. In the Stansted valve design (which is made of high-pressure-resistant ceramic material), the fluid streams axially under high pressure along the mobile part of the valve and flows with high velocity through the radial narrow gap formed between the valve seat and the piston, before leaving the valve seat at atmospheric pressure (Figure 19.1B). The size of the slit and the resulting stream velocity and pressure of the liquid ahead of the valve depend on the force acting on the valve piston, which can be adjusted to regulate the homogenizing intensity. The pressure drop of the liquid in the valve is called the 'homogenizing pressure' (Floury *et al.*, 2004). Effluent from the homogenizer is normally chilled to minimize thermal damage to the product caused by friction heat generated due to the high fluid velocity that elevates the product temperature by about 2.0–2.5°C/10 MPa (Popper and Knorr, 1990) (Figure 19.2).

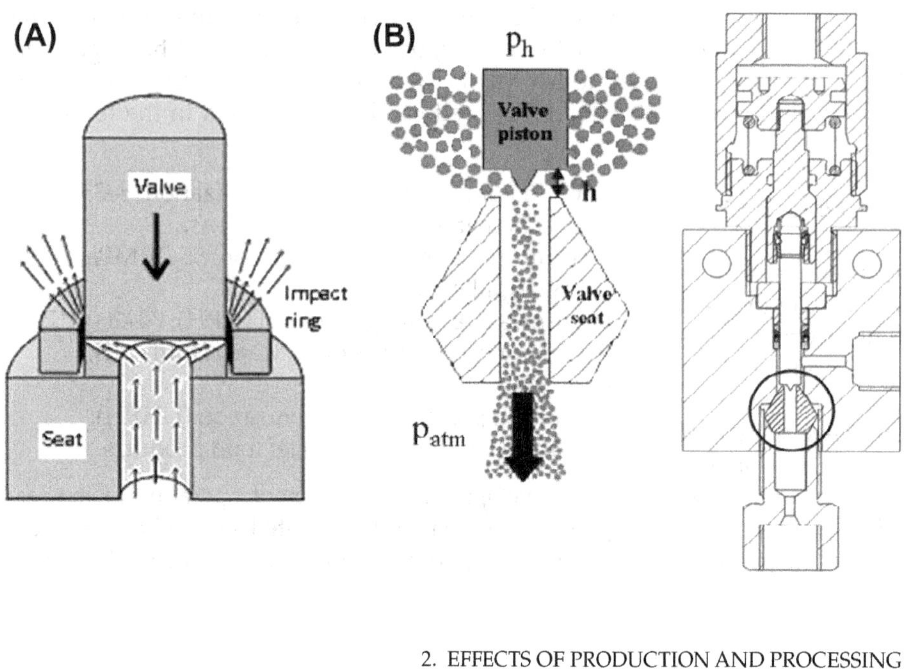

FIGURE 19.1 Homogenization valve geometries: (a) APV-Gaulin valve; (b) Stansted Fluid Power valve.

FRUIT JUICES PRESERVED BY HPH AND UHPH

The consumer demand for fresh, safe, high-quality, functional and convenience fruit and vegetable juices is increasing. According to a recent survey, new healthier juice products with low/no additives, preservatives, or colors are considered to be very important to consumers (Dodds, 2008). Fruit and vegetable juices are popular drinks consumed by people of all ages for their sensory and nutritional qualities, thus the application of UHPH in these kinds of products represents an interesting opportunity to reduce the microbiological activity while preserving the quality characteristics (nutritional, physicochemical, and organoleptic properties) and contributing to an extension of its shelf-life without pasteurization.

Homogenization of orange juice to reduce the pulp mean particle size and the viscosity of orange juice concentrate is a well-known technique, consisting of exposing the orange juice to high shear in a high-pressure homogenizer, pumped at pressures up to 50 MPa. The high-pressure differential between the inlet of the homogenizer valve and the outlet results in high shear and cavitation, which may alter the size and/or the properties of the suspended/insoluble pulp in the juice. Pressure homogenizations (14 MPa) in tomato pomaced juice increased the degree of product uniformity, improved smell and taste, and substantially lengthened the storage period, without the occurrence of juice separation (Komsta *et al.*, 1999). Thakur *et al.* (1995) also observed an increased consistency and reduction of serum separation in pressure-homogenised tomato juice at room temperature (28°C). Consistency in all the samples increased with increasing pressure of homogenization up to a pressure of 20 MPa, above which it became constant. In general, the juice with higher initial efflux and serum viscosities showed a larger increase in consistency and greater reduction in serum separation.

In 1993, the Coca-Cola® Company used 'ultra-high-pressure homogenization' (as it was called) to preserve freshly squeezed orange. Pressure of about 100 MPa was used and, surprisingly, the treated juice exhibited an increased shelf-life and decreased microbiological activity compared to homogenized juice under conventional pressures. The microbiological activity present in the juice was significantly reduced despite the fact that no thermal pasteurization step was conducted. In addition, juice processed by this technology maintained good flavor and palatability for 40 days at a storage temperature of 4°C, while untreated juice exhibited deterioration in 20 days (Clark *et al.*, 1993).

Recently, HPH an UHPH technologies have been applied to preserve fruit juices, for example the UHPH inactivation of endogenous microbiota in apple juice (Suárez-Jacobo *et al.*, 2009, 2010, 2012; McKay *et al.*, 2011; Donsì *et al.*, 2009a), banana juice (Calligaris *et al.*, 2012), and orange juice (Tahiri *et al.*, 2006; Velázquez-Estrada, 2011) was achieved with promising results. Application of multi-pass HPH-processing (100–200 MPa) was necessary to reduce microorganism load in orange, carrot, and apricot juices (Maresca *et al.*, 2011).

A complete inactivation and 5 log reduction of *Escherichia coli* O157:H7 was achieved in orange juice at 200 MPa after five and three passes at 25°C, respectively, by using an Emulsiflex C5 (Avestin®, Canada). Lower inactivation was obtained with *Penicillium* ssp. (4 log units), *Saccharomyces cerevisiae* (2.5 log units), *Lactobacillus plantarum* (2.3 log units) and *Leuconostoc mesenteroides* (1.6 log units) (Tahiri *et al.*, 2006; Maresca *et al.*, 2011).

Briñez *et al.* (2006a,b) investigated the inactivation after a two-stage UHPH treatment (300 MPa in the first valve, 30 MPa in the second valve; inlet temperature ranging between 6 and 20°C) of inoculated strains of *E. coli* (O58:H21 ATCC 10536 and O157:H7 CCUG 44857), *Listeria innocua*, *Staphylococcus aureus*, and *Staphylococcus carnosus* in orange juice and the evolution of survival over up to 2 weeks of storage at 4°C. Significant differences in lethality values between the two *E. coli* strains (approximately 3.9–4.7 log reduction at 2°C inlet temperature, respectively) were observed. However, after 27 days of storage, a decreasing tendency of 3.0 log units was observed at 4°C.

Analogously, *L. innocua*, *S. aureus*, and *S. carnosus*, which were initially reduced by ~3 log units, constantly

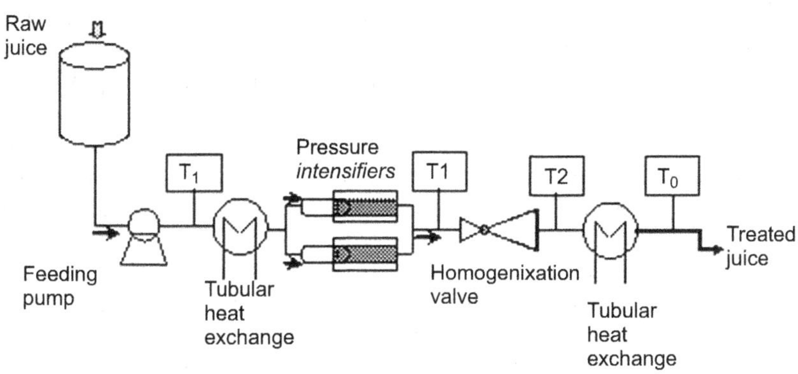

FIGURE 19.2 Scheme of the UHPH homogenizer. The actual DGR No. FPG 11300-400 Hygienic Homogenizer, from Stansted Fluid Power, has two high-pressure homogenization valves, but only the first one was used in the present study. T_i, inlet temperature; T1, temperature just before the high-pressure valve; T2, temperature just after the high-pressure homogenization valve; T_o, the outlet temperature. *Reproduced with permission from Suárez-Jacobo* et al. *(2010).*

decreased during the storage period, by additional reduction due to juice acidic conditions (Briñez *et al.*, 2006b). Yeasts and lactic acid bacteria are the usual contaminants in acid fruit juices and the agents responsible for decreasing product shelf-life. *L. plantarum* and *S. cerevisiae* were chosen by Campos and Cristianini (2007) as common contaminants of acid juice, in order to evaluate the sensitivity of these types of microorganisms to UHPH. Pressures higher than 250 MPa were able to completely destroy initial loads of 1.2×10^7 cfu/ml of *L. plantarum* and 2.9×10^5 cfu/ml of *S. cerevisiae* in orange juice, confirming this technology as a promising way to preserve acid juices.

The inactivation of fungal spores has been studied in apple juice. The UHPH treatment at 300 MPa reduced more than 5-log units of ascospores of *S. cerevisiae*, conidiospores of filamentous fungi, and sporulated black yeasts. Ascospores of *Talaromyces macrosporus* and *Neosartorya spinosa* were found to be resistant to the UHPH treatment (McKay, 2009).

Conidia from a heat-resistant *Aspergillus niger* were inoculated into mango nectar, and subjected to UHPH (300–100 MPa) (Tribst *et al.*, 2009). A heat shock (80°C for 5–20 min) was applied before or after UHPH treatment in order to improve the effectiveness of the treatment. Results indicated that UHPH inactivated *A. niger* in mango nectar at 300 MPa (>6.24-log cycles) and that, with pressure (200 MPa) combined with post-heat shock, it was possible to obtain the same decimal reduction, showing a synergistic effect. Conversely, pre-heat shock associated with UHPH resulted in an additive effect (Tribst *et al.*, 2009). Corbo *et al.* (2010) aimed to investigate the possibility of using UHPH to inactivate *Fusarium oxysporum*, *Emericella nidulans*, and *Penicillium italicum*. Pressure was an important variable as the number of conidia decreased to the progressive increase of the pressure applied (from 0 to 150 MPa). The reduction of conidia also increased while increasing the number of passes, as reported for other microorganisms (Dumay *et al.*, 2013). In particular, multi-step processing seemed to influence the effectiveness of treatment.

As a preliminary study Bevilacqua *et al.* (2007) investigated the effectiveness of HPH against *Alicyclobacillus acidoterrestris*. The treatment (50–170 MPa) caused a significant reduction of the initial cell number (1–2 log units) at the highest pressures and it was also concluded that the application of several successive rounds of HPH could have an additive effect on the reduction of viability and increase of susceptibility of *A. acidoterrestris* spores.

In general, the literature review shows that fruit juices are being considered as an interesting field of application for HPH and UHPH technologies because a reduction of microbial load is reached while preserving the quality attributes of the fresh products.

EFFECT OF HPH AND THERMAL TREATMENT ON HEALTH-RELATED COMPOUNDS AND ANTIOXIDANT CAPACITY OF FRUIT JUICES

Extensive scientific literature exists dealing with the beneficial effects that the consumption of fruit juice offers to human health. Increasing attention is being paid to the role of bioactive compounds in fruit juices, because they possess a possible health-protecting effect in the form of antioxidant capacity. The majority of the antioxidant effects of a fruit may be attributed to compounds such as vitamin C (L-ascorbic acid and dehydroascorbic acid), vitamin E, β-carotene, and phenolic compounds such as hydrocinnamic acids, flavanols or catechins, flavonols, and anthocyanins (Kaur and Kapoor, 2001). These compounds are of interest for human nutrition because they are assumed to inhibit oxidative damage to biomolecules (such as DNA, lipids, and proteins) thereby preventing cardiovascular diseases and certain cancers (Eberhardt *et al.*, 2000; Hollman, 2001).

Due to increasing consumer health consciousness, the importance of these biocompounds has oriented the fruit juice industry towards the manufacture of healthy and convenient foods by using new technologies that are able to preserve or increase the health-related compound content. Heat processing, commonly applied to preserve liquid food products, has presented some drawbacks such as undesirable biochemical, sensory, and nutritional changes in processed products (color changes, flavor and aroma decreases, and vitamin losses) that affect the quality of the final product and make it unattractive to the consumer.

Effect on Vitamin C Content

During heat treatment, ascorbic acid loss is one important problem associated with orange juice quality (Manso *et al.*, 2001) and temperature is the most critical factor in preserving ascorbic acid (Nagy, 1980). After thermal treatment of orange juice, vitamin C was reduced significantly (approximately 80–90%) (Elez-Martínez and Martín-Belloso, 2007; Sanchez-Moreno *et al.*, 2003b).

As previously described, UHPH processing has been introduced as an alternative non-thermal technology that causes inactivation of microorganisms. However, few data on the effect of HPH on ascorbic acid content for different juices have been published. Hence, it is important to study the effect of UHPH processing on the retention of vitamin C in fruit juices and kinetic modeling to predict the losses during and after processing.

Regarding this issue, Welti-Chanes *et al.* (2009) evaluated the effect of HPH on freshly squeezed orange juice, at pressures ranging from 0 to 250 MPa with three inlet temperatures (22–45°C) and up to five passes (1–5) on the vitamin C content. In just-squeezed juice, the vitamin concentration was 57 ± 7 mg/100 ml, and remained stable after homogenization at 100 MPa (66 ± 3 mg/100 ml) or 250 MPa (54 ± 3 mg/100 ml), irrespective of the number of passes. These results were in disagreement with those published by Velázquez-Estrada *et al.* (2011) when orange juice was treated at 100, 200 MPa and 300 MPa and losses in 2%, 5% and 11%, respectively, were found.

These differences could be associated with the different inlet temperature and the short thermal increment (less than 0.7 s) during the UHPH processing, which causes an increase in the outlet temperature of the product after the UHPH processing. Floury *et al.* (2000) described an increase of 0.164°C per MPa, whereas Hayes and Kelly (2003) reported that the temperature increase was 0.176°C per MPa. However, Floury *et al.* (2000) also estimated the exposure time of the fluid for about 10^{-4} s, meaning that the thermal effect could be negligible in comparison with the conventional heat treatments. Apple juice, another important juice, was also subject to UHPH treatment (100, 200, and 300 MPa) by Suárez- Jacobo *et al.* (2011); the result showed that ascorbic acid concentration was found at low levels (0.46 mg/l), without being affected by the different pressure intensities or by thermal treatment applied. However, presence of dehydroascorbic acid, a compound that can be converted back into ascorbic acid, was found in fresh and UHPH-treated samples but in low levels in pasteurized samples (12.24 mg/l and 1.01 mg/l, respectively). In this study, no significant differences were found in vitamin C (ascorbic acid plus dehydroascorbic acid) content when comparing UHPH-treated samples with samples which had not been treated, but their content was significantly higher than in pasteurized samples (1.51 mg/l). In the same research, the thermal effect on vitamin C, observed in the pasteurized samples, reduced its content to around 88% of fresh apple juice, being inversely correlated (97%, $P < 0.01$) with the hydroxymethylfurfural content (a good index for thermal treatment) (Saldo *et al.*, 2009).

Isostatic high pressure is known to cause vitamin C degradation during the pressure treatment, and the severity of this effect is dependent on the amount of dissolved oxygen, and the final level achieved can be regarded as a result on the molar ratio of vitamin and oxygen concentration (Oey *et al.*, 2008). It is therefore acceptable to consider that this oxidation can occur in an analogous way during UHPH treatment and once the major part of soluble oxygen has been eliminated by reacting with vitamin C, the increase in pressure level and/or prolonged treatment has almost no effect on further degradation if temperature and time are kept at moderate values.

Long storage time can produce a further degradation of vitamin C, and this effect is increased with temperature. Some research studies reported that the rate of ascorbic acid degradation increased with the storage temperature (Ros-Chumillas *et al.*, 2007). Esteve *et al.* (1996) studied the stability of ascorbic acid in fresh and commercial orange juice maintained at 4 and 10°C, finding that at 4°C the loss of ascorbic acid was less than 10% after 7 days of storage. At 30°C and 15 days, ascorbic acid was totally degraded.

On a combined study of UHPH (300 MPa, 4°C inlet temperature) and aseptically packaged storage of apple juice (4, 10, 20, and 30°C), Suárez-Jacobo *et al.* (2012) described the vitamin C changes during 3 months storage (Figure 19.3). In this research work the original vitamin C content (13.19 mg/l) decreased and its final losses ranged between 75 and 98% at 4–20°C temperature of storage.

Effect on Carotenoid Content

The impact of UHPH on carotenoids in fruit juices has not been studied in depth, because studies have been focused on evaluating the technology as a preservation method.

In addition to vitamin C, orange juice is also a good source of carotenoids such as β-carotene, α-carotene and β-cryptoxanthin, as well as zeaxanthin and lutein (Melendez-Martinez *et al.*, 2003; Sánchez-Moreno *et al.*, 2003a). In fact, its color is mainly due to these kinds of compounds.

The effect of UHPH-treatments on these compounds result in significant differences between different intensities of homogenization pressures and inlet temperatures applied. In orange juice the content ranged from 57 and 67 μg/l after treatments at 100 and 200 MPa. A significant decrease (approximately 30%) from the original content was observed in samples treated at 300 MPa (Velázquez-Estrada, 2011).

In apple juice, β-carotene content was also evaluated due to lack of information on the effect of this technology on its content. In comparison with the UHPH treatments, fresh apple juice had the highest β-carotene content (11.38 μg/l). Values from 9.41 μg/l (100–200 MPa) to 7.59 μg/l (100–200–300 MPa at 20°C) were reported when juices was subjected to UHPH treatment. The decrease in carotene content after UHPH treatment of apple juice could be due to oxidation reactions that could have occurred during the processing. Rodríguez-Amaya (2003) reported that the oxidation of this compound depends on: availability of oxygen, water activity, presence of antioxidants, presence of metals, peroxidases and severity of treatments (destruction of the cellular structure that protects the carotenoids, increase of surface area, duration and temperature of heat treatment).

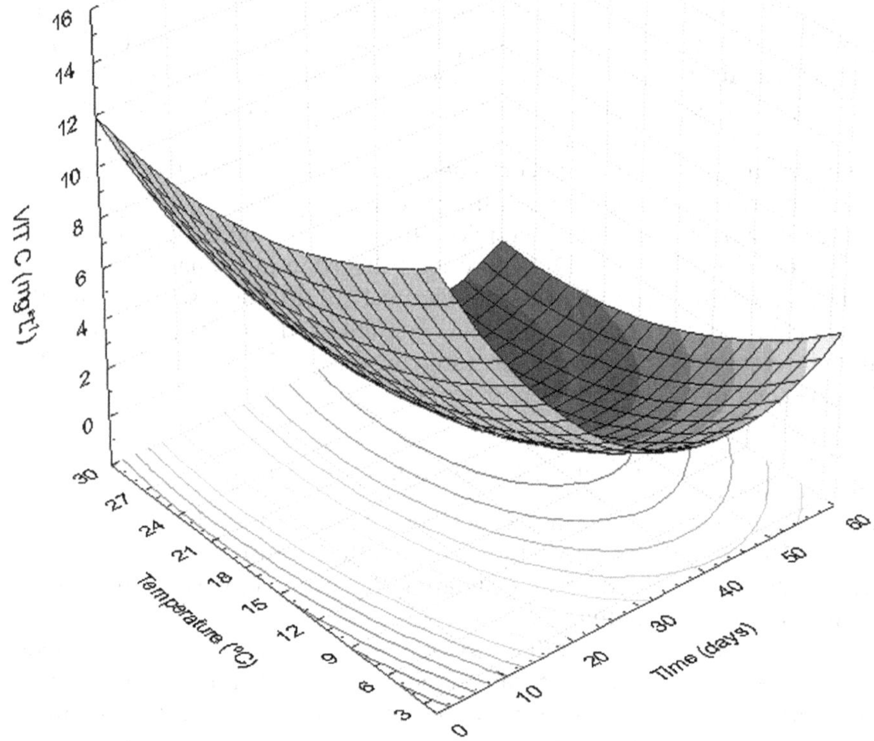

FIGURE 19.3 Changes total vitamin C content during storage of UHPH-treated apple juice over 60 days in Tetra Brik containers. X, temperature (°C); Y, days. *Source: Unpublished data.*

In the studies of apple and orange juices, thermal pasteurization of juice had the lowest value of only 6.82 μg/l (60% of losses) and 870 μg/l (35% of losses), respectively (Velázquez-Estrada, 2011; Suárez-Jacobo *et al.*, 2011). These results were in agreement with the reported values; after pasteurization (90°C, 20 s) of orange juice the *cis/trans* carotenoid contents decreased by 12.6% with respect to untreated fresh orange juice (Cortés *et al.*, 2006). Torregrosa *et al.* (2005) also observed a decrease in the carotenoid content of the orange–carrot juice mixture (80:20, v/v) when juice was subjected to a conventional heat treatment (98°C, 21 s). Lee *et al.* (2003) observed significant carotenoid content losses after thermal pasteurization at 90°C for 30 s. The thermal effect was especially observed with violaxanthin (−46.4%) and antheraxhantin (−24.8%). Contrary to these results, it has been reported that traditional thermal treatment (90°C, 1 min) did not exert any effect on total carotenoid content or vitamin A value in orange juice (Sanchez-Moreno *et al.* 2005).

Effect on Phenolic Content and Antioxidant Capacity

Phenolic compounds are secondary plant metabolites that contribute to the flavor and color characteristics of fruit juices. Ranging from simple phenolic molecules to highly polymerized compounds with molecular weights above 30,000 Da, the occurrence of this complex group of substances is extremely variable (Bravo, 1998).

In apple juice, several phenolic compounds such as hydroxycinnamic acids, flavanols, dihydrochalcones. and flavonols have been found. Clear apple juice has been subject to UHPH processing and analysis of the effect on polyphenol content (Suárez-Jacobo *et al.* 2011). High-pressure liquid chromatography coupled to mass spectrometry (HPLC-MS) and high pressure liquid chromatography with diode array detection (HPLC-DAD) were used to identify and quantify the polyphenol components. The major content of polyphenolic compounds was ascribed to hydroxycinnamic acids with 64.9%, followed by flavonols with 21.6%, and dihydrochalcones with 13.6%; chlorogenic acid was the dominant constituent of hydroxycinnamic acids (26.5% of total quantity) compared with other compounds analyzed.

Polyphenol content was not affected by the UHPH technology or thermal treatments; total polyphenol content obtained ranged from 11.7 mg/l in non-treated and 14.5 mg/l in pasteurized juice. However, it was observed that highest phenolic compound content was quantified in pasteurized and UHPH-treated juice at 300 MPa, being related to the preservation of phenolic compounds as a result of HPH impact on enzyme inactivation (polyphenoloxidase) at high pressure in comparison to non-treated and UHPH-treated juices at low pressure levels (Saldo *et al.*, 2009; Schilling *et al.* 2008). The stability of polyphenols in UHPH-treated and aseptically packaged apple juice have been evaluated; the total polyphenolic content (which ranged from 22.75 to 14.11 mg of GAE/l) was affected by time and temperature of storage, with reductions up to 38% at highest storage temperature (38°C) (Betoret *et al.*, 2009). The stability of various classes of polyphenolic antioxidants in an apple juice enriched with these compounds as a function of storage temperature has been evaluated (van der Sluis *et al.*, 2004, 2005). Results reveal that high storage temperatures result in a decrease in the polyphenolic antioxidants, which are more stable at ambient or refrigerated storage conditions.

The effect of homogenization and UHPH on flavonoids in orange juice has been evaluated. The content of the flavonoids was not affected by homogenization pressure (which ranged from 5 up to 30 MPa), but after 5 months of storage the content of flavonoid hesperidin in juice was affected (Betoret *et al.*, 2009). In addition, Velázquez-Estrada *et al.* (2011) observed that neither pasteurization nor UHPH treatment in orange juice affect the flavonoid content. However, high concentrations of hesperidin were found in juices treated at 200 and 300 MPa. The authors pointed out that UHPH could improve the flavanoid content due to forces such as cavitation, friction, and shear; and, further, turbulence suffered by the juice during the processing could diminish the cloud fraction size of juice, improving the extractability of flavonoids from this fraction and producing an increase in the soluble flavanone content.

Regarding the antioxidant capacity of fruit juices treated by UHPH, this test has been measured using different antioxidant assays [2,2-diphenyl-1-picrylhydrazyl radical (DPPH), Trolox-equivalent antioxidant capacity (TEAC), ferric-reducing antioxidant power (FRAP), and oxygen radical absorbance capacity antioxidant test (ORAC)].

The antioxidant capacity of apple juices has been mainly ascribed to phenolic compounds (hydroxycinnamic acids, flavanols, dihydrochalcones, and flavonols) and to a lesser extent to ascorbic acid content. UHPH treatment in apple juice did not significantly affect their total phenolic content (Figure 19.4); values obtained from the Folin–Ciocalteu assay ranged from 49.32 to 52.39 mg GAE/l. Furthermore, the antioxidant capacity of non-treated, pasteurized and UHPH-treated apple juice was measured by FRAP, TEAC, 2,2′-azino-bis (3-ethylbenzothiazoline-6-sulphonic acid) radical ($ABTS^+$), and ORAC assays. No significant differences were found between non-treated (R) and UHPH-treated (100, 200, and 300 MPa) juices analyzed by DPPH and FRAP assays; however, significant differences were detected in comparison with pasteurized juice. ORAC and TEAC assays showed more sensitive responses, and again significant differences between UHPH treatments and pasteurized juice were found. The antioxidant capacity in pasteurized apple juice (PA) was higher by between 5 and 19% than in all the other assays. It was observed that some polyphenol compounds are decomposed during heating at high temperatures (above 100°C). However, some decomposition products still have radical-scavenging activity. This result suggests that the radical-scavenging activity is more stable than the original content of polyphenolic compounds in foods during cooking or heat processing (Murakami *et al.*, 2004). In UHPH-treated apple juice the cooking effect was not observed but a high value was reported

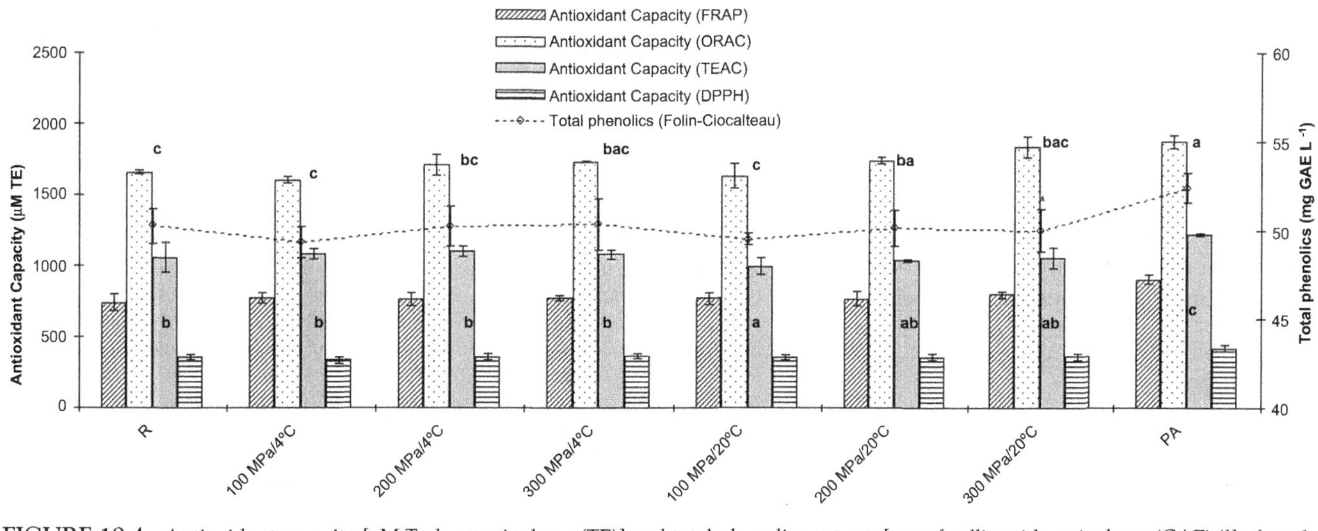

FIGURE 19.4 Antioxidant capacity [μM Trolox equivalents (TE)] and total phenolic contents [mg of gallic acid equivalents (GAE)/l] of apple juices, raw (R), pasteurized (PA), and treated by ultra-high-pressure homogenization (UHPH) at two different inlet temperatures (4 and 20°C). Treatments with the same letter index for ORAC and TEAC series did not differ significantly ($P < 0.05$). *Source: Suárez-Jacobo* et al. *(2011); reproduced with the permission of the editor.*

in the pasteurized samples (Saldo *et al.*, 2009). This could explain the effect of cooking and their relationship with the result of antioxidant capacity of pasteurized juices in comparison with the UHPH-treated juices.

Contrary to these results, pasteurization processing (90°C, 1 min) of orange juice resulted in high impact by decreasing the polyphenol content in comparison with the UHPH-treated and non-treated juices (Velázquez-Estrada, 2011). In the same study, the TEAC and FRAP analysis demonstrated no significant differences in the UHPH-treated juices' antioxidant capacities. Nevertheless, significant differences were reported at non-treated and pasteurized juices.

Gardner *et al.* (2000), when studying the relative contributions of vitamin C, carotenoids, and phenolics to the antioxidant potential of fruit juices (including orange, grapefruit, pink grapefruit, apple, pineapple and vegetable juices), found that both vitamin concentrations and total phenolic contents strongly correlated with antioxidant capacity. This could explain the differences obtained in UHPH-treated apple and orange juices. In apple juices, a strong correlation between polyphenol content and antioxidant capacity (86%, $P<0.01$) (Suárez-Jacobo *et al.*, 2011) and inversely correlated with β-carotene content was found (−0.63%, $P<0{,}05$); while, in orange juice, vitamin C, carotenoids, and flavonoids are the most important health-related compounds that exhibit antioxidant capacity by multiple different mechanisms at different concentrations.

CONCLUSIONS

UHPH-treated juices retained better the health-related compounds (such as vitamin C, carotenoid, phenolic, and antioxidant capacity content) than thermally treated juices. Operational parameters such as the pressure and inlet temperatures used during the treatment could impact the health-related compound content in fruit juices. As observed, highest inlet and processing temperatures could induce the degradation of these biocompounds that exhibit antioxidant capacity.

UHPH above 200 MPa seems to be a quite suitable technology to improve the preservation of fruit juices by helping to retain their nutritional and antioxidant capacities while achieving sufficient microbiological inactivation without the additional thermal damage.

References

Betoret, E., Betoret, N., Carbonell, J.V., Fito, P., 2009. Effects of pressure homogenisation on particle size and the functional properties of citrus juices. J. Food Eng. 92 (1), 18–23.

Bevilacqua, A., Cibelli, F., Corbo, M., Sinigaglia, M., 2007. Effects of high-pressure homogenisation on the survival of Alicyclobacillus acidoterrestris in a laboratory medium. Lett. Appl. Microbiol. 45 (4), 382–386.

Bravo, L., 1998. Polyphenols: Chemistry, Dietary Sources, Metabolism, and Nutritional Significance. Nutr. rev. 56 (11), 317–333.

Briñez, W.J., Roig-Sagués, A.X., Hernández Herrero, M.M., Guamis, B., 2006a. Inactivation by Ultrahigh-Pressure Homogenisation of Escherichia coli strains inoculated into orange juice. J. Food Prot. 69 (5), 984–989.

Briñez, W.J., Roig-Sagués, A.X., Hernández Herrero, M.M., Guamis, B., 2006b. Inactivation of Listeria Innocua in milk and orange juice by ultrahigh-pressure homogenisation. J. Food Prot. 69 (1), 86–92.

Calligaris, S., Foschia, M., Bartolomeoli, I., Maifreni, M., Manzocco, L., 2012. Study on the applicability of high-pressure homogenisation for the production of banana juices. LWT Food Sci. Technol. 45 (1), 117–121.

Campos, F.P., Cristianini, M., 2007. Inactivation of Saccharomyces cerevisiae and Lactobacillus plantarum in orange juice using ultra high-pressure homogenisation. Innovative Food Sci. Emerg. Technol. 8 (2), 226–229.

Clark, A.V., Rejimbal, T.R., Gomez, C.M., 1993. Ultra-high pressure homogenisation of unpasteurised juice.

Corbo, M.R., Bevilacqua, A., Campaniello, D., Ciccarone, C., Sinigaglia, M., 2010. Use of high pressure homogenisation as a mean to control the growth of foodborne moulds in tomato juice. Food Control 21 (11), 1507–1511.

Cortés, C., Esteve, M.J., Rodrigo, D., Torregrosa, F., Frígola, A., 2006. Changes of colour and carotenoids contents during high intensity pulsed electric field treatment in orange juices. Food Chem. Toxicol. 44 (11), 1932–1939.

Diels, A.M.J., Michiels, C.W., 2006. High-pressure homogenisation as a non-thermal technique for the inactivation of microorganisms. Crit. Rev. Microbiol. 32 (4), 201.

Dodds, A., 2008. Future Convenience Food And Drinks.

Donsì, F., Esposito, L., Lenza, E., Senatore, B., Ferrari, G., 2009a. Production of shelf-stable annurca apple juice with pulp by high pressure homogenisation. Int. J. Food Eng. 5. pp. Article 12.

Donsì, F., Ferrari, G., Maresca, P., 2009b. High-Pressure Homogenization for food sanitization in *Global Issues in Food Science and Technology*. In: Gustavo Barbosa- Cánovas, Alan, Mortimer (Eds.), Academic Press, San Diego, pp. 309–352.

Dumay, E., Chevalier-Lucia, D., Picart-Palmade, L., Benzaria, A., Gràcia-Julià, A. & Blayo, C., 2013. Technological aspects and potential applications of (ultra) high-pressure homogenisation. Trends Food Sci. Technol. 31, 13–26.

Eberhardt, M., Lee, C., Liu, R.H., 2000. Antioxidant activity of fresh apples. Nature 405, 903–904.

Elez-Martínez, P., Martín-Belloso, O., 2007. Effects of high intensity pulsed electric field processing conditions on vitamin C and antioxidant capacity of orange juice and gazpacho, a cold vegetable soup. Food Chem. 102 (1), 201–209.

Esteve, M.J., Farré, R., Frígola, A., 1996. Stability of ascorbic acid in orange juices after initial use at home begins. J. Food Quality 19 (3), 243–249.

Floury, J., Bellettre, J., Legrand, J., Desrumaux, A., 2004. Analysis of a new type of high pressure homogeniser. A study of the flow pattern. Chem. Eng. Sci. 59 (4), 843–853.

Floury, J., Desrumaux, A., Lardières, J., 2000. Effect of high-pressure homogenisation on droplet size distributions and rheological properties of model oil-in-water emulsions. Innovative Food Sci. Emer. Technol. 1 (2), 127–134.

Gardner, P.T., White, T.A.C., McPhail, D.B., Duthie, G.G., 2000. The relative contributions of vitamin C, carotenoids and phenolics to the antioxidant potential of fruit juices. Food Chem. 68 (4), 471–474.

Hayes, M.G., Kelly, A.L., 2003. High pressure homogenisation of milk (b) effects on indigenous enzymatic activity. J. Dairy Res. 70 (03), 307–313.

Hollman, P.C., 2001. Evidence for health benefits of plant phenols: local or systemic effects? J. Sci. Food Agric. 81 (9), 842–852.

Kaur, C. & Kapoor, C.H. 2001, Antioxidants in fruits and vegetables—the millennium's health, Int. J. Food Sci. Technol. 36, 703–725.

Komsta, H., Popko, R., Popko, H. & Hys, L. 1999. Pressure homogenisation in fruit-vegetable processing industry. Int. Agrophysics, 13, 363–368.

Lee, J., Lin, Y., Chang, H., Chen, W., Wu, M., 2003. Temperature-time relationships for thermal inactivation of pectinesterases in orange juice, 96. J. Sci. Food Agri. 83 (7), 681–684.

Manso, M.C., Oliveira, F.A.R., Oliveira, J.C., Frías, J.M., 2001. Modelling ascorbic acid thermal degradation and browning in orange juice under aerobic conditions. Int. J. Food Sci. Technol. 36 (3), 303–312.

Maresca, P., Donsì, F., Ferrari, G., 2011. Application of a multi-pass high-pressure homogenisation treatment for the pasteurisation of fruit juices. J. Food Eng. 104 (3), 364–372.

McKay, A.M., 2009. Inactivation of fungal spores in apple juice by high pressure homogenisation. J. Food Prot. 72 (12), 2561–2564.

McKay, A.M., Linton, M., Stirling, J., Mackle, A., Patterson, M.F., 2011. A comparative study of changes in the microbiota of apple juice treated by high hydrostatic pressure (HHP) or high pressure homogenisation (HPH). Food Microbiol. 28 (8), 1426–1431.

Melendez-Martinez, A.J., Vicario, I.M., Heredia, F.J., 2003. A Routine High-Performance Liquid Chromatography Method for Carotenoid Determination In Ultrafrozen Orange Juices. J. Agric. Food Chem. 51 (15), 4219–4224.

Middelberg, A.P.J., 1995. Process-scale disruption of microorganisms. Biotechnol. Adv. 13 (3), 491–551.

Murakami, M., Yamaguchi, T., Takamura, H., Atoba, T.M., 2004. Effects of thermal treatment on radical-scavenging activity of single and mixed polyphenolic compounds. J. Food Sci. 69 (1), FCT7–FCT10.

Nagy, S., 1980. Vitamin C contents of citrus fruit and their products: a review. J. Agric. Food Chem. 28, 8–18.

Oey, I., Van der Plancken, I., Van Loey, A., Hendrickx, M., 2008. Does high pressure processing influence nutritional aspects of plant based food systems? Trends Food Sci. Technol. 19 (6), 300–308.

Pandolfe, W.D., 1999. Homogeniser in *Encyclopedia of Food and Science and Technology*. In: Francis, F.J. (Ed.), second ed. Wiley, New York, USA, pp. 1289–1294.

Pandolfe, W.D., 1983. Homogenising apparatus and method.

Popper, L., Knorr, D., 1990. Applications of high-pressure homogenisation for food preservation. Food Technol. 44 (7), 84–89.

Rodriguez-Amaya, D.B., 2003. Carotenoids: occurrence, properties, and determination in *Encyclopedia of Food Sciences and Nutrition*. In: Caballero, B. (Ed.), Academic Press, Oxford, pp. 927–936.

Ros-Chumillas, M., Belissario, Y., Iguaz, A., López, A., 2007. Quality and shelf life of orange juice aseptically packaged in PET bottles. J. Food Eng. 79 (1), 234–242.

Saldo, J., Suárez-Jacobo, Á., Gervilla, R., Guamis, B., Roig-Sagués, A.X., 2009. Use of ultra-high-pressure homogenisation to preserve apple juice without heat damage. High Press. Res. 29 (1), 52.

Sanchez-Moreno, C., Plaza, L., Elez-Martinez, P., De Ancos, B., Martin-Belloso, O., Cano, M.P., 2005. Impact of High Pressure and Pulsed Electric Fields on Bioactive Compounds and Antioxidant Activity of Orange Juice in Comparison with Traditional Thermal Processing. J. Agric. Food Chem. 53 (11), 4403–4409.

Sánchez-Moreno, C., Plaza, L., de Ancos, B., Cano, M.P., 2003a. Quantitative bioactive compounds assessment and their relative contribution to the antioxidant capacity of commercial orange juices. J. Sci. Food Agric. 83 (5), 430–439.

Sanchez-Moreno, C., Plaza, L., de Ancos, B., Cano, M.P., 2003b. Vitamin C, provitamin A carotenoids, and other carotenoids in high-pressurized orange juice during refrigerated storage. J. Agric. Food Chem. 51 (3), 647–653.

Schilling, S., Schmid, S., Jäger, H., Ludwig, M., Dietrich, H., Toepfl, S., Knorr, D., Neidhart, S., Schieber, A., Carle, R., 2008. Comparative study of pulsed electric field and thermal processing of apple juice with particular consideration of juice quality and enzyme deactivation. J. Agric. Food Chem. 56 (12), 4545–4554.

Suárez-Jacobo, Á., Gervilla, R., Guamis, B., Roig-Sagués, A.X., Saldo, J., 2010. Effect of UHPH on indigenous microbiota of apple juice: A preliminary study of microbial shelf-life. Int. J. Food Microbiol. 136 (3), 261–267.

Suárez-Jacobo, Á., Gervilla, R., Guamis, B., Roig-Sagués, A.X., Saldo, J., 2009. Microbial inactivation by ultra high-pressure homogenisation on fresh apple juice. High Press. Res. 29 (1), 46.

Suárez-Jacobo, Á., Rüfer, C.E., Gervilla, R., Guamis, B., Roig-Sagués, A.X., Saldo, J., 2011. Influence of ultra-high pressure homogenisation on antioxidant capacity, polyphenol and vitamin content of clear apple juice. Food Chem. 127 (2), 447–454.

Suárez-Jacobo, Á., Saldo, J., Rüfer, C.E., Guamis, B., Roig-Sagués, A.X., Gervilla, R., 2012. Aseptically packaged UHPH-treated apple juice: Safety and quality parameters during storage. J. Food Eng. 109 (2), 291–300.

Tahiri, I., Makhlouf, J., Paquin, P., Fliss, I., 2006. Inactivation of food spoilage bacteria and Escherichia coli O157:H7 in phosphate buffer and orange juice using dynamic high pressure. Food Res. Int. 39 (1), 98–105.

Thakur, B.R., Singh, R.K., Handa, A.K., 1995. Effect of homogenisation pressure on consistency of tomato juice. J. Food Qual. 18 (5), 389–396.

Torregrosa, F., Cortes, C., Esteve, M.J., Frigola, A., 2005. Effect of High-Intensity Pulsed Electric Fields Processing and Conventional Heat Treatment on Orange and Carrot Juice Carotenoids. J. Agric. Food Chem. 53 (24), 9519–9525.

Tribst, A.A., Franchi, M.A., Cristianini, M., de Massaguer, P.R., 2009. Inactivation of *Aspergillus niger* in mango nectar by high-pressure homogenisation combined with heat shock. J. Food Sci. 74 (9), M509–14.

van der Sluis, A.A., Dekker, M., Skrede, G., Jongen, W.M.F., 2004. Activity and concentration of polyphenolic antioxidants in apple juice. 2. Effect of novel production methods. J. Agric. Food Chem. 52 (10), 2840–2848.

van der Sluis, A.A., Dekker, M., van Boekel, Martinus A.J. S., 2005. Activity and Concentration of Polyphenolic Antioxidants in Apple Juice. 3. Stability during Storage. J. Agric. Food Chem. 53 (4), 1073–1080.

Velázquez-Estrada, R.M., 2011. Evaluation of the efficacy of Ultra-High Pressure Homogenisation Technology to improve the safety and quality of liquid food and especially orange juice. Universitat Autònoma de Barcelona.

Welti-Chanes, J., Ochoa-Velasco, C.E., Guerrero-Beltrán, J.Á., 2009. High-pressure homogenisation of orange juice to inactivate pectinmethylesterase. Innovative Food Sci. Emerg. Technol. 10 (4), 457–462.

CHAPTER

20

Enzymatic Debittering on Antioxidant Capacity of Grapefruit Juice

María D. Busto, Mónica Cavia-Saiz, Natividad Ortega, Pilar Muñiz

Department of Biotechnology and Food Science, Area of Biochemistry and Molecular Biology, University of Burgos, Burgos, Spain

CHAPTER POINTS

- Flavonoids, limonoids, and their glycosides have potential biological functions and contribute to antioxidant capacity of citrus juices.
- Conventional methods to reduce the bitterness in citrus juices involve the removal of not only the bitter compounds, but also bioactive components from the juice.
- Biotechnological approaches for removing the bitterness of citrus juices are focused on the development of enzymes or whole cells that modify naringin or limonin in other flavonoids or limonoids, respectively, to remove bitterness.
- Enzymatic technology is an interesting alternative way to remove bitterness, without affecting the nutritional and healthy properties of the processed juice, due to the substrate specificity and mild operating conditions of the enzymes.
- Some practical problems in the use of soluble enzymes in the juice debittering process, e.g., inactivation at pH conditions of juice or their high cost, can be overcome by the immobilization technology.
- Enzymatic debittering with naringinase immobilized by entrapment in a biocompatible and non-toxic polymer poly(vinyl alcohol) preserves the antioxidant and biomolecule protection capacity of freshly squeezed grapefruit juice.
- Much research is still required before considering a future commercial application of immobilized enzymes for debittering of juices, mainly on the design of the immobilized biocatalyst, and in particular on the effects of the immobilization material in the antioxidant properties of treated juice.

INTRODUCTION

Some citrus juices, such as from grapefruits and sour oranges, have encountered some commercial restrictions due to their inherent undesirable bitter taste (Yusof *et al.*, 1990). The bitterness of grapefruit juice is caused mainly by naringin, a flavanone diglycoside. As in all citrus fruits, the grapefruit also contains the bitter triterpenoid, limonin (Hasegawa *et al.*, 1985a). Reduction in the level of these substances during the processing of juice has been the focus of many studies. A number of physico-chemical methods and materials (macroporus resins, affinity columns, cyclodextrin polymer, activated diatomaceous earths,...) have been applied to minimize the flavonoid and limonoid bitterness problem (Singh *et al.*, 2008; Ribeiro *et al.*, 2002). However, these techniques involve not only the removal of the bitter compounds, but also nutrients/flavor/color from the juice (Kranz *et al.*, 2011; Puri *et al.*, 2008; Lee and Kim, 2003). Faced with these technologies, biotechnological approaches for removing the bitterness of citrus juices are focused on the development of enzymes that specifically modify naringin or limonin, in other flavonoids (naringenin) or limonoids (e.g., 17-dehydrolimonoate or glycosides) to remove bitterness.

Numerous reports have shown that flavonoids, limonoids and their glycosides have a wide range of

Processing and Impact on Antioxidants in Beverages
http://dx.doi.org/10.1016/B978-0-12-404738-9.00020-9

biological and pharmacological actitivities, and contribute to the antioxidant capacity of citrus juices (Xu *et al.*, 2008; Manners, 2007). Hence, it would be very interesting to find a method to remove bitterness without affecting the level of flavonoids, limonoids, or their metabolites.

In this context, researchers are putting emphasis on enzyme technology as an alternative way to remove bitterness with maximum preservation of the sensorial, nutritional, and healthy qualities in processed fruit juices.

In this chapter, the current state of knowledge of the potential application of enzymes in the debittering of citrus juices and their effects on the antioxidant properties of the grapefruit juice are reviewed and discussed.

ENZYMES IN DEBITTERING OF CITRUS JUICES

Nowadays, enzyme technology plays an important role in the food industry as an integral part of processing. The field of enzymes for food applications has expanded rapidly, particularly in the last two decades, aiming to optimize processing parameters, design novel functional foods, provide alternative applications for several agricultural products and/or improve the safety, health impact, and the quality of foods. In the sector of fruit juice production, enzymes are specifically used for fruit maceration, extraction yield improvement, clarification, and enhancement of stability of the final product (Ribeiro *et al.*, 2010b; Busto and Ortega, 2008). In addition to these well-established industrial applications of enzymes, numerous studies have demonstrated the potential of some enzymes for removing the bitterness of citrus juices. Based on these studies, the debittering can be achieved by treatment of juices with one of the following enzymes: (i) *naringinase*, (ii) *limonoate dehydrogenase* (LD*ase*) or (iii) *limonoid glucosyl transferase* (LGT*ase*). Naringinase is by far the main focus of research.

The substrates and reaction products of these enzymes are described below and summarized in Figure 20.1.

Naringinase

Naringinase, an enzyme complex containing α-L-rhamnosidase and β-D-glucosidase, first catalyzes the hydrolysis of naringin (4,5,7-trihydroxy flavanone 7-rhamnoglucoside), the main bittering component in grapefruit, into rhamnose and prunin (responsible for one-third of the bitterness of naringin). Then, prunin is hydrolyzed to tasteless naringenin (4,5,7-trihydroxyflavanone) and glucose (Prakash *et al.*, 2002) (Figure 20.1A).

Naringinase occurs widely in nature and is produced by many microorganisms (Puri, 2012). Nevertheless, *Aspergillus niger* is considered as the most

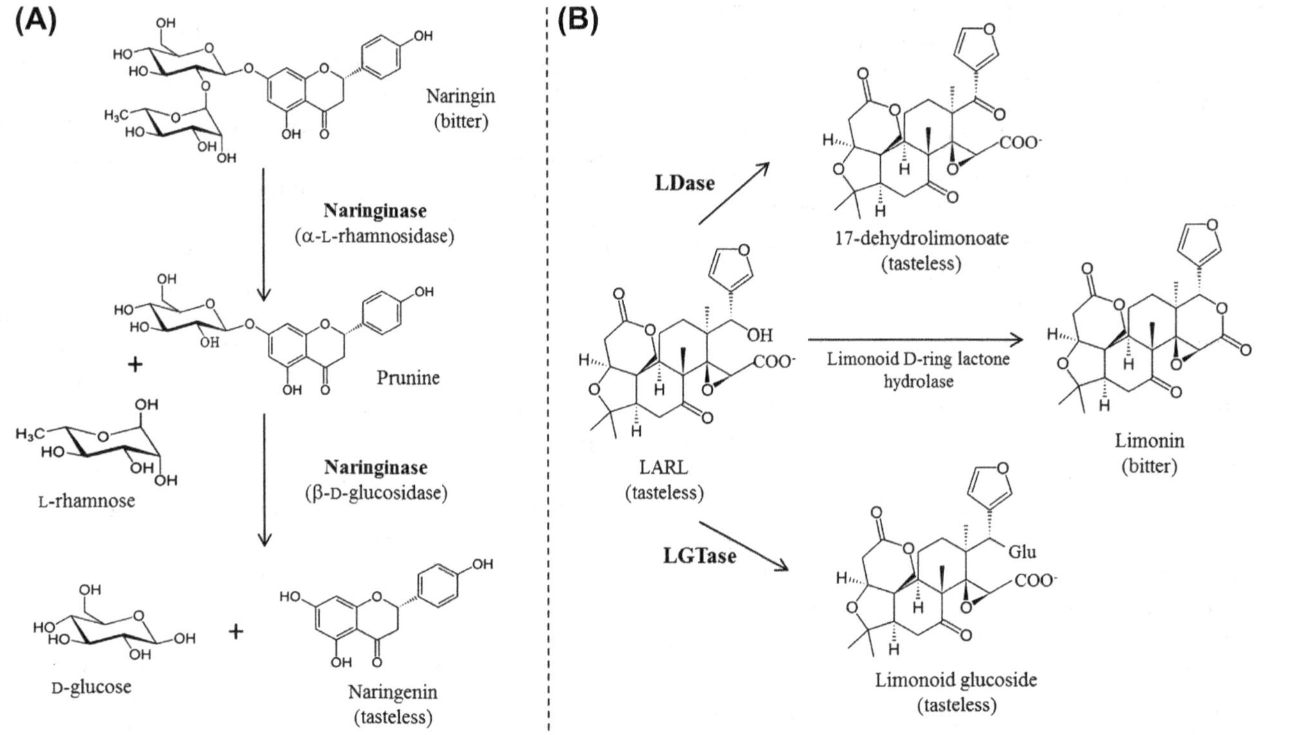

FIGURE 20.1 Enzymatic debittering of citrus juices: A, naringinase; B, limonoate dehydrogenase (LDase) and limonoid glucosyl transferase (LGTase).

important source of naringinase for food use regarding safety characteristics and feasible culture processes (Ni *et al.*, 2012).

Several reviews about this enzyme have been published recently (Puri, 2012; Ribeiro, 2011; Yadav *et al.*, 2010). These reviews focus on the sources and production of naringinase, structural aspects, biochemical characteristics, immobilization, and the potential biotechnological applications.

Limonoate Dehydrogenase

After citrus juice extraction, and also in the fruits after freezing or mechanical damage, acidic pH conditions facilitate the conversion of limonoate A-ring lactone (LARL) (non-bitter limonoid) to limonin, a process known as delayed bitterness (Manners, 2007). Limonoate dehydrogenase, an enzyme detected in different species of microrganisms, can prevent limonin production by catalyzing the oxidation of LARL to corresponding 17-dehydrolimonoate form, a non-bitter derivative which can not be converted into limonin (Manners, 2007) (Figure 20.1B).

This enzyme has been isolated from *Athrobacter globiformis* (Hasegawa *et al.*, 1972), *Pseudomonas* sp. (Hasegawa *et al.*, 1974), and *Rhodococcus fascians* (Puri *et al.*, 2002; Humanes *et al.*, 1997). However, the optimal pH of LDase (8.0–9.0) prevents the use of the isolated enzyme in debittering processes due to the acidic pH of the juices (Hasegawa *et al.*, 1983). In contrast, when whole bacterial cells are employed for citrus-juice debittering, the enzyme is protected from the external pH and, therefore, the problem is minimized (Martínez-Madrid *et al.*, 1989).

Limonoid Glucosyl Transferase

Limonoid UDP-glucosyltransferase has been reported to catalyze the conversion of limonoid aglycones to limonoid 17-O-glucosides (Hasegawa *et al.*, 1997). A limonoid glucosyltransferase has been isolated and purified from the navel orange albedo tissue (Hasegawa *et al.*, 1997). Karim and Hashinaga (2002) also isolated and characterized a LGTase from pummel albedo tissue. Glucosylation of LARL hampers lactonization and thus prevents formation of limonin (Daniel *et al.*, 2011) (Figure 20.1B).

Limonoid glucosyltransferase seems to be a key enzyme for creating transgenic citrus without limonoid bitterness, increasing specific limonoid glucoside molecules (Zaare-Nahandi *et al.*, 2008). Karim and Hashinaga (2002) proposed that immobilized LGTase might give promising results in reducing bitterness in juice by conversion of free limonoids into glucosides.

IMMOBILIZED BIOCATALYSTS IN DEBITTERING OF JUICES

The use of soluble enzymes in juice processing to obtain different technological effects is well established. Nevertheless, in spite of the unquestionable advantages, there exist a number of practical problems in the use of soluble enzymes in the juice debittering process, such as their inactivation at pH conditions of citrus juices and their high cost (Karim and Hasinaga, 2002). Immobilizing the enzyme on solid supports allows the re-use of catalytic activity, higher enzyme concentrations and higher stability, prevention of contamination of the processed product, modification of catalytic properties, better control of reactions, and a design of bioreactors that can be easily incorporated into a continuous processing line (Puri, 2008; Karim and Hasinaga, 2002).

An up-to-date review of immobilized biocatalysts developed for debittering of citrus juices reported in the literature is presented in Table 20.1. With the exception of one publication about the immobilization of LGTase (Karim and Hashinaga, 2002) and a few reports on immobilized microorganisms capable of metabolizing limonoids (Martínez-Madrid *et al.*, 1989; Hasegawa *et al.*, 1985b), the rest of the research has focused on the immobilization of naringinase. In fact, this enzyme has been immobilized on very different supports such as hollow fibers (Gay and Olson, 1981), glycophase-coated porous glass (Manjón *et al.*, 1985), cellulose triacetate (Tsen, 1989), glutaraldehyde-coated hen egg white (Puri *et al.*, 2001), celite (Sekeroglu *et al.*, 2006), poly(vinyl alcohol) cryogels (Busto *et al.*, 2007), κ-carrageenan beads (Ribeiro *et al.*, 2008), calcium alginate beads (Puri *et al.*, 2010), mesoporous molecular sieve MCM-41 (Lei *et al.*, 2011), magnetic particles (Soria *et al.*, 2012), and coating films of poly(vinyl alcohol) (PVA) with glutaraldehyde for active packaging (Del Nobile *et al.*, 2003). Most of these studies deal with kinetic parameters, thermal and operational stability, and reusability of the immobilized naringinase.

Data regarding the effectiveness of the immobilized enzymes for reducing bitterness and the nature of citrus juice treated have been included in Table 20.1.

EFFECTS OF PRODUCTION AND PROCESSING

Among the various techniques that are used to reduce bitterness, adsorption on macroporous resin is the process that is applied most frequently in the citrus industry (Kranz *et al.*, 2011). This and other techniques for debittering routinely remove bioactive compounds, such as ascorbic acid, phenolics, flavonoids, isoflavonoids, isoflavones, and terpenes from fruit juices (Puri *et al.*, 2008;

TABLE 20.1 Overview of Immobilized Biocatalysts Developed for Debittering of Citrus Juices

Enzyme/Cells	Support	Juice/Debittering Efficiency[a]	References
NARINGINASE FROM			
A. niger	Hollow fiber	Grapefruit juice/67.8%	Gray and Olson (1981)
A. niger	Chitin with glutaraldehyde	—	Tsen (1984)
Penicillium sp.	Glycophase-coated porous glass	—	Manjón *et al.* (1985)
Penicillium sp.	Cellulose triacetate fiber	Grapefruit juice/80%	Tsen *et al.* (1989)
Penicillium sp.	Alginate	Kinnow mandarin juice/56%	Puri *et al.* (1996)
Penicillium sp.	Cellulose triacetate (films)	Grapefruit juice/60%	Soares and Hotchkiss (1998)
Penicillium sp.	Glutaraldehyde-coated hen egg white	Kinnow mandarin juice/68%	Puri *et al.* (2001)
P. decumbems	PVA with glutaraldehyde (film)	—	Del Nobile *et al.* (2003)
A. niger	Alginate	Grapefruit juice/84%	Mishra and Kar (2003)
Penicillium sp.	Glutaraldehyde-coated wood chip	Kinnow mandarin juice/76%	Puri *et al.* (2005)
P. decumbems	Celite	—	Sekeroglu *et al.* (2006)
A. niger	Poly(vinyl alcohol) hydrogel	Simulated juice/34%	Busto *et al.* (2007)
P. decumbems	Alginate	—	Pedro *et al.* (2007)
P. decumbems	κ-Carrageenan beads	Grapefruit juice/95%	Ribeiro *et al.* (2008)
P. decumbems	PVA-alginate beads	—	Nunes *et al.* (2010)
P. decumbems	Calcium alginate beads	Grapefruit juice/82%	Ribeiro *et al.* (2010a)
Recombinant *E. coli* BL21 (DE3)	Calcium alginate beads	Kinnow mandarin juice/67%	Puri *et al.* (2010)
P. decumbems	Mesoporous molecular sieve MCM-41	Grapefruit juice/95%	Lei *et al.* (2011)
MICROORGANISM			
A. globiformis II	Acrylamide gel	Navel orange juice serum/81%	Hasegawa *et al.* (1983)
C. fascians	Acrylamide gel	Citrus juice serum	Hasegawa *et al.* (1985b)
R. fascians	Alginate, κ-carrageenan, agarose, polyacrilamide gel	—	Martínez-Madrid *et al.* (1989)
R. fascians	κ-karrageenan	—	Manjón *et al.* (1991)
A. globiformis II	Alginate	Kinnow mandarine juice	Premi *et al.* (1998)
R. fascians	Polyurethane foam	Synthetic citrus juice/85%	Cánovas *et al.* (1998)
R. fascians	PVA-PEG cryogels	Orange juice	Lopez-Fouz *et al.* (2007)
LGTASE FROM			
Pummelo albedo tissue	DEAE-Toyopearl, chitosan, cellulose carbonate	—	Karim and Hashinaga (2002)

[a]*Percentage of bitter compounds removed, in the conditions described in references.*
PVA, poly(vinyl alcohol); PEG, polyethylene glycol; DEAE, diethylaminoethanol.

Lee and Kim, 2003). In contrast, biological methods, either isolated enzymes or whole cells, are a very interesting alternative to eliminate the bitterness with preservation of natural qualities of juice, due to the substrate specificity and mild operating conditions.

It has been reported that the biological activity of non-bitter limonoid glucosides is equal or similar to that of bitter limonoids (Karim and Hashinaga, 2002). In fact, limonoids and their glycosides have been shown to inhibit chemically induced carcinogenesis and possess hypocholesterolemic and antioxidant properties (Breksa and Manners, 2006; Kurowska *et al.*, 2000; Miller *et al.*, 1994).

Both naringin and naringenin have been shown to have pharmacological applications (Jayachitra and Nalini, 2011) and free-radical scavenging activities (Cavia-Saiz *et al.*, 2010; Kumar *et al.*, 2003). Moreover, both flavanones were equally effective in reducing DNA damage (Cavia-Saiz *et al.*, 2010). Thus, the debittering by naringinase, which transforms naringin into naringenin, does not reduce the health-promoting effects of juice.

The interest in immobilized biocatalysts in comparison to their free counterparts for reducing bitterness has already been described. However, when the enzyme or cells are immobilized, some retention of bioactive components of the juice by the support of immobilization can occur. In this regard, references that study the effect of the immobilization matrix on the juice constituents (e.g., sugars, organic acids, ascorbic acid) are very scarce. For instance, the use of a bioreactor with *Corynebacterium fascians* cells, capable of metabolizing limonoids, immobilized in acrylamide gel, caused no significant adverse effects on the composition of citric, malic, ascorbic acids, fructose, glucose, and sucrose (Hasegawa *et al.*, 1985b).

With respect to the effect of enzymatic debittering on pharmacological properties of the juice, Ribeiro *et al.* (2008) described that the enzymatic processing with naringinase immobilized in κ-carrageenan beads did not affect the anti-inflammatory properties of the grapefruit juice, which was demonstrated in comparability studies in rats administered orally with juice before and after debittering.

Regarding the effects of enzymatic debittering on the antioxidant properties of juice, recent studies have found that debittering of grapefruit juice with free naringinase was more effective than traditional adsorption technology (Amberlite®IRA-400 ion exchange resin) in preserving the antioxidant and biomolecule protection capacity of fresh grapefruit (Cavia-Saiz *et al.*, 2011). In fact, this is the only study published to date about the enzymatic treatment and its effect on the antioxidant properties of juice.

Effects of Enzymatic Debittering on Antioxidant Capacity of Grapefruit Juice

The authors' group has studied the influence of the debittering with immobilized naringinase into a polymeric matrix consisting of poly(vinyl alcohol) (PVA) cryogels (Busto *et al.*, 2007) on total antioxidant capacity of grapefruit juice. The results obtained by 2,2′-azinobis(3-ethylbenzothiazoline-6-sulphonic acid) diammonium salt (ABTS) method (Pellegrini *et al.*, 1999) showed (Figure 20.2A) that pasteurized juices treated with free or immobilized naringinase (N-PJ or IN-PJ) exhibited a higher capacity than fresh juice (FJ) samples to inhibit 50% of ABTS radical cation generation, no significant differences ($p < 0.005$) being detected between the debittered juices. These results are in agreement with those reported in a previous study (Cavia-Saiz *et al.*, 2010), in which the bitter flavanone naringin exhibited

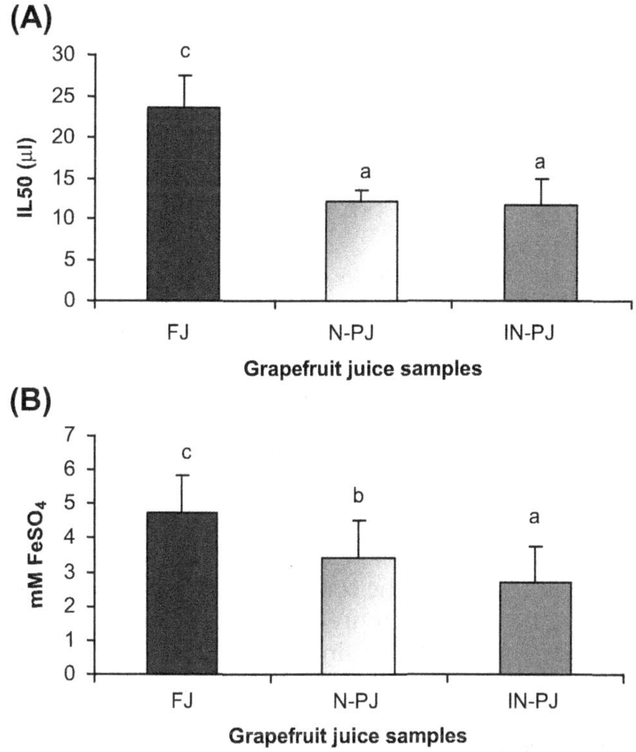

FIGURE 20.2 Total antioxidant capacity of grapefruit juice by ABTS [2,2′-azinobis(3-ethylbenzothiazoline-6-sulphonic acid) diammonium salt] (A) and FRAP (ferric reducing antioxidant power) (B) methods. (A) Capacity of juice samples to inhibit the generation of $ABTS^{•+}$ radical. IL_{50}, juice volume giving 50% inhibition of $ABTS^{•+}$ radical production. (B) FRAP method analyzes the ability of the juice samples to reduce Fe^{3+} to Fe^{2+}. FJ, fresh juice; N-PJ, pasteurized[a] and naringinase-treated juice; IN-PJ, pasteurized and immobilized naringinase-treated juice. Results are the mean ± SD of 10 experiments. Different letters for each column correspond to statistically different values ($p < 0.005$). [a]Tatum, J. H., and Berry, R. E. (1973) Method for determining naringin content in grapefruit juice. *J. Food Sci.* 38, 340–341.

lower capacity to scavenge $ABTS^{•+}$ than naringenin. Heim *et al.* (2002) also described that naringenin significantly increased the antioxidant capacity of the flavanone towards the $ABTS^{•+}$ stable radical.

In addition, all juices analyzed had the ability to reduce Fe^{3+} to Fe^{2+} (Figure 20.2B). The FRAP (ferric reducing antioxidant power) values of juice samples were 4.73 ± 1.09 mM of $FeSO_4$ for FJ and 3.41 ± 1.10 mM and 2.70 ± 1.07 mM for N-PJ and IN-PJ, respectively. These data showed that the juice debittered with the immobilized enzyme had the lowest levels of antioxidant activity by the FRAP method (Benzie and Strain, 1999). The making of a control test with fresh juice and the immobilization matrix (PVA beads) without enzyme, effectively removed 12.5% of total phenolic content. Therefore, the cryogel of PVA partly retains some of the compounds responsible for reducing the capacity of the sample. In this respect, a recent study by Xu *et al.* (2009) reported the intermolecular hydrogen-bonded interaction between PVA polymer

TABLE 20.2 Scavenger Effect of Grapefruit Samples on Superoxide Radicals

	IL_{50}[a] (μl)		
Superoxide Radical Generation Method	FJ[b]	N-PJ[c]	IN-PJ[d]
No enzymatic (NBT)[e]	27 ± 0,5†	19.1 ± 2,8*	21 ± 6,3*†
Enzymatic (Xanthyne oxidase)	120 ± 18*	146 ± 47*†	139 ± 55*†

[a]*IL_{50}: juice volume giving 50% inhibition of radical production.*
[b]*FJ, fresh juice.*
[c]*N-PJ, pasteurized and naringinase-treated juice.*
[d]*IN-PJ, pasteurized and immobilized naringinase-treated juice.*
[e]*Nitroblue tetrazolium.*
Each value is presented as the mean ± SD of 10 independent experiments. Different symbols for each row correspond to statistically different values (p < 0.005).

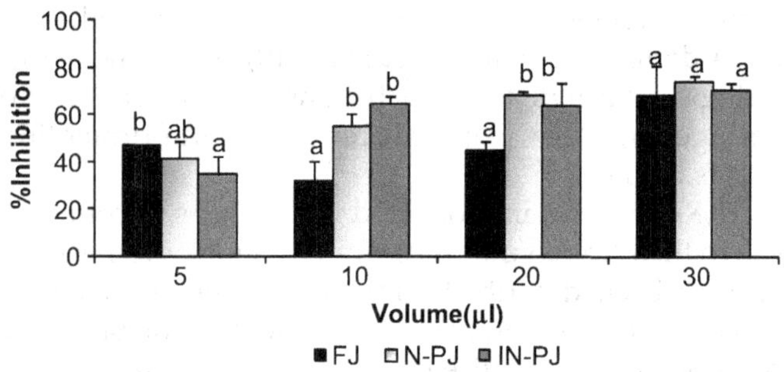

FIGURE 20.3 Scavenger effect of grapefruit samples on hydroxyl radical. %Inhibition, percentage of inhibition in the production of radical, by juice samples at different volumes. FJ, fresh juice; N-PJ, pasteurized and naringinase-treated juice; IN-PJ, pasteurized and immobilized naringinase-treated juice. Results are the mean ± SD of 10 experiments. Different letters for each column correspond to statistically different values (p < 0.005).

and plant polyphenolic compounds. Conversely, vitamin C is also one of the components of the citrus juices with antioxidant activity that most contributes to reducing capacity (Abeysinghe *et al.*, 2007) and it is also a typically heat-sensitive compound (Yeom *et al.*, 2000), so it can produce losses during pasteurization of debittering samples.

Effects of Enzymatic Debittering on Superoxide and Hydroxyl Radical Scavenger Activity of Grapefruit Juice

Reactive oxygen species (ROS), e.g., hydroxyl and superoxide radicals, represent the most important class of radical species generated in living systems. The unpaired electron(s) usually gives a considerable degree of reactivity to the free radicals, making them highly toxic (Valko *et al.*, 2004). Therefore, the study of the radical scavenger capacity of different juice samples may be of interest when testing their antioxidant status.

Radicals can be generated either enzymatically or non-enzymatically (e.g., by a transition metal), or in combination with a reducing agent such as ascorbate. The authors' group evaluated the superoxide scavenging capacity measuring the ability of the samples to scavenge the free radicals produced by an enzymatic (xanthine oxidase/cytochrome c system) (McCord and Fridovich, 1969) and a non-enzymatic [nitroblue tetrazolium (NBT) method] (Liu *et al.*, 1997) assay. In these assays, the components of the juices can influence not only the propagation reactions of the free-radicals but also their formation, either by inhibiting the enzymes involved in the initiation reaction or by chelating the transition metals (Yoo *et al.*, 2009).

According to the IL_{50} values (volumes giving 50% inhibition of radical production) reported in Table 20.2, the juice samples debittered with free o immobilized naringinase showed higher scavenger efficiency (NBT method) when comparing with the fresh juice sample. Previous researchers reported that the presence of both 7- and 5-hydroxyl groups in the aglycone (naringenin) makes it an excellent candidate for the superoxide scavenging reaction compared to the bitter compound naringin (Yu *et al.*, 2005). Cavia-Saiz *et al.* (2010) also provided evidence that the o-glycosylation in naringin decreased the efficiency in inhibiting superoxide radical generation by the xanthine oxidase system. No significant differences were observed between juice samples when using xanthine oxidase/cytochrome c system (Table 20.2).

Previous studies have shown that grapefruit processed with free naringinase exhibits the highest superoxide scavenging activity in comparison to juice treated with an exchange resin (Cavia-Saiz *et al.*, 2011).

The capacity of the citrus juices to inhibit the hydroxyl radical is largely attributed to the action of flavonoids, by their high scavenger capacity against these radicals, their ability to chelate transition metals, and by the synergistic action with other antioxidants (Marti *et al.*, 2009). Our group has observed that the enzyme technology provided great effectiveness in the juices for neutralizing the hydroxyl radical in a dose-dependent manner. In fact, the volume of grapefruit juice samples debittered with free or immobilized naringinase required to reach 50% inhibition of radical production was 2.8- and 1.8-fold lower, respectively, as compared to the fresh juice sample at a volume of 10 μl (Figure 20.3). However, no significant differences were observed between fresh and debittered juice at the highest volume tested.

Effects of Enzymatic Debittering on Protective Role Against Oxidative Stress of Grapefruit Juice

The capacity of debittered grapefruit juice to protect DNA from oxidation has also been assessed by the authors' group. The method used consisted of

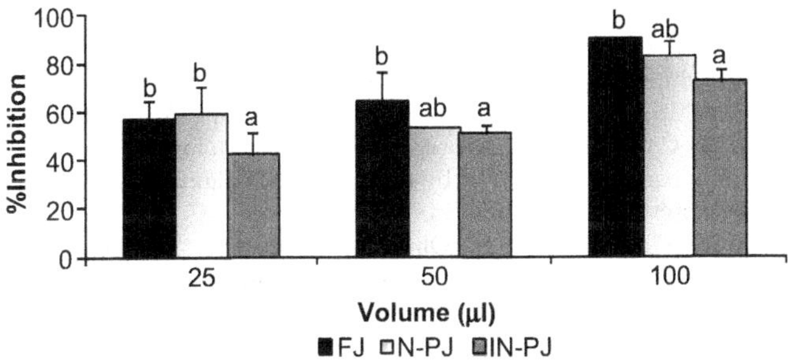

FIGURE 20.4 Effect of grapefruit juice samples on the formation of 8-hydroxy-deoxyguanosine (8-OH-dG)/10^5dG in calf thymus DNA induced by ascorbic acid and copper (II). The method used consists of quantifying the capacity of juice samples to inhibit the level of modified base (8-OH-dG) formed during attack of the •OH on the DNA. FJ, fresh juice; N-PJ, pasteurized and naringinase-treated juice; IN-PJ pasteurized and immobilized naringinase-treated juice. Results are the mean ± SD of six experiments. Different letters for each volume correspond to statistically different values ($p < 0.005$).

quantifying the capacity to inhibit the level of modified base 8-hydroxy-deoxyguanosine (8-OH-dG) formed during attack of the •OH on the DNA (Muñiz *et al.*, 1995). Our findings indicated (Figure 20.4) that fresh juice showed the highest inhibition percentage. Nevertheless, samples of juice debittered with naringinase immobilized in PVA also inhibited oxidative DNA damage induced by copper(II)-ascorbic acid in a dose-dependent manner. No significant differences were observed between using free or immobilized enzyme.

References

Abeysinghe, D.C., Li, X., Sun, C.D., Zhang, W.S., Zhou, C.H., Chen, K.S., 2007. Bioactive compounds and antioxidant capacities in different edible tissues of citrus fruit of four species. Food Chem. 104, 1338–1344.

Benzie, I.F.F., Strain, J.J., 1999. Ferric reducing/antioxidant power assay: Direct measure of total antioxidant activity of biological fluids and modified version for simultaneous measurement of total antioxidant power and ascorbic acid concentration. Methods Enzymol. 299, 15–27.

Breksa III, A.P., Manner, K., 2006. Evaluation of antioxidant capacity of limonin, nomilin and limonin glucoside. J. Agric. Food Chem. 54, 3827–3831.

Busto, M.D., Ortega, N., 2008. Immobilized enzymes in fruit juice processing and wine-making. In: Busto, M.D., Ortega, N. (Eds.), Food Enzymes Applications of New Technologies. Transworld Research Network, Trivandrum, pp. 59–90. Transworld Research Network, Trivandrum.

Busto, M.D., Meza, V., Ortega, N., Perez-Mateos, M., 2007. Immobilization of naringinase from *Aspergillus niger* CECT 2088 in poly(vinyl alcohol) cryogels for the debittering of juices. Food Chem. 104, 1177–1182.

Cánovas, M., García-Cases, L., Iborra, J.L., 1998. Limonin consumption at acidic pH values and absence of aeration by *Rhodococcus fascians* cells in bath and immobilized continuous systems. Enzyme Microb. Technol. 22, 111–116.

Cavia-Saiz, M., Busto, M.D., Pilar-Izquierdo, M.C., Ortega, N., Perez-Mateos, M., Muñiz, P., 2010. Antioxidant properties, radical scavenging activity and biomolecule protection capacity of flavonoid naringenin and its glycoside naringin: a comparative study. J. Sci. Food Agric. 90, 1238–1244.

Cavia-Saiz, M., Muñiz, P., Ortega, N., Busto, M.D., 2011. Effect of enzymatic debittering on antioxidant capacity and protective role against oxidative stress of grapefruit juice in comparison with adsorption on exchange resin. Food Chem. 125, 158–163.

Daniel, J.J., Owens, D.K., McInctosh, C.A., 2011. Secondary product glucosyltransferase and putative glucosyltransferase expression during *Citrus paradise (c.v. Duncan)* growth and development. Phytochemistry 72, 1732–1738.

Del Nobile, M.A.D., Piergiovanni, L., Buonocore, G.G., Fava, P., Puglisi, M.L., Nicolais, L., 2003. Naringinase immobilization in polymeric films intended for food packaging application. J. Food Sci. 68, 2046–2049.

Gray, G.M., Olson, A.C., 1981. Hydrolysis of high levels of naringin of grapefruit juice using a hollow fibre naringinase reactor. J. Agric. Food Chem. 29, 1298–1301.

Hasegawa, S., Herman, Z., Orme, E., Ou, P., 1985a. Biosynthesis of limonoids in citrus: Sites and translocation. Phytochemistry 25, 2783–2785.

Hasegawa, S., Vandercook, C.E., Choi, G.Y., Herman, Z., Ou, P., 1985b. Limonoid debittering of citrus juice sera by immobilized cells of *Corynebacterium fascians*. J. Food Sci. 50, 330–332.

Hasegawa, S., Maier, V.P., King, A.D., 1974. Isolation of new limonoate dehydrogenase from *Pseudomonas*. J. Agric. Food Chem. 22, 523–526.

Hasegawa, S., Nennett, R.D., Maier, V.P., King, A.D., 1972. Limonoate dehydrogenase from *Arthrobacter globiformis*. J. Agric. Food Chem. 20, 1031–1034.

Hasegawa, S., Pelton, V.A., Bennett, R.D., 1983. Metabolism of limonoids by *Arthrobacter globiformis* II: basis for a practical means of reducing the limonin content of orange juice by immobilized cells. J. Agric. Food Chem. 31, 1002–1004.

Hasegawa, S., Suhayda, C.G., Hsu, W., Robertson, G.H., 1997. Purification of limonoid glucosyltransferase from navel orange albedo tissues. Phytochemistry 46, 33–37.

Heim, K.E., Tagliafero, A.R., Bobyl, D.J., 2002. Flavonoid antioxidants: chemistry, metabolism and structure–activity relationships. J. Nutr. Biochem. 13, 572–584.

Humanes, L., López-Ruiz, A., Merino, M.T., Roldán, J.M., Diéz, J., 1997. Purification and characterization of limonoate dehydrogenase from *Rhodococcus fascians*. Appl. Environ. Microbiol. 63, 3385–3389.

Jayachitra, J., Nalini, N., 2011. Effect of naringinin (citrus flavanone) on lipid in ethanol-induced toxicity in rats. J. Food Biochem. 36, 502–511.

Karim, M.R., Hashinaga, F., 2002. Preparation and properties of immobilized pummel limonoid glucosyltransferase. Process Biochem. 38, 809–814.

Kranz, P., Adler, P., Kunz, B., 2011. Sorption of citrus flavour compounds on XAD-7HP resin during the debittering of grapefruit juice. Int. J. Food Sci. Technol. 46, 30–36.

Kumar, M.S., Unnikrishnan, M.K., Patra, S., Murthy, K., Srinivasan, K.K., 2003. Naringin and naringenin inhibit nitrite-induced methemoglobin formation. Pharmazie 58, 564–566.

Kurowska, E., Spence, J.D., Jordan, J., Wetmore, S., Freeman, D.J., Piche, L.A., Serratore, P., 2000. HDL cholesterol raising effect of orange juice in subjects with hypercholesterolemia. Am. J. Clin. Nutr. 72, 1095–1100.

Lee, H.S., Kim, J.G., 2003. Effects of debittering on red grapefruit juice concentrate. Food Chem. 82, 177–180.

Lei, S., Xu, Y., Fan, G., Xiao, M., Pan, S., 2011. Immobilization of naringinase on mesoporousmolecular sieve MCM-41 and its application to debittering of white grapefruit. Appl. Surf. Sci. 257, 4096–4099.

Liu, F., Ooi, V.E.C., Chang, S.T., 1997. Free radical scavenging activities of mushroom polysaccharide extracts. Life Sci. 60, 763–771.

Lopez-Fouz, M., Pilar-Izquierdo, M.C., Martinez-Mayo, I., Ortega, N., Perez-Mateos, M., Busto, M.D., 2007. Immobilization of *Rhodococcus fascians* cells in poly(vinyl alcohol) cryogels for the debittering of citrus juices. J. Biotechnol. 131 S104–S104.

Manjón, A., Bastida, J., Romero, C., Jimeno, A., Iborra, J.L., 1985. Immobilization of naringinase on glycophase-coated porous glass. Biotecnol. Lett. 7, 477–482.

Manjón, A., Iborra, J.L., Martínez-Madrid, C., 1991. pH control of limonin debittering with entrapped *Rhodococcus fascians* cells. Appl. Microbiol. Biotechnol. 35, 176–179.

Manners, G.D., 2007. Citrus limonoids: analysis, bioactivity, and biomedical prospects. J. Agric. Food Chem. 55, 8285–8294.

Marti, N., Mena, P., Canovas, J.A., Micol, V., Saura, D., 2009. Vitamin C and the role of citrus juices as functional food. Natural Product Commun. 4, 677–700.

Martínez-Madrid, C., Manjón, A., Iborra, J.L., 1989. Degradation of limonin by entrapped *Rhodococcus fascians* cells. Biotechnol. Lett. 2, 653–658.

McCord, J.M., Fridovich, I., 1969. Superoxide dismutase: an enzymic function for erythrocuprein (hemocuprein). J. Biol. Chem. 244, 6049–6055.

Miller, E.G., Gonzales-Sanders, A.P., Couvillon, A.M., Wright, J.M., Hasegawa, S., Lam, L.K.T., 1994. Inhibition of oral carcinogenesis by green coffee beans and limonoid glucosides. In: Huang, M.T., Osawa, T., Ho, C.-T., Rosen, R.T. (Eds.), Food phytochemicals for cancer prevention I, vol. 546. American Chemical Society, Washinton, DC, pp. 220–229.

Mishra, P., Kar, R., 2003. Treatment of grapefruit juice for bitterness removal by Amberlite IR 120 and Amberlite IR 400 and alginate entrapped naringinase enzyme. J. Food Sci. 68, 1229–1233.

Muñiz, P., Valls, V., Perez-Broseta, C., Iradi, A., Climent, J.V., Oliva, M., Sáez, G.T., 1995. The role of 8-hydroxy-2′deoxyguanosine in Rifamycin-induced DNA damage. Free Radic. Biol. Med. 18, 747–755.

Ni, H., Chen, F., Cai, H., Xiao, A., You, Q., Lu, Y., 2012. Characterization and preparation of *Aspergillus niger* naringinase for debittering citrus juice. J. Food Sci. 71, 1–7.

Nunes, M.A.P., Vila-Real, V., Fernades, P.C.B., Ribeiro, M.H.L., 2010. Immobilization of naringinase in PVA–alginate matrix using an innovative technique. Appl. Biochem. Biotechnol. 160, 212–2147.

Pedro, H.A.L., Alfaia, A.J., Marques, J., Vila-Real, H.J., Calado, A., Ribeiro, M.H.L., 2007. Design of an immobilized enzyme system for naringin hydrolysis at high-pressure. Enzyme Microb. Technol. 40, 442–446.

Pellegrini, N., Proteggente, A., Pannala, A.S., Rice-Evans, C.A., 1999. Antioxidant activity applying an improved ABTS radical cation decolorization assay. Free Radic. Biol. Med. 26, 1231–1237.

Prakash, S., Singhal, R.S., Kulkarmi, P.R., 2002. Enzymic debittering of Indian grapefruit (*Citrus paradise*) juice. J. Sci. Food Agric. 82, 394–397.

Premi, R.B., Joshi, V.K., Lal, B.B., 1998. Physico-chemical and sensory qualities of kinnow juice debittered by immobilized cells of *Arthrobacter globiformis II*. J. Sci. Ind. Res. 57, 698–702.

Puri, M., 2012. Updates on naringinase: structural and biotechnological aspects. Appl. Microbiol. Biotechnol. 93, 49–60.

Puri, M., Kaur, A., Signh, R.S., Schwarz, W.H., Kaur, A., 2010. One-step purification and immobilization of His-tagged rhamnosidase for naringin hydrolysis. Process Biochem. 45, 451–456.

Puri, M., Kaur, A., Singh, R.S., Kanwar, J.R., 2008. Immobilized enzyme technology for debittering citrus juices. In: Busto, M.D., Ortega, N. (Eds.), Food Enzymes Applications of New Technologies. Transworld Research Network, Trivandrum, pp. 91–103.

Puri, M., Kaur, H., Kennedy, J.F., 2005. Covalent immobilization of naringinase for the transformation of a flavonoid. J. Chem. Technol. Biotechnol. 80, 1160–1165.

Puri, M., Lakhwinder, K., Marwaha, S.S., 2002. Partial purification and characterization of limonoate dehydrogenase from *Rhodococcus fascians* for the degradation of limonin. J. Microbiol. Biotechnol. 12, 669–673.

Puri, M., Marwaha, S.S., Kothari, R.M., 1996. Studies on the applicability of alginate-entrapped naringinase for the debittering of kinnow juice. Enzyme Microb. Technol. 18, 281–285.

Puri, M., Seth, M., Marwaha, S.S., Kothari, R.M., 2001. Debittering of Kinnow mandarin juice by covalently bound naringinase on hen egg white. Food Biotechnol. 15, 13–23.

Ribeiro, M.H.L., Afonso, C., Vila-Real, H.J., Alfaia, A., Ferreira, L., 2010a. Contribution of response surface methodology to the modeling of naringin hydrolysis by naringinase Ca-alginate beads under high pressure. LWT Food Sci. Technol. 43, 482–487.

Ribeiro, D.D., Henrique, S.M., Oliveira, L.S., Macedo, G.A., Fleuri, L.F., 2010b. Enzymes in juice processing: a review. Int. J. Sci. Technol. 45, 635–641.

Ribeiro, I.A., Rocha, J., Sepodes, B., Mota-Filipe, H., Ribeiro, M.H., 2008. Effect of naringin enzymatic hydrolysis towards naringenin on the anti-inflammatory activity of both compounds. J. Mol. Catal. B Enzym. 52–53, 13–18.

Ribeiro, M.H., 2011. Naringinases: occurrence, characteristics, and applications. Appl. Microb. Biotechnol. 90, 1883–1895.

Ribeiro, M.H.L., Silveira, D., Ferreira-Diaz, S., 2002. Selective adsorption of limonin and naringin from orange juice to natural and synthetic adsorbents. Eur. Food Res. Technol. 215, 462–471.

Sekeroglu, G., Fadiloglu, S., Gogus, F., 2006. Immobilization and characterization of naringinase for the hydrolysis of naringin. Eur. Food Res. Technol. 224, 55–60.

Singh, S.V., Gupta, A.K., Jain, R.K., 2008. Adsorption of naringin on nonionic (netural) macroporus adsorbent resin from its aqueous solutions. J. Food Eng. 86, 259–271.

Soares, N.F.F., Hotchkiss, J.H., 1998. Naringinase immobilization in packaging films for reducing naringinase concentration in grapefruit juice. J. Food Sci. 63, 61–65.

Soria, F., Ellenrieder, G., Bezerra, G., Cabrera, M., Bezerra, L., 2012. α-L-Rhamonosidase of *Aspergillus terreus* immobilized on ferromagnetic supports. Appl. Microbiol. Biotechnol. 93, 1127–1134.

Tsen, H.-Y., 1984. Factors affecting the inactivation of naringinase immobilized on chitin during debittering of fruit juice. J. Ferment. Technol. 62, 263–267.

Tsen, H.Y., Tsai, S.Y., Gee-Kaite, Y., 1989. Fiber entrapment of naringinase from *Penicillium* species and application to fruit juice debittering. J. Ferment. Bioeng. 67, 186–189.

Valko, M., Izakovic, M., Mazur, M., Rhodes, C.J., Telser, J., 2004. Role of oxygen radicals in DNA damage and cancer incidence. Mol. Cell. Biochem. 266, 37–56.

Xu, G., Liu, D., Chen, J., Ye, X., Ma, Y., Shi, J., 2008. Juice components and antioxidant capacity of citrus varieties cultivated in China. Food Chem. 106, 545–551.

Xu, X.J., Huang, S.M., Zhang, L.H., 2009. Biodegradability, antibacterial properties, and ultraviolet protection of polyvinyl alcohol–natural polyphenol blends. Polym. Compos. 30, 1611–1617.

Yadav, V., Pramod, P.K., Yadav, S., Yadav, K.D.S., 2010. α-L-Rhamnosidase: A review. Process Biochem. 45, 1226–1235.

Yeom, H.W., Streaker, C.B., Zhang, Q.H., Min, D.B., 2000. Effects of pulsed electric field on the quality of orange juice and comparison with heat pasteurization. J. Agric. Food Chem. 48, 4597–4605.

Yoo, K.M., Lee, K.W., Park, J.B., Lee, H.J., Hwang, I.K., 2009. Variation in major antioxidants and total antioxidant activity of Yuzu (Citrus junos Sieb ex Tanaka) during maturation and between cultivars. J. Agric. Food Chem. 52, 5907–5913.

Yu, J., Wang, L.M., Walzem, R.L., Miller, E.G., Pike, L.M., Patil, B.S., 2005. Antioxidant activity of citrus limonoids, flavonoids, and coumarins. J. Agric. Food Chem. 53, 2009–2014.

Yusof, S., Gazali, H.M., King, G.S., 1990. Naringin content in local citrus fruit. Food Chem. 37, 113–121.

Zaare-Nahandi, F., Hosseinkhani, S., Zamani, Z., Asadi-Abkenar, A., Omidbaigi, R., 2008. Delay expression of limonoid UPD-glucosyltransferase makes delayed bitterness in citrus. Biochem. Biophys. Res. Comm. 371, 59–62.

CHAPTER

21

Production Processes of Orange Juice and Effects on Antioxidant Components

Gianni Galaverna, Chiara Dall'Asta

Department of Food Science, University of Parma, Parma, Italy

CHAPTER POINTS

- Orange juice (OJ) antioxidant components are affected by thermal stabilizing treatments.
- Thermal treatments negatively affect carotenoids, anthocyanins, and color.
- Thermal treatments decrease vitamin C in OJ.
- High-pressure processing increases carotenoid and flavanone content, while preserving other components.
- Pulsed electric fields only slightly affect vitamin C, while preserving other components.
- Thermally accelerated short-time evaporation is very effective in maintaining antioxidant components.
- Innovative membrane-based concentration processes are promising techniques for obtaining high-quality concentrates.

INTRODUCTION

Orange juice (OJ) (*Citrus sinensis*) is the most important fruit juice consumed worldwide: its market share is continuously increasing and orange production has reached 68.4 million tons worldwilde (FAO Statistics, 2010), with Brazil as the top producer. In western Europe, OJ accounts for 60% of all consumed fruit juice and juice-based drinks (Yeom *et al.*, 2000): both blond and blood varieties are fairly common, with the latter experiencing an ever increasing popularity in recent years.

This widespread success is due to OJ's particularly appreciated organoleptic properties, but consumers' interest is also based on a general trend towards healthier eating and lifestyle in which fruit and vegetables play a major role, as they are natural rich sources of vitamins, dietary fiber, and health-promoting bioactive compounds. Indeed, from the nutritional viewpoint, a serving of OJ (240 ml) would cover the reference daily intake (RDI) amount of vitamin C. Moreover, OJ is very rich in natural antioxidants (Figure 21.1), mainly phenolic compounds, being an important source of flavanones, especially hesperidin and naringenin, as well as hydroxycinnamic acids (particularly in the blood varieties). Orange juice is also very rich in carotenoids, being an important dietary source of vitamin A carotenoids (β-carotene, α-carotene and β-cryptoxanthin) and antioxidant carotenoids (β-carotene, β-cryptoxanthin, zeaxanthin, and lutein). In the blood varieties, the red color is due to the presence of anthocyanins. The relative amounts of these different components are reported in Table 21.1.

OJ's excellent organoleptic properties are due to a very rich and fresh flavor, on account of the sweet and acid taste and of the rich volatile profile mainly characterized by terpenes and aldehydes (Sandhu *et al.*, 2006).

Preserving these fundamental properties is one of the main goals of production technology; indeed, as most parts of the consumed orange juice are produced by industrial integrated processes, researchers are continuously seeking innovative technological solutions to extend shelf-life without affecting quality.

In this quest, the main problems to be faced during production and storage are the inhibition of microbiological spoilage and enzyme inactivation. In particular, enzymes such as pectin methyl esterases (PME) are responsible for gelling phenomena during storage; indeed, inducing the hydrolysis of pectins, they

Processing and Impact on Antioxidants in Beverages
http://dx.doi.org/10.1016/B978-0-12-404738-9.00021-0

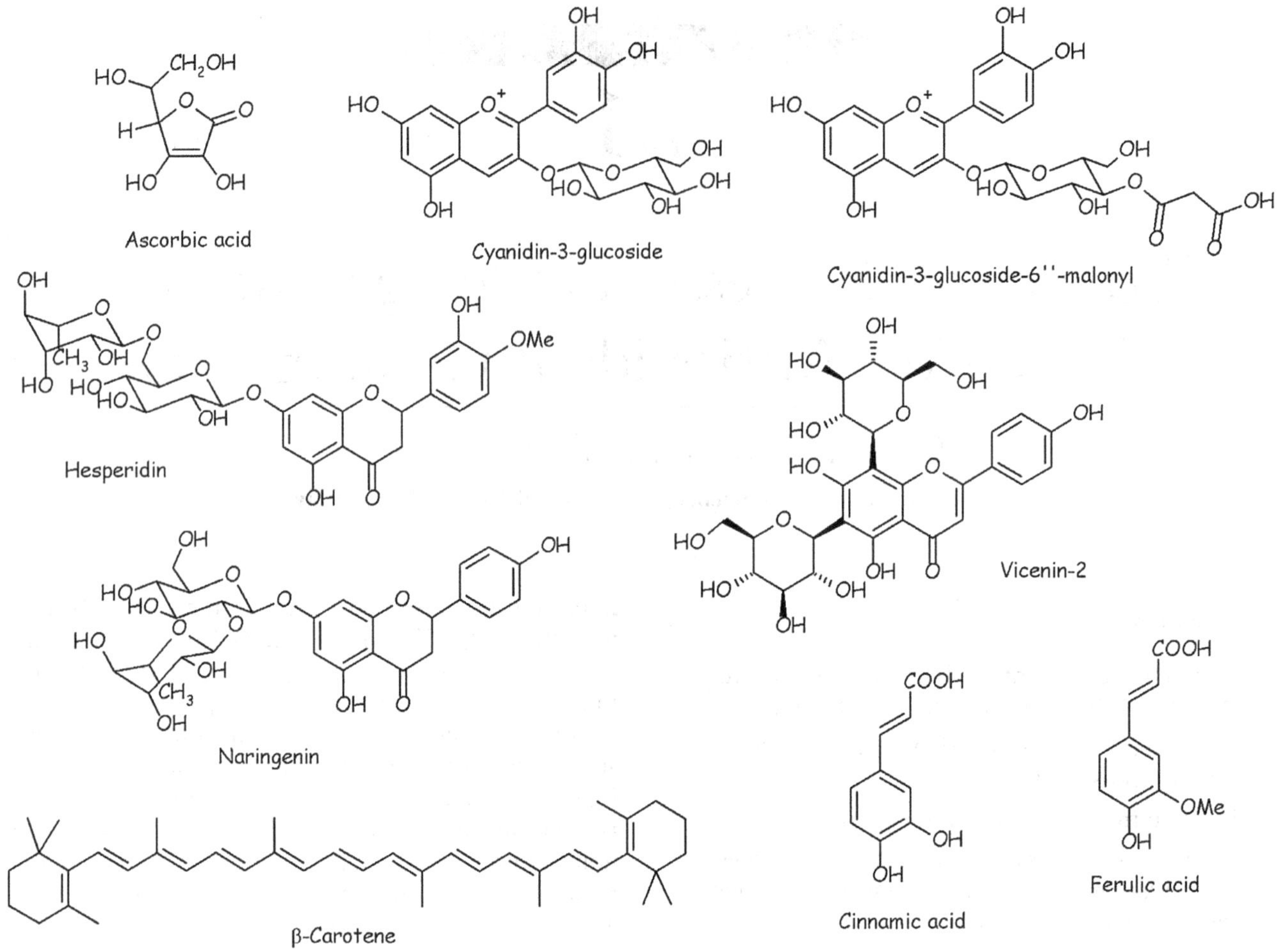

FIGURE 21.1 Antioxidant components of orange juice.

TABLE 21.1 Relative Amounts of the Different Orange Juice's Bioactive Antioxidants (mg/l)

Component	Blond Orange Juice	Blood Orange Juice
Ascorbic acid	150–350	450–650
Hydroxycinnamates	60–70	70–80
Flavanones	55–150	80–110
Carotenoids	4–9	3–8
Anthocyanins	—	50–100

destabilize the cloud fraction and induce juice clarification (Wicker *et al.*, 2003). Moreover, polyphenoloxidases (PPO) and peroxidases (POD) induce oxidation reaction of ascorbic acids, polyphenols, and carotenoids, thus being responsible for browning and color changes.

The most effective and common way to inhibit these phenomena are thermal stabilization treatments, although several undesired degradation reactions involving both the organoleptic and nutritional aspects are often encountered.

THE ORANGE JUICE PRODUCTION PROCESS

The orange juice production process is characterized by several consecutive steps (Figure 21.2) (Sandhu *et al.*, 2006).

First of all, fruits are washed and sorted by size in order to obtain proper feeding to the juicing machines (extractors). Two types of extraction methods are mostly used by industry. In the first one, the fruit is placed between two metal cups equipped with sharpened metal tubes. When the fruit is pressed between the two cups, the juice is pushed out from the tubes penetrating the fruit and recovered. The peel pressed between the cups is washed by a water spray in order to recover the oil, as this is a valuable raw material for the aroma and essential oil industry. In the second type of extraction, oranges are cut in half and the juice is extracted by moving the two halves against rotating reamers. Finishers, hydrocyclones, and separators are then used to remove the excessive pulp and seeds which, along with the peel, are to be used as by-products. After filtering through a stainless

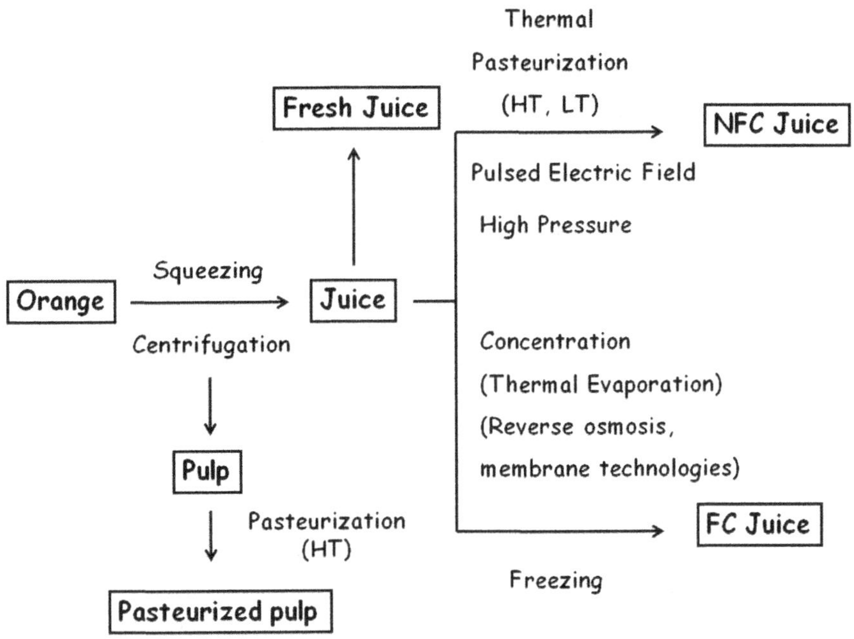

FIGURE 21.2 General scheme of production process of orange juice.

steel screen, the juice can be chilled or concentrated, leading to the production of two type of products: the not from concentrated (NFC) and the from concentrated (FC) juices. NFC is generally de-oiled to 0.02–0.04% oil levels with a centrifuge, then pasteurized in order to destroy microorganisms and to deactivate enzymes. NFC orange juice may be stored frozen or chilled. Pasteurization methods are similar to those applied to other liquid foods such as milk and are performed using plate heat exchangers or steam heating (typically 94–85°C for about 15–60 s (Nagy *et al.*, 1993). High temperature pasteurization allows us to obtain stable products at room temperature (20°C). Alternative stabilization methods are currently under study, such as the application of high-pressure treatment (HPP) and ohmic heating with pulsed electric fields (PEF), with the aim of limiting thermal damage to the juice.

As an alternative, juice may be concentrated for easier preservation, storage, and transport: reconstitution to the original concentration is generally made before marketing the product (reconstituted from concentrated, RFC). Concentrated juice is enzymatically and microbially stable: this product can be frozen or used to prepare ready-to-drink beverages by proper dilution with water. Concentration is usually performed to obtain a 60–65°Brix final concentration, which is a 7-to-1 strength ratio to normal single-strength juice (the original concentration is about 11.5–12°Brix). Concentrated juice is mostly obtained using Thermally Accelerated Short-Time Evaporators (TASTE): The juice is heated under vacuum with hot steam in order to evaporate water without causing excessive thermal damage both from the nutritional as well as from the organoleptic viewpoint. Alternative innovative methods for juice concentration are currently under study such as those based on application of membrane technology. As an example, reverse osmosis (after pulp separation by ultrafiltration) allows us to obtain intermediate concentration of about 30–35°Brix without heat application. Innovative membrane operations may allow us to obtain highly concentrated juice, mainly based on selective water removal without heat application (membrane and osmotic distillation). In any case, juice concentrates are stored in refrigerated stainless steel bulk tanks until they are ready to be packaged or reconstituted. Once produced, aseptic packaging may afford a shelf-life of 6–8 months at room temperature. Reconstitution is performed by adding water and flavor oils and essences to obtain a final concentration of 11.8°Brix, then pasteurized and packaged.

EFFECTS OF PRODUCTION AND PROCESSING ON ANTIOXIDANT COMPONENTS OF ORANGE JUICE

As far as the bioactive antioxidant components of orange juice are concerned, several investigations have been performed to evaluate their fate during production and storage of the products. The most recent studies are summarized in Table 21.2.

Mostly, investigations have regarded the fate of vitamin C (L-ascorbic acid and L-dehydroascorbic acid), individual phenolic components (flavonoids, anthocyanins, and hydroxycinnamic acids), carotenoids, and total antioxidant capacity. Most of the bioactive components

TABLE 21.2 Recent Studies on the Effect of Processing on Antioxidant Components

Treatment	Processing Conditions	Storage Conditions/ Period	Nutrient/Component	Reference
Concentration	TASTE	2°C–20°C/60 days	Ascorbic acid, flavanones, hydroxycinnamic acids, TAA	Arena *et al.* (2001)
HPP	30–350 Mpa, 30°C and 60°C	4°C/15–30 days	Carotenoids, antiradical activity	de Ancos *et al.* (2002)
Squeezing Mild pasteurization Standard pasteurization Concentration Freezing	Domestic/industrial 75°C/30 s 95°C/30 s TASTE –40°C	— — — — 1 month	Flavanones, flavones, hydroxycinnamates, free radical scavenging capacity	Gil-Izquierdo *et al.* (2002)
HPP	100–400 Mpa, 30°C–60°C, 1–5 min	1–10 days, 4°C	Flavanones, radical scavenging activity	Sanchez-Moreno *et al.* (2003)
HPP *versus* thermal pasteurization	500 Mpa, 5 min *versus* 80°C, 20 s	1–2 months, 0, 5, 10, 15°C	Total antioxidant activity, ascorbic acid	Polydera *et al.* (2004)
Thermal pasteurization	80°C, 1 min	—	Anthocyanins, hydroxycinnamates, ascorbic acid, radical scavenging activity	Lo Scalzo *et al.* (2004)
HP PEF Mild pasteurization Standard pasteurization Standard pasteurization plus freezing Freezing	400 Mpa, 40°C, 1 min 35 kVcm^{-1}/750 μs 70°C, 30 s 90°C/1 min –38°C/15 min	—	Ascorbic acid, flavanones, carotenoids and radical scavenging capacity	Sánchez-Moreno *et al.* (2005)
HPP *versus* thermal pasteurization	600 Mpa, 40°C, 1 min	1–3 months, 0, 5, 10, 15, 30°C	Antioxidant activity, Ascorbic acid	Polydera *et al.* (2005)
PEF *versus* thermal pasteurization	35 kV/cm for 1000 μs; bipolar 4-μs pulses at 200 Hz *versus* 90°C, 1 min	0–2 months, 4 and 22°C	Ascorbic acid, Radical scavenging activity	Elez-Martinez *et al.* (2006)
PEF *versus* thermal pasteurization	25, 30, 35, 40 kV/cm for 30–340 μs *versus* 90°C, 20 s	—	Carotenoids	Cortes *et al.* (2006)
HP PEF Mild pasteurization	400 MPa, 40°C, 1 min 35 kv cm^{-1}, 750 μs 70°C, 30 s	40 days, 4°C	Ascorbic acid, Radical scavenging activity	Plaza *et al.* (2006)
Clarification by UF		—	Antioxidant activity, ascorbic acid, anthocyanins, flavanones	Cassano *et al.* (2007)
Pasteurization	85°C, 10 s	1–3 months, 4 and 20°C	Flavanones, Flavones	Sentandreu *et al.* (2007)
PEF *versus* thermal pasteurization	15, 25 and 35 kV/cm, 100, 400 and 1000 μs, 4 μs pulses at 200 Hz *versus* 90°C, 1 min	—	Ascorbic acid, Radical scavenging activity	Elez-Martinez *et al.* (2007)
Membrane concentration *versus* thermal concentration	Reverse osmosis, membrane and osmotic distillation *versus* TASTE	—	Ascorbic acid, anthocyanins, flavanones, hydroxycinnamates, total antioxidant activity	Galaverna *et al.* (2008)
HPP PEF Thermal pasteurization	400 MPa/40°C/1 min 35 kV cm^{-1}/750 μs 70°C/30 s	40 days, 4°C	Carotenoids, flavanones	Plaza *et al.* (2011)

TASTE, Thermally Accelerated Short-Time Evaporators; TAA, total antioxidant activity; HPP, high-pressure processing; HP, high pressure; PEF, pulsed electric fields; UF, ultrafiltration.

were determined by chromatographic methods (HPLC), whereas antioxidant activity was mainly determined by spectrophotometric measurement of the radical scavenging capacity against the well known chromophoric radicals ABTS [2,2′-azinobis-(3-ethylbenzothiazoline-6-sulfonic acid)] and DPPH (2,2-diphenylpicrylhydrazyl) or with more biologically feasible assay such as the evaluation of the inhibition of linoleic acid peroxidation. In the following paragraphs, a survey of these studies grouped on the basis of the applied technological processes is presented.

Effect of Standard Processing

Among the different technological parameters of standard processing, the most important is certainly the temperature applied during thermal stabilization or concentration. Many studies compare the quality of the industrial products with hand-made squeezed juice, which is the most widely used alternative when consuming OJ at home. Effects of processing have been evaluated per single step and during storage in different conditions, taking into consideration that the shelf-life and the storage conditions are strictly determined by the intensity and the efficacy of the stabilizing treatment, the most important problems being the microbiological spoilage and the enzymatic degradation of organoleptic and nutritional components.

Ascorbic acid is the main antioxidant component of orange juice and it is the most prone to degradation reaction: its degradation is frequently used as a measure of the quality of the juice, also because it is related to other parameters, such as color retention (Roig *et al.*, 1999), although other important antioxidant components have received considerable attention in recent years.

The main processing steps of orange juice production have been studied at industrial scale using oranges from the blond variety 'Navel.' The effects of squeezing, mild pasteurization (75°C for 30s), standard pasteurization (95°C for 30s), concentration (up to 60°Brix), and freezing (−40°C) on the nutritional value of the juice (vitamin C content, polyphenols and antioxidant activity) were studied (Gil-Izquierdo *et al.*, 2002) and the results compared with those obtained for hand-squeezed juice. Generally, the industrial single-strength extractor is more effective in extracting bioactive substances into the juice: indeed, a higher amount of phenolics (>22%) and ascorbic acid (>25%) was obtained. This effect is partially due to higher amounts of phenolics in the cloud fraction of the commercial juice, and also due to a major contribution from the albedo portion, which is more pressed by the industrial squeezing method, and for a probably more efficient extraction of ascorbic acid from the solid parts of the fruits. The considered phenolics are flavanones (mainly narirutin and hesperidin), flavones (mainly vicenin 2) and hydroxycinnamic acids (mainly caffeic acid and p-coumaric acid derivatives): particularly interesting is the evaluation of the partition of the different compounds between the soluble and the cloud fractions, which can be performed after centrifugation. The total content of phenolics in both fractions did not change upon application of mild and standard pasteurization techniques, nor with the concentration process, although a different partition between the two phases was observed. Indeed, partial precipitation of these compounds to the cloud fraction was observed during concentration. The phenolic content of the juice was much more affected by freezing, particularly in the liquid phase (global loss >35%): this effect could also be due to the thawing phase. The most important effects were recorded for pasteurization of the pulp fraction, which induced significant degradation of both phenolic compounds and ascorbic acid. Pasteurization seems to slightly affect ascorbic acid, whereas no variations were observed during concentration and freezing. Accordingly, the total antioxidant capacity of the juice, which is mainly due to the ascorbic acid contribution, was not affected by the considered treatments, whereas antioxidant capacity of pulps was affected and reduced by 47%. Thus, it appears that immediately after treatment, pasteurization does not negatively influence the nutritional and antioxidant content, whereas pulp pasteurization has negative effects on vitamin C and phenolic compounds. Concentration and freezing generally decrease the phenolic content in comparison with the untreated juice.

The effect of thermal treatments was also evaluated in the case of orange juice obtained from the blood variety 'Moro' from Italy, taking into consideration the variation of antioxidant and antiradical activity of the juice obtained from orange segments that had been squeezed directly or after blanching (80°C, 6min), eventually coupled to pasteurization (80°C for 1min) of the obtained juice (Lo Scalzo *et al.*, 2004). Different results were obtained according to the applied antioxidant activity assay; in particular, upon thermal treatments an increase of the inhibition of enzymatically mediated linoleic acid peroxidation was observed. Conversely, hydroxyl radical and DPPH-scavenging activity decreased. The increase of antioxidant capacity upon thermal treatment was related to the thermally induced extraction of antioxidant molecules previously complexed or polymerized and the retention of active principles caused by the inactivation of catabolyc enzymes. It appears evident that thermal treatments induced a decrease in free radical scavenging activity as they are responsible for the degradation of ascorbic acid in blood orange juice, as already observed for blond orange juice. On the basis of these findings, the authors speculated that phenolic compounds (anthocyanins and hydroxycinnamates) were able to protect against the oxidation of lipophilic substances whereas the highly hydrophilic ascorbic acid acts as an hydroxyl radical or DPPH scavenger.

In the case of blood orange juice (Fiore *et al.*, 2005), the study of the antioxidant activity by three different tests [ABTS, DPPH, and ferric reducing antioxidant power (FRAP)] showed that in pasteurized and sterilized juices the antioxidant activity is positively related to the content of anthocyanins and that the reduction of anthocyanin content, typical of commercial sterilized long shelf-life juices, leads to a remarkable loss of antioxidant power. Indeed, thermal degradation is highly significant for anthocyanins, the typical natural pigments of blood orange juice (Cao *et al.*, 2011; Kirca *et al.*, 2003). Thermal degradation followed first-order reaction kinetics, with a high Ea value (55.81 kJ/mol), indicating a general high sensitivity to thermal treatment (Choi *et al.*, 2002).

On account of the degradation observed with harsh treatment, interest has arisen around minimally processed products that can be more closely related to freshly squeezed juice. Thus, mild pasteurization is a valid approach in order to better preserve the 'freshness' of the juice, although with the need for refrigerated storage and with a shorter shelf-life. As an example, the effect of storage time and temperature on the degradation of ascorbic acid was measured for Spanish orange juices lightly pasteurized (77°C, 20 s) and stored at 4°C or 10°C (Esteve *et al.*, 2005). Degradation kinetics of ascorbic acid concentration against time follows a straight line whose slope indicates the degradation rate: a shelf-life period of at least 42 days at 4°C and 35 days at 10°C was calculated.

The study of the variation of another class of antioxidant components, namely flavanone-7-O-glycosides (FGs) and fully methoxylated flavones (FMFs), and of antiradical activities have been performed on Valencia late orange juice which has been industrially squeezed, pasteurized, concentrated, and stored refrigerated (4°C) or at room temperature (20°C) (Sentandreu *et al.*, 2007). In particular, the total antioxidant activity (TAA) of orange juice, measured by the DPPH assay, has been related to the contribution of three different types of compound: those characterized by fast-kinetics (FAA), such as ascorbic acid, those characterized by fast + slow kinetics, such as hydroxycinnamic acids, and those characterized by slow-kinetics (SAA) antiradical activities, such as flavanones and flavones. The variation of all parameters was measured after each processing step and the results were compared with those obtained for fresh hand-squeezed juices. Neither industrial squeezing, nor pasteurization or concentration significantly affected FG and FMF contents and antiradical activities. Storage caused a slight decrease of the FMF content but a significant reduction of both soluble hesperidin content and cumulative fast-kinetics antiradical activities in all juices. These decreases were dependent on storage temperature: At lower temperature a slower decrease was observed. These results are in agreement with general findings reporting the relative stability of this class of antioxidant compounds as well as the decrease of hesperidin due to its low water solubility. The decrease of fast-kinetics antiradical activity is related to the degradation of ascorbic acid during storage.

As a confirmation of these results, the variation of multiple parameters (content of vitamin C, total polyphenols, and individual phenolic compounds, as well as antioxidant activity) was measured in commercial orange juices as a function of storage time and temperature, taking into account fresh juices and juices after storage periods of 2, 4, and 6 months at 18, 28, and 38°C (Klimczak *et al.*, 2007). Also in this case, the most affected compounds were vitamin C and free and conjugated hydroxycinnamic acids as well as antioxidant activity, whereas small changes in flavanone content were observed.

On the basis of these observations, a predictive model has been proposed using the kinetics of thermal degradation in order to design optimal processing conditions for minimizing thermal damage to citrus juice components, particularly vitamin C, carotenoids, and polyphenols (Dhuique-Mayer *et al.*, 2007). Provitamin A carotenoids (β-carotene and β-cryptoxanthin) revealed a higher heat stability compared to xanthophylls: degradation fitted first-order reaction kinetics. Also the degradation of vitamin C follows a first-order reaction and it is strongly influenced by the concentration of dissolved oxygen. Conversely, in agreement with the literature, flavanones such as hesperidin showed very high stability during thermal treatment. The developed model may be used to predict losses of bioactive components during isothermal treatment such as pasteurization.

As a general comment, it seems that storage conditions more than thermal treatment are responsible for the degradation of the bioactive components of orange juice.

Color, carotenoid content, and carotenoid bioaccessibility were assessed comparing industrially processed and hand-squeezed orange juices: pasteurization induces significant changes in terms of color (darker and more reddish *versus* brighter, more yellowish, and colorful) and total carotenoid content. Nevertheless, bioaccessibility of bioactive carotenoids increased upon industrial extraction on account of reduced particle size distribution (Stinco *et al.*, 2011).

The effects of different parameters (organic *versus* traditional agricultural practice, pasteurization, storage) on the content of vitamin C, color, carotenoids, and antioxidant capacity of orange juices was recently performed (Navarro *et al.*, 2011). After pasteurization at 98°C (20 s), a 30% loss of vitamin C content was measured as well as loss of carotenoids (lutein loss of 16% for organic and 8% for traditional, and β-cryptoxanthin loss of 30%) along 145 days storage period at 20–25°C. Pasteurization and storage induced significant color changes, and a 40% loss of antioxidant capacity was found along the storage period.

Beside thermal treatment, the second important type of OJ is the concentrate, produced upon thermal evaporation of water in order to obtain easier preservation and more economically feasible transport and storage. Concentration is usually performed by TASTE evaporator, applying controlled heating under partial vacuum. Arena *et al.* (2001) studied the effect of the thermal concentration process of blond and blood orange juice on antioxidant capacity as well as changes occurring during storage. The concentration process has a negative impact on anthocyanins and increases the content of carotenoid pigment in the serum. These variation could partly explain the higher TAA measured for RFC juices in comparison with NFC ones. Antioxidant activity is relatively unaffected by storage at 2°C (stability up to 60 days), whereas a significant decrease was measured at 20°C. As the major contribution to the total antioxidant activity of the juice is due to ascorbic acid, its degradation is mainly responsible for the observed behavior.

Innovative Stabilizing Treatments: High-Pressure Processing (HPP) and Pulsed Electric Fields (PEF)

High-pressure (HP) treatments are innovative stabilizing treatments applied to fruit juice, which are able to better preserve the organoleptic properties and bioactive compounds, while assuring microbiological stability (Smelt, 1998) and enzyme inactivation (Cano *et al.*, 1997; Devlieghere *et al.*, 2004). Different pressure ranges may be applied up to 800 MPa with the juice inserted in polyethylene plastic bags and immersed in water as pressure driving liquid inside hydrostatic pressure units (Figure 21.3). Pressure is generally applied and released in a few minutes (typical range, 2–10 min). Very often, as spores and some enzymes are highly resistant towards pressure, HPP is applied in combination with thermal treatment (up to 95°C) (Nienaber *et al.*, 2001; Parish, 1998). It is noteworthy that upon pressure application, temperature increase ca. 3°C per every 100 MPa and adjabatic heating occurs; nevertheless, the effect is absolutely temporary and temperature decreases immediately after pressure release (Senorans *et al.*, 2003).

The use of pulsed electric fields is another very active field of research for stabilization of orange juice as a valuable alternative to heat treatments. (Jeyamkondan *et al.*, 1999; Espachs-Barroso *et al.*, 2003).

Typically, high voltage pulses are applied in the range of 20–80 kV in a time frame of microseconds. The stabilizing effect is based on the ability of the electric field to inactivate microbial cells without applying high temperature treatment: the effect is due to the ability of high-voltage pulses to break cell membranes by electroporation phenomena (Figure 21.4).

The effect of high-pressure treatments on orange juice components (in particular carotenoids) as well as on antioxidant activity was tested using different pressure values (50–350 MPa) combined with different temperatures (30 and 60°C) and times (2.5, 5, and 15 min) (de Ancos *et al.*, 2002). High-pressure treatments at 350 MPa induced a significant increase (20–43%) in the carotenoid content of orange juice, although no correlation was found between this increase and the free radical-scavenging activity. The latter result was more related to the contribution of polyphenols and vitamin C. Nevertheless, the increase in the concentration of the carotenoids due to an higher extraction efficiency from the vegetal cells, probably caused by a degradation of carotenoids-proteins or cell wall interactions, are very interesting from the organoleptic and nutritional point of view. The effects were maintained during refrigerated storage up to 30 days.

A similar higher extraction effect was also observed in the case of flavanones. Indeed, HP treatment in the

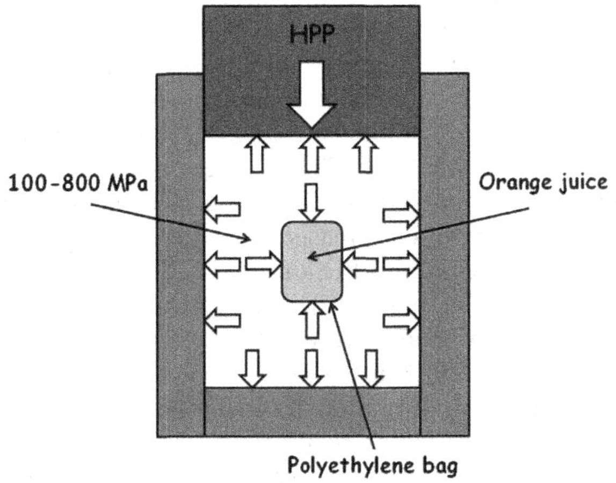

FIGURE 21.3 High-pressure processing (HPP).

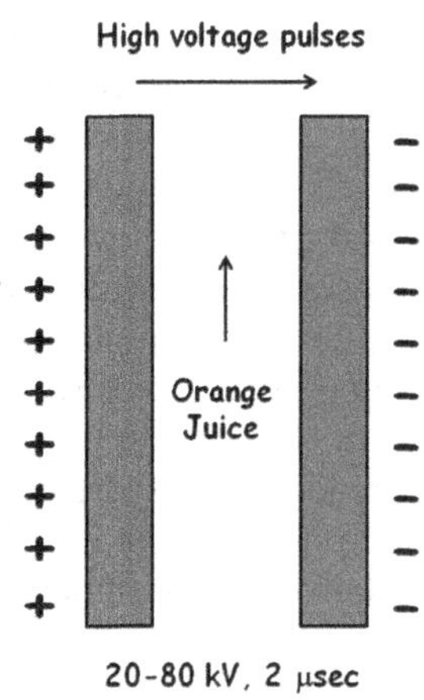

FIGURE 21.4 Pulsed electric field processing.

range of 350–450 MPa at 30–40°C led to an increase in the extraction of flavanones (Sánchez-Moreno *et al.*, 2003). Flavanone content is preserved during chilled storage (10 days at 4°C), in agreement with the observed general stability of these compounds. Antioxidant activity was unchanged upon treatment and during chilled storage. However, the application of higher temperature (60°C) was detrimental for preserving antioxidant activity, probably on account of the thermal degradation of ascorbic acid.

A comparison between thermal treatment (75–120°C) and high-pressure treatment (100–800 MPa; 30–65°C), performed on a kinetic basis, showed that the water-soluble antioxidant capacities increased by a first-order reaction model for the first 60 min and 90 min of thermal treatment, respectively (Indrawati *et al.*, 2004). The observed increase applying thermal treatment may be related to Maillard-type reactions taking place in the juice, although these effects could be relevant only in sterilized long-shelf-life products. Conversely, HPP induced a slight decrease of the antioxidant capacity of orange juice, and this decrease was higher as the pressure was increased. Nevertheless, applying high-pressure treatment with moderate temperature may allow preservation of water-soluble antioxidant capacities. Analogously, no significant differences in antioxidant capacity and vitamin C and carotene content were found in HPP-treated OJ in comparison with untreated juice during a storage period of up to 21 days at 4°C (Fernandez-García *et al.*, 2001).

Again, the variation of antioxidant activity during storage (1–2 months) of reconstituted orange juice that had been pasteurized by heat or high pressure was studied at different temperatures (between 0–15°C) (Polydera *et al.*, 2004). As a general behavior, the total antioxidant activity of orange juice decreased during storage, mainly depending on ascorbic acid degradation, although orange juice treated by high pressure retained higher overall antioxidant activity during storage than the conventionally pasteurized juice. These results are related to the rates of ascorbic acid degradation, which were lower for orange juice treated with high pressure when compared with conventionally pasteurized juice. As far as the behavior of flavonoids and other polyphenolic compounds is concerned, they show a time-dependent antioxidant reaction, which may be described by the sum of two exponential decay functions of different rates. Thus, variation of total antioxidant activity during storage is the result of different simultaneous or consecutive effects, a fast reaction of ascorbic acid followed by a slower reaction of polyphenols: in this respect, HPP-treated juice shows better performance.

A very deep investigation of the variation of vitamin C, carotenoids, and flavanones and of antioxidant activity was performed on juices treated by HPP (400 MPa/40°C per 1 min), pulsed electric fields (PEF) (35 kV/cm per 750 μs), pasteurization at low (LPT) (70°C/30 s) and high temperature (HPT) (90°C/1 min), eventually followed by freezing (−38°C/15 min) (Sanchez-Moreno *et al.*, 2005). As already observed in other studies, loss of vitamin C was generally higher at higher temperature. However, losses were generally lower than 9%: HP did not exert any change while a 6% reduction was measured in the case of PEF. Moreover, the property of HP treatment of inducing a higher release of carotenoids, vitamin A, and also flavanones in the serum of the juice was confirmed, whereas the other treatments showed no effects or a slight decrease for some individual flavanones (pasteurization and freezing). The radical scavenging activity was maintained by HP and PEF treatments, whereas a decrease was observed for traditional thermal technologies, mainly due to vitamin C degradation.

The better performance of HPP compared to thermal pasteurization in retaining the total antioxidant activity of the juice is also confirmed by a mathematical description of the reaction rate constant of nth order kinetics of the decolorization of ABTS radical cation solution as a function of storage temperature and time (Polydera *et al.*, 2005). Variation of antioxidant activity was studied during storage at different temperatures in the range of 0–15°C. The mathematical description takes into account the different antioxidant behavior and reactivity of ascorbic acid and polyphenols, which show fast and slow reaction kinetics. Again, ascorbic acid loss was the main factor governing the variation of TAA.

In another work (Plaza *et al.*, 2006), the extent of vitamin C degradation observed during refrigerated storage (40 days at 4°C) of HP-treated juices was 14% whereas for LTP juices it was 18%, along with a concomitant reduction of antioxidant activity; nevertheless, these variations were slower than for LTP juices. A similar preserving capacity was measured in the case of PEF-treated juices.

The stability of the other main bioactive compounds of orange juice, i.e. carotenoids and flavanones, was also investigated during refrigerated storage after LPT, HP, or PEF treatments (Plaza *et al.*, 2011). HP juice showed a significant increase in total carotenoid and flavanone content extracted (45.19 and 15.46%, respectively) and in vitamin A value (30.89%), whereas no significant changes were observed for PEF and LPT juices. During storage, flavanones decreased (around 50%), probably as a result of residual PPO and POD activity, as well as carotenoids (less than 11%), although in general carotenoids and flavanones were better retained in HP juice than in LPT and PEF juices, confirming the effectiveness of these innovative treatments in preserving the bioactive components of orange juice.

As far as PEF processing is specifically concerned, several studies have investigated the effects of processing on vitamin C in orange juice in comparison with untreated or thermally treated juices.

As an example, concentration of ascorbic acid in PEF-treated orange juice (35 kV/cm, 59 µs) after 47 days at 4°C was significantly higher than in heat-pasteurized orange juice (94.6°C, 30 s) (Yeom *et al.*, 2000). Analogous results were obtained by Hodgins *et al.* (2002).

Considering also the nutritional consequences, an interesting *in vivo* study on the impact of orange juice drinking on 8-epiPGF2 isoprostane concentrations (biomarker of lipid peroxidation) was performed (Sánchez-Moreno *et al.*, 2004). In particular, the effect of freshly squeezed orange juice and pulsed electric field-treated orange juice was compared analyzing plasma levels of vitamin C and 8-epiPGF2, the two parameters being inversely correlated. Both juices were able to induce an increase in plasma vitamin C and a decrease in 8-epiPGF2, thus showing that pulsed electric field-preservation of orange juice retains the vitamin C bioavailability and antioxidant properties of fresh juice with a longer shelf-life.

Comparing traditional heat pasteurization (90°C for 1 min) and high-intensity pulsed electric fields (HIPEF) processing (35 kV/cm for 1000 µs; vbipolar 4-µs pulses at 200 Hz) with an unprocessed juice, it was demonstrated that HIPEF treatment ensures better color retention and much higher vitamin C retention for a longer period (still 49.2% after 56 days storage at 4°C) than obtained for thermal pasteurization, which exhibited poor vitamin C retention beyond 14 days' storage at 4°C (43%) (Elez-Martınez *et al.*, 2006). Also, antioxidant capacity as free radical scavenging activity, although decreasing during storage for all juices, is higher for the HIPEF-processed juice and similar to the unprocessed one. This better performance is related to a higher inactivation of POD as well as an efficient microbial inactivation. PME is better inactivated by heat treatment, although sufficient inactivation was also obtained by HIPEF (80–90%) and, more importantly, its activity is not recovered during storage.

As far as color change is concerned, in thermally treated products it has been thought to be linked to ascorbic acid degradation, which takes part in Maillard-type reaction during storage: better color retention in HIPEF-treated OJ may be linked to better preservation of ascorbic acid.

The effects of HIPEF on carotenoid content of orange juice and of their cis/trans ratio was evaluated applying different electric field intensities (25, 30, 35, and 40 kV/cm) and different treatment times (30–340 µs) in comparison with a conventional heat treatment (90°C, 20 s) (Cortes *et al.*, 2006). Pasteurization treatment causes a greater decrease in the concentration of most of the carotenoids identified and of the carotenoids with vitamin A activity. The total carotenoid concentration decreased by 12.6% in pasteurized orange juice with respect to untreated fresh orange juice, as opposed to decreases of 9.6, 6.3, or 7.8% when fields of 25, 30, or 40 kV/cm were applied. HIPEF treatment induces the appearance of a more intense yellow component if compared with the more intense red tonality of the untreated juice, although this effect is less pronounced than for the pasteurized juice. Carotenoids with vitamin A activity decreased by 15% in comparison with untreated juice, whereas a 36% reduction was observed in thermally treated juice.

Again, the effects of electric field strength, treatment time, pulse frequency, width, and polarity, as process parameters on vitamin C retention and antioxidant capacity of orange juice were evaluated and compared to those obtained by heat pasteurization (Elez-Martınez *et al.*, 2007). Orange juice retained 87.5–98.2% of vitamin C after HIPEF treatments. Pulses applied in bipolar mode, as well as a lower electric field strength, treatment time, pulse frequency, and width, led to higher levels of vitamin C retention. HIPEF-treated orange juice always showed a vitamin C retention higher than that of the heat-pasteurized products. There were no differences in antioxidant capacity between HIPEF-treated and untreated products, whereas heat-treated foods showed lower values of antioxidant capacity.

Minimally Processed Products and Alternative Stabilizing Treatment

Several studies were recently performed in order to clarify the behavior of minimally processed fruit products. As an example, Salustiana and Shamouti oranges were minimally processed as juices and stored for up to 12 or 15 days at refrigeration temperature, respectively (Del Caro *et al.*, 2004). Minimal processing consists of fruit preliminary selection operations, washing and drying, then squeezing using a household electric juice extractor: the juices are put into 50-ml glass test tubes, leaving only 5 ml empty space at the top; after hermetically sealing, the juices were refrigerated at 4°C. Chemical analyses were performed at 5, 10, and 15 days of storage, taking 12 samples of juice. Minimal processing had almost no effect on the main chemical constituents, but ascorbic acid decreased in Salustiana juices as well as flavonoid content and antioxidant capacity. The main problems limiting the shelf-life of these products are the enzymatic activity and the potential microbiological spoilage.

The effects of very low temperature and storage time on the antioxidant capacity, ascorbic acid and soluble and hydrolyzable polyphenol content of orange extracts (Polinati *et al.*, 2010) was evaluated, taking into account also the effect of ascorbic acid (AA) addition to the fruit extracts. High stability of polyphenol contents and antioxidant capacity was measured as well as ascorbic acid content at both freezing temperatures of −18°C and −70°C for up to 10 days.

Very interesting innovative stabilizing methods were recently proposed based on the application of high-pressure carbon dioxide treatment (HPCD) to freshly squeezed blood orange juice (Fabroni *et al.*, 2010). In this new system, CO_2 and the liquid food are pumped through the system and mixed together through a high-pressure pump; applied pressure range varies between 230–130 bar. At the end, CO_2 is released by an expansion valve. Effects of high-pressurized CO_2 on microbial and enzymatic inactivation are probably based on simultaneous effects of solubilized CO_2, based on pH decrease, negative effects on electrolyte balance, and on metabolism, also on account of cell membrane modifications. In the present work, applying three different operating conditions (temperature, pressure, and CO_2/juice ratio), physicochemical, antioxidant, and microbiological parameters were measured in order to find out the best operative parameters for the application to shelf-life of the treated juice. Best operating conditions were found to be at 130 bar with a 0.385 g_{CO2}/g_{juice} ratio: this treatment better preserves antioxidant components than thermal pasteurization, although it is not an alternative for an extended shelf-life product. It allows for obtaining a stabilized blood orange juice with a shelf-life of 20 days at 4°C.

Ultraviolet (UV) treatment (254 nm) was also tried as a stabilizing treatment for orange juice (Thi Tran *et al.*, 2004), analogously to that applied to water treatment. Due to the presence of high-suspended solids, a thin-film glass reactor was built up. Limited exposure afforded a shelf-life of 5 days, whereas higher intensity treatment was more effective, but partial reduction of vitamin C similar to that usually found in thermal stabilization was measured (17%) and pectin methylesterase activity was not affected by the treatment.

Innovative Concentration Techniques

Novel approaches for the concentration of fruit juices were recently proposed based on membrane technologies (Cassano *et al.*, 2007), which are able to strip out water at lower temperatures thus strongly limiting thermal damages to the juice components. Processes are generally based on clarification by ultrafiltration (UF), followed by other membrane technologies to concentrate the juice up to almost 65°Brix. The effect of UF process was evaluated in terms of several parameters: total soluble solids (TSS), suspended solids (SS), total antioxidant activity (TAA), ascorbic acid, flavanone and anthocyanin content. The clarified orange juice obtained by UF was highly similar to the initial juice except for insoluble solids, which were concentrated in the retentate stream. In the permeate a low reduction of the TAA (1.5%) was observed with respect to the fresh juice, mainly related to anthocyanins (− 9.4%) and vitamin C (−8.4%) variation, whereas flavanones were totally unaffected.

On the basis of these results for UF clarification, a new integrated membrane process was designed and evaluated in terms of preservation of the total antioxidant activity and of the bioactive antioxidant components of the juice (Galaverna *et al.*, 2008). The overall set up involved a clarification step by UF, followed by two consecutive concentration processes: first, reverse osmosis (RO), then osmotic distillation.

Reverse osmosis makes it possible to concentrate juices up to 25–30°Brix at room temperature (Silva *et al.*, 1998; Singh *et al.*, 2000), using composite polyamide membrane. OD makes it possible to obtain the desired final concentration of about 60°Bx and it is based on the use of a microporous hydrophobic polypropylene membrane. In order to strip out water from the juice, a hypertonic salt solution is oppositely circulated on the other side of the membrane. A vapor transfer is induced from the dilute solution (the juice) towards the stripping solution (the salt solution), due to the vapor pressure difference induced by the different water activities (Figure 21.5).

The overall process induced only a slight decrease of total antioxidant activity (15%), which was related to the loss of ascorbic acid (15%) and anthocyanins (20%), whereas flavanones and hydroxycinnamic acids were totally unaffected. Reduction of ascorbic acid mainly occurred during UF and RO treatment, whereas anthocyanins were mostly affected by the concentration step. These preliminary results seem promising and probably would be even better using a fully integrated plant. It is worth noting that in the case of thermal concentration a higher solubilization of carotenoid pigments in the serum coupled to partial degradation of anthocyanins led to an increase of the yellow color component with a red-brick color appearance of the juice, while in the present case a total preservation of the original red brilliant color was obtained. It has to be considered also that these mild treatments have great influence on the organoleptic properties, preserving most of the freshness irreversibly lost during thermal concentration processes. Further studies are needed to fully exploit the potentiality of membrane concentration technologies in this field.

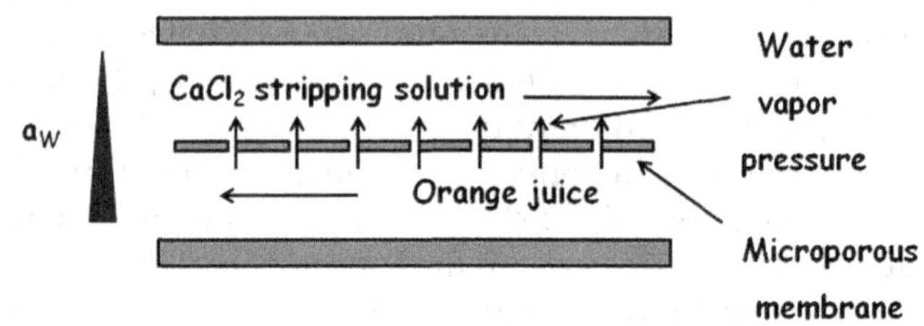

FIGURE 21.5 Membrane distillation process.

TABLE 21.3 Effect of Processing on Orange Juice's Antioxidant Components

Treatment	Vitamin C	Flavanones	Flavones	Carotenoids	Hydroxycinnamic Acids	Anthocyanins	TAA
HPT (95°C, 30 s)	Decrease	No change	No change	Decrease	Decrease	Decrease	Decrease
LPT (75°C, 30 s)	Slight decrease	No change	No change	Slight decrease	Slight decrease	Decrease	Slight decrease
HPP	No change	Increase	—	Increase	No change	No change	No change or slight decrease
PEF	Slight decrease	No change		Slight decrease	—		No change
Concentration by TASTE	Slight decrease	No change	No change	Slight decrease	Slight decrease	Decrease	Decrease
Concentration by UF-MD	Slight decrease	No change	No change	No change	No change	Slight decrease	No change

TAA, total antioxidant activity; HPT, high pasteurization temperature; LPT, low pasteurization temperature; HPP, high-pressure processing; PEF, pulsed electric fields; TASTE, Thermally Accelerated Short-Time Evaporators; UF-MD, ultrafiltration and and membrane distillation.

CONCLUSIONS

Stabilizing treatments aimed at inhibiting microbial spoilage as well as inactivating degrading enzymes are mandatory to obtain OJ with an acceptable shelf–life. The main effects on the bioactive components of OJ are summarized in Table 21.3.

Thermal processes are very effective, but may cause detrimental effects on organoleptic/nutritional properties. Many new technological approaches are under study, mainly based on the application of high-pressure processing as well as pulsed electric fields, which better preserve OJ's bioactive components. The alternative approach is based on the production of concentrates, which are actually obtained by evaporation of water under vacuum. Membrane technologies able to strip out water without applying heating are under consideration for future industrial applications and, in the near future, may afford juices of higher nutritional and organoleptic quality.

References

Arena, E., Fallico, B., Maccarrone, E., 2001. Evaluation of antioxidant capacity of blood orange juices as influenced by constituents, concentration process and storage. Food Chem. 74, 423–427.

Cao, S.-q., Liu, L., Pan, S.-y., 2011. Thermal Degradation Kinetics of Anthocyanins and Visual Color of Blood Orange Juice. Agric. Sci. Chin. 10 (12), 1992–1997.

Cano, M.P., Hernandez, A., De Ancos, B., 1997. High pressure and temperature effects on enzyme inactivation in strawberry and orange products. J. Food Sci. 62, 85–88.

Cassano, A., Marchio, M., Drioli, E., 2007. Clarification of blood orange juice by ultrafiltration: analyses of operating parameters, membrane fouling and juice quality. Desalination 212, 15–27.

Choi, M.H., Kim, G.H., Lee, H.S., 2002. Effects of ascorbic acid retention on juice color and pigment stability in blood orange (Citrus sinensis) juice during refrigerated storage. Food Res.Int. 35, 753–759.

Cortes, C., Esteve, M.J., Rodrigo, D., Torregrosa, F., Frıgola, A., 2006. Changes of color and carotenoids contents during high intensity pulsed electric field treatment in orange juices. Food Chem. Toxicol. 44, 1932–1939.

de Ancos, B., Sgroppo, S., Plaza, L., Cano, M.P., 2002. Possible nutritional and health-related value promotion in orange juice preserved by high-pressure treatment. J. Sci. Food Agric. 82, 790–796.

Del Caro, A., Piga, A., Vacca, V., Agabbio, M., 2004. Changes of flavonoids, vitamin C and antioxidant capacity in minimally processed citrus segments and juices during storage. Food Chem. 84, 99–105.

Devlieghere, F., Vermeiren, L., Debevere, J., 2004. New preservation technologies: Possibilities and limitations. Int. Dairy J. 14, 273–285.

Dhuique-Mayer, C., Tbatou, M., Carail, M., Caris-Veyrat, C., Dornier, M., Amiot, M.J., 2007. Thermal Degradation of Antioxidant Micronutrients in Citrus Juice: Kinetics and Newly Formed Compounds. J. Agric. Food Chem. 55, 4209–4216.

Elez-Martınez, P., Martın-Belloso, O., 2007. Effects of high intensity pulsed electric field processing conditions on vitamin C and antioxidant capacity of orange juice and gazpacho, a cold vegetable soup. Food Chem. 102, 201–209.

Elez-Martınez, P., Soliva-Fortuny, R.C., Martın-Belloso, O., 2006. Comparative study on shelf life of orange juice processed by high intensity pulsed electric fields or heat treatment. Eur. Food Res. Technol. 222, 321–329.

Espachs-Barroso, A., Barbosa-Canovas, G.V., Martın-Belloso, O., 2003. Microbial and enzymatic changes in fruit juice induced by high intensity pulsed electric fields. Food Rev. Int. 19, 253–273.

Esteve, M.J., Frıgola, A., Rodrigo, C., Rodrigo, D., 2005. Effect of storage period under variable conditions on the chemical and physical composition and color of Spanish refrigerated orange juices. Food Chem. Toxicol. 43, 1413–1422.

Fabroni, S., Amenta, M., Timpanaro, N., Rapisarda, P., 2010. Supercritical carbon dioxide-treated blood orange juice as a new product in the fresh fruit juice market. Innovative Food Sci. Emerg. Technol. 11, 477–484.

FAO Statistics. www.fao.org Statistics for 2010.

Fernandez-Garcıa, A., Butz, P., Bognar, A., Tauscher, B., 2001. Antioxidative capacity, nutrient content and sensory quality of orange juice and an orange–lemon–carrot juice product after high pressure treatment and storage in different packaging. Eur. Food Res. Technol. 213 (4–5), 290–296.

Fiore, A., La Fauci, L., Cervellati, R., Guerra, M.C., Speroni, E., Costa, S., Galvano, G., De Lorenzo, A., Bacchelli, V., Fogliano, V., Galvano, F., 2005. Antioxidant activity of pasteurized and sterilized commercial red orange juices. Mol. Nutr. Food Res. 49, 1129–1135.

Gil-Izquierdo, A., Gil, M.I., Ferreres, F., 2002. Effect of processing techniques at industrial scale on orange juice antioxidant and beneficial health compounds. J. Agric. Food Chem. 50, 5107–5114.

Kirca, A., Cemeroglu, B., 2003. Degradation kinetics of anthocyanins in blood orange juice and concentrate. Food Chem. 81, 583–587.

Klimczak, I., Maecka, M., Szlachta, M., Gliszczynska-Swigo, A., 2007. Effect of storage on the content of polyphenols, vitamin C and the antioxidant activity of orange juices. J. Food Composition Anal. 20, 313–322.

Galaverna, G., Di Silvestro, G., Cassano, A., Sforza, S., Dossena, A., Drioli, E., Marchelli, R., 2008. A new integrated membrane process for the production of concentrated blood orange juice: Effect on bioactive compounds and antioxidant activity. Food Chem. 106, 1021–1030.

Hodgins, A.M., Mittal, G.S., Griffiths, M.W., 2002. Pasteurization of fresh orange juice using low-energy pulsed electrical field. J. Food Sci. 67, 2294–2298.

Indrawati, Van Loey, A., Hendrickx, M., 2004. Pressure and temperature stability of water-soluble antioxidants in orange and carrot juice: a kinetic study. Eur. Food Res. Technol. 219, 161–166.

Jeyamkondan, S., Jayas, D.S., Holley, R.A., 1999. Pulsed electric field processing of foods: A review. J. Food Prot. 62, 1088–1096.

Lo Scalzo, R., Iannoccari, T., Summa, C., Morelli, R., Rapisarda, P., 2004. Effect of thermal treatments on antioxidant and antiradical activity of blood orange juice. Food Chem. 85, 41–47.

Nagy, S., Chen, C.S., Shaw, P.E., 1993. Fruit juice processing technology. Auburndale, FL: Agscience, pp. 110–165.

Navarro, P., Melendez-Martinez, A.J., Heredia, F., Gabaldon, J.A., Carbonell-Barrachina, A.A., Soler, A., Perez-Lopez, A.J., 2011. Effects of β-cyclodextrin addition and farming type on vitamin C, antioxidant activity, carotenoids profile, and sensory analysis in pasteurised orange juices. Int. J. Food Sci. Technol. 46, 2182–2190.

Nienaber, U., Shellhammer, T.H., 2001. High-pressure processing of orange juice: kinetics of pectinmethylesterase inactivation. J. Food Sci. 66, 328–331.

Parish, M.E., 1998. Orange juice quality after treatment by thermal pasteurization or isostatic high pressure. Lebensmittel Wiss. und Technol. Food Sci. Technol. 31, 439–442.

Plaza, L., Sanchez-Moreno, C., Elez-Martınez, P., de Ancos, B., Martın-Belloso, O., Cano, M.P., 2006. Effect of refrigerated storage on vitamin C and antioxidant activity of orange juice processed by high-pressure or pulsed electric fields with regard to low pasteurization. Eur. Food Res. Technol. 223, 487–493.

Plaza, L., Sánchez-Moreno, C., De Ancos, B., Elez-Martínez, P., Martín-Belloso, O., Cano, M.P., 2011. Carotenoid and flavanone content during refrigerated storage of orange juice processed by high-pressure, pulsed electric fields and low pasteurization. LWT Food Sci. Technol. 44, 834–839.

Polinati, R.M., Kremer Faller, A.L., Fialho, E., 2010. The effect of freezing at −18°C and −70°C with and without ascorbic acid on the stability of antioxidant in extracts of apple and orange fruits. Int. J. Food Sci. Technol. 45, 1814–1820.

Polydera, A.C., Stoforos, N.G., Taoukis, P.S., 2004. The effect of storage on the antioxidant activity of reconstituted orange juice which had been pasteurized by high pressure or heat. Int. J. Food Sci. Technol. 39, 783–791.

Polydera, A.C., Stoforos, N.G., Taoukis, P.S., 2005. Effect of high hydrostatic pressure treatment on post processing antioxidant activity of fresh Navel orange juice. Food Chem. 91, 495–503.

Roig, M.G., Bello, J.F., Rivera, Z.S., Kennedy, J.F., 1999. Studies on the occurrence of non-enzymatic browning during storage of citrus juice. Food Res. Int. 57, 235–242.

Sánchez-Moreno, C., Cano, M.P., de Ancos, B., Plaza, L., Olmedilla, B., Granado, F., Elez-Martinez, P., Martin-Belloso, O., Martin, A., 2004. Pulsed electric fields—processed orange juice consumption increases plasma vitamin C and decreases F2-isoprostanes in healthy humans. J. Nutr. Biochem. 15, 601–607.

Sánchez-Moreno, C., de Ancos, L.P.B., Cano, M.P., 2003. Effect of high-pressure processing on health-promoting attributes of freshly squeezed orange juice (Citrus sinensis L.) during chilled storage. Eur. Food Res. Technol. 216, 18–22.

Sánchez-Moreno, C., Plaza, L., Elez-Martinez, P., De Ancos, B., Martin-Belloso, O., Cano, M.P., 2005. Impact of High Pressure and Pulsed Electric Fields on Bioactive Compounds and Antioxidant Activity of Orange Juice in Comparison with Traditional Thermal Processing. J. Agric. Food Chem. 53, 4403–4409.

Sandhu, K.S., Minhas, K.S., 2006. Oranges and Citrus Juices. In: Hui, Y.H. (Ed.), Handbook of Fruits and Fruit Juices. Blackwell Publishing, Ames, Iowa, USA, pp. 309–357.

Senorans, F.J., Ibanez, E., Cifuentes, A., 2003. New trends in food processing. Crit. Rev. Food Sci. Nutr. 43, 507–526.

Sentandreu, E., Navarro, J.L., Sendra, J.M., 2007. Effect of technological processes and storage on flavonoids content and total, cumulative fast-kinetics and cumulative slow-kinetics antiradical activities of citrus juices. Eur. Food Res. Technol. 225, 905–912.

Silva, F.T., Jardine, J.G., Matta, V.M., 1998. Concentration of orange (Citrus sinensis) juice by reverse osmosis. Ciencia e Tecnologia des Alimentos 18, 99–104.

Singh, N.G.I., Eipeson, W.E., 2000. Concentration of clarified orange juice by reverse osmosis. J. Food Sci. Technol. 37, 363–367.

Smelt, J.P.P.M., 1998. Recent advances in the microbiology of high pressure processing. Trend Food Sci. Technol. 9, 152–158.

Stinco, C.M., Fernández-Vázquez, R., Escudero-Gilete, M.L., Heredia, F.J., Meléndez-Martínez, A.J., Vicario, I.M., 2011. Antioxidant Activity of Commercial Ready-to-Drink Orange Juice and Nectar. J. Food Sci. 76 (3), C392–C397.

Thi Tran, M.T., Farid, M., 2004. Ultraviolet treatment of orange juice. Innovative Food Sci. Emerg. Technol. 5, 495–502.

Yeom, H.W., Streaker, C.B., Zhang, Q.H., Min, D.B., 2000. Effects of pulsed electric field on the quality of orange juice and comparison with heat pasteurization. J. Agric. Food. Chem. 48, 4597–4605.

Wicker, L., Ackerley, J.L., Hunter, J.L., 2003. Modification of pectin by pectinmethylesterase and the role in stability of juice beverages. Food Hydrocolloids 17 (6), 809–814.

CHAPTER

22

Total Antioxidant Capacity of Flavored Waters

M. Fátima Barroso, Alexandra Plácido*, Luísa Correia-Sá*, Valentina F. Domingues*, M.B.P.P. Oliveira†, Cristina Delerue-Matos**

*REQUIMTE, Instituto Superior de Engenharia do Porto, Instituto Politécnico do Porto, Porto, Portugal,
†REQUIMTE, Departamento de Ciências Químicas, Faculdade de Farmácia, Universidade do Porto, Porto, Portugal

CHAPTER POINTS

- Food processing converts plain water into so-called 'flavored waters.'
- Antioxidant compounds are transferred from the flavors to the flavored waters.
- The full antioxidant profile of flavored waters was assessed.
- Flavored waters have measurable antioxidant capacity.
- The volatile fraction of these waters is mainly contributed by esters, terpenes, terpenoids, and ketones.
- Flavored waters can be a good dietary source of antioxidants and phenolic compounds.

INTRODUCTION

To meet consumers' preferences, the beverage industry has been developing new non-alcoholic beverages. In fact, the beverage market comprises a wide variety of soft drinks formats, from carbonated drinks, squashes, and nectars, through to energy and sports drinks. The drink production business is focused on providing appealing (i.e., pleasant taste and aroma), high quality, and safe products (Heaven and Nash, 2012) to satisfy the increasing consumer demand for safety and nutritional excellence (Balasubramanian and Panigrahi, 2011). The industry offers a range of choice for every palate and occasion. This includes different flavors, packaging formats and sizes, from single servings through to larger sizes for sharing. It also includes a full range of no- and low-calorie products.

According to the Union of European Soft Drinks Associations (UNESDA), the consumption of non-alcoholic drinks has moderately increased over time. In the European Union (EU), in 2011, around 119-thousand-million liters of these types of beverages were marketed. Figure 22.1 shows the statistics for the consumption (liters per capita) of non-alcoholic beverages in 24 countries from the EU.

Analyzing the data provided by UNESDA (UNESDA, 2012), it is possible to verify that the average intake of marketed non-alcoholic drinks in the EU in 2011, was 103.81 per capita of packaged water (which includes flavored waters), 96.01 per capita of soft drinks, 21.21 per capita of juice and nectars, and 15.81 per capita of dilutables. However, analyzing individual data from each of the 24 EU countries, one possible conclusion is that consumption habits of non-alcoholic beverages differ from country to country. The factors that affect beverage choice decisions are sensory appeal, familiarity and habit, social interactions, cost, availability, time constraints, personal ideology, social class and gender, media and advertising, and health (Pollard *et al.*, 2002).

With regard to soft drinks, dilutables, juices, and nectars, a mixed behavior emerges across the EU: the minimum consumption of soft drinks is in Latavia and Lithuania (around 491 per capita) and the maximum in Germany (143.31 per capita). The minimum consumption of dilutables is in Germany, Spain, and Romania (around 0.51 per capita) and the maximum is in Denmark (35.21 per capita). Regarding the consumption of juice and nectars, the minimum is from Romania (2.71 per capita) and the maximum is from Germany (33.51 per capita) (all these data are provided from the UNESDA website).

Evaluating the consumption of packaged water, the lower intake consumption is related to Finland (18.31 per capita) and the highest consumption is from Germany and Italy (170.9 and 179.11 per capita, respectively).

Processing and Impact on Antioxidants in Beverages
http://dx.doi.org/10.1016/B978-0-12-404738-9.00022-2

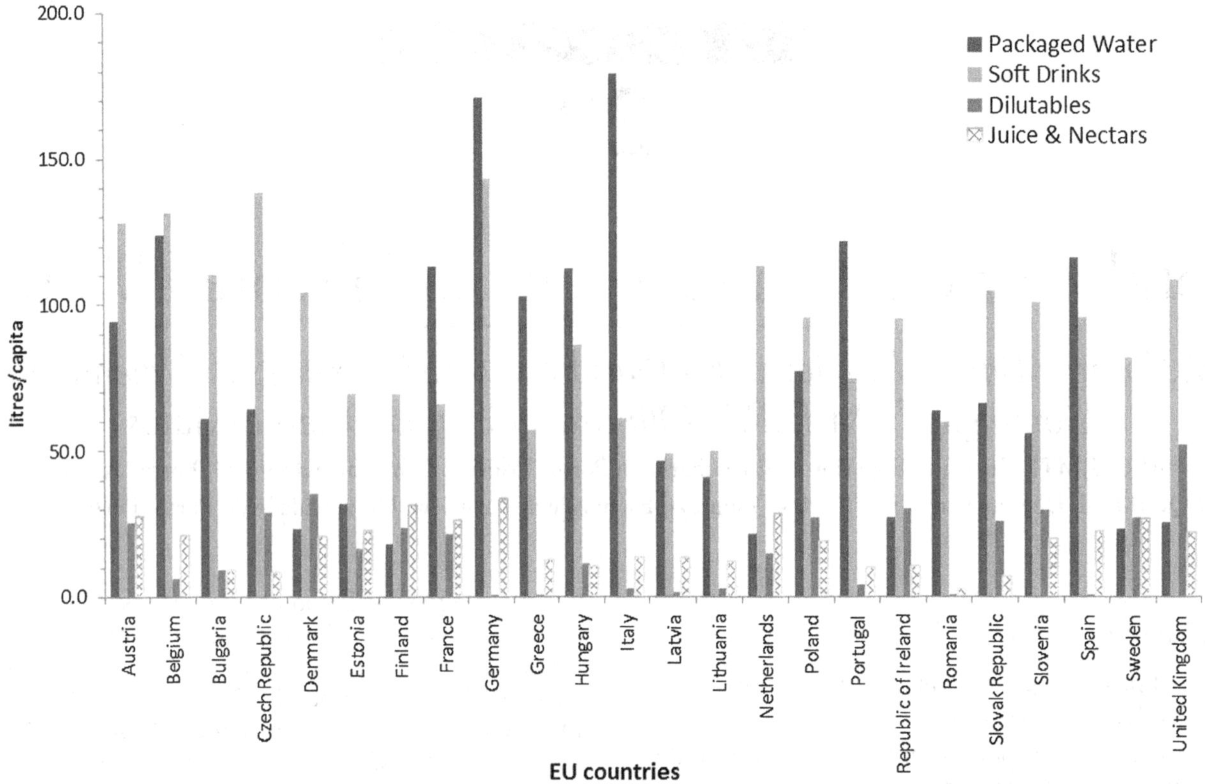

FIGURE 22.1 Statistics of consumption of non-alcoholic beverages in EU. *(adapted from UNESDA, 2012).*

A number of factors have contributed to the growth of the consumption of packaged water, including the fact that many people dislike the taste of tap water. Also, given that some people dislike natural water, the food industry has applied several technical improvements to plain water to produce new beverages.

FLAVORED WATERS: EFFECTS OF PRODUCTION AND PROCESSING

Modern consumers expect products that are ready and easy to consume, and that can be related to health and well-being. They also expect that processed foodstuffs have the same or more nutrients than natural foods (Costa *et al.*, 2012).

The basic ingredient of non-alcoholic drinks is plain water (still or carbonated water). To produce different types of beverages, several additives (natural or synthetic) are added to the water. Food additives are chemical substances used to ensure the desired quality and safety of the product, for the producer and the consumer. These compounds are useful to preserve flavor or enhance taste and appearance. For that, preservatives (potassium sorbate, potassium benzoate, dimethyl dicarbonate, and sodium benzoate), sweeteners (acessulfame-K, aspartame, and sucralose) and acidifying regulators (citric acid, sodium citrate) are added to the water.

Recently, a new beverage product was developed by the industry, a beverage termed 'flavored water.' The manufacture of flavored water is a very simple process. To produce this type of beverage, the required ingredients are mixed and homogenized, in a correct order. After checking the final volume, this mixture is added to the water and then pasteurized and transferred to commercial bottles.

The production of these kind of waters consists of the addition of flavors (essences and/or aromatic substances), juices, and sweeteners that provide to the water singular tastes and smells that are appreciated by consumers. In general, flavored waters present a strong and sweet odor. Flavors (or essential oils) added to the water are volatiles and complex natural mixtures characterized by a strong odor, which can contain about 20–60 components at different concentrations.

Flavored water can be made using only one flavor or by using a mix of flavors (Barroso *et al.*, 2011a; Correia-Sá *et al.*, 2012). Examining the variety and quantity of flavored waters on the market, it is possible to verify that the predominant market flavor is lemon fruit (Barroso *et al.*, 2009a). This flavor is conventional and assuring, being more acceptable and appreciated by the majority of consumers; hence, it offers less economical risks to producers. Other citrus flavors can be found in the marketed flavored waters, such as lime, orange, bergamot, clementine, tangerine), but also berries (wild berries, yellow berries, strawberries, raspberries, gooseberries, cranberries, elderberries, blackberries, pomegranates), exotic fruits (pineapple, passion fruit, mango, guava), apple, green apple, melon, peach, and apricot.

Today, a significant part of marketed water is flavored. Nevertheless, little information regarding the effects of the different processing steps used on the production of flavored water is available. However, the different ingredients added to natural waters influence its mineral and antioxidant composition. It has been reported that drink flavored waters could represent advantages due to the presence of certain minerals, vitamins, antioxidants, and bioactive compounds (Barroso *et al.*, 2009a,b).

FLAVORED WATERS AS AN ANTIOXIDANT SOURCE

Knowing the preference of consumers for high quality and appealing flavored water products, the flavors used in these formulations are mostly fruit extract or fruit juice. The use of synthetic flavors is becoming residual.

Considering that fruit extract and/or fruit juice are important sources of antioxidants such as minerals, vitamins, phenolics, and flavonoids (Gulcin, 2012; Ignat *et al.*, 2011) it is expected that flavored waters, which are prepared with fruits, contain natural antioxidants (Barroso *et al.*, 2011a). So, drinking these waters can increase the daily intake of natural exogenous antioxidants and may help one's health (Barroso *et al.*, 2011a; Correia-Sá *et al.*, 2012).

Until now, very few studies of the antioxidant capacity of flavored waters have been performed. The evaluation of the total antioxidant capacity (TAC) of several different types of flavored waters (more than 39) and flavor extracts (six) provided by a producer was carried out by using ultraviolet–visible (UV-vis) spectrometry (Barroso *et al.*, 2011a; Correia-Sá *et al.*, 2012; Moreira *et al.*, 2012), gas chromatography (Barroso *et al.*, 2011a, Correia-Sá *et al.*, 2012) and electrochemistry techniques (Barroso *et al.*, 2012).

Total Phenolic Content and Antioxidant Capacity of Flavored Waters

Assessing the reported literature, it was possible to verify that these flavored waters contained a moderate TAC, but, obviously, less than the fruit extract. Therefore, it was concluded that the antioxidants from the fruit were transferred into the water bottle. However, considering that labels sometimes indicate the presence of other added ingredients, it is necessary to do a preliminary evaluation to predict some synergetic interaction between the various ingredients (including flavor and fruit) added to the water.

With regard to the measurement of the TAC of the flavored waters by applying conventional optical techniques, nine analytical methods were carried out, including the determination of the total phenolic contents (TPC), total flavonoid contents (TFC), reducing power (RP), DPPH radical scavenging activity (DPPH-RSA) (Barroso *et al.*, 2011a), ferric reducing antioxidant potential (FRAP), oxygen radical absorbance capacity (ORAC), Trolox equivalent antioxidant capacity (TEAC), and total reactive antioxidant potential (TRAP) (Moreira *et al.*, 2012). By using all these analytical methods it was possible to obtain a full TAC profile of these beverages.

Depending on the chemical principle of antioxidant capacity assays, different antioxidant standards have been used. In the TFC and DPPH-RSA assays, the antioxidant epicatechin and Trolox, respectively, were used. In the case of the TPC and RP assays, the gallic acid (GA) antioxidant was used. Ascorbic acid (AA) was the antioxidant used in the FRAP, ORAC, TEAC, and TRAP assays.

The range of TAC values for different kinds of marketed flavored waters is presented in Table 22.1. The different flavors studied were lemon, orange, apple, berries, mango, pineapple, peach, guava, and melon. Wide variations on the TAC contents were observed in these different flavored waters. Literature data vary widely according to several factors: it is important to take into account the processing, storage, all the other ingredients added in the water (vitamins, sweeteners), the water brand, flavor formulation, and the amount and concentration of flavors added to the plain water. It is important to note that according to the labeled information, the added amounts are very different amongst brands and some of them only refer to their presence (Barroso *et al.*, 2009a). In addition, these beverages can provide a variety of antioxidants that may play a part in synergistic mixtures. For instance, it was reported that some labeled flavored waters have the indication of several different types of fruit juices, minerals (magnesium), ascorbic acid (a powerful antioxidant), and bioactive compounds (white and green teas, ginseng, *Ginkgo biloba*, and L-carnitine). These compounds, with biological activity in the presence of flavors, presented synergic interaction resulting in an increase of TAC (Barroso *et al.*, 2011a; Correia-Sá *et al.*, 2012).

Fruits, vegetables, and beverages are major sources of phenolic compounds in the diet that contribute to *in vitro* antioxidant activity (Frankel, 2007). As mentioned earlier, by using fruit extract and essential oils extracted from fruits, these phenolic compounds are transferred to the water bottle.

Phenolic and flavonoid compounds are correlated with antioxidant activity and seem to have an important role in stabilizing lipid oxidation. Generally, the antioxidant mechanism of phenolic compounds inactivates lipid free radicals and prevents decomposition of hydroperoxides into free radicals (Rapisarda *et al.*, 2009).

TABLE 22.1 Total antioxidant capacity (TAC) and total phenolic content (TPC) Range Values of Flavored Waters

Beverage	TPC	TAC					
Flavored Water	(mg GA/l)	RP (mg GA/l)	DPPH-RSA (mg Trolox/l)	FRAP (mg AA/l)	ORAC (mg AA/l)	TEAC (mg AA/l)	TRAP (mg AA/l)
Lemon	0.29–40	0.28–9.7	12–45	2.0–39	0.88–100	0.45–47	0.10–21
Orange	6.2–18	1.1–6.0	14–44	2.5–3.8	2.0–38	0.43–1.2	0.10–1.0
Apple	0.54–284	3.3–154	17–269	1.8–35	1.0–69	0.74–62	0.10–58
Strawberry	5.9–9.3	3.9–6.1	12–43	1.2–5.0	4.6–28	1.7–2.5	0.14–0.60
Mango	7.7–37	3.7	13	1.9–6.7	20–40	1.6–2.4	0.5–3.0
Pineapple	1.5–18	0.14–6.0	8.1–15	2.0–3.8	33–51	0.13–0.74	0.26
Raspberry	6.2–38	1.1–13.8	17–49	3.7–6.9	28–32	0.93–3.8	1.0–2.4
Peach	1.5–147	0.14–8.3	15–133	2.5–33	2.0–34	0.13–37	0.10–2.1
Guava	8.6	5.5	15	2.8	7.9	1.7	1.1
Melon	20	10.1	42	6.4	40	–	1.7

Adapted from Barroso et al. *(2011a) and Moreira* et al. *(2012).*
DPPH-RSA, DPPH radical scavenging activity; FRAP, ferric reducing antioxidant potential; ORAC, oxygen radical absorbance capacity; RP, reducing power; TEAC, Trolox equivalent antioxidant capacity; TRAP, total reactive antioxidant potential.

The TPC of several flavors and flavored waters was measured by using the Folin–Ciocalteu method (Barroso *et al.*, 2011a; Correia-Sá *et al.*, 2012). However, this methodology should always be considered as an indicative value of the total phenolic compounds available in a real sample. This method can be affected by possible interferences (reducing sugars and some amino acids) that can lead to overestimation of evaluated amounts. Nevertheless this assay is simple, economical, sensitive, and a precise technique (Prior *et al.*, 2005).

According to the data obtained, flavored waters presented polyphenolic compounds in their composition. TPC values of the waters ranged between 0.29 and 284 mg of GA/l. A wide variation on the TPC values was found in the same flavored water but belonging to a different brand. It is important to note that different technologies, processing, flavor formulations, and ingredients are used by producers to make similar flavored waters.

Taking into account apple-flavored water, TPC values ranged from 0.54 to 284 mg GA/l. The high variation in the TPC of apple waters can be explained by analyzing the bottle label. Independently of apple flavors used in beverages, the highest TPC value (284 mg GA/l) was obtained from a beverage also containing vitamin C (12 mg/250 ml) and other ingredients that can obviously increase the TPC value. Conversely, the least TPC value (0.54 mg GA/l) was found in a beverage that only contains apple flavor and sweeteners.

Compared to other flavored water, for instance mango water, TPC values of this water varied from 7.7 to 36.5 mg GA/l. In this case, the addition of a bioactive compound, *G. biloba*, and fruit juice (mango, apple, and pear) seems to produce a dual phenolic protective effect by increasing the TPC level.

Relative to the remaining water samples, the same analytical behavior was verified, namely, the increasing of the TPC by adding other ingredients with antioxidant interest, besides the flavors.

With regard to flavonoid measurement by using optical methods (formation of a flavonoid–aluminin complex), no flavonoids were found in detectable amounts, neither in flavors nor in flavored waters (Barroso *et al.*, 2011a; Correia-Sá *et al.*, 2012). Considering that flavors contain about 1–15% of non-volatile compounds, where flavonoids are included (Fisher and Philips, 2008), it was expected that flavonoids would be present in the beverages.

With regard to the measurement of the RP in flavored waters (Table 22.1), it was verified that the antioxidant compounds included in these waters had a moderate ability to reduce the iron complex present in this analytical assay.

In the case of food and beverages, it is important to determine the efficacy of natural antioxidants for food and health protection against oxidative damage promoted by free radicals, to avoid deleterious changes, loss of commercial and nutritional value, and diseases. Consequently, different analytical methods are employed in screening tests to evaluate the effectiveness of antioxidants; there is evidence that many different methods are necessary to measure the free radical scavenging ability of antioxidants (Sánchez-Moreno, 2002). In this context, the utilization of five antiradical analytical methods for the determination of TAC of flavored waters were reported (Barroso *et al.*, 2011a; Moreira *et al.*, 2012). These different analytical methods are based on monitoring of the inhibition of the oxidation of a suitable substrate by using different free radical generation systems. In these reports, the DPPH-RSA assay (against the DPPH radical), the FRAP (against the TPTZ radical), the ORAC (against the AAPH radical), the TEAC (against the ABTS radical) and TRAP (against the ABTS radical) were applied. It is important to indicate that free radicals used in these procedures are artificially generated and do not have a direct correlation with the real biological system.

The evaluation of data indicated that flavored waters presented a moderate antiradical capacity promoted by the antioxidant compounds added to the plain water. The range of TAC values (Table 22.1) varied from 8.1 to 269 mg Trolox/l when the DPPH-RSA assay was used, between 2.0 and 32.8 mg AA/l with the FRAP assay, 0.9–100 mg AA/l in the ORAC, and 0.1 to around 60 mg AA/l in the TEAC and TRAP assays.

Several factors can influence the TAC values of flavored waters, such as the methodology, the antioxidant standard used in the assay, the flavors, ingredients added, and the commercial brand of the water (Moreira *et al.*, 2012).

With regard to the analytical methods, several authors agree that TAC values provided by these five antiradicals methods are not statistically correlated, in that the lowest TAC values were obtained with the TRAP method and highest TAC levels with the ORAC methods (Moreira *et al.*, 2012; Prior *et al.*, 2005; Pulido *et al.*, 2000). Analyzing Table 22.1, it is possible to observe that polyphenolic and antioxidant compounds presented different effectiveness in promoting the scavenging activity of the free radicals generated in the chemical assay. Indeed, flavored waters showed a good ability to promote inhibition of the peroxyl radical (ORAC method) reflecting a classical radical chain-breaking antioxidant activity by H atom transfer. Conversely, less efficacy was found on the ability to quench radical propagation as demonstrated in the TAC values obtained in DPPH-RSA, TEAC, FRAP, and TRAP assays.

According to statistical data (Table 22.1), flavors also presented different radical scavenging capacity. Higher TAC values were obtained in apple-, peach-, and lemon-flavored waters and lower values in guava and raspberry flavors.

As expected and mentioned by authors, the vitamins (C and B complex) presented in the waters (by transferring from flavor/juice or as an added ingredient) also increase the TAC and the ability of the real samples to eliminate or inhibit radical propagation and generation.

The optical methodologies used for the assessment of TAC are useful, simple to use, reliable, and reproductive; however they only give indication of the total antioxidants present in a sample. Considering that flavored waters possessed a strong odor, it is of maximum importance to know what kind of compound can give water its smell. Hence, the identification of the individual volatile compounds present in flavors and flavored waters was performed (Barroso *et al.*, 2011a; Correia-Sá *et al.*, 2012).

Volatile Composition of Flavors and Flavored Waters

Flavor is one of the most important quality criteria of fruit and related products; therefore, a considerable amount of research has been devoted to developing analytical techniques for characterizing aroma-producing compounds. Both qualitative and quantitative information is desired in order to monitor flavor quality and to provide quality control for fresh and processed products (Song *et al.*, 1997).

Volatile compounds present in the beverages directly affect the sensorial quality. The qualitative composition of the volatile fraction is formed by a complex group of chemical substances (e.g., aldehydes, alcohols, ketones, esters, lactones, terpenes). The concentration of these volatile compounds is generally low (μg/l) and can be affected by a number of agronomic (variety, climatological conditions, ripening stage) and technological factors (harvest, post-harvest treatments, storage and processing conditions) (Riu-Aumatell *et al.*, 2004). For extraction of flavors, the microextraction techniques are often used due to their potential in flavor analysis, fast sample preparation, high sensitivity, high enrichment factor, and often possibility of automation. Of these techniques, solid phase microextraction (SPME) is the most commonly used (Jelen *et al.*, 2012). The extraction can be performed in two common ways: (i) headspace-SPME (HS-SPME) where fiber is exposed in the vapor phase above a gaseous, liquid, or solid sample; (ii) direct immersion-SPME (DI-SPME) where the fiber is directly immersed in liquid samples (Balasubramanian and Panigrahi, 2011).

The headspace sampling methods have the advantages of being solvent free and non-invasive for the samples. The HS-SPME is currently the most widely used and has been reported to be cheap, fast, and reproducible, and it eliminates problems associated with chemically and thermally unstable samples, thus avoiding generation of artefacts (Gamero and Jong, 2013).

Microextraction techniques, mainly SPME and stir bar sorptive extraction (SBSE) have been used for food flavor compounds analysis for two decades. With SBSE, several applications were found for food, such as flavors/volatiles analyses, wine and other beverages analyses, and monitoring metabolism of flavor compounds. Due to the high volume of phase in SBSE stirrer, which results in low detection limits, this is a good tool therefore for the analysis of flavor (Jelen *et al.*, 2012). However, the main drawback of SBSE is the necessity of a specific device for desorption. Other microextraction techniques can also be applied for flavor detection, such as liquid phase microextraction (LPME). LPME is usually divided into three approaches: single drop microextraction (SDME), hollow fiber protected microextraction (HF-LPME), and dispersive liquid–liquid microextraction (DLLME) (Jelen *et al.*, 2012).

Nonetheless, comparing these several extraction techniques, the DI-SPME showed the highest sensitivity, as expressed by detection of the highest percentage of total aroma compounds at low levels of concentration (Gamero and Jong, 2013).

Depending on the food matrix and the aroma compounds to be determined, different aroma extraction methods can be applied in combination with chromatographic techniques (Gamero and Jong, 2013). Gas chromatography in combination with mass spectrometry (GC-MS) is the method of choice for the analysis of volatile fractions (Bicchi *et al.*, 2011).

HS-SPME–GC/MS was selected for the analysis of lime, lemon, and tangerine flavor extracts. It was concluded that the monoterpenes represented >99% of the lime volatile fraction. From this group, the citral was the major ingredient, followed by α-terpineol, carveol and α-limonene. In lemon flavor, it was also found that, once more, the monoterpenes were the most abundant group, followed by the sesquiterpenes. Therefore, the terpenes citral, carveol and α-limonene, and the sesquiterpene β-caryophyllene were the major compounds. With regard to tangerine flavor, methyl aminobonzoate, α- limonene, and α-terpineol were the major compounds, followed by the sesquiterpene α-farnesene (Barroso *et al.*, 2011a) (Table 22.2A).

The analysis of three flavors (strawberry, gooseberry, and apple) and two flavored waters (one with apple and green tea, and the other with gooseberry) applying HS-SPME–GC/MS, showed that the strawberry flavor extract was predominantly composed of esters (Table 22.2B). In this compound group it was found that the 2-propenoic acid, 3-phenyl-, and methyl ester (methyl cinnamate) are considered to be responsible for the typical strawberry note found in fresh syrups (Ducruet *et al.*, 2001). Conversely, for the gooseberry flavor extract, the major identified aroma compounds were the ketones and the terpene compounds.

This study also presented the analysis of the aroma profile of two flavored waters. In one particular case (gooseberry-flavored water) the aroma profile of the flavor extract and the water sample was almost identical, as can be seen in Figure 22.2B. This similarity was because the water sample had been produced by adding the previous gooseberry flavor extract to sparkling mineral water. Nevertheless, some differences were determined; namely the 4-methyl-1,3-oxazinane-2-thione was more abundant and the other compound groups, esters, and terpenes were present in lower

TABLE 22.2A Relative Abundance of Compound Groups Detected in Aroma Extracts

Sample	Achieved Aroma Compounds (Peak Number)	Relative Abundance (%)
Apple	*Esters*	79
	Hexanoic acid, ethyl ester	
	Hexanoic acid, 2-methyl-, 1-methylpropyl ester	
	Octanoic acid, hexylester	
	Menthol, fumarate	
	Terpenes/Terpenoids	15
	Isocitronellol	
	Thio Ketone	3.8
	4-Methyl-1,3-oxazinane-2-thione	
Gooseberry	*Esters*	12
	Hexanoic acid, ethyl ester (1)	
	Butanedioic acid, hydroxy-, dipentyl ester (3)	
	Terpenes/Terpenoids	24
	Cis-myrtanol (4)	
	β-Damascenone (5)	
	α- Isomethyl ionone (6)	
	β-ionone (7)	
	Thio Ketone	25
	4-Methyl-1,3-oxazinane-2-thione (2)	
Strawberry	*Esters*	69
	Hexanoic acid,-ethyl ester	
	6.43 Propyl heptanoate	
	Butanol, 3-methyl-, acetate	
	Pentyloctanoate	
	4-Hexen-1-ol, acetate	
	Butanedioic acid, hydroxy-, dipentyl ester	
	Hexanoic acid, 3-hexenyl ester	
	2-Propenoic acid, 3-phenyl-, methyl ester	
	Sulfur compounds	23
	Decyldisulfide	
Citrus	*Terpenes/Terpenoids*	100
Lime	β-pinene	
Lemon	1,4-Cineole	
Tangerine	o-cymene	
	α-limonene	
	β-caryophyllene	
	α-bergamotene	
	β-bisabolene	
	α-terpineol (p-menth-1,8-dien-6-ol)	
	carveol (p-mentha-1,8-dien-6-ol)	
	citral (geranial)	
	methyl aminobenzoate	

Adapted from Barroso et al. *(2011a) and Correira-Sá* et al. *(2012).*

TABLE 22.2B Relative Abundance of Compound Groups Detected in Flavored Waters

Sample	Achieved Aroma Compound (Peak Number)	Relative Abundance (%)
Gooseberry-flavored water	*Esters* Hexanoic acid, ethyl ester (1) Butanedioic acid, hydroxy-, dipentyl ester (3)	8.8
	Terpenes/Terpenoids Cis-myrtanol (4) β-Damascenone (5) α- Isomethyl ionone (6) β-Ionone (7) τ-Elemene (8)	20
	Thio Ketone 4-Methyl-1,3-oxazinane-2-thione (2)	53
Apple-flavored water	*Esters* Butanoic acid, 2-methyl	5.4
	Terpenes/Terpenoids ϱ-Cymene τ-Terpinene Trans-anethole	5.9
	Thio Ketone 4-Methyl-1,3-oxazinane-2-thione	58
	Aldehydes Hexanal	9.9

Adapted from Correia-Sá et al. *(2012).*

abundance. Additionally, τ-elemene was detected in the drink sample but not in the corresponding flavor extract sample (Correia-Sá *et al.*, 2012) (Table 22.2A and B). The detected ionones were similar to other studies related to gooseberry volatile analysis (Leino and Kallio, 1993; Varming *et al.*, 2006). The different steps used in the processing of beverages, namely the thermal treatment, can have an effect on the volatile profile. An increase in the proportion of β-ionone and damascenone in heated juices was reported (Leino and Kallio, 1993).

Regarding an apple flavor extract sample (Correia-Sá *et al.*, 2012), it was concluded that the most abundant compounds were esters. This flavor also revealed the presence of a terpene. The analyzed apple-flavored water had in its composition apple flavor and white tea. In this case some difference in the individual composition was found comparatively with apple flavor. In this water sample, the predominant compounds were ketones and, in lesser quantity, compounds from the other groups (aldehydes, terpenes/terpenois, esters, etc.) (Table 22.2A and A).

In summary, the aroma is a complex mixture of compounds and although hundreds of compounds can be detected for a specific flavor, only a few of these chemicals are character- impact compounds. Therefore, the analysis of flavored water aroma profiles can be a challenge. One must bear in mind that the aroma is crucial for consumers' acceptance and that several factors can affect the volatile compounds profile that make it almost unique. Therefore, aroma volatile compounds analysis can be crucial for quality control.

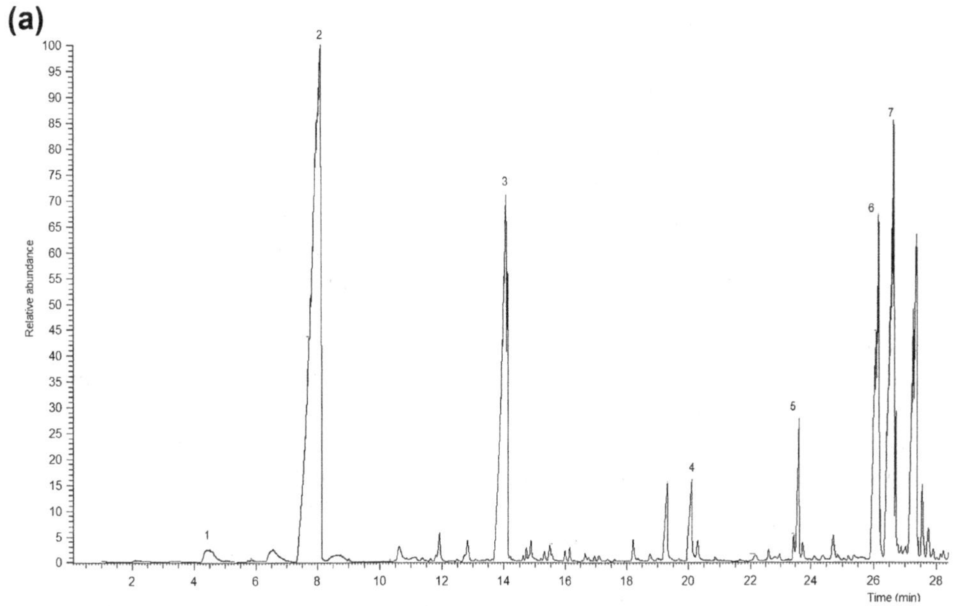

FIGURE 22.2A GC-MS chromatogram of gooseberry flavor extract.

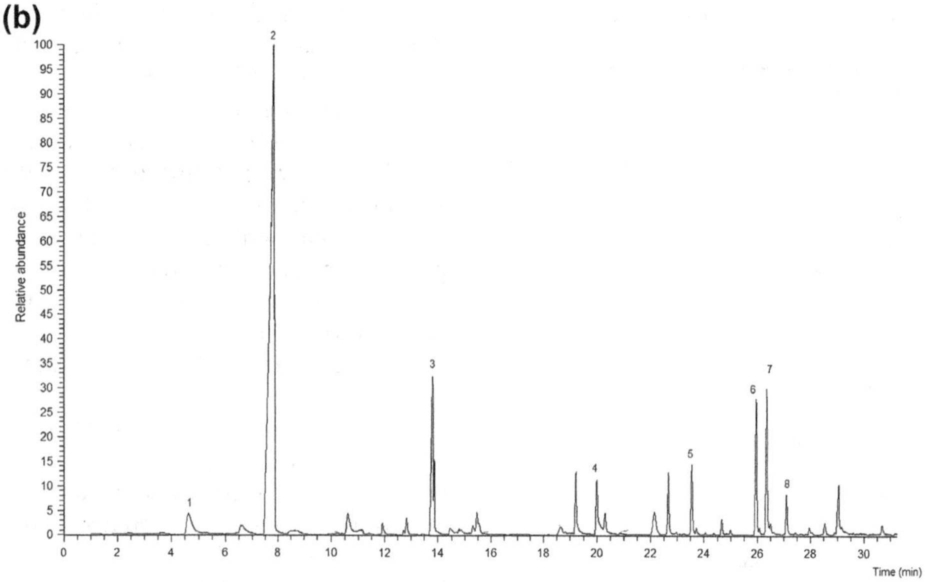

FIGURE 22.2B GC-MS chromatogram of gooseberry-flavored water.

TAC of Flavored Waters by Electrochemical DNA-Based Biosensor

The use of electrochemical biodevices for the assessment of TAC in beverages is a good alternative to optical methods. In general, these devices are very sensitive and easy to miniaturize, which paves the way to portability and lacks interference from colored samples, avoiding time consumption (Barroso *et al.*, 2011b).

For antioxidant quantification, cyclic voltammetry (CV) is the most widely used electrochemical technique although differential pulse voltammetry (DPV) and square wave voltammetry (SWV) are advantageous in terms of sensitivity for individual detection.

In general, these voltammetric techniques are based on the principle that an antioxidant tends to be easily oxidized at a bare electrode, presenting an anodic peak potential and a measurable electrochemical signal. The accomplishment of electrochemical measurement is highly influenced by the working electrode material, which can be made of carbon, gold, platinum, and silver, among several others.

In particular, the use of electrochemical DNA-based biosensors for the determination of TAC in beverages seems to be a reasonable option because the protective role of antioxidants at a cellular level could be properly studied by monitoring DNA integrity. In this sense, a simulation to an *in vivo* assay can be performed by the exposition of DNA layers to a biological free radical and an antioxidant.

Purine-based biosensors for the assessment of the TAC in flavored waters have been developed.

For this propose, purine bases, guanine and adenine, and adenine-rich oligonucleotide (dA_{21}) were immobilized on glassy carbon (GCE) and carbon paste (CPE) electrode. After this step, the modified electrode was placed in contact with radicals such as hydroxyl ($HO^{\bullet}$), and superoxide ($O_2^{\bullet -}$) radical that can be found in biological systems (Barroso *et al.*, 2012; Barroso *et al.*, 2011c,d,e).

The chemical production of these free radicals was similar to *in vivo* radical generation. Indeed, in living systems, the hydroxyl radicals are mostly generated in the presence of ferrous or cupric ions and hydrogen peroxide (Eq. 22.1). On the other-hand, biological superoxide radical can be generated enzymatically via xanthine and xanthine oxidase in the presence of oxygen (Eq. 22.2).

$$H_2O_2 + Fe^{2+} \rightarrow OH^{\bullet} + OH^{-} + Fe^{3+} \qquad \text{(Eq. 22.1)}$$

$$\text{Xanthine} + H_2O + O_2 \underset{\text{XOD}}{\longrightarrow} \text{uric acid} + 2H^{+} + O_2^{\cdot -} \qquad \text{(Eq. 22.2)}$$

Radicals induced oxidative damage in the purine bases and in the DNA (Barroso *et al.*, 2011b). This observation was confirmed by analyzing the oxidation signal of guanine and adenine that dramatically decreased in the presence of radicals, when compared to their native electroactivity. When an antioxidant is added to the radical-containing solution, the peak current is partially recovered, demonstrating the protective role promoted by antioxidants on the DNA by scavenging the free radicals (Barroso *et al.*, 2011b). Antioxidants (ascorbic acid, gallic acid, caffeic acid, coumaric acid, and resveratrol) exhibited hydroxyl- and superoxide -scavenger activity with efficiencies ranging from 47–79% when it was used with the purine-based biosensor and 19–59% when it was applied with the DNA-based biosensor. Different trends in the protection of DNA were observed depending on the source of radicals used, suggesting a different mode of action against free radicals. In particular, when it was used with DNA-CPE, the order of the protective efficacy carried out by the antioxidants was the following: resveratrol > ascorbic acid > gallic acid > Trolox ~ coumaric acid. Using this methodology the TAC of flavors and flavored waters was measured. Data provided in the literature indicated these beverages present a moderate TAC.

CONCLUSION

Knowing the preferences of consumers, beverage producers have been developing new non-alcoholic beverages, such as flavored water. Flavored water consists of the addition of flavors/fruit extract, juice, and sweeteners that provide tastes and smells appreciated by consumers.

Several analytical methodologies, based on optical, chromatographic, and electrochemical techniques, have been used to discover the full antioxidant profile of flavors and flavored waters.

Using colorimetric and electrochemical methods, the TAC of beverages was evaluated. Considering the chemical complexity of the food composition and the possibility of synergic interactions among the antioxidant compounds, the determination of the total antioxidants is highly desired. Moreover, the operating principle of the electrochemical DNA-based biosensors is closer to the activity of antioxidants in biological systems.

The use of GC-MS makes it possible to identify the individual volatile compounds and the volatile profile of the flavors and flavored waters. Generically, the volatile fraction is constituted by esters, terpenes, terpenoids, and ketones.

Flavored waters can be an adequate alternative to consumers that dislike tap/packaged water. The ingredients added to the plain water insert antioxidants into its composition that can have a positive effect on health. Flavored waters could represent some advantages due to the presence of certain minerals, some vitamins, antioxidants, and bioactive compounds. However, consumers are advised to read the label in order to avoid some health problems that can occur with some other ingredients (e.g., sweeteners, preservatives).

Acknowledgments

M. F. Barroso and Luísa Correia-Sá are grateful for the post-doc fellowship (SFRH/BPD/78845/2011) and doc fellowship financed by (SFRH/BD/87019/2012), respectively, POPH - QREN - Tipologia 4.1 - Formação Avançada, subsidized by Fundo Social Europeu and Ministério da Ciência, Tecnologia e Ensino Superior. This work has also been supported by Fundação para a Ciência e a Tecnologia (FCT) through grant no. PEst-C/EQB/LA0006/2011. The authors acknowledge the Frize (Portuguese farmers) for providing flavor samples.

References

Balasubramanian, S., Panigrahi, S., 2011. Solid-Phase Microextraction (SPME) Techniques for Quality Characterization of Food Products: A Review. Food Bioprocess. Technol. 4 (1), 1–26.

Barroso, M.F., Delerue-Matos, C., Oliveira, M.B.P.P., 2012. Electrochemical evaluation of total antioxidant capacity of beverages using a purine-biosensor. Food Chem. 132 (2), 1055–1062.

Barroso, M.F., Noronha, J.P., Delerue-Matos, C., Oliveira, M.B.P.P., 2011a. Flavored Waters: Influence of Ingredients on Antioxidant Capacity and Terpenoid Profile by HS-SPME/GC-MS. J. Agric. Food Chem. 59 (9), 5062–5072.

Barroso, M.F., de-los-Santos-Álvarez, N., Delerue-Matos, C., Oliveira, M.B.P.P., 2011b. Towards a reliable technology for antioxidant capacity and oxidative damage evaluation: electrochemical (bio) sensors. Biosens. Bioelectron. 30 (1), 1–12.

Barroso, M.F., de-los-Santos-Álvarez, N., Lobo-Castañón, M.J., Miranda-Ordieres, A.J., Delerue-Matos, C., Oliveira, M.B.P.P.P., Tuñón-Blanco, P., 2011c. DNA-based biosensor for the electrocatalytic determination of antioxidant capacity in beverages. Biosens. Bioelectron. 26 (5), 2396–2401.

Barroso, M.F., de-los-Santos-Álvarez, N., Lobo-Castañón, M.J., Miranda-Ordieres, A.J., Delerue-Matos, C., Oliveira, M.B.P.P.P., Tuñón-Blanco, P., 2011d. Electrocatalytic Evaluation of DNA Damage by Superoxide Radical for Antioxidant Capacity Assessment. J. Electroanal. Chem. 659 (1), 43–49.

Barroso, M.F., Delerue-Matos, C., Oliveira, M.B.P.P., 2011e. Electrochemical DNA-sensor for evaluation of total antioxidant capacity of flavors and flavored waters using superoxide radical damage. Biosens. Bioelectron. 26 (9), 3748–3754.

Barroso, M.F., Silva, A., Ramos, S., Oliva-Teles, M.T., Delerue-Matos, C., Sales, M.G.F., Oliveira, M.B.P.P., 2009a. Flavored versus natural waters: Macromineral (Ca, Mg, K, Na) and micromineral (Fe, Cu, Zn) contents. Food Chem. 116 (2), 580–589.

Barroso, M.F., Ramos, S., Oliva-Teles, M.T., Delerue-Matos, C., Sales, M.G.F., Oliveira, M.B.P.P., 2009b. Survey of trace elements (Al, As, Cd, Cr, Co, Hg, Mn, Ni, Pb, Se, and Si) in retail samples of flavored and bottled waters. Food Addit. Contam. Part B-Surveill. 2 (2), 121–130.

Bicchi, C., Cagliero, C., Rubiolo, P., 2011. New trends in the analysis of the volatile fraction of matrices of vegetable origin: a short overview. A review. Flavor Frag. J. 26 (5), 321–325.

Correia-Sá, L., Barroso, M.F., Bragança, I., Domingues, V.F., Oliveira, M.B.P.P., Delerue-Matos, C., 2012. Antioxidant capacity and flavor profile of juice and flavored waters by HS-SPME/GC-MS. Submitted.

Costa, A.S.G., Nunes, A., Almeida, A.M.C., Carvalho, M.R., Barroso, M.F., Alves, R.C., Oliveira, M.B.P.P., 2012. Teas, dietary supplements and fruit juices: a comparative study regarding antioxidant activity and bioactive compounds. LWT Food Sci. Technol. 49 (2), 324–328.

Ducruet, V., Fournier, N., Saillard, P., Feigenbaum, A., Guichard, E., 2001. Influence of Packaging on the Aroma Stability of Strawberry Syrup during Shelf Life. J. Agric. Food Chem. 49 (5), 2290–2297.

Fisher, K., Phillips, C., 2008. Potential antimicrobial uses of essential oils in food: is citrus the answer? Trends Food Sci. Technol. 19, 156–164.

Frankel, E.N., 2007. Antioxidants in food and biology facts and fiction-vol. 2. Publisher Oily Press, Bridgwater, England, p. 231.

Gamero, A., Wesselink, W., Jong, C., 2013. Comparison of the sensitivity of different aroma extraction techniques in combination with gas chromatography–mass spectrometry to detect minor aroma compounds in wine. J. Chromatogr. A 1272 (0), 1–7.

Gulcin, I., 2012. Antioxidant activity of food constituents: an overview. Arch. Toxicol. 86 (3), 345–391.

Heaven, M.W., Nash, D., 2012. Recent analyses using solid phase microextraction in industries related to food made into or from liquids. Food Control 27 (1), 214–227.

Ignat, I., Volf, I., Popa, V.I., 2011. A critical review of methods for characterisation of polyphenolic compounds in fruits and vegetables. Food Chem. 126 (4), 1821–1835.

Jelen, H.H., Majcher, M., Dziadas, M., 2012. Microextraction techniques in the analysis of food flavor compounds: A review. Anal. Chim. Acta. 738, 13–26.

Leino, M., Kallio, H., 1993. Volatile compounds of blackcurrant juice and wine. Zeitschrift für Lebensmitteluntersuchung und -Forschung A 196 (5), 410–414.

Moreira, F.T.C., Guerreiro, J.R.L., Barros, R., Sales, M.G.F., 2012. The effect of method, standard and sample components on the total antioxidant capacity of commercial waters assessed by optical conventional assays. Food Chem. 134 (1), 564–571.

Pollard, J., Kirk, J.S.F., Cade, J.E., 2002. Factors affecting food choice in relation to fruit and vegetable intake: a review. Nutr. Res. Rev. 15, 373–387.

Prior, R.L., Wu, X., Schaich, K., 2005. Standardized methods for the determination of antioxidant capacity and phenolic in foods and dietary supplements. J. Agric. Food Chem. 53, 4290–4302.

Pulido, R., Bravo, L., Saura-Calixto, F., 2000. Antioxidant activity of dietary polyphenols as determined by a modified ferric reducing/antioxidant power assay. J. Agric. Food Chem. 48, 3396–3402.

Rapisarda, P., Fabroni, S., Peterek, S., Russo, G., Mock, H.P., 2009. Juice of new citrus hybrids (Citrus clementina Hort. Ex Tan. X C. sinensis L. Osbeck) as a source of natural antioxidants. Food Chem. 117, 212–218.

Riu-Aumatell, M., Castellari, M., López-Tamames, E., Galassi, S., Buxaderas, S., 2004. Characterisation of volatile compounds of fruit juices and nectars by HS/SPME and GC/MS. Food Chem. 87 (4), 627–637.

Sánchez-Moreno, C., 2002. Review: Methods used to evaluate the free radical scavenging activity in foods and biological systems. Food Sci. Technol. Int. 8, 121–137.

Song, J., Gardner, B.D., Holland, J.F., Beaudry, R.M., 1997. Rapid Analysis of Volatile Flavor Compounds in Apple Fruit Using SPME and GC/Time-of-Flight Mass Spectrometry. J. Agric. Food Chem 45 (5), 1801–1807.

UNESDA, Union of European Soft Drinks Associations, 2012. Statistics on the consumption of non-alcoholic beverages in Europe. Available in http://www.unesda.org/industry.

Varming, C., Andersen, M.L., Poll, L., 2006. Volatile monoterpenes in black currant (Ribes nigrum L.) juice: Effects of heating and enzymatic treatment by beta-glucosidase. J. Agric. Food Chem. 54 (6), 2298–2302.

CHAPTER

23

Antioxidant Properties of Soy-Based Drinks and Effects of Processing

José Manuel Silván, Miryam Amigo-Benavent*, Mª. Dolores del Castillo†*

*Department of Nutrition and Metabolism, Institute of Food Science and Technology and Nutrition (ICTAN-CSIC), Madrid, Spain, †Food Bioscience Group, Department of Food Bioactivity and Analysis, Institute of Food Science Research (CSIC-UAM), Madrid, Spain

CHAPTER POINTS

- Soy isoflavones, proteins, and minerals contribute to the overall antioxidant properties of soy-based drinks.
- Many variables related and non-related to food processing influence the antioxidant properties of soy-based drinks.
- All processing steps involving thermal treatments of different intensity dramatically affect the overall antioxidant properties of soy-based drinks.
- Non-thermal packing processes [high-intensity pulsed electric fields (HIPEF) and high-pressure processing (HPP)] induce small changes in isoflavone profile and total content.
- The overall antioxidant properties of soy-based drinks are due to natural antioxidants surviving food processing and those neo-formed antioxidants generated by different chemical events.
- The fermentation process greatly modifies composition, antioxidant properties, and sensorial quality of soy milk.
- In fermented soy milk isoflavones may not be the main antioxidant.
- The analysis of the contribution of each individual antioxidant to the overall antioxidant properties has not been performed and is of interest.
- Most studies pinpoint the isoflavones as the principal antioxidant of soy-based drinks.
- Little information regarding changes in soy milk antioxidant properties is currently available.

INTRODUCTION

Soy beverages constitute one of the food industry sectors with the highest worldwide growth. The consumption of soy-based drinks has experienced a huge increase in the last few years (Sloan, 2005). These drinks are made either from fresh or reconstituted soy milk, soy protein isolate, or a combination of traditional soy milk augmented with soy protein isolate technologies. Soy milk is considered the most popular soy-based drink. Soy milk is basically the aqueous extract of whole soybeans and has gained much popularity as a healthy drink because it is an important source of isoflavones with high antioxidant properties. Table 23.1 shows the typical soy milk composition.

A positive correlation between antioxidant properties in soy-based drinks and total isoflavone content has been well described (Ishihara *et al.*, 2007). Isoflavones are a group of naturally occurring heterocyclic phenols found mainly in soybeans. There are three main types of isoflavones in raw soybeans, which are also present in soy-food products, such as soy beverages. The parent isoflavones are the aglycone structure of daidzein, genistein, and glycitein (Friedman and Brandon, 2001), which are conjugated to form the β-glucosides (daidzin, genistin, and glycitin), the malonyl-glucosides (6″-*O*-malonyldaidzin, 6″-*O*-malonylgenistin, and 6″-*O*-malonylglycitin), and the acetyl-glucosides (6″-*O*-acetyldaidzin, 6″-*O*-acetylgenistin, and 6″-*O*-acetylglycitin) (Figure 23.1). The conjugated forms have an additional glucose moiety, which could be free of other groups (β-glucosides) or could be bound to either a malonyl group (malonyl-glucosides) or an acetyl group (acetyl-glucosides). Analyses of the

Processing and Impact on Antioxidants in Beverages
http://dx.doi.org/10.1016/B978-0-12-404738-9.00023-4

different isoflavone isomers in soybean foods indicate that 97–98% of total isoflavone in whole soybeans are in the esterified forms, glucoside, malonyl-glucoside or acetyl-glucoside. In raw, unprocessed soybeans, the 6″-*O*-malonyl-β-glucosides are the predominant forms, but malonyl forms are unstable and they are converted to acetyl, β-glucosides and aglycone forms during soybean processing (Huang *et al.*, 2006).

TABLE 23.1 Soy Milk Composition (Per 100g)

Component	Amount	Component	Amount
Water	93.00 (g)	Potassium, K	141.0 (mg)
Energy	138.00 (kJ)	Sodium, Na	12.00 (mg)
Protein	2.80 (g)	Zinc, Zn	0.23 (mg)
Fat (total lipid)	2.00 (g)	Copper, Cu	0.12 (mg)
Fatty acids, saturated	0.21 (g)	Manganese, Mn	0.17 (mg)
Fatty acids, mono-unsaturated	0.32 (g)	Selenium, Se	1.30 (μg)
Fatty acids, poly-unsaturated	0.83 (g)	Thiamin (Vit. B1)	0.16 (mg)
Carbohydrates	1.80 (g)	Riboflavin (Vit. B2)	0.07 (mg)
Fiber	1.30 (g)	Niacin (Vit. B3)	0.14 (mg)
Ash	0.27 (g)	Panthotenic acid (Vit. B5)	0.04 (mg)
Isoflavones	8.80 (mg)	Vitamin B6	0.04 (mg)
Calcium, Ca	4.00 (mg)	Folic acid	1.50 (μg)
Iron, Fe	0.58 (mg)	Vitamin A	3.00 (μg)
Magnesium, Mg	19.00 (mg)	Vitamin E	0.01 (mg)
Phosphorous, P	49.00 (mg)		

Source: Soya – information about soy and soya products (2012). Nutritional values of soy milk. http://www.soya.be/nutritional-valuesof-soy-milk.php.

The extent of the antioxidant capacity of isoflavones is positively correlated to the number of hydroxyl groups in the isoflavone nucleus. Most antioxidant properties of isoflavones are mainly attributed to their aglycone forms, especially genistein, followed by daidzein (Ishihara *et al.*, 2007). Positive correlations were found between antioxidant capacity and total isoflavone content, aglycone conjugates, glucoside conjugates, genistein series, and daidzein series. The great variability in the antioxidant capacity of soy beverages depends on the chemical form and concentration of isoflavone compounds present, which depends on raw material, processing, and storage conditions (Monje *et al.*, 2006).

Soy protein isolate and defatted soy flour are ingredients used for soy-based drinks (Deshpande *et al.*, 2008) and contain proteins that constitute one of the main ingredients in soy drinks. Soy proteins are composed of storage proteins, enzymes, and protease inhibitors, such as Bowman–Birk inhibitor (BBI) (Garcia *et al.*, 1997). The most abundant storage proteins are β-conglycinin (7S) and glycinin (11S), which are present in soy drinks and foodstuffs (Amigo-Benavent *et al.*, 2008). Peptides derived from these two proteins have been described as antioxidants in animal models (Yu *et al.*, 2008) and *in vitro* assays (Darmawan *et al.*, 2010). Antioxidant peptides from soybean storage proteins described by different authors are shown in Table 23.2. Soy enzymatic hydrolyzates are used as ingredients in nutritious beverages, infant food, sport nutrition, and patient food formulae (Sun, 2011). Depending on the quantitative and qualitative composition of soy proteins in soy-based drinks, products are expected to display variation in their antioxidant properties (Lee *et al.*, 2002).

BBI is a minor component of soy proteins present in soy drinks. BBIs are soluble low-molecular-weight (MW ~8000 Da) proteins, consisting of a chain of 71 amino acids cross linked by seven disulfide bonds (Birk, 1985). Huang *et al.* (2010) have demonstrated that BBI shows

Aglycones	R_1		R_2	Compound
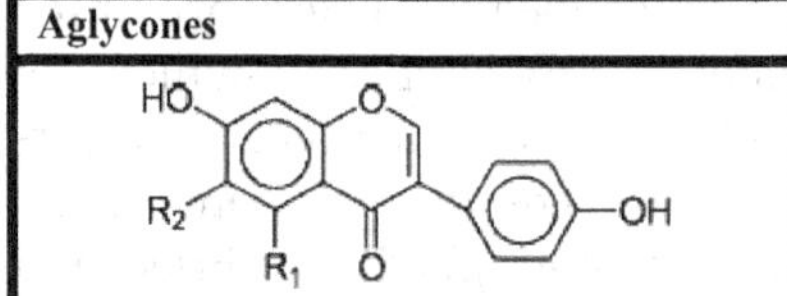	H		H	Daidzein
	OH		H	Genistein
	H		OCH_3	Glycitein
Glucosides	R_3	R_4	R_5	**Compound**
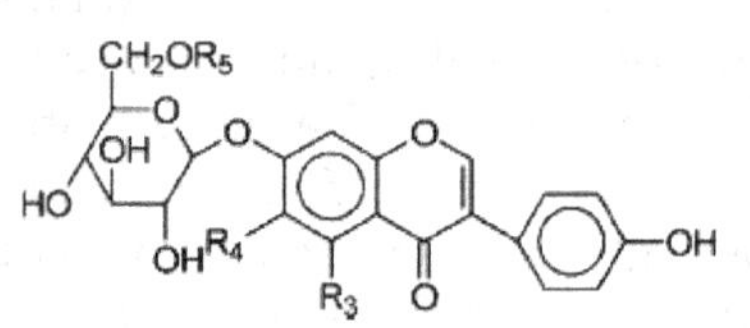	H	H	H	Daidzin
	OH	H	H	Genistin
	H	OCH_3	H	Glycitin
	H	H	$COCH_3$	Acetyldaidzin
	OH	H	$COCH_3$	Acetylgenistin
	H	OCH_3	$COCH_3$	Acetylglycitin
	H	H	$COCH_2COOH$	Malonyldaidzin
	OH	H	$COCH_2COOH$	Malonylgenistin
	H	OCH_3	$COCH_2COOH$	Malonylglycitin

FIGURE 23.1 Soybean isoflavones (Shimoni, 2004).

a dose-dependent antioxidant activity in 2,2′-azino-bis (3-ethylbenzothiazoline-6-sulfonic acid) (ABTS), 2,2-diphenyl-1-picrylhydrazyl (DPPH), reducing power, and chelating ability assays, and protection against hydroxyl radicals. BBI may resist the processing and storage conditions (Losso, 2008).

Soy milk and other soy-based drinks also contain minerals such as copper and zinc (Table 23.1), which are constituents of the natural antioxidant enzyme defence of the organism (copper/zinc superoxide dismutase, cytochrome c oxidase) and can regulate different gene expression in relation to aging and inflammation (Mocchegiani *et al.*, 2012). Health claims of these micronutrients in relation to protection of cells from oxidative stress have been approved by EFSA (2009a, b).

During processing, several chemical events affecting soy-based drink composition and functionality take place. Chemical events such as isomerization, esterification, hydrolysis, and Maillard reaction may generate novel compounds possessing antioxidant characters that affect the overall antioxidant properties of the ready-to-drink beverage. The impact of different processing steps in the antioxidant profile and overall antioxidant capacity of soy milk, as a fundamental ingredient for formulation of soy-based-drinks, is discussed in this chapter.

TABLE 23.2 Soybean Antioxidant Peptides

Protein	Amino Acid Sequence	Free Radical Scavenging
β-Conglycinin	VNPHDHQN, LVNPHDHQN, LLPHH, LLPHHADADY, VIPAGYP, LQSGDALRVPSGTTYY[1]	Peroxidation of linoleic acid
β-Conglycinin	FESFFL, FQTLF, SYLQGF, TTYY, LFF, LY, IY[2]	$ABTS^{\bullet+}$
Glycinin	WNLNAN, SLDFPALW, LY, IY[2]	$ABTS^{\bullet+}$
Glycinin	GVAW, GVAWW, GVAWWM, GVAWWMY, WWMY, SWWPF, WTY[3]	$ABTS^{\bullet+}$, peroxyl radicals, prevent oxidation of linoleic acid, superoxide ion
Bowman–Birk inhibitor[4]		$ABTS^{\bullet+}$, DPPH, reducing power, Fe^{2+}-chelating activity (•OH)

[1]*Chen* et al., *1995.*
[2]*Beermann, C., Euler, M., Herzberg, J., and Stahl, B. (2009). Anti-oxidative capacity of enzymatically released peptides from soybean protein isolate.* Eur. Food Res. Technol. *229, 637–644.*
[3]*Gao* et al., *2004.*
[4]*Huang* et al., *2010.*
$^{\bullet+}$*ABTS, 2,2′-azino-bis (3-ethylbenzothiazoline-6-sulfonic acid); DPPH, 2,2-diphenyl-1-picrylhydrazyl*

EFFECTS OF PRODUCTION AND PROCESSING

The content of isoflavones and proteins in soybean seeds is significantly higher compared to their content in soy milk and soy-based drinks. Generally, the processed soy drink products showed lower levels of isoflavones and proteins than the raw material, and the main reasons for this effect are the soaking and heating processes, which caused significant losses of these antioxidant compounds (Ishihara *et al.*, 2007; Jackson *et al.*, 2002). It is known that the antioxidant properties are more pronounced in soybean seed extracts, whereas the processed products, such as soy milk, show reduced antioxidant activity; thus, free radical-scavenging activity might be affected during processing (Lee *et al.*, 2007). The basic manufacturing steps of soy milk preparation include selection and dehulling of soybeans, soaking in water, wet grinding and filtering of soy milk from fibre (okara), cooking to inactivate lipoxygenase and trypsin inhibitors, aseptic packaging, and storage. All these processing operations will determine the final content and profile of isoflavones and proteins and thus, its final antioxidant properties. Figure 23.2 describes the traditional manufacturing process for the soy milk.

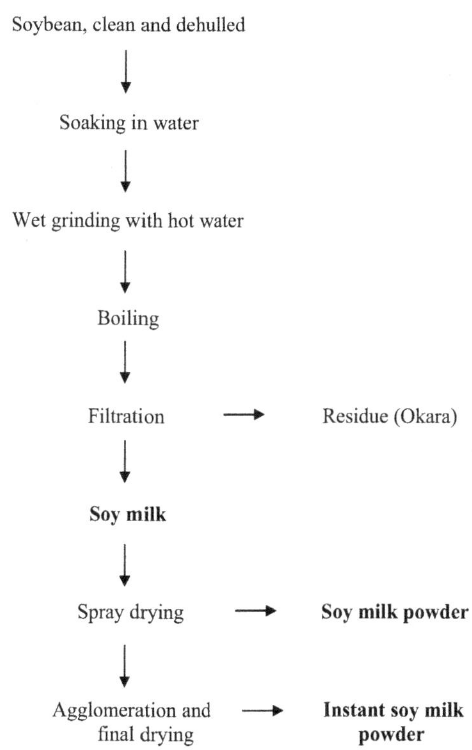

FIGURE 23.2 Traditional soy milk process (Giri and Mangaraj, 2012).

The isoflavone content of soy milk can be highly variable, and thus appears to be partially responsible for the large variability of isoflavone contents and antioxidant properties in soy beverages. It is known that different commercial soy milks present different antioxidant properties. These differences are probably the result of variabilities in the isoflavone contents of soybeans used for soy milk production and the amount of other ingredients added to instant soy milk. In raw soy milk the predominant isoflavone forms are the malonyl-glucosides, but after manufacturing steps, predominant forms are the β-glucoside conjugates due to the de-esterification of malonyl- and acetyl-glucosides (Kao *et al.*, 2004), while aglycones appear in only very small proportions; however, the different heat processing methods impact the transformation of isoflavones. Figure 23.3 shows the chemical conversions of isoflavones during soy food processing.

Raw Material

The protein and isoflavone content in soy-based drinks and thus their antioxidant properties depend on the raw materials, which are influenced by cultivation conditions, environmental factors, variations in genotype, crop location, harvesting time, storage period, humidity, temperature, etc. (Kim *et al.*, 2012; Lee *et al.*, 2003). The content of individual and total isoflavones and proteins in soy milk is higher when it comes from soybean cultivars with higher protein content and therefore these cultivars will be more suitable for soy milk production (Šertović *et al.*, 2012). Some studies of the different commercial soy beverages showed that similar products from different manufacturers often contained different amounts of isoflavones and thus different antioxidant properties (Rau De Almeida Callou *et al.*, 2010). The explanation could be that the soybean raw materials used might be different, with variation in the content of isoflavone glucoside and aglycone forms. The mineral content is also affected by soybean genotype and growing location (Bhardwaj *et al.*, 2007), and therefore this could affect the final antioxidant properties of the soy drink.

Soaking Process

During processing of soy milk, soybeans have to be soaked and heated for a certain period of time. It is known that the concentration of total isoflavones is decreased during the soaking process, the soaking temperature being the main factor in this effect (Kao *et al.*, 2004). However, during this process the concentrations of the three aglycone forms, which are known to possess better antioxidant properties, increase with the soaking temperature and time, while a reversed trend is produced for the other glucoside forms. This effect can be explained because of the conversion of β-glucosides (Kao *et al.*, 2004). In fact, different studies have demonstrated that during soaking process, malonyl-glucosides can be transformed to acetyl-glucosides, which could be further converted to β-glucosides or aglycones depending on soaking temperature and time (Jackson *et al.*, 2002). These inter-conversion reactions of isoflavones during soaking process can be accelerated by heat, acidic or alkaline conditions, and enzymes (Vaidya *et al.*, 2007).

Grinding Process

This process of soaked-soybean in boiling water seems to result in the main loss of isoflavones during processing of soy milk. During grinding malonyl-glucosides are de-esterified to aglycone forms. Beforehand, to obtain soy milk the soy flakes are washed in water or alcohol to remove soluble carbohydrates during soy protein concentrate production. Alcohol washing results in the loss of most of the isoflavones, whereas a substantial amount

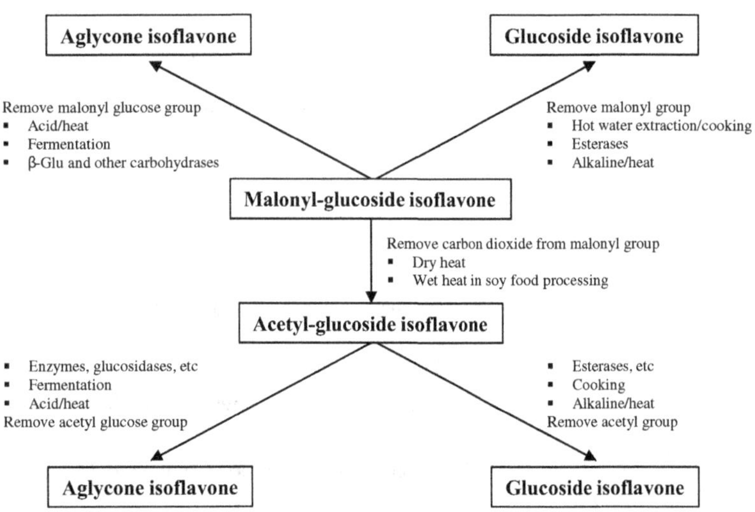

FIGURE 23.3 Chemical conversions of isoflavones during soy food processing (Morales-de la Peña *et al.*, 2010a).

of isoflavones are retained after water washing. Finally, when the slurry is filtered to extract the soy beverage, approximately two-thirds of the isoflavones are retained in the soy beverage (Jackson *et al.*, 2002).

Heating Process

Soy milk production includes heat treatment to inactivate lipoxygenase and trypsin inhibitors. In general, isoflavone compounds are quite heat stable and are not destroyed during the heating process. Therefore, conventional thermal processing operations generally have no impact on the total isoflavone content on soy milk; however, the composition of the isoflavone profile may be altered during heat processing and consequently their antioxidant properties. When the heating process is excessive a decrease in the total isoflavone content may occur.

Heat treatment promotes the conversion of predominant malonyl-glucoside conjugates to more stable acetyl groups and β-glucosides (Choi and Rhee, 2006; Genovese, 2002). Therefore, malonyl conjugates are decreased significantly during thermal processing due to the de-esterification into acetyl and glucosidic conjugates, which are significantly increased. Genistein and daidzein, known for their high antioxidant properties, are degraded when exposed to high temperatures, but daidzein is generally more labile to heat treatments in soy milk (Eisen *et al.*, 2003; Ungar *et al.*, 2003). Thus, the degradation of these isoflavones present in soy beverages during heat processing will affect their antioxidant properties. Interestingly, daidzein degradation products may exhibit important antioxidant activity, whereas genistein degradation products show only minor antioxidant activity (Ungar *et al.*, 2003).

During thermal treatment, proteolysis, cross-linking, oxidation, and Maillard reaction may cause changes in the protein profile of soy milk and affect its final antioxidant properties (Amigo-Benavent *et al.*, 2008). However, chemical events such as Maillard reaction and caramelization can occur simultaneously during soy milk thermal processing and might give rise to neo-antioxidants. In simplified model systems with soy protein isolate and carbohydrates (glucose, fructose, and fructo-oligosaccharides), ingredients that can be found in soy beverages, different authors have reported an increase in antioxidant activity due to Maillard reaction (Amigo-Benavent *et al.*, 2010; Mesa *et al.*, 2008). Therefore, it has been suggested that both remaining natural antioxidants and those formed during processing could contribute to the overall antioxidant capacity of the soy-based drink final product.

Spray-drying, a common thermal technological approach employed for producing powdered soy milk, causes a higher reduction of the antioxidant properties of soy foods than that provoked by freeze-drying (Wang *et al.*, 2006).

Aseptic Packing Process

Usually, an additional thermal treatment is applied to prolong the shelf-life of soy beverages through inactivation of pathogenic microorganisms and anti-nutritional factors; however, the severity of heat may alter the isoflavone and protein profile of soy beverages and accordingly their final antioxidant properties. A variety of heating methods and time–temperature conditions are used by soy milk manufacturers, such as, pasteurization, in-container sterilization and ultra-high-temperature (UHT) treatment (Deshpande *et al.*, 2008).

The isoflavones are rather stable compounds and there is little evidence for thermal degradation of these compounds. The conversion of the isoflavone to aglycones in soy milk during these heating processes is of particular interest, because there is evidence that the aglycones, especially genistein, show the greatest biological activity (such as antioxidant properties) of the soy isoflavones. Pasteurization refers to a heat treatment, normally below 100°C and generally at 95°C for 60 min. This process in soy milk causes de-esterification of malonyl- and acetyl-glucosides to their β-glucoside and aglycone forms. UHT processing involves the use of high temperatures (135–150°C) for short times (a few seconds) to achieve a product that is commercially sterile. Commercial soy beverages in cans or glass bottles are sterilized at 121°C for 5 min to achieve commercial sterility. This process destroys about 87% of daidzein, 72% of glycitein and 17% of genistein in soy milk (Huang *et al.*, 2006). It can be seen that high-temperature, short-time heat treatment greatly minimizes the degradation of the isoflavone aglycones in soy milk.

Recently, different non-thermal processing technologies have been employed in order to avoid the negative effects of heat on the overall antioxidant quality of soy beverages. High-intensity pulsed electric fields (HIPEF) treatment has been shown to be a potential process capable of preserving soy beverages and does not cause significant changes in the total isoflavone content (Morales-de la Peña *et al.*, 2010a). The application of HIPEF can provide soy beverages with similar isoflavone content to the fresh beverage, as well as better quality products than those produced by heat pasteurization processes. Moreover, the antioxidant capacity of treated beverages by HIPEF process is not modified with respect to the untreated sample (Morales-de la Peña *et al.*, 2010b).

There are few studies reporting results about the effects of HIPEF or thermal processes on mineral content of soy beverages, but recently, it has been demonstrated that the initial content of some minerals in the soy beverages can be affected in different ways depending on the treatments applied, thus while the concentration of Ca and Mg remain with no significant changes after HIPEF or thermal processing, the Cu content is slightly diminished and the concentration of Zn and Fe augmented after HIPEF processes (Morales-de la Peña *et al.*, 2010b).

High-pressure processing (HPP), a novel food processing and preservation technology, has also been tried as an alternative to thermal processing for soy milk elaboration. This methodology applied in soy milk production overall induced small changes in isoflavone profile and total content (Jung *et al.*, 2008).

Storage Process

Storage conditions are also important factors that can affect the antioxidant properties of soy isoflavones present in soy-based drinks. In general, soy isoflavones are stable compounds during cold storage, but the soluble isoflavones and antioxidant capacity may be reduced by storage of soy beverages. No degradation of total isoflavones in soy food has been observed throughout storage time; nevertheless, they can be subjected to several inter-conversions (Uzzan and Labuza, 2004). The least stable compounds of the isoflavones (malonyl- and acetyl-conjugates) continue to de-esterify into the more stable β-glucoside forms during storage, and the degree of change is affected mostly by storage temperature and is more pronounced at higher storage temperatures. Some authors have observed a decrease of radical scavenging capacity and total phenolic content during soy beverages storage, and this effect is more pronounced after 9 months of storage (Rau de Almeida *et al.*, 2010). Up to now, very little information has been available on the effects of storage on mineral content of soy-based drinks; however, according to Morales-de la Peña *et al.* (2010b) minerals are generally stable in soy beverages throughout storage time irrespective of the treatment applied.

Fermentation Process

Fermentation can enhance nutritional value, sensory properties, and functional qualities of soy milk and soy-based drinks. In fact, during this technological process, the antioxidant properties of soy beverages can be improved by the conversion of isoflavones to their aglycone forms, peptides releasing from soy proteins, and the intracellular antioxidants of starter microorganisms which can be released to the medium (Lin *et al.*, 2012; Wang *et al.*, 2006). Different probiotic organisms have been used as starter cultures to ferment soy milk: *Lactobacillus plantarum*, *L. casei*, *L. bulgaricus*, *L. fermentum*, and *Bifidobacterium* spp. (Gobbetti *et al.*, 2010).

Non-fermented soy milk contains β-glucoside conjugates, whereas fermented soy milk contains the unconjugated isoflavone, aglycones. This manufacturing process decreases the isoflavone malonyl-glucoside, acetyl-glucoside and β-glucoside content with a corresponding increase of the aglycone form content. Thus, fermentation process of soy milk with lactic acid bacteria and/or bifidobacteria with high β-glucosidase activity, can improve the antioxidant activity (Cheng *et al.*, 2011; Wang *et al.*, 2006). Therefore, it is possible after fermentation to obtain isoflavone aglycone-rich soy-based drinks. For example, fermented black soybean milk possesses high antioxidant activity as well as cytotoxic activity toward cancer cells due to the higher daidzein and genistein (aglycone forms) content compared with the unfermented soy milk (Cheng *et al.* 2011). Moreover, it is well established that the isoflavone aglycones formed during fermentation process have a higher antioxidant activity than the β-glucosides (Marazza *et al.*, 2012).

Traditionally, authors have related protein fermentation to an improvement in food nutritional value (Lin *et al.*, 2012). Bioactive peptides can be formed from soy proteins due to microbial protease action, but little information regarding to the effect of protein fermentation on antioxidant properties of soy beverages has been published so far. Previous studies have reported the relationship between the degree of hydrolysis of soy proteins and their antioxidant capacity (Moure *et al.*, 2006; Peña-Ramos and Xiong, 2002). In general, it is accepted that antioxidant capacity of soy proteins is correlated positively with degree of hydrolysis to some extent and, moreover, the amino acid composition (Peña-Ramos and Xiong, 2002). Table 23.2 shows the sequence of the main antioxidant peptides described in soy proteins. These peptides can be produced by microbial fermentation or enzymatic hydrolysis in the production of soy-based drinks. For example, soybean protein hydrolyzates obtained by *Bacillus subtilis* fermentation show antioxidant capacity against superoxide and hydroxyl radicals (Yu *et al.*, 2008). These peptides and other hydrolyzates might be used in soybean ready-to-drink and powered beverages for commercial sports nutrition to improve performance (Paul, 2009). Peptides derived from soy glycinin protein, one of the most abundant storage proteins in soybean, have shown a high degree of antioxidant activity (ABTS, superoxide and alkyl peroxyl radicals, and prevent the oxidation of linoleic acid) after the enzymatic hydrolysis with alcalase/corolase (3h at 50°C) and have been patented for food uses (Gao *et al.*, 2004). Hydrolysis of soy β-conglycinin protein, the other most abundant storage protein in soybean, with different proteolytic microorganisms releases LLPHH peptides, which show more antioxidant capacity than butylated hydroxyanisole (BHA), a common food antioxidant additive (Chen *et al.*, 1995).

Acknowledgments

The authors would like to thank Projects AGL2010-17779 (NATURAGE), CONSOLIDER INGENIO 2010 CSD2007-00063 FUN-C-FOOD (Ministerio de Educación y Ciencia) and ALIBIRD, project S-505/AGR-0153, (Comunidad Autónoma de Madrid) for economical support. J.M.S and M.A. would like to thank CSIC for their JAE-DOC research contracts.

References

Amigo-Benavent, M., del Castillo, M.D., Fogliano, V., 2010. Are the major antioxidants derived from soy protein and fructo-oligosaccharides model systems colored aqueous soluble or insoluble compounds? Eur. Food Res. Technol. 231, 545–553.

Amigo-Benavent, M., Silván, J.M., Moreno, F.J., Villamiel, M., del Castillo, M.D., 2008. Protein quality, antigenicity, and antioxidant activity of soy-based foodstuffs. J. Agric. Food Chem. 56, 6498–6505.

Bhardwaj, H.L., Hamama, A.A., Rangappa, M., Joshi, J.M., Sapra, V.T., 2007. Mineral composition of tofu made from twelve soybean genotypes grown at three locations in USA. World J. Agric. Sci. 3, 140–144.

Birk, Y., 1985. The Bowman-Birk inhibitor. Int. J. Pept. Protein Res. 25, 113–131.

Chen, H.M., Muramoto, K., Yamauchi, F., 1995. Structural-analysis of antioxidative peptides from soybean beta-conglycinin. J. Agric. Food Chem. 43, 574–578.

Cheng, K.C., Lin, J.T., Liu, W.H., 2011. Extracts from fermented black soybean by immobilized *Rhizopus oligosporous* exhibit both antioxidative and cytotoxic activities. Food Technol. Biotechnol. 49, 111–117.

Choi, M.S., Rhee, K.C., 2006. Production and processing of soybeans and nutrition and safety of isoflavone and other soy products for human health. J. Med. Food 9, 1–10.

Darmawan, R., Bringe, N.A., de Mejia, E.G., 2010. Antioxidant capacity of alcalase hydrolysates and protein profiles of two conventional and seven low glycinin soybean cultivars. Plant Foods Hum. Nutr. 65, 233–240.

Deshpande, R.P., Chinnan, M.S., Phillips, R.D., 2008. Process development of a chocolate-flavored peanut-soy beverage. Int. J. Food Sci. Technol. 43, 886–894.

EFSA Panel on Dietetic Products Nutrition and Allergies (NDA), 2009a. Scientific Opinion on the substantiation of health claims related to copper and protection of DNA, proteins and lipids from oxidative damage (ID 263, 1726), function of the immune system (ID 264), maintenance of connective tissues (ID 265, 271, 1722), energy-yielding metabolism (ID 266), function of the nervous system (ID 267), maintenance of skin and hair pigmentation (ID 268, 1724), iron transport (ID 269, 270, 1727), cholesterol metabolism (ID 369), and glucose metabolism (ID 369) pursuant to Article 13(1) of Regulation (EC) No 1924/2006. EFSA J. 7, 1211.

EFSA Panel on Dietetic Products Nutrition and Allergies (NDA), 2009b. Scientific Opinion on the substantiation of health claims related to zinc and function of the immune system (ID 291, 1757), DNA synthesis and cell division (ID 292, 1759), protection of DNA, proteins and lipids from oxidative damage (ID 294, 1758), maintenance of bone (ID 295, 1756), cognitive function (ID 296), fertility and reproduction (ID 297, 300), reproductive development (ID 298), muscle function (ID 299), metabolism of fatty acids (ID 302), maintenance of joints (ID 305), function of the heart and blood vessels (ID 306), prostate function (ID 307), thyroid function (ID 308), acid-base metabolism (ID 360), vitamin A metabolism (ID 361) and maintenance of vision (ID 361) pursuant to Article 13(1) of Regulation (EC) No 1924/2006. EFSA J. 7, 1229.

Eisen, B., Ungar, Y., Shimoni, E., 2003. Stability of isoflavones in soy milk stored at elevated and ambient temperatures. J. Agric. Food Chem. 51, 2212–2215.

Friedman, M., Brandon, D.L., 2001. Nutritional and health benefits of soy proteins. J. Agric. Food Chem. 49, 1069–1086.

Gao, S., Miller, M., Han, X.Q., 2004. Peptide antioxidants from soy protein. U.S. patent 2004/0192611.

Garcia, M.C., Torre, M., Marina, M.L., Laborda, F., 1997. Composition and characterization of soyabean and related products. Crit. Rev. Food Sci. Nutr. 37, 361–391.

Genovese, M.I., Lajolo, F.M., 2002. Isoflavones in soy-based foods consumed in Brazil: levels, distribution, and estimated intake. J. Agric. Food Chem. 50, 5987–5993.

Giri, S.K., Mangaraj, S., 2012. Processing influences on composition and quality attributes of soymilk and its powder. Food Eng. Rev. 4, 149–164.

Gobbetti, M., Di Cagno, R., De Angelis, M., 2010. Functional microorganisms for functional food quality. Crit. Rev. Food Sci. Nutr. 50, 716–727.

Huang, G.J., Chiu, C.S., Wu, C.H., Huang, S.S., Hou, W.C., Amegaya, S., Sheu, M.J., Liao, J.C., Lin, Y.H., 2010. Redox status of Bowman-Birk inhibitor from soybean influence its *in vitro* antioxidant activities. Bot. Stud. 51, 431–437.

Huang, H., Liang, H., Kwok, K.C., 2006. Effect of thermal processing on genistein, daidzein and glycitein content in soymilk. J. Sci. Food Agric. 86, 1110–1114.

Ishihara, M., Singh, H., Chung, G., Tam, C., 2007. Content composition and antioxidant activity of isoflavones in commercial and homemade soymilk and tofu. J. Sci. Food Agric. 87, 2844–2852.

Jackson, C.J.C., Dini, J.P., Lavandier, C., Rupasinghe, H.P.V., Faulkner, H., Poysa, V., Buzzell, D., DeGrandis, S., 2002. Effects of processing on the content and composition of isoflavones during manufacturing of soy beverage and tofu. Process Biochem. 37, 1117–1123.

Jung, S., Murphy, P.A., Sala, I., 2008. Isoflavone profiles of soymilk as affected by high-pressure treatments of soymilk and soybeans. Food Chem. 111, 592–598.

Kao, T.H., Lu, Y.F., Hsieh, H.C., Chen, B.H., 2004. Stability of isoflavone glucosides during processing of soymilk and tofu. Food Res. Int. 37, 891–900.

Kim, E.H., Kim, S.L., Kim, S.H., Chung, I.M., 2012. Comparison of isoflavones and anthocyanins in soybean [*Glycine max* (l.) Merrill] seeds of different planting dates. J. Agric. Food Chem. 60, 10196–10202.

Lee, S.J., Chung, I.M., Ahn, J.K., Lee, S.K., Kim, S.H., Yoo, N.H., 2002. Variation in antioxidative activity of soybean (*Glycine max* L.) varieties with crop year and duration of storage time. Food Sci. Biotechnol. 11, 649–653.

Lee, S.J., Yan, W., Ahn, J.K., Chung, I.M., 2003. Effects of year, site, genotype and their interactions on various soybean isoflavones. Field Crops Res. 81, 181–192.

Lee, Y.W., Kim, J.D., Zheng, J., Row, K.H., 2007. Comparisons of isoflavones from Korean and Chinese soybean and processed products. Biochem. Eng. J. 36, 49–53.

Lin, C.C., Wu, P.S., Liang, D.W.M., Kwan, C.C., Chen, Y.S., 2012. Quality, antioxidative ability, and cell proliferation-enhancing activity of fermented black soybean broths with various supplemental culture medium. J. Food Sci. 77, C95–C101.

Losso, J.N., 2008. The biochemical and functional food properties of the Bowman-Birk inhibitor. Crit. Rev. Food Sci. Nutr. 48, 94–118.

Marazza, J.A., Nazareno, M.A., Savoy de Giori, G., Garro, M.S., 2012. Enhancement of the antioxidant capacity of soymilk by fermentation with *Lactobacillus rhamnosus*. J. Funct. Foods 4, 594–601.

Mesa, M.D., Silván, J.M., Olza, J., Gil, A., del Castillo, M.D., 2008. Antioxidant properties of soy protein-fructooligosaccharide glycation systems and its hydrolyzates. Food Res. Int. 41, 606–615.

Mocchegiani, E., Costarelli, L., Giacconi, R., Piacenza, F., Basso, A., Malavolta, M., 2012. Micronutrient (Zn, Cu, Fe)-gene interactions in ageing and inflammatory age-related diseases: implications for treatments. Ageing Res. Rev. 11, 297–319.

Monje, M.C., Berger, M., Farines, V., Reybier, K., Verger, A., Dayde, J., Theodorou, V., Nepveu, F., 2006. Antioxidant capacity of cotyledons and germs of soybean in relation to their isoflavone content. Food Technol. Biotechnol. 44, 493–498.

Morales-de la Peña, M., Salvia-Trujillo, L., Rojas-Graü, M.A., Martín-Belloso, O., 2010a. Isoflavone profile of a high intensity pulsed electric field or thermally treated fruit juice-soymilk beverage stored under refrigeration. Innovative Food Sci. Emerg. Technol. 11, 604–610.

Morales-de la Peña, M., Salvia-Trujillo, L., Rojas-Graü, M.A., Martín-Belloso, O., 2010b. Impact of high intensity pulsed electric field on antioxidant properties and quality parameters of a fruit juice-soymilk beverage in chilled storage. Food Sci. Technol. 43, 872–881.

Moure, A., Dominguez, H., Parajo, J.C., 2006. Antioxidant properties of ultrafiltration-recovered soy protein fractions from industrial effluents and their hydrolysates. Process Biochem. 41, 447–456.

Paul, G.L., 2009. The rationale for consuming protein blends in sports nutrition. J. Am. Coll. Nutr. 28, 464S–472S.

Peña-Ramos, E.A., Xiong, Y.L.L., 2002. Antioxidant activity of soy protein hydrolysates in a liposomal system. J. Food Sci. 67, 2952–2956.

Rau De Almeida Callou, K., Sadigov, S., Lajolo, F.M., Genovese, M.I., 2010. Isoflavones and antioxidant capacity of commercial soy-based beverages: effect of storage. J. Agric. Food Chem. 58, 4284–4291.

Šertović, E., Mujić, I., Jokić, S., Alibabić, V., Sarić, Z., 2012. Effect of soybean cultivars on the content of isoflavones in soymilk. Rom. Biotechnol. Lett. 17, 7151–7159.

Shimoni, E., 2004. Stability and shelf life of bioactive compounds during food processing and storage: soy isoflavones. J. Food Sci. 69, R160–R166.

Sloan, A.E., 2005. Top 10 global food trends. Food Technol. 59, 20–32.

Sun, X.D., 2011. Enzymatic hydrolysis of soy proteins and the hydrolysates utilisation. Int. J. Food Sci. Technol. 46, 2447–2459.

Ungar, Y., Osundahunsi, O.F., Shimoni, E., 2003. Thermal stability of genistein and daidzein and its effect on their antioxidant activity. J. Agric. Food Chem. 51, 4394–4399.

Uzzan, M., Labuza, T.P., 2004. Critical issues in R&D of soy isoflavone-enriched foods and dietary supplements. J. Food Sci. 69, R77–R86.

Vaidya, N.A., Mathias, K., Ismail, B., Hayes, K.D., Corvalan, C.M., 2007. Kinetic modeling of malonylgenistin and malonyldaidzin conversions under alkaline conditions and elevated temperatures. J. Agric. Food Chem. 55, 3408–3413.

Wang, Y.C., Yu, R.C., Chou, C.C., 2006. .Antioxidative activities of soymilk fermented with lactic acid bacteria and bifidobacteria. Food Microbiol. 23, 128–135.

Yu, B., Lu, Z.X., Bie, X.M., Lu, F.X., Huang, X.Q., 2008. Scavenging and anti-fatigue activity of fermented defatted soybean peptides. Eur. Food Res. Technol. 226, 415–421.

SECTION 3

Selective Assays for Antioxidants

CHAPTER

24

The CUPRAC Methods of Antioxidant Measurement for Beverages

Reşat Apak, Mustafa Özyürek, Kubilay Güçlü, Burcu Bekdeşer, Mustafa Bener

Istanbul University, Istanbul, Turkey

CHAPTER POINTS

- The CUPRAC reagent is selective, because it has a lower redox potential than that of the ferric/ferrous couple in the presence of *ortho*-phenanthroline type or *batho*-phenanthroline type ligands. The standard potential of the Cu(II,I)-Nc redox couple is about 0.6 V, close to that of $ABTS^{+}$/ABTS (E^o = 0.68 V) and FRAP (E^o = 0.70 V). Simple sugars and citric acid are not oxidized with the CUPRAC reagent.
- The CUPRAC reagent is much more stable and easily accessible than the chromogenic radical reagents (e.g., ABTS and DPPH).
- The redox reaction producing colored species is carried out at nearly physiological pH (pH7 of ammonium acetate buffer) as opposed to the unrealistic acidic conditions (pH3.6) of the FRAP or alkaline conditions (pH10, with carbonate) of the Folin assay required for phenolics to dissociate their protons. Under more acidic conditions than the physiological pH, the reducing capacity may be suppressed due to protonation on antioxidant compounds, whereas in more alkaline conditions proton dissociation of phenolics would enhance a sample's reducing capacity. Thus, CUPRAC gives a better estimate of *in vivo* antioxidant behavior.
- Sensor-based CUPRAC method offers advantages over the solution-based assay, including the potential of being used as a rapid, low-cost, diversely, and easily applicable screening tool for total antioxidant capacity (TAC) of plant and food extracts of unknown composition without pretreatment.
- In the most common UV-method for determination of hydrogen peroxide scavenging (HPS) activity, scavenging ability depends on the change of the absorbance value at 230 nm when H_2O_2 concentration is decreased by scavenger compounds. As this UV-method suffers from the interference of some phenolics in real samples having strong absorption in the UV-region, HPS activity of polyphenols was alternatively determined without interference by directly measuring the concentration of undegraded H_2O_2 using the HPS-CUPRAC method in the presence of a Cu(II) catalyst. The proposed methodology is also superior to the rather non-specific horseradish peroxidase-based assays.
- Ascorbic acid in complex matrices could be sensitively determined over a wide concentration range, without interference from flavones and flavonols with La-CUPRAC method.
- CUPRAC methodology, comprising many assays, has evolved into an 'antioxidant-measurement package' in food and biochemistry, and by maintaining the CUPRAC reagent and related chemicals in the laboratory, one can simultaneously measure the TAC and reactive oxygen species (ROS) scavenging activity of antioxidants, most easily in fruit juices and beverages.

INTRODUCTION

Reactive oxygen species (ROS) have gained considerable attention due to their role in the progression of a number of human diseases and carcinogenesis. When the

Processing and Impact on Antioxidants in Beverages
http://dx.doi.org/10.1016/B978-0-12-404738-9.00024-6

organism defence systems (of enzymatic, nonenzymatic, or dietary origin) are overwhelmed by an excessive generation of ROS, a situation of oxidative stress occurs, in which cellular and extracellular macromolecules (i.e., proteins, lipids, and nucleic acids) can suffer oxidative damage (Halliwell and Gutteridge, 1989; Halliwell and Aruoma, 1991).

The term 'antioxidant' encompasses a vast array of substances that are divided into endogenous and exogenous antioxidants. Huang *et al.* (2005) defined dietary antioxidants as "substances which can (sacrificially) scavenge reactive oxygen/nitrogen species (ROS/RNS) to stop radical chain reactions, or can inhibit the reactive oxidants from being formed in the first place." Beverages are predominantly rich sources of phenolic acids, flavonoids, vitamins, and minerals. Orange juices are a rich source of vitamin C, which is an important antioxidant in these juices. Concentration of vitamin C is a significant indicator of orange juice quality (Miller and Rice-Evans, 1997; Rapisarda *et al.*, 1999; Gardner *et al.*, 2000; Arena *et al.*, 2001). Pomegranate juice is an important source of anthocyanins and anthocyanidins. It also contains significant levels of citric acid and ascorbic acid (Du *et al.*,1975; El-Nemr *et al.*, 1990). Grapes are a good source of polyphenols such as anthocyanins, procyanidins, gallic acid, cathechin, and epicathechin (Palma and Taylor, 1999; Souquet *et al.*, 1996; Yamakoshi *et al.*, 1999). In apple juices there are four classes of polyphenol (phenolic acid, dihydrochalcones, catechins, procyanidins) (Spanos *et al.*, 1990; Oleszek *et al.*, 1988). Cherries are rich in phenolics known to contain high levels of anthocyanins, and show high radical scavenging activity (Proteggente *et al.*, 2002). Apricot juice is an important source of provitamin A, carotenoids, β-carotene, chlorogenic and neochlorogenic acids, (+)-catechin, (−)-epicathechin, and rutin (Bolin and Stafford, 1974; Ruiz *et al.*, 2005; Radi *et al.*, 1997). Flavonoids are secondary plant metabolites occurring widely in plant foods such as fruits, vegetables, cereals, and beverages (Harborne, 1989; Cook and Samman, 1996). Fruits and vegetables contain a variety of flavonoids and phenolic compounds in amounts ranging from traces to several grams per kilogram of fresh weight (Bravo, 1998). Beverages make a significant contribution to the polyphenol intake of the diet. Flavonoids and other plant phenolics have a significant antioxidant activity with respect to vitamins C and E (Scott *et al.*, 1993; Rice-Evans *et al.*, 1997). The antioxidant activity of this diverse group of compounds depends on the individual structure and number of hydroxyl groups (Richelle *et al.*, 2001).

Commercial fruit juice products are produced by processing techniques (high pressure, thermal, and physical) that may have profound effects on the levels of antioxidants within the plant and the separate parts such as peel and seeds. Processing (heating, separation, and storage techniques etc.) causes both increases and decreases in antioxidant activity.

Beverages have an important role to deliver high concentrations of dissolved functional ingredients (i.e., in sports and performance beverages, ready-to-drink teas, vitamin-enhanced water, and energy beverages). In addition, most consumers prefer beverages that are convenient and provide widely accepted methods of taking functional components. Beverages may contribute to increasing the consumption of fruits to the balance between recommendations and actual intake, although juice intake does not equate to intake of whole fruits. (Wootton-Beard and Ryan, 2011).

The consumption of fruits has been inversely associated with morbidity from degenerative diseases (La Vecchia *et al.*, 2001; Terry *et al.*, 2001; Cohen *et al.*, 2000). Epidemiologic studies show that the health implications of dietary components rely on the intake estimates in sample populations found in databases that list the commonly consumed beverage components. Thus, the availability of a complete beverage composition database is important. Because of the chemical diversity of antioxidant compounds in beverages, complete databases on beverage antioxidant content are not yet available (Pellegrini *et al.*, 2003).

Consumption of antioxidant-rich beverages naturally bearing antioxidant activity is the most efficient way of combating tissue injuries, undesired transformations, and health risks. The chemical diversity of antioxidants makes it difficult to separate and quantify antioxidants from the food matrix. There are a number of commercial ready-to-drink polyphenol-rich beverages, which base their marketing strategies on antioxidant potency. It is of great interest to the general public to know the antioxidant capacity of the beverages that they consume. A variety of molecular spectroscopic assays to determine the total antioxidant capacity (TAC) of food and plant extracts, beverages, and biological fluids (Huang *et al.*, 2005) have been developed, which may be broadly classified as electron transfer (ET)-based [cupric reducing antioxidant capacity (CUPRAC), Folin–Ciocalteu, ferric reducing antioxidant power (FRAP), ferricyanide], hydrogen atom transfer (HAT)-based [oxygen radical absorbance capacity (ORAC)], and mixed ET/HAT-based (2,2′-azino-bis(3-ethylbenzothiazoline-6-sulfonic acid: ABTS-persulfate/Trolox equivalent antioxidant capacity: TEAC, 2,2′-diphenyl-1-picrylhydrazyl: DPPH) radical assays (Prior *et al.*, 2005; Özyürek *et al.*, 2011). However, it should be cautioned that because of the inherent complexity of food matrices, the use of a single antioxidant capacity method to determine antioxidant potency is ineffective. This is because antioxidants respond to different reactive species in different tests, which is partially attributed to different redox potentials of TAC reagents, multiple reaction mechanisms and reaction phases (Seeram *et al.*, 2008). However, the CUPRAC methodology is reliable in determining the antioxidant capacity and activity of a wide variety of antioxidants. So this methodology is evolving into an 'antioxidant measurement package' in food chemistry and biochemistry.

THE MAIN CUPRAC METHOD

The CUPRAC method (Apak *et al.*, 2004) was first developed as a total antioxidant capacity assay and its scope has been expanded with some modifications. The chromogenic redox reagent used for the CUPRAC assay is bis(neocuproine)copper(II) [Cu(II)-Nc] chelate. The copper(II)neocuproine reagent is useful at pH 7, easily accessible, stable, selective, and capable of responding to all types of biologically important antioxidants [e.g., ascorbic acid, α-tocopherol, β-carotene, reduced glutathione (GSH), uric acid, and bilirubin] and all types of food and beverage antioxidants (e.g., flavonoids, simple phenolic and hydroxycinnamic acids, and polymeric phenols) regardless of chemical type or hydrophilicity. CUPRAC method was used for TAC determination of various food extracts [fresh and dried apricot (Güçlü *et al.*, 2006), Turkish hazelnuts (Altun *et al.*, 2013), Van herby cheese (Çelik *et al.*, 2008)], beverages [fruit juices and red wine (Güçlü *et al.*, 2005), herbal teas (Apak *et al.*, 2006)], biological samples [tissue and serum (Apak *et al.*, 2005)], and natural dyes containing flavonoids (Bener *et al.*, 2010a).

In the CUPRAC reaction, the absorbance of the orange–yellow colored Cu(I)-Nc chelate formed as a result of the redox reaction of cupric neocuproine reagent with antioxidants (acting as reducing agents) is measured at 450 nm (see Figure 24.1 for the CUPRAC reaction and the chromophore). The reaction conditions (e.g., reagent concentration, pH, and oxidation time at room and elevated temperatures) were optimized (Apak *et al.*, 2004, 2005).

The chromogenic oxidizing reagent of the CUPRAC method (i.e., cupric neocuproine) reacts with n-electron-reductant antioxidants (AOX) in the following manner:

$$n\ Cu(Nc)_2^{2+} + \text{n-electron reductant (AOX)} \rightarrow n\ Cu(Nc)_2^{+} + \text{n-electron oxidized product} + n\ H^+$$

In this reaction, the reactive Ar-OH groups of polyphenolic antioxidants are usually oxidized to the corresponding quinones (Ar=O), and $Cu(Nc)_2^{2+}$ is reduced to the highly colored $Cu(Nc)_2^{+}$ chelate. Although the concentration of Cu^{2+} ions is in stoichiometric excess of that of neocuproine in the CUPRAC reagent for driving the redox equilibrium reaction to the right, the actual oxidant is the $Cu(Nc)_2^{2+}$ species and not just Cu^{2+}, because the standard redox potential of the Cu(II/I)-neocuproine complex is about 0.6 V, much higher that of the single Cu^{2+}/Cu^{+} couple (0.17 V) (Tütem and Apak, 1991). As a result, polyphenols are oxidized much more rapidly and efficiently with Cu(II)-Nc than with Cu^{2+}, and the amount of colored product [i.e., Cu(I)-Nc chelate] emerging at the end of the redox reaction is equivalent to that of reacted Cu(II)-Nc. The liberated protons are buffered in ammonium-acetate medium. The CUPRAC reagent (as an ET-reagent) can oxidize suitably situated phenolic –OH groups to the corresponding aryl oxyl radicals or quinones as long as the corresponding conditional aryl oxyl/phenol or quinone/phenol potential is less than or close to that of cupric/cuprous-neocuproine in neutral medium (Özyürek *et al.*, 2011).

As a distinct advantage over other ET-based and mixed-mechanism TAC assays (e.g., Folin, FRAP, ABTS, DPPH), CUPRAC is superior in regard to its realistic pH close to the physiological pH, favorable redox potential, accessibility and stability of reagents, flexibility, simplicity, low cost, and applicability to lipophilic antioxidants as well as hydrophilic ones.

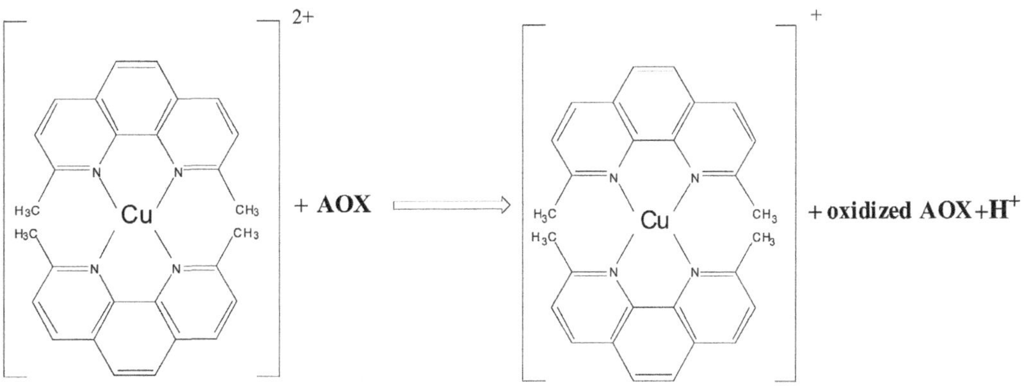

FIGURE 24.1 The CUPRAC reaction and chromophore: Bis(neocuproine) copper(I) chelate cation (protons liberated in the reaction are neutralized by the NH_4Ac buffer).

THE MODIFIED CUPRAC METHODS

The CUPRAC assay does not merely measure the TAC of an antioxidant sample, but has opened the way to many other modified assays for scavenging activity measurement of reactive oxygen species (hydroxyl and superoxide radicals, and hydrogen peroxide).

Sensor Based-CUPRAC Assay

A low-cost optical sensor was developed using a membrane-immobilized CUPRAC reagent, Cu(II)-Nc complex, for the assessment of antioxidant capacity of non-enzymatic antioxidants, their synthetic mixtures, and real samples (Bener *et al.*, 2010b). The Cu(II)-Nc reagent was immobilized onto a cation-exchange polymer (Nafion, a sulfonated tetrafluoroethylene–based copolymer) membrane matrix, and the absorbance changes on the membrane associated with the formation of the colored Cu(I)-Nc chelate as a result of reaction with antioxidants was measured at 450 nm (Figure 24.2). The TEAC (Trolox equivalent antioxidant capacity) coefficients, a unitless quantity defined as the reducing potency, in Trolox mM equivalents, of 1 mM antioxidant solution under investigation measured for various antioxidant compounds, suggest that the reactivity of the Nafion-immobilized reagent is comparable to that of the standard solution-based CUPRAC assay. Testing of synthetic ternary mixtures of antioxidants yielded the theoretically expected CUPRAC antioxidant capacities, in accordance with the principle of additivity of absorbances of mixture constituents obeying Beer's law. This assay was validated through linearity, additivity, precision, and recovery, demonstrating that the optical sensor is reliable and robust. The sensor was used to screen the TAC of some commercial fruit juices such as orange, cherry, peach, and apricot juices, and proved to be an effective tool in measuring the TAC values of food and plant samples without pretreatment. The optical sensor-based CUPRAC assay has been shown not to be adversely affected by common food ingredients (that are not true antioxidants) such as citrate, oxalate, fruit acids, and reducing sugars, and offers good prospects of providing a versatile antioxidant sensor in food industries.

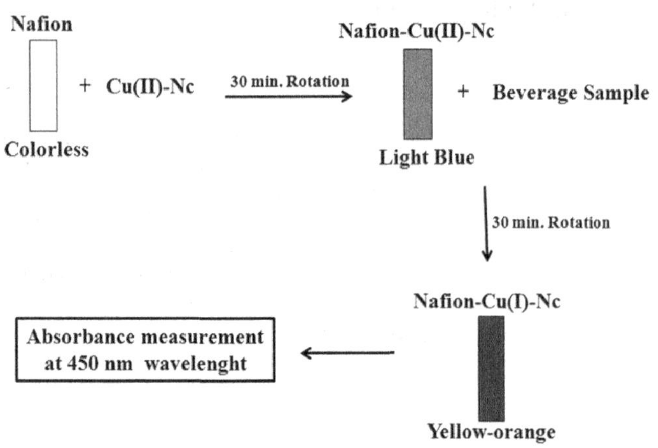

FIGURE 24.2 Procedures for utilization of sensor-based CUPRAC method.

Hydroxyl Radical Scavenging Assay

The CUPRAC procedure can also detect hydroxyl radicals (•OH), and measure the activity of •OH scavengers (Bekdeşer *et al.*, 2012). The terephthalate (TP) probe is used for capturing the hydroxyl radicals (•OH) generated from an equivalent mixture of Fe(II)-EDTA (ethylenediaminetetraacetic acid) with hydrogen peroxide, known as the Fenton system. Hydroxyl radicals attacked both the probe and the •OH scavengers in 37°C-incubated solution for 30 min, and formed a single type of hydroxylated product from the probe responsive to the CUPRAC reagent. Since less hydroxylated probe was formed in the presence of antioxidants, the hydroxyl radical scavenging activity was measured using the decrease in CUPRAC absorbance at 450 nm, arising from the reduction of Cu(II)-Nc reagent to the Cu(I)-Nc chelate by the hydroxylated probe (Figure 24.3).

Under optimal conditions, about half of the probe (TP) was converted into 2-hydroxy terephthalate, and this mono-hydroxylated derivative, being the only product of hydroxylation, was a more specific marker of •OH than the non-specific malondialdehyde end-product of the thiobarbituric acid reactive substances (TBARS) test.

Determination of Ascorbic Acid in the Presence of Flavonoids [Lathanum (La)-CUPRAC Assay]

Ascorbic acid (AA) determination is based on the oxidation of ascorbic acid to dehydroascorbic acid with the CUPRAC reagent, i.e., Cu(II)-Nc, in ammonium acetate-containing medium at pH 7, where the absorbance of the formed Cu(I)-Nc chelate is measured at 450 nm (Özyürek *et al.*, 2007). The flavonoids (essentially flavones and flavonols) normally interfering with the

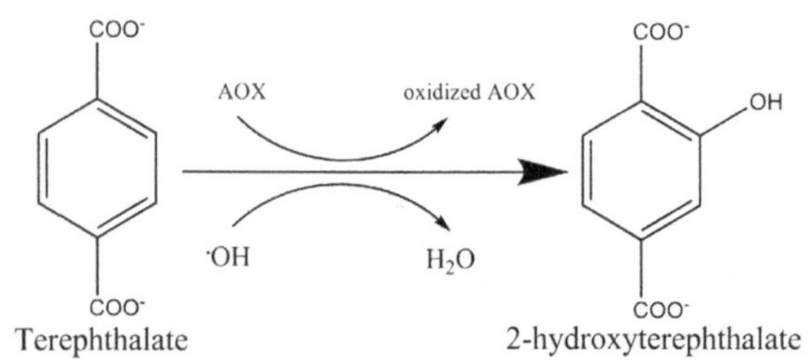

FIGURE 24.3 Schematic representation of the competition between a scavenger compound and terephthalate probe for •OH radicals.

CUPRAC procedure were separated with preliminary extraction as their La(III) chelates into ethyl acetate. The Cu(I)-Nc chelate responsible for color development was formed immediately with AA oxidation. The molar absorptivity for AA in the CUPRAC method was $\varepsilon = 1.60 \times 10^4 M^{-1}cm^{-1}$.

Superoxide Radical Scavenging Activity (SRSA) Assay

A novel probe, *tert*-butylhydroquinone (TBHQ), is used for superoxide radical (SR) nonenzymatically generated in the phenazine methosulfate-β-nicotinamide adenine dinucleotide system. SR attacks both TBHQ and SR scavengers incubated in solution for 30 min where scavengers compete with TBHQ for the $O_2^{\bullet -}$ produced. TBHQ, but not its $O_2^{\bullet -}$ oxidation product, *tert*-butyl-1,4-benzoquinone (TBBQ), is responsive to the CUPRAC spectrophotometric assay (Figure 24.4) (Bekdeşer *et al.*, 2011). The CUPRAC absorbance of the ethyl acetate (EtAc) extract of the incubation solution arising from the reduction of Cu(II)-Nc reagent by the remaining TBHQ was higher in the presence of $O_2^{\bullet -}$ scavengers (due to less conversion to TBBQ), the difference being correlated to the SR scavenging activity of the analytes. With the use of this reaction, a kinetic approach was adopted to assess the superoxide radical scavenging activity (SRSA) of amino acids, vitamins, plasma, and thiol antioxidants.

Hydrogen Peroxide Scavenging (HPS) Assay

HPS activity of antioxidants was determined without interference by directly measuring the concentration of H_2O_2 using the modified CUPRAC spectrophotometric method at 450 nm in the presence of a Cu(II) salt (since H_2O_2 is a relatively stable compound, not scavenged unless transition metal compounds are present as catalysts) (Özyürek *et al.*, 2010). The CUPRAC absorbance of the incubation solution due to the reduction of Cu(II)-Nc reagent by remaining H_2O_2 decreased in the presence of polyphenolics, the difference being proportional to the H_2O_2 scavenging ability of the tested compound.

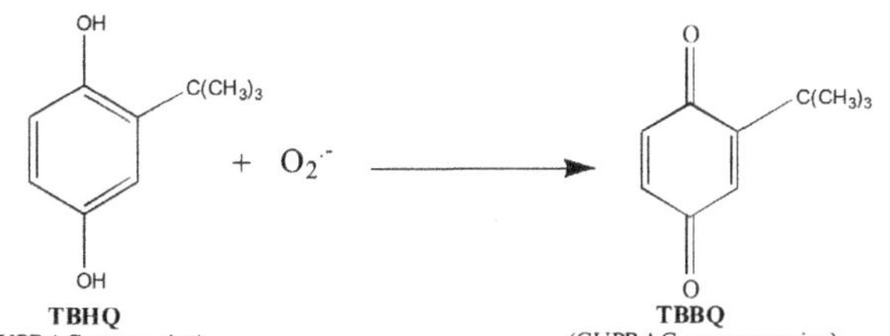

FIGURE 24.4 Schematic representation of the competition between a scavenger compound and TBHQ probe for $O_2^{\bullet -}$ radicals.

PREPARATION OF SOLUTIONS

Preparation of CUPRAC Assay Solutions

$CuCl_2$ solution, 1.0×10^{-2} M, is prepared by dissolving 0.4262 g $CuCl_2.2H_2O$ in water, and diluting to 250 ml. Ammonium acetate buffer at pH = 7.0, 1.0 M, is prepared by dissolving 19.27 g NH_4Ac in water and diluting to 250 ml. Neocuproine solution, 7.5×10^{-3} M, is prepared daily by dissolving 0.039 g Nc in 96% ethanol, and diluting to 25 ml with ethanol.

Preparation of CUPRAC Sensor

The commercial Nafion membrane was sliced into 4.5 × 0.5 cm pieces, immersed into a tube containing 2 ml of 1.0×10^{-2} M Cu(II) + 2 ml of 7.5×10^{-3} M Nc + 2 ml of 1 M NH_4Ac + 2.2 ml of H_2O, and agitated for 30 min in a rotator.

Preparation of Hydroxyl Radical Scavenging Assay Solutions

The Na_2-terephthalate at 10 mM concentration was prepared by dissolving 0.105 g in distilled water and diluting to 50 ml. Fe(II) at 20 mM concentration was prepared by dissolving 0.1988 g $FeCl_2.4H_2O$ with 2 ml of 1 M HCl, and diluting to 50 ml with distilled water. Na_2-EDTA at 20 mM concentration was prepared by dissolving 0.372 g of the salt in distilled water and diluting to 50 ml. Hydrogen peroxide at 10 mM concentration was prepared from a 0.5 M intermediary stock solution, the latter being prepared from 35% H_2O_2 and standardized with permanganate. The NaH_2PO_4–Na_2HPO_4 buffer solution (pH 7.4) at 200 mM was prepared in distilled water.

Preparation of La-CUPRAC Assay Solutions

La(III) aqueous stock solution at 3.0×10^{-3} M concentration (for chelation of phenolics and flavonoids) was prepared from $LaCl_3.7H_2O$.

Preparation of Superoxide Radical Scavenging Assay Solutions

The stock TBHQ solution at 10 mM concentration was prepared by dissolving 0.0831 g in distilled water and diluting to 50 ml. The working solution of TBHQ was prepared by diluting to 1.0 mM with distilled water. β-Nicotinamide adenine dinucleotide (NADH) at 468 μM concentration was prepared by dissolving 0.0173 g of the salt in distilled water and diluting to 50 ml. Phenazine methosulfate (PMS) at 600 μM concentration was prepared by dissolving 0.0183 g of the salt in distilled water and diluting to 100 ml. The working solution of PMS was prepared by diluting the main solution to 60 μM with distilled water. The solutions were prepared daily.

Preparation of Hydrogen Peroxide Scavenging Assay Solutions

Hydrogen peroxide at 1.0 mM concentration was prepared from a 0.5 M intermediary stock solution, the latter being prepared from 30% H_2O_2 and standardized with permanganate titration. The NaH_2PO_4–Na_2HPO_4 buffer solution (pH 7.4) at 0.2 M total phosphate concentration was prepared in water. The original catalase solution of initial activity 1340 U/mg solid was diluted with 0.2 M phosphate buffer (pH 7.4) to a concentration of 268 U/ml.

Preparation of Fruit Juices for CUPRAC Assay

Orange, red grape, pomegranate, apple, cherry, and apricot juices were purchased from a local supermarket. Homogenates were diluted and filtered through a 0.45-μm membrane filter (Chromafil GF/PET-45/25) before analysis.

PROCEDURES FOR THE MAIN AND MODIFIED CUPRAC METHODS

Procedure for the Main CUPRAC Method

To a test tube were added 1 ml each of Cu(II), Nc, and NH_4Ac buffer solutions. Beverage sample solution (x ml) and H_2O ($1.1-x$) ml were added to the initial mixture so as to make the final volume 4.1 ml. The tubes were stoppered, and after 30 min, the absorbance at 450 nm (A_{450}) was recorded against a reagent blank. The UV-Vis spectrophotometer used was Varian CARY 1E, equipped with matched quartz cuvettes. The standard calibration curves of each antioxidant compound were constructed by plotting absorbance *versus* molar concentration, and the molar absorptivity of the CUPRAC method for each antioxidant was found from the slope of the calibration line concerned. The scheme for normal measurement (at room temperature) of hydrophilic antioxidants is summarized as: 1 ml Cu(II) + 1 ml Nc + 1 ml buffer + x ml antioxidant solution + ($1.1-x$) ml H_2O; total vol. = 4.1 ml, measure A_{450} against a reagent blank after 30 min of reagent addition (Apak *et al.*, 2004).

Procedure for the CUPRAC Sensor

The reagent-impregnated membrane was taken out and immersed in a tube containing 8.2 ml of standard antioxidant or real solutions (x ml of sample + ($8.2-x$) ml of EtOH). The tube was placed in a rotator and agitated for 30 min so as to enable color development. The colored membrane was taken out and placed in a 1 mm-width optical cuvette containing H_2O, and its absorbance at 450 nm was read against a blank membrane prepared under identical conditions excluding analyte (Bener *et al.*, 2010b).

Procedures for Hydroxyl Radical Scavenging Assay

To a test tube were added 1.5 ml of phosphate buffer (pH 7.4), 0.5 ml of 10 mM TP (terephthalate probe), 0.25 ml of 20 mM Na_2–EDTA, 0.25 ml of 20 mM $FeCl_2$ solution, ($2-x$) ml H_2O, (x) ml beverage sample solutions at various concentrations, and 0.5 ml of 10 mM H_2O_2 rapidly in this order. The mixture in a total volume of 5 ml was incubated for 30 min in a water bath kept at 37°C. At the end of this period, the reaction was stopped by adding 0.5 ml of 1.0 M HCl solution and 5 ml EtAc vortexed for 30 s (Bekdeşer *et al.*, 2012). To 1.0 ml of the EtAc extract, the solution-based CUPRAC method was applied in the following manner:

1 ml Cu(II) + 1 ml Nc + 1 ml NH_4Ac buffer + 1 ml EtAc extract + 0.5 ml EtOH $\xrightarrow{A_{450}}$ the absorbance at 450 nm of the final solution at 4.5 ml total volume was recorded 30 min later against a reagent blank.

The inhibition ratio of scavengers and beverages (%) was calculated using the following formula:

$$\text{Inhibition ratio }(\%) = 100\,[(A_0 - A)/A_0] \quad \text{(Eq. 24.1)}$$

where A_0 and A are the CUPRAC absorbances of the system in the absence and presence of scavenger, respectively.

Procedure for La-CUPRAC Assay

To (x) ml of beverage sample solutions, ($2x$) ml La(III) and ($10-3x$) ml H_2O were added, and extracted with two successive portions of 10 ml EtAc (x was usually taken as 2 ml). After the separation of phases, 1.0 ml of the lower aqueous phase was taken for analysis with the modified CUPRAC procedure (Özyürek *et al.*, 2007).

Procedure for Superoxide Radical Scavenging Activity (SRSA) Assay

Superoxide anion radicals are generated nonenzymatically by a PMS-NADH system. To a test tube were added 0.5 ml of 1.0 mM TBHQ (probe), 0.5 ml of beverage solution samples at a suitable concentration or water (for reference), 2.0 ml of 468 μM NADH, and 1.0 ml of 60 μM PMS, rapidly in this order. The reaction was started by adding PMS solution. The mixture in a total volume of 4.0 ml was incubated for 30 min in a water bath kept at 25°C. At the end of this period, the remaining TBHQ together with its oxidation product (TBBQ) was extracted by adding 3.0ml of EtAc (extraction of probe) and vortexed for 20s (Bekdeşer *et al.*, 2011). To 1.0ml of the EtAc extract, the modified CUPRAC method was applied in the following manner:

1.0ml of Cu(II) + 1.0ml of Nc + 1.0ml of NH_4Ac buffer + 1.0ml of EtAc extract + 0.1ml of EtOH.

The absorbance at 450nm of the final solution at 4.1ml of total volume was recorded 30min later against a reagent blank.

The IC_{50} (50% inhibitive concentration) values of the scavengers were determined with the use of the NADH-PMS system by means of a linear plot of inhibition percentage as a function of $C_{scavenger}$, where A_0 is the initial CUPRAC absorbance of the original TBHQ probe solution, A_1 and A_2 are those of the TBHQ probe subjected to PMS-NADH action in the absence and presence of SR scavenger, respectively, and C is the molar concentration of relevant scavenger. The IC_{50} values were then compared with those found with the reference NBT method. The SRSA of various scavengers were calculated using the following equation:

$$\text{SRSA}\,(\%) = 100\,[(A_2 - A_1)\,/\,(A_0 - A_1)] \qquad \text{(Eq. 24.2)}$$

Procedure for Hydrogen Peroxide Scavenging (HPS) Assay

To a test tube were added 0.7ml of phosphate buffer (pH7.4), 0.4ml of 1mM H_2O_2, 0.4ml of 0.1mM $CuCl_2.2H_2O$ in this order (hydrogen peroxide incubation solution, used as reference). To two other test tubes were added 0.5ml of phosphate buffer (pH7.4), 0.4ml of 1.0mM H_2O_2, 0.2ml scavenger sample solution, and 0.4ml of 0.1mM $CuCl_2.2H_2O$ solution rapidly in this order (scavenger solutions-I and -II, identical at this stage). The mixtures in a total volume of 1.5ml were incubated for 30min in a water bath kept at 37°C. At the end of this period, to reference and scavenger solution-I was added 0.4ml H_2O and to scavenger solution-II was added 0.4ml of 268Uml^{-1} catalase solution (the contents of scavenger solution-I and -II became different at this stage), and vortexed for 30s. To 1.0ml of the final incubation solutions, the HPS–CUPRAC method was applied in the following manner: 1ml Cu(II) + 1ml Nc + 2ml NH_4Ac buffer + 1ml final incubation solution. The absorbance at 450nm of the final solution at 5.0ml total volume was recorded 30min later against a reagent blank (Özyürek *et al.*, 2010).

The HPS activity of antioxidants and beverages (%) were calculated using the following formula:

$$\text{HPS}\,(\%) = 100\left[\frac{A_0 - (A_1 - A_2)}{A_0}\right] \qquad \text{(Eq. 24.3)}$$

where A_0 is the CUPRAC absorbance of reference hydrogen peroxide incubation solution, A_1 and A_2 the CUPRAC absorbances of scavenger solutions-I and -II, respectively.

APPLICATION OF THE MAIN AND MODIFIED CUPRAC METHODS TO BEVERAGES

Total Antioxidant Capacity (TAC) of Beverages due to Phenolics

Beverages contain many different antioxidant components (flavonoids, vitamins, etc.) demonstrating strong antioxidant capacity/activity. Therefore, the authors aimed to measure the TAC of some commercial ready-to-drink and freshly squeezed fruit juices by using the main (i.e., solution-based) and sensor-based CUPRAC methods. In Figure 24.5, the main and optical sensor-based TAC_{CUPRAC} values of ready-to-drink fruit juices were reported as Trolox equivalents (mM TR). The hierarchy for TAC of ready-to-drink fruit juices with respect to optical sensor-based CUPRAC method was

apple < apricot < orange < cherry < redgrape < pomegranate.

Linear regression analysis of capacity data presented in Figure 24.5 found with the main CUPRAC method showed that this method correlated well with the optical sensor-based CUPRAC method.

An investigation of antioxidant capacity of fruit juices was studied by Wang *et al.* (1996) who found that the order of antioxidant effectiveness was: apple < orange < grape juices using the ORAC method. In a similar study, Tosun

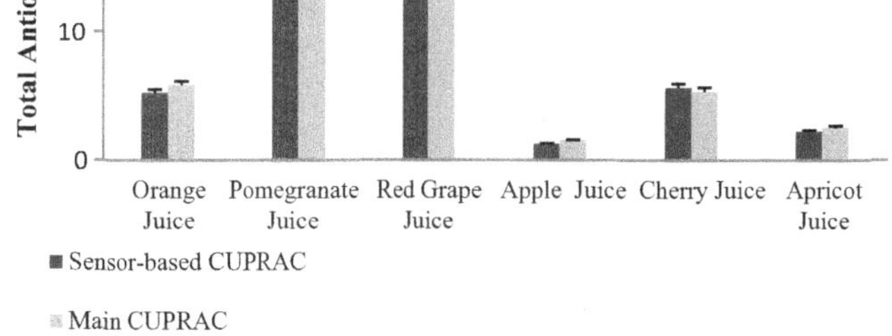

FIGURE 24.5 Comparative total antioxidant capacity (TAC) values (mM Trolox equivalents (TR)) of some commercial beverages measured by the solution- and sensor-based CUPRAC assays. Data are presented as (mean ± SD) (error bars), $N=3$ ($P=0.05$, $F_{exp}=3.27$, $F_{crit(table)}=5.78$, $F_{exp} < F_{crit(table)}$).

and Ustun (2003) reported that the order of antioxidant capacity was: apricot<orange<cherry with FRAP method. The TAC order of fruit juices in both reports agreed with our results found by the CUPRAC method.

Total antioxidant capacities of freshly squeezed orange juice were determined as 7.07 and 6.24 mM TR with sensor-based CUPRAC and main CUPRAC methods, respectively. The corresponding TAC values for freshly squeezed apple juice were found to be 1.80 and 1.62 mM TR using the sensor-based CUPRAC and main CUPRAC methods, respectively. Total antioxidant capacities found with both CUPRAC methods of freshly squeezed fruit juices were higher than those of ready-to-drink fruit juices, probably due to undesired oxidative degradation, polymerization and interconversion reactions of antioxidants resulting in TAC losses during processing.

Reactive Oxygen Species (ROS) Scavenging Activities of Beverages

The hydroxyl radical, superoxide anion radical and hydrogen peroxide scavenging activities of ready-to-drink fruit juices were determined with modified CUPRAC methods. The ROS (mainly •OH, $O_2^{\bullet-}$, and H_2O_2) scavenging activity of some fruit juices as inhibition % is shown in Table 24.1. Table 24.1 displays the effectiveness of all juices as ROS scavengers. These results show that all juice samples have the capacity to scavenge ROS, with the efficiency dependent on the type of juice. The most effective scavengers were determined as apple, orange, and red grape juices as potential inhibitors of •OH, $O_2^{\bullet-}$, H_2O_2, respectively.

On the one hand, •OH, $O_2^{\bullet-}$, and H_2O_2 scavenging activities of freshly squeezed orange juice were found to be 62.78, 58.36, and 10.97%, respectively. On the other hand, the corresponding scavenging activities of freshly squeezed apple juice for the same ROS were 70.48, 63.91, and 14.69%, respectively. It may be deduced that freshly squeezed apple juice is a better scavenger of ROS than freshly squeezed orange juice. It is apparent that ROS scavenging assays give quite different results from TAC assays (Figure 24.5 and Table 24.1). These findings confirm the different underlying mechanisms of TAC and ROS scavenging assays, because TAC assays generally measure the reduction or H-atom donation capability of (mostly) phenolic antioxidants in these juices while ROS scavenging assays have other structural requirements for effective phenolics.

TABLE 24.1 The Percentage ROS Scavenging Activity of Various Beverages (Fruit Juices)

Beverages	•OH Scavenging Activity	SO Scavenging Activity	H_2O_2 Scavenging Activity
Orange juice	47.45 ± 2.21	97.60 ± 2.76	30.42 ± 2.08
Pomegranate juice	42.21 ± 2.28	71.97 ± 3.69	38.43 ± 2.44
Red grape juice	44.47 ± 0.89	16.61 ± 1.79	44.11 ± 0.74
Apple juice	55.55 ± 3.35	21.68 ± 3.63	19.72 ± 0.30
Cherry juice	43.26 ± 0.92	50.30 ± 3.36	24.52 ± 1.16
Apricot juice	53.89 ± 2.70	19.50 ± 2.64	24.62 ± 1.44

Individual Determination of Ascorbic Acid with a La-CUPRAC Method

Ascorbic acid, commonly known as vitamin C, is one of the most important water-soluble vitamins. Fruit juices are one of the main sources of ascorbic acid in the human diet. The ascorbic acid contents (mM AA equivalents) of ready-to-drink fruit juices were found with the use of La-CUPRAC method. The hierarchy for AA content of ready-to-drink fruit juices was: apple < apricot < orange < cherry < pomegranate < red grape. Inspection of Figure 24.5 and Figure 24.6 shows that the hierarchy for TAC and AA content of fruit juices were essentially parallel. The possible two reasons for this observation may be listed as: (i) ascorbic acid content contributes to a relatively high percentage of the

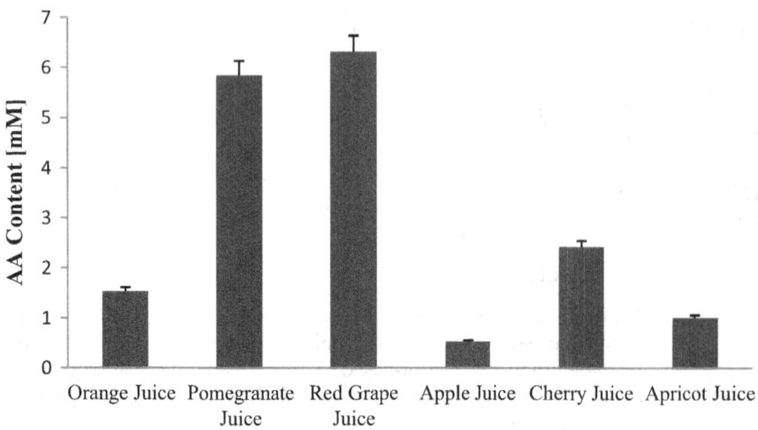

FIGURE 24.6 Ascorbic acid (AA) contents of some commercial beverages measured by the La-CUPRAC method.

observed TAC in some (especially citrus) fruit juices; (ii) ascorbic acid is well protected from oxidative degradation by other phenolics acting as synergic co-antioxidants. The amounts of ascorbic acid present in fruit juices listed in Figure 24.6 show significant variations among themselves, and range from 6.32 mM in red grape juice to 0.53 mM in apple juice. For red grape juice, having an ascorbic acid concentration of 6.32 mM, •OH scavenging activity (as inhibition percentage) was found to be 44.47%. Interestingly, orange juice, having only 1.53 mM of ascorbic acid, yielded an •OH scavenging activity of 47.45%. While orange juice contained much less ascorbic acid, it had comparable effectiveness for •OH scavenging. This may indicate that ascorbic acid is not the sole factor in determining radical scavenging ability. Ascorbic acid contents of freshly squeezed orange and apple juice were determined as 2.58 and 0.73 mM, respectively, confirming the well-known fact that the AA content of fruit juices decreases during processing and storage.

References

Altun, M., Çelik, S.E., Güçlü, K., Özyürek, M., Erçağ, E., Apak, R., 2013. Total antioxidant capacity and phenolic contents of Turkish hazelnut (*Corylus Avellana* L.) kernels and oils. J. Food Biochem. 37, 53–61.

Apak, R., Güçlü, K., Özyürek, M., Karademir, S.E., 2004. A novel total antioxidant capacity index for dietary polyphenols, vitamin C and E, using their cupric ion reducing capability in the presence of neocuproine: CUPRAC method. J. Agric. Food Chem. 52, 7970–7981.

Apak, R., Güçlü, K., Özyürek, M., Karademir, S.E., Altun, M., 2005. Total antioxidant capacity assay of human serum using copper(II)-neocuproine as chromogenic oxidant: The CUPRAC method. Free Radic. Res. 39, 949–961.

Apak, R., Güçlü, K., Özyürek, M., Karademir, S.E., Erçağ, E., 2006. The cupric ion reducing antioxidant capacity (CUPRAC) and polyphenolic content of some herbal teas. Int. J. Food Sci. Nutr. 57, 292–304.

Arena, E., Fallico, B., Maccarone, E., 2001. Evaluation of antioxidant capacity of blood orange juices as influenced by constituents, concentration process and storage. Food Chem. 74, 423–427.

Bekdeşer, B., Özyürek, M., Güçlü, K., Apak, R., 2011. *Tert*-Butylhydroquinone (TBHQ) as a novel spectroscopic probe for the superoxide radical scavenging activity assay of biological samples. Anal. Chem. 83, 5652–5660.

Bekdeşer, B., Özyürek, M., Güçlü, K., Apak, R., 2012. Development of a novel spectroscopic sensor for the hydroxyl radical scavenging activity measurement of biological samples. Talanta 99, 689–696.

Bener, M., Özyürek, M., Güçlü, K., Apak, R., 2010a. Polyphenolic contents of natural dyes produced from industrial plants assayed by HPLC and novel spectrophotometric methods. Ind. Crop. Prod. 32, 499–506.

Bener, M., Özyürek, M., Güçlü, K., Apak, R., 2010b. Development of a low-cost optical sensor for cupric reducing antioxidant capacity measurement of food extracts. Anal. Chem. 82, 4252–4258.

Bolin, H.R., Stafford, A.E., 1974. Effect of processing and storage on provitamin A and vitamin C in apricots. J. Food Sci. 39, 1034–1036.

Bravo, L., 1998. Polyphenols: Chemistry, dietary sources, metabolism and nutritional significance. Nutr. Rev. 56, 317–333.

Cohen, J.H., Kristal, A.R., Stanford, J.L., 2000. Fruit and vegetable intakes and prostate cancer risk. J. Natl. Cancer Inst. 92, 61–68.

Cook, N.C., Samman, S., 1996. Flavonoids: Chemistry, metabolism, cardioprotective effects an dietary sources. J. Nutr. Biochem. 7, 66–76.

Çelik, S.E., Özyürek, M., Bektaşoğlu, B., Güçlü, K., Berker, K.I., Özgökçe, F., Apak, R., 2008. Antioxidant capacities of herbal plants used in the manufacture of Van herby cheese: "Otlu peynir". Int. J. Food Prop. 11, 747–761.

Du, C.T., Wang, P.L., Francis, F.J., 1975. Anthocyanins of pomegranate, Punica granatum. J. Food Sci. 40, 417–418.

El-Nemr, S.E., Ismail, I.A., Ragab, M., 1990. Chemical composition of juice and seeds of pomegranate fruit. Nahrung 7, 601–606.

Gardner, P.T., White, T.A.C., McPhail, D.B., Duthie, G.G., 2000. The relative contributions of vitamin C, carotenoids and phenolics to the antioxidant potential of fruit juices. Food Chem. 68, 471–474.

Güçlü, K., Altun, M., Özyürek, M., Karademir, S.E., Apak, R., 2006. Antioxidant capacity of fresh, sun- and sulfited-dried Malatya apricot (*prunus armeniaca*) assayed by CUPRAC, ABTS/TEAC and Folin methods. Int. J. Food Sci. Technol. 41, 76–85.

Güçlü, K., Sözgen, K., Tütem, E., Özyürek, M., Apak, R., 2005. Spectrophotometric determination of ascorbic acid using copper(II)-neocuproine reagent in beverages and pharmaceuticals. Talanta 65, 1226–1232.

Halliwell, B., Aruoma, O.I., 1991. DNA damage by oxygen-derived species: Its mechanisms and measurement in mammalian systems. FEBS Lett. 281, 9–19.

Halliwell, B., Gutteridge, J.M.C., 1989. Free Radicals in Biology and Medicine. Oxford University Press, Oxford, U.K.

Harborne, J.B., 1989. General procedures and measurement of total phenolics. In: Harborne, J.B. (Ed.), Methods in Plant Biochemistry, vol. 1. Academic Press, London, pp. 1–28.

Huang, D., Ou, B., Prior, R.L., 2005. The chemistry behind antioxidant capacity assays. J. Agric. Food Chem. 53, 1841–1856.

La Vecchia, C., Altieri, A., Tavani, A., 2001. Vegetables, fruit, antioxidants and cancer: A review of Italian studies. Eur. J. Nutr. 40, 261–267.

Miller, N.J., Rice-Evans, C.A., 1997. The relative contributions of ascorbic acid and phenolic antioxidants to the total antioxidant activity of orange and apple fruit juices and blackcurrant drink. Food Chem. 60, 331–337.

Oleszek, W., Lee, C.Y., Jaworski, A.W., Price, K.R., 1988. Identification of some phenolic compounds in apples. J. Agric. Food Chem. 36, 430–436.

Özyürek, M., Güçlü, K., Apak, R., 2011. The main and modified CUPRAC methods of antioxidant measurement. Trends Anal. Chem. 30, 652–664.

Özyürek, M., Güçlü, K., Bektaşoğlu, B., Apak, R., 2007. Spectrophotometric determination of ascorbic acid by the modified cuprac method with extractive separation of flavonoids-La(III) complexes. Anal. Chim. Acta 588, 88–95.

Özyürek, M., Bektaşoğlu, B., Güçlü, K., Güngör, N., Apak, R., 2010. A novel hydrogen peroxide scavenging assay of phenolics and flavonoids using cupric reducing antioxidant capacity (CUPRAC) methodology. J. Food Compos. Anal. 23, 689–698.

Palma, M., Taylor, L.R., 1999. Extraction of polyphenolic compounds from grape seeds with near critical carbon dioxide. J. Chromatogr. A 849, 117–124.

Pellegrini, N., Serafini, M., Colombi, B., Rio, D.D., Salvatore, S., Bianchi, M., Brighenti, F., 2003. Total antioxidant capacity of plant foods, beverages and oils consumed in Italy assessed by three different in vitro assays. J. Nutr. 133, 2812–2819.

Prior, R.L., Wu, X., Schaich, K., 2005. Standardized methods for the determination of antioxidant capacity and phenolics in foods and dietary supplements. J. Agric. Food Chem. 53, 4290–4302.

Proteggente, A.R., Pannala, A.S., Paganga, G., Van Buren, L., Wagner, E., Wiseman, S., Van de Put, F., Dacombe, C., Rice-Evans, C.A., 2002. The antioxidant activity of regularly consumed fruit and vegetables reflects their phenolic and vitamin C composition. Free Radic. Res. 36, 217–233.

Radi, M., Mahrouz, M., Jaouad, A., Tacchini, M., Aubert, S., Hugues, M., Amiot, M.J., 1997. Phenolic composition, browning susceptibility, and carotenoid content of several apricot cultivars at maturity. HortScience 32, 1087–1091.

Rapisarda, P., Tomaino, A., Cascio, R.L., Bonina, F., de Pasquale, A., Saija, A., 1999. Antioxidant effectiveness as influenced by phenolic content of fresh orange juices. J. Agric. Food Chem. 47, 4718–4723.

Rice-Evans, C.A., Miller, N.J., Paganga, G., 1997. Antioxidant properties of phenolic compounds. Trends Plant Sci. 2, 152–159.

Richelle, M., Tavazzi, I., Offord, E., 2001. Comparison of the antioxidant activity of commonly consumed polyphenolic beverages (coffee, cocoa, and tea) perepared per cup serving. J. Agric. Food Chem. 49, 3438–3442.

Ruiz, D., Egea, J., Tomás-Barberán, F.A., Gil, M.I., 2005. Carotenoids from new apricot (Prunus armeniaca L.) varieties and their relationship with flesh and skin color. J.Agric. Food Chem. 53, 6368–6374.

Scott, B.C., Butler, J., Halliwell, B., Aruoma, O.I., 1993. Evaluation of the antioxidant actions of ferulic acid and catechins. Free Radic. Res. Commun. 19, 241–253.

Seeram, N.P., Aviram, M., Zhang, Y., Henning, S.M., Feng, L., Dreher, M., Heber, D., 2008. Comparison of antioxidant potency of commonly consumed polyphenol-rich beverages in the United States. J. Agric. Food Chem. 56, 1415–1422.

Souquet, J., Cheynier, V., Brossaud, F., Moutounet, M., 1996. Polymeric proanthocyanidins from grape skins. Phytochemistry 43, 509–512.

Spanos, G.A., Wrolstad, R.E., Heatherbell, D.A., 1990. Influence of processing and storage on the phenolic composition of apple juice. J. Agric. Food Chem. 38, 1572–1579.

Tosun, I., Ustun, N.S., 2003. An investigation about antioxidant capacity of fruit nectars. Pakistan J. Nutr. 2, 167–169.

Terry, P., Terry, J.B., Wolk, A., 2001. Fruit and vegetable consumption in the prevention of cancer: An update. J. Int. Med. 250, 280–290.

Tütem, E., Apak, R., 1991. Simultaneous spectrophotometric determination of cystine and cysteine in amino acid mixtures using copper(II)-neocuproine reagent. Anal. Chim. Acta 255, 121–125.

Wang, H., Cao, G., Prior, R.L., 1996. Total antioxidant capacity of fruits. J. Agric. Food Chem. 44, 701–705.

Wootton-Beard, P.C., Ryan, L., 2011. Improving public health?: The role of antioxidant-rich fruit and vegetable beverages. Food Res. Int. 44, 3135–3148.

Yamakoshi, J., Kataoka, S., Koga, T., Ariga, T., 1999. Proanthocyanidin-rich extract from grape seeds attenuates the development of aortic atherosclerosis in cholesterol-fed rabbits. Atherosclerosis 142, 139–149.

CHAPTER

25

The Use of Oxygen Radical Absorbance Capacity (ORAC) and Trolox Equivalent Antioxidant Capacity (TEAC) Assays in the Assessment of Beverages' Antioxidant Properties

Simona Carmen Litescu, Sandra A.V. Eremia, Andreia Tache, Ioana Vasilescu, Gabriel-Lucian Radu

National Institute for Biological Sciences, Bucharest, Romania

CHAPTER POINTS

- The appropriate ascription of the antioxidant efficiency of a compound can not be performed solely based on an experiment consisting in determination of radical scavenging properties; it should be completed by the evaluation of the compound efficacy with respect to biological systems and relevant oxidative markers.
- In setting up an experimental design for antioxidant assessment, it is necessary to consider sample treatment specificity and consequently the solvent influence on final results values, antioxidant solubility characteristics, and the reaction pathway (type of antioxidant action).
- Comparison between obtained antioxidant capacity values must be performed using the same methods (assays), similar/identical solvents, and identical experimental conditions.
- At least two different assays have to be performed in order to suitably ascribe the antioxidant properties, of which one must be a method involving an oxidizable substrate.

INTRODUCTION

Antioxidants represent living organisms' front line of defense against major health conditions such as neurodegenerative diseases, cardiovascular diseases, cancer, and macular degeneration. There is scientific consensus that the key factor initiating such pathologies is cellular damage induced by free radical attack, generally by reactive oxygen species (ROS) radicals, so-called 'oxidative stress.' According to Sies oxidative stress takes place when an imbalance between ROS and antioxidants exists (Sies, 1991).

The definition of an 'antioxidant compound' refers to a substance able to end, suspend, or retard the damaging reaction between an oxidative substrate and ROS. But the reaction itself is not sufficient to properly define an antioxidant, the concentration level of the presumed antioxidant being the main element in deciding the compound effectiveness, since according to Halliwell (Halliwell and Gutteridge, 2007), the proper definition of an antioxidant is "a substance that, when present at low concentrations compared to those of an oxidizable substrate, significantly delays or prevents oxidation of that substrate." It has to be emphasized that a proper antioxidant (AOxH) has to generate as result of a chain reaction a corresponding radical ($AOx^{\bullet}$) less reactive than the

Processing and Impact on Antioxidants in Beverages
http://dx.doi.org/10.1016/B978-0-12-404738-9.00025-8

free radical initiating the oxidative stress, otherwise the radical reaction is not quenched.

When dealing with antioxidants assays, several issues related to free radical structure, oxidative substrate, and mechanism of action have to be considered to avoid both misunderstandings of the real effect and experimental misinterpretation. Accordingly, it has to be mentioned that there are not overall applicable antioxidants; rather, their effect depends on the type of ROS initiating oxidative stress that generates specific damage to a certain substrate.

As Prior and co-workers emphasized in one review (Huang, 2005), paradoxically, despite the huge interest in antioxidants, a major problem has been the absence of validated assay protocols able to perform a reliable measure of the antioxidant capacity of foods and biological samples. At the same time a lot of research dealing with antioxidant assays from foods was published, but opinions on the results are quite often significantly different (Niederlander *et al.*, 2008; Litescu *et al.*, 2010; Prior and Cao, 1999; Sanchez-Moreno, 2002; Laguerre *et al.*, 2007). These paradoxical opinions are mainly due to the fact that the antioxidant concept is very comprehensive, frequently even covering other effects (e.g., anti-inflammatory). The lack of validated protocols and validated methods devoted to antioxidant assessment either from food, beverages, biological samples, or pharmaceutical formulations is another factor contributing to the previously mentioned paradox of reported results variability.

The proper assessment of beverage antioxidant capacity became essential for suitable labeling of a particular beverage as a potential contributor to antioxidant intake, considering the importance of beverages in ensuring the daily intake of antioxidants. This is due not only to increased use of such products in daily diet, but also to the improved bioavailability that beverages supply for the active antioxidant compounds compared with solid products, which may need additional metabolic steps prior to antioxidant release.

The present chapter, based on critical evaluation of a few experimental key points, attempts to set out the argument for an overall evaluation of foods or beverages from the point of view of their antioxidant capacity, oxygen radical absorbance capacity (ORAC) method being one of the compulsory assays to be performed in order to provide reliable information with Trolox equivalent antioxidant capacity (TEAC) being the required complementary assay.

BACKGROUND OF EXPERIMENTAL SET-UP FOR ANTIOXIDANT ASSAYS

Any experimental set-up of antioxidant assessment, no matter what the nature of the sample, involves three key points: (i) sample treatment—including the extraction procedures for antioxidants, if necessary; (ii) antioxidant capacity measurement; and (iii) results calculation and reporting (Perez-Jimenez *et al.*, 2008).

Sample Treatment

When the sample matrix is a beverage, the sample treatment is easier, generally consisting simply of dilution of the sample in the appropriate solvent (water; mixture of water and alcohol; ethyl acetate). At this point it should be mentioned that the solvent polarity has a significant influence on the values of measured antioxidant capacity, of course depending on the method used. The highest influence was noticed for the methods involving hydrogen atom transfer (HAT) mechanisms and polar antioxidants, e.g., phenols, polyphenols, flavonoids, meaning frequently occurring active principles in beverages and known as efficient antioxidants. Addressed methods are numerous, either the DPPH (2,2-diphenylpicrylhydrazyl) or ORAC methods or even the TEAC method, etc. An example is the influence of alcoholic solvents on the results obtained by the DPPH method (Litwinienko and Ingold, 2003, 2005): the rate constants for hydrogen atom abstraction for various hindered (thus powerful antioxidants) and unhindered polyphenols is abnormally increased in alcoholic solvents, the demonstrated explanation being the occurrence of a slight ionization of the phenols followed by their implication in a sequential proton loss electron transfer reaction involving the solvent. When measurements of the values of antioxidant capacity were performed for a mixture of standard antioxidants, catechin and gallic acid, in different solvents (methanol, water, and acetone/water 50:50, v/v), it was shown that the reaction medium influences the obtained results; the effect was greater in ORAC and TEAC assays (Perez-Jimenez *et al.*, 2008).

It should be concluded at this point that the sample treatment is highly important; the solvent influence on the reported results is significant and, accordingly, every comparison performed between obtained antioxidant capacity values has to use an identical reaction medium and similar assays.

Moreover, antioxidant solubility characteristics have to be considered when sample treatment is performed, and assessment of antioxidant properties has to comprise both hydrophilic and lypophilic conditions for an accurate result.

Antioxidant Capacity Measurement

In order to provide reliable results with respect to antioxidant efficiency measurement in a manner applicable to a wide range of presumed antioxidant compounds, a four-step draft-protocol was proposed (Becker *et al.*, 2004) for the drawing up of an experiment of antioxidant efficacy assessment.

This draft-protocol involves:

1. the quantification and identification of potential antioxidant compounds from the raw sample;
2. the quantification of the radical scavenging properties and the determination of formal redox potential;
3. the evaluation of the degree of lipid oxidation inhibition or determination of the oxidation end-point using biological systems; and
4. the study of the antioxidant effectiveness against relevant oxidative markers.

The antioxidant effect is related to the compound structure; accordingly, the identification of the compounds with potential antioxidant effect (step 1) is important to predict the outcome of food or beverage intake with respect to certain free radicals. For example one class of antioxidants mostly used as food and beverage natural flavors and colorants are anthocyans and anthocyanidins, which are highly effective in quenching lipid peroxides, especially those resulted from low-density lipoproteins peroxidation (Denev *et al.*, 2010); another example are carotenoids, which are very efficient in scavenging hydroxyl radicals and hydrogen peroxide (Parker *et al.*, 2010). Consequently, it might be presumed that a beverage enriched with anthocyans and carotenoids will be effective in preventing degenerative diseases (Kahkonen and Heinonen, 2003; Ghosh and Konishi, 2007; Bell and Gochenaur 2006). At the same time, it has to be noted that quantification of the antioxidant amount (step 1) is important in order to comply with the given definition of the antioxidant (Halliwell and Gutteridge, 2007). However, when dealing with sample antioxidant effectiveness it has to be considered that the antioxidant efficiency is really not an additive property, since a higher amount of antioxidants does not necessarily lead to better efficiency against oxidative stress; this behavior is due to either the synergetic or antagonist effect that may occur when various antioxidants are mixed.

As is clear from the above-presented step-sequence, the appropriate ascription of the antioxidant efficiency of a compound cannot be performed solely based on an experiment consisting in determination of radical scavenging properties (step 2); it should be completed by the evaluation of the compound efficacy with respect to biological systems (step 3) and relevant oxidative markers (step 4). This experimental approach is highly important when food and beverages have to be assessed as potential antioxidant sources, since it is supposed that a proper diet, rich in antioxidants, might prevent oxidative stress and further, the occurrence of the related pathologies.

A rationale for this assertion is given below, in order to provide a basis for understanding the authors' plea for a complex assessment of a food or beverage sample before making claims about its antioxidant properties. Accomplishing experimental step 2 involves an assay based on the use of a relatively stable reagent or a long lifetime free radical [e.g. Folin–Ciocalteu; DPPH·, 2,2-azinobis-(3-ethylbenzothiazoline-6-sulfonate) cation radical ($ABTS^{+\bullet}$)]. The information provided by subsequent experiments concerns only the radical scavenging capability of the tested compound, which does not automatically correlate with proper antioxidant activity. Not all samples presenting a high radical scavenger effect demonstrate equally good antioxidant properties with respect to an oxidizable substrate. This apparent discrepancy has its origin in the reaction pathway, which commonly is the same for both effects—radical scavenging effect and antioxidant effect. The reaction pathway involves either hydrogen atom transfer (HAT) or electron transfer (ET), or even both HAT and ET (Figure 25.1).

Hydrogen Atom Transfer:

$$AOxH + ROO\bullet \rightarrow ROOH + AOx\bullet$$

$$AOxH + HO\bullet \rightarrow H_2O + AOx\bullet$$

Electron Transfer:

$$ROO\bullet + AOxH \rightarrow ROO^- + AOxH^+$$

$$AOxH^+ + H_2O \rightarrow AOx\bullet + H_3O^+$$

$$ROO^- + H_3O^+ \rightarrow ROOH + H_2O$$

FIGURE 25.1 Examples of mechanism of antioxidants (AOxH) action.

While the radical scavenging property means reaching chemical stability by coupling hydrogen atoms, electrons (or both) between antioxidant and free radicals, without the participation of a substrate subjected to oxidative damage, the antioxidant property means the substrate protection as a result of a competitive reaction gained by the antioxidant against the ROS damaging the substrate.

Our assertion complies with Laguerre's conclusion, based on the Halliwell definition of antioxidant:

> *"by this definition, any method that does not involve such a substrate could not measure antioxidant activity. These are called indirect methods, which are generally used to measure the capacity of a molecule to reduce a stable artificial free radical, or a transition metal"*
>
> *(Laguerre et al., 2008).*

An experimental example supporting this rationale is Katsube and co-workers, study on antioxidant activity of several edible plant products, which proved to have a higher efficiency in preventing LDL oxidation than estimated by subsequent DPPH• assay and Folin–Ciocalteu analysis (Katsube *et al.*, 2004).

Results Calculation and Reporting

Another important contribution in the above-mentioned paradox of the reported variability is given by the manner of expressing the obtained values after antioxidant capacity measurement. The reports of the obtained results differ from method to method; e.g., in TEAC and ORAC methods the results obtained are reported as reference to a standard antioxidant, Trolox—the hydro-soluble correspondent of vitamin E, while total radical-trapping antioxidant parameter (TRAP) assay reports the results as associated lag-phase in radical quenching, and DPPH assay reports the results as the equivalent concentration able to quench 50% from initial DPPH radical. Trolox is nowadays generally accepted as the reference compound in an attempt to support a common value to be used to compare results from different laboratories for various samples having similar effects, despite all the associated drawbacks of using it. (It is not the most efficient, has no physiological relevance, has unsuitable solubility characteristics for assessment of oils, etc.)

OXYGEN RADICAL ABSORPTION CAPACITY ASSAY

Oxygen radical absorbance capacity assay (ORAC) is a method that was first developed by Cao, Sofic, and Prior (Cao *et al.*,1993,1996) and relies on assessing the effect of presumed antioxidants by measuring fluorescence quenching.

Fluorescence measurements were used because the technique is highly sensitive, allowing limits of detection lower than nanomolar.

In the initial method, a fluorescent hydrosoluble protein, β-phycoerythrin (β-PE), was used as a fluorescent probe, with an excitation wavelength at 540nm and an emission at 565nm. β-PE absorbs the visible light and possesses high fluorescence yield, proving highly sensitive to ROS (Cao *et al.*, 1993). Further, due to inconsistency of β-PE fluorescent emission (it is photo-sensitive so that it loses its fluorescence even in the absence of the free radicals) and in order to avoid the interaction between β-PE and the samples containing phenols and polyphenols, different fluorescent probes were proposed. The new probes had to have high molar absorption coefficients, considerable fluorescent yields, and photochemical stabilities, the most frequently used being fluorescein and 6-carboxyfluorescein (Naguib, 2000; Ou *et al.*, 2001; Huang *et al.*, 2002).

The ORAC assay is based on *in situ* production of peroxyl free radicals generated via an azo-compound, 2,2,-azobis(2-methylpropionamidine) dihydrochloride (AAPH), according to the reaction presented in Figure 25.2. AAPH thermally generates C centred free radicals which, in the presence of oxygen generates the peroxy-free radicals interacting with fluorescent probe.

The peroxyl radical is able to react with the oxidizable substrate/fluorescent probe, changing the fluorescence intensity and generally increasing the rate of fluorescence decay. In the presence of an antioxidant compound (mixture of antioxidants) the fluorescence decay is inhibited.

The result is obtained by calculating the area under the fluorescence curve, and it is expressed as equivalent of micromoles of Trolox per mL (for beverages) or per g (for food products from plant), since prior to sample assessment, the proper calibration is performed for Trolox.

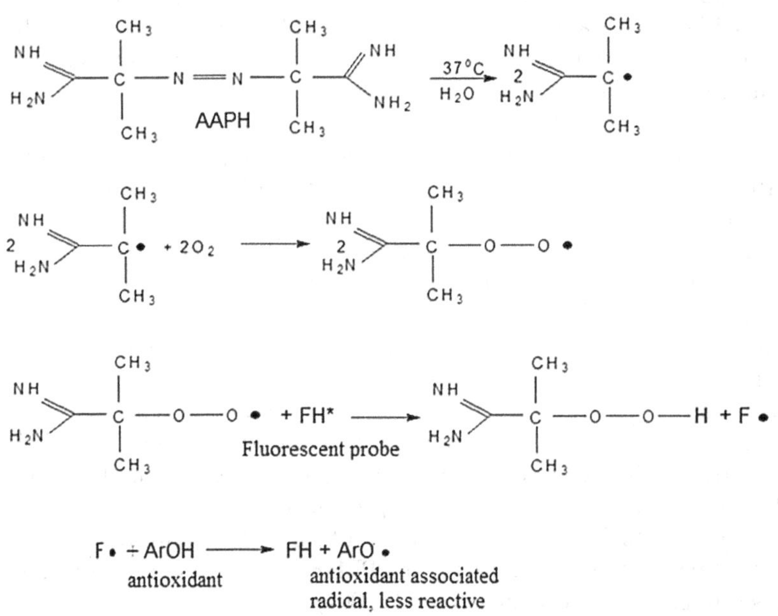

FIGURE 25.2 ORAC assay measuring principle.

The following formula is used for results expression:

$$AC_{Fl} = \frac{AUC_{sample} - AUC_{blank}}{AUC_{Trolox} - AUC_{blank}} \cdot C_{Trolox}$$

where AC_{Fl} represents the antioxidant capacity; AUC_{sample} represents the net area under the curve of the mixture: sample, AAPH, fluorescent probe; the AUC_{blank} is the net area under the blank curve buffer, AAPH, fluorescent probe; AUC_{Trolox} is the net area under the curve of Trolox, AAPH, fluorescent probe and C_{Trolox} represents the molar concentration of Trolox.

Usually, when the fluorescent probe is fluorescein (excitation at 490 nm, emission at 514 nm), the working conditions involve buffer at pH 7.40; millimolar concentration of AAPH and nanomolar concentration of fluorescein in the measuring cell). The Trolox concentration normally is 1 μmol/l.

The ORAC method is frequently used to evaluate the antioxidant capacity of water-soluble antioxidants, juices, wines, teas, etc., despite the fact that it has several drawbacks: long time of measurement (30–80 min), and reproducibility is highly influenced by pH values and ionic strength (since fluorescent emission is affected). Its large applicability is generally based on the fact that ORAC assay deals with the ubiquitous free radical from biological systems, peroxyl radical.

TROLOX EQUIVALENT ANTIOXIDANT CAPACITY (TEAC) ASSAY

The TEAC assay was originally developed and optimized by Miller and Rice Evans, and was based on production of a colored cation radical, ABTS cation radical ($ABTS^{•+}$), as a result of the reaction between ABTS and ferrylmyoglobin radical generated *in situ* by the methmyoglobin activation with hydrogen peroxide (Rice-Evans and Miller, 1994). In this first approach the sample to be tested had to be added before the formation of $ABTS^{•+}$ and consequently, the lag phase of radical formation induced by tested samples was measured.

This assay proved to be tedious and moreover it was affected by the drawback that the antioxidant cannot discriminate between $ABTS^{•+}$ and ferrylmyoglobin radicals, the reaction being similar. As a consequence, Re and Rice-Evans improved the method by developing a new technique for $ABTS^{•+}$ radical cation generation, through the reaction between ABTS and potassium persulfate (Re *et al.*, 1999). In order to produce the blue–green $ABTS^{•+}$ cation radical chromophore, 7 mmol of ABTS were dissolved in water then treated with 2.45 mmol of potassium persulfate and the mixture was allowed to react for 12–16 h at room temperature, in the dark, up to a stable blue–green solution. The so-formed radical cation has maximum absorption wavelengths at 415 nm (the most commonly used), 645 nm, 734 nm, and 815 nm, and it is stable for more than two days when stored in the dark, at room temperature. Further, this last solution is diluted before experiments with ethanol or buffer (pH 7.40) until the absorbance reaches 0.7 a.u., at 734 nm. The sample to be tested is added to the $ABTS^{+•}$ solution (blue–green), which is reduced by the antioxidants from the sample to ABTS (colorless) and the decrease of the maximum absorbency is measured spectrometrically. The decrease of the maximum absorbency is proportional with the antioxidant efficacy. The result is expressed as equivalent of micromoles of Trolox (Re *et al*, 1999). In Figure 25.3 the ABTS cation radical reaction and the formulas to be applied for antioxidant capacity calculation are given.

In the formula, A_{Trolox} represents the maximum absorbency, measured at end-point addition after addition of an established volume of Trolox (μl) in a cell containing $ABTS^{+•}$; A_{sample} represents the maximum absorbency, measured at end-point addition after addition of an established volume of sample (μl, same as Trolox volume) in a cell containing $ABTS^{+•}$; A_{blank} represents the maximum absorbency, measured at end-point addition after addition of an established volume of solvent (μl, same as Trolox volume) in a cell containing $ABTS^{+•}$; *f* is the dilution factor of the sample; and C_{Trolox} is the effective concentration of Trolox, in μmol/l.

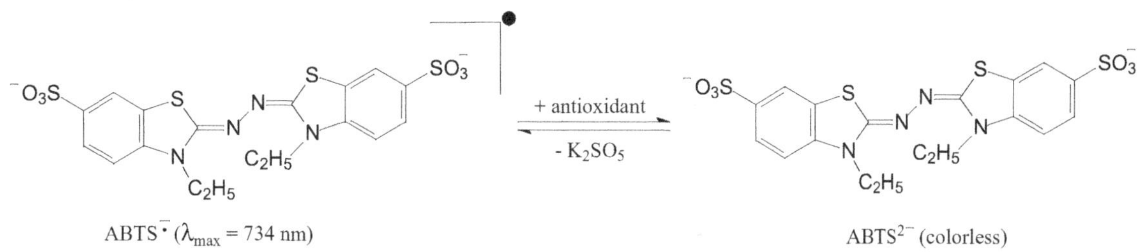

$$TEAC_{sample} = \frac{A_{sample} - A_{blank}}{A_{Trolox} - A_{blank}} \cdot f \cdot C_{Trolox}$$

FIGURE 25.3 TEAC measuring principle and results expression.

TEAC is one of the assays widely applied to the study of both water-soluble and lipid-soluble antioxidants, despite the existing drawback given mainly by the fact that measuring time—generally 3–6 min—does not allow reaching the end-point (reaction completion). An automated TEAC assay, using a sequential injection analysis system, was reported for the evaluation of the antioxidant capacity of white and red wines (Pinto *et al.*, 2005).

TABLE 25.1 Romanian Red Wine Antioxidant Capacity Assessment by TEAC and ORAC Methods

Wine Sample	TEAC μmol/l Trolox	ORAC μmol/l Trolox
Feteasca Neagra	428.03 ± 1.09	90.07 ± 3.20
Merlot	359.26 ± 0.92	88.28 ± 2.08
Pinot	282.77 ± 0.72	87.33 ± 2.02
Burgundy	364.50 ± 0.93	89.02 ± 3.20

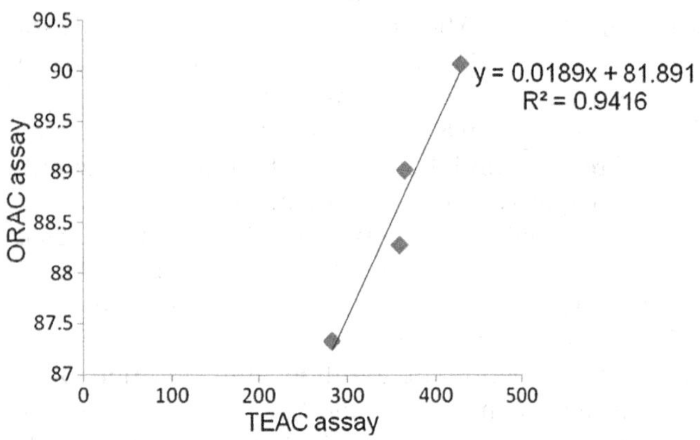

FIGURE 25.4 Comparison of TEAC and ORAC antioxidant capacity values for red wines samples.

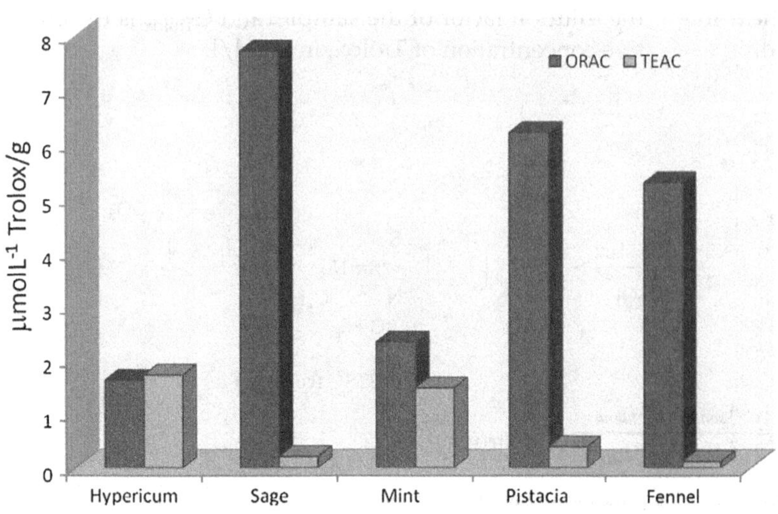

FIGURE 25.5 Comparison between ORAC values and TEAC values for essential oils from edible plants frequently used in infusion, teas, and spice formulations.

CONSISTENCY OF ORAC AND TEAC RESULTS BEYOND THE PARADOX OF VALUES VARIABILITY

As previously mentioned the expression of results is of high importance in order to accurately ascribe the antioxidant properties of a sample (beverage). The resulting reports and, further, the product labeling with regard to antioxidant intake have to consider the obtained antioxidant capacity values strictly related to the composition of the sample and to the solubility characteristics of the presumed antioxidants (hydrophilic or lypophylic) since the antioxidant action is highly influenced by the solvent, as formerly stated. All these precautions have to be taken in order to avoid considering the obtained results as artifacts, if there are significantly different values when two different methods are used for antioxidant assessment. Further, we provide several examples supporting the results consistency beyond the apparent paradoxical values obtained for antioxidant capacities.

Experiments performed in the authors' laboratories to assess antioxidant capacity of different Romanian red wines of controlled origin led, apparently, to different results, depending on the method used to evaluate the antioxidant capacity, ORAC or TEAC assay (see Table 25.1). At a glance, only seeing the values, result discharge might be considered as an option. However, a deeper analysis of the obtained value leads to accepting the results. It is obvious that the ORAC equivalent values are in a very narrow range, which is consistent with the associated anthocyan and anthocyanidin effectiveness against lipoperoxide formation and the composition profile of the red wine (all having an equivalent content of anthocyans of 6.70–8.18 μg/ml), while TEAC equivalent values are higher and more widely spread. The TEAC values are consistent to the sample wines chemical profile too, since ABTS does not discriminate between OH phenolics providing a response related to total groups able to quench the radical reaction.

The results consistency is proven even by the quite accurate fitting between the results obtained for the investigated wines by the two employed assays, since the antioxidant capacities matching values is linear, with a correlation coefficient of $R = 0.9703$ (see Figure 25.4).

The same paradox was noticed even for plant extracts or essential oils from edible plants mainly used either as teas, infusions, flavors in beverages, or even as spices. One accessible example is that provided by the results obtained for ORAC and TEAC analysis of five essential oils from some of the most commonly used edible plants: sage, *Hypericum*, mint, fennel, and *Pistacia*.

As shown in Figure 25.5, it seems that ORAC assay was more appropriate for comprehensive antioxidant assessment of essential oils while TEAC gave only partial information, considering even solubility characteristics. It is important to underline that TEAC clearly provides the antioxidant capacity value associated to the few compounds extracted in essential oil with structures compatible to $ABTS^{+\bullet}$ quenching.

References

Becker, E.M., Nissen, L.R., Skibsted, L.H., 2004. Antioxidant evaluation protocols: Food quality or health effects. Eur. Food Res. Technol. 219, 561–571.

Bell, D.R., Gochenaur, K., 2006. Direct vasoactive and vasoprotective properties of anthocyanin-rich extracts. J. Appl. Physiol. 100, 1164–1170.

Cao, G., Alessio, H.M., Cutler, R.G., 1993. Oxygen-radical absorbance capacity assay for antioxidants. Free Radic. Biol. Med. 14, 303–311.

Cao, G., Sofic, E., Prior, R.L., 1996. Antioxidant capacity of tea and common vegetables. J. Agric. Food Chem. 44, 3426–3431.

Denev, P., Ciz, M., Ambrozova, G., Lojek, A., Yanakieva, I., Kratchanova, M., 2010. Solid-phase extraction of berries' anthocyanins and evaluation of their antioxidative properties. Food Chem. 123, 1055–1061.

Ghosh, D., Konishi, T., 2007. Anthocyanins and anthocyanin-rich extracts: Role in diabetes and eye function. Asia Pac. J. Clin. Nutr. 16 (2), 200–208.

Halliwell, B., Gutteridge, J.M.C., 2007. Antioxidant defences: endogenous and diet derived in Free Radicals in Biology and Medicine, fourth ed. Oxford University Press, Clarendon 79–81.

Huang, D., Ou, B., Hampsch-Woodill, M., Flanagan, J.A., Prior, R.L., 2002. High-throughput assay of oxygen radical absorbance capacity (ORAC) using a multichannel liquid handling system coupled with a microplate fluorescence reader in 96-well format. J. Agric. Food Chem. 50, 4437–4444.

Huang, D., Ou, B., Prior, R.L., 2005. The Chemistry behind Antioxidant Capacity Assays. J. Agric. Food Chem. 53, 1841–1856.

Kahkonen, M., Heinonen, M., 2003. Antioxidant activity of anthocyanins and their aglycons. J. Agric. Food Chem. 51, 628–633.

Katsube, T., Tabata, H., Ohta, Y., Yamasaki, Y., Anuurad, E., Shiwaku, K., Yamane, Y., 2004. Screening for antioxidant activity in edible plant products: comparison of low-density lipoprotein oxidation assay, DPPH radical scavenging assay, and Folin-Ciocalteu assay. J. Agric. Food Chem. 52 (8), 2391–2396.

Laguerre, M., Lecomte, J., Villeneuve, P., 2007. Evaluation of the ability of antioxidants to counteract lipid oxidation: Existing methods, new trends and challenges. Prog. Lipid Res. 46, 244–282.

Litescu, S.C., Eremia, S.A.V., Radu, G.L., 2010. Methods for the Determination of Antioxidant Capacity in Food and Raw Materials in Bio-Farms for Nutraceuticals: Functional Food and Safety Control by Biosensors. In: Teresa Giardi, Maria, Rea, Giuseppina, Berra, Bruno (Eds.), Landes Bioscience & Springer Science. Texas, Austin.

Litwinienko, G., Ingold, K.U., 2003. Abnormal Solvent Effects on Hydrogen Atom Abstractions. 1. The Reactions of Phenols with 2,2-Diphenyl-1-picrylhydrazyl (dpph) in Alcohols. J. Org. Chem. 68, 3433–3438.

Litwinienko, G., Ingold, K.U., 2005. Abnormal Solvent Effects on Hydrogen Atom Abstraction. 3. Novel Kinetics in Sequential Proton Loss Electron Transfer Chemistry J. Org. Chem. 70, 8982–8990.

Naguib, Y.M.A., 2000. A Fluorometric method for measurement of oxygen radical-scavenging activity of water-soluble antioxidants. Anal. Biochem. 284, 93–98.

Niederlander HAG,Van Beek, T.A., Bartasiute, A., Koleva, I.I., 2008. Antioxidant activity assays on-line with liquid chromatography. J. Chromatogr. A 1210, 121–134.

Ou, B., Hampsch-Woodill, M., Prior, R., 2001. Development and Validation of an Improved Oxygen Radical Absorbance Capacity Assay Using Fluorescein as the Fluorescent Probe. J. Agric. Food Chem. 49, 4619–4626.

Parker, T., Esgro, S., Miller, S., Myers, L., Rustin, Meister, Toshkov, S., Engeseth, N., 2010. Development of an optimised papaya pulp nectar using a combination of irradiation and mild heat. Food Chem. 118, 861–869.

Perez-Jimenez, J., Arranz, S., Tabernero, M., Dıaz- Rubio, M.E., Serrano, J., Goni, I., Saura-Calixto, F., 2008. Updated methodology to determine antioxidant capacity in plant foods, oils and beverages: Extraction, measurement and expression of results. Food Res. Int. 41, 274–285.

Pinto, P.C.A.G., Saraiva, M.F.S.L., Reis, S., Lima, R., 2005. Automatic sequential determination of the hydrogen peroxide scavenging activity and evaluation of the antioxidant potential by the 2,2′-azinobis(3-ethylbenzothiazoline-6-sulfonic acid) radical cation assay in wines by sequential injection analysis. Anal. Chim. Acta 531, 25–32.

Prior, R.L., Cao, G., 1999. In vivo antioxidant capacity: comparison of different analytical methods. Free Radic. Biol. Med. 27 (11–12), 1173–1181.

Re, R., Pellegrini, N., Proteggente, A., Pannala, A., Yang, M., Rice-Evans, C., 1999. Antioxidant activity applying an improved ABTS radical cation decolorization assay. Free Radic. Biol. Med. 26, 1231–1237.

Rice-Evans, C., Miller, N.J., 1994. Total antioxidant status in plasma and body fluids. Methods Enzymol. 234, 279–293.

Sanchez-Moreno, C., 2002. Methods Used to Evaluate the Free Radical Scavenging Activity in Foods and Biological Systems. Food Sci. Technol. Int. 8 (3), 121–137.

Sies, H., 1991. Oxidative stress: from basic research to clinical application. Am. J. Med. 91 (3C), 31S–38S.

CHAPTER

26

Methodology for the Measurement of Antioxidant Capacity of Coffee: A Validated Platform Composed of Three Complementary Antioxidant Assays

Sebastian E.W. Opitz, Samo Smrke†, Bernard A. Goodman‡, Chahan Yeretzian**

*Zurich University of Applied Sciences, Institute of Chemistry and Biological Chemistry, Wädenswil, Switzerland, †National Institute of Chemistry, Laboratory for Food Chemistry, Ljubljana, Slovenia, ‡State Key Laboratory for Conservation and Utilization of Subtropical Agro-Bioresources, Guangxi University, Nanning, Guangxi, China

CHAPTER POINTS

- A platform consisting of three assays was developed to determine the antioxidant capacity in coffee brew.
- A complementary set of assays has been chosen, consisting of the Folin–Ciocalteu (FC) assay, ABTS [2,2′-azinobis-(3-ethylbenzothiazoline-6-sulphonate)] assay and oxygen radical absorbance capacity (ORAC) assay.
- The three assays have been thoroughly validated and are suitable for routine measurements.
- Analysis of coffee brew has been performed using flow injection analysis (FIA) and sequential injection analysis (SIA), which allow a fast and reliable assessment of antioxidant activity.
- The GAE (gallic acid equivalents) values of the dark roast produced by the three methods were 710 for the ABTS assay, 5868 for the ORAC assay, and 2103 for the FC assay.
- These different values illustrate fundamental differences in the chemistry of the assays, and collectively provide a fuller picture of the antioxidant properties of the beverage than could be obtained from any single assay.

INTRODUCTION

Epidemiological studies have revealed increasingly positive benefits of coffee consumption to our health and wellbeing (Dorea and da Costa, 2005; Goodman, 2012), and there is evidence that moderate consumption of three to five cups of coffee per day has protective effects against the development of certain neurological disorders, such as Parkinson's disease (Saaksjarvi *et al.*, 2007), inflammatory and cardiovascular diseases (Wu *et al.*, 2009), type 2 diabetes (Iso *et al.*, 2006; van Dam and Hu, 2005; Yamaji *et al.*, 2004), liver cirrhosis (Ruhl and Everhart, 2005; Tverdal and Skurtveit, 2003), various types of cancers (Tao *et al.*, 2008), and other diseases and medical conditions. There is a strong belief that these effects arise from the presence of components in coffee with antioxidant properties.

ANTIOXIDANTS—DEFINITION

An antioxidant is defined as "any substance that, when present at low concentration compared with that of an oxidizable substrate, significantly delays or prevents oxidation of that substrate" (Halliwell and Gutteridge, 1999). Pro-oxidants are regarded as chemicals that induce oxidative stress, usually through either creating reactive

Processing and Impact on Antioxidants in Beverages
http://dx.doi.org/10.1016/B978-0-12-404738-9.00026-X

oxygen species (ROS) or inhibiting antioxidant systems. Antioxidant capacity itself can be defined as the ability of a compound to decrease oxidation, or pro-oxidant activity. Thus antioxidants and pro-oxidants can be treated as biological terms, whereas reductants and oxidants are chemical terms (Prior and Cao, 1999).

ASSAYS FOR ANTIOXIDANTS

A wide variety of assays have been reported for measuring total antioxidant capacity (TAC) (for reviews on methodology and chemistry, see Huang *et al.*, 2005; Magalhaes *et al.*, 2008; Prior and Cao, 1999). Many publications claim to measure TAC, but the chemical bases of the various assays differ fundamentally, and the numbers generated correspond to specific chemical properties and do not necessarily relate to biological antioxidant capacity. Thus, measurement of TAC with only one method seems to be unrealistic (Huang *et al.*, 2005), and a combination of a few assays that involve different chemical reactions gives a more realistic assessment of the overall antioxidant properties of a sample such as a coffee brew.

In general, antioxidants can be classified as preventive or chain-breaking. Preventive antioxidants, such as superoxide dismutase, catalase or peroxidase inhibit the formation of ROS, whereas chain-breaking antioxidants, such as vitamin C, vitamin E, uric acid, bilirubin, polyphenols, etc., are compounds that scavenge ROS and thereby break radical chain sequences (Huang *et al.*, 2005). There are two possible reaction mechanisms through which chain-breaking antioxidants can function, namely (i) hydrogen atom transfer (HAT), where an oxygen-derived radical abstracts a hydrogen atom from the antioxidant, resulting in formation of a (relatively) stable antioxidant-derived radical, or (ii) single electron transfer (SET), in which an electron is first transferred and then followed by a proton (Huang *et al.*, 2005). Both mechanisms occur in parallel at different rates and the net result from HAT and SET reactions is the same, although SET is solvent dependent due to the need to stabilize the charged species (Wright *et al.*, 2001). It is important to note that all antioxidant assays are based on chemical reactions carried out *in vitro*, and values may not be biologically relevant since they do not measure bioavailability, *in vivo* stability, retention by tissues, and reactivity in tissues (Huang *et al.*, 2005).

Magalhaes *et al.* (2008) have classified *in vitro* analytical assays according to whether they involve competitive or non-competitive reaction schemes. In the competitive scheme, the target species (e.g., biomolecules or other compounds) and the antioxidant compound(s) compete for the oxidizing agent, whereas in non-competitive assays, antioxidants interact with oxidizing agents without competing molecules.

General guidelines for TAC assays have been suggested by Prior *et al.* (2005), who stressed the importance of understanding the chemistry that actually occurs in potential applications. Also, utilization of biologically relevant compounds [ROS, reactive nitrogen species (RNS)] is important, but this is not the case in most SET reactions, e.g., *in vitro* quenching of stable radicals such as 2,2-diphenyl-1-picrylhydrazyl (DPPH) or ABTS [2,2′-azinobis-(3-ethylbenzothiazoline-6-sulfonate)] (Frankel and Meyer, 2000). Assays should have a defined endpoint, known chemical mechanism, and exhibit good reproducibility. For routine use, assays should be technically simple, measurable on readily available instrumentation, and adaptable to high-throughput analyses. Furthermore, information should also be provided on both lipophilic and hydrophilic antioxidants, although for coffee only the assessment of the hydrophilic antioxidants in the brew is feasible. In standardization of an assay, consideration should be given to the analytical range, its linearity, recovery, repeatability, reproducibility, recognition of interfering substances, and robustness. All these considerations set the stage for the following development of the analytical platform to assess TAC in coffee brews.

DEVELOPMENT OF A PRACTICAL ANALYTICAL PLATFORM FOR MEASUREMENT OF ANTIOXIDANT CAPACITY OF COFFEE

One of the main target groups of antioxidants for assays in coffee is the chlorogenic acids (CGAs), a group of esters formed between quinic acid and various phenolic acids (primarily caffeic acid, ferulic acid, p-coumaric acid) (Jaiswal *et al.*, 2010). According to the European Food Safety Authority (EFSA), CGAs are active food constituents in coffee (EFSA Panel on Dietetic Products, Nutrition and Allergies, 2011), which is a major dietary source of antioxidants (Clifford, 1997). This panel reported that antioxidants from dietary sources provide protection of cells from free radical damage, protect body tissues, lipids, cells and DNA from oxidative damage, and help strengthen the body's natural defenses against oxidative stress. They concluded that CGAs in coffee extracts contribute to the maintenance of normal blood glucose levels and have a beneficial effect on glucose metabolism/insulin metabolism. However, they also concluded that a cause and effect has not been established between consumption of CGAs and the claimed beneficial health effects, mainly because most effects have not been established *in vivo*. In addition to their 'direct' beneficial effects, CGAs can minimize the formation of potentially mutagenic pyrolysis products, and CGAs or their transformation products also minimize oxidative deterioration in the roasted coffee (Clifford, 1997).

Other coffee components, such as caffeine (Nawrot *et al.*, 2003) and Maillard reaction products (Yanagimoto *et al.*, 2002), have also been reported to exhibit antioxidant activity. The high-molecular-weight Maillard reaction end-products are often referred to as melanoidins (Borrelli *et al.*, 2002; Delgado-Andrade and Morales, 2005; Delgado-Andrade *et al.*, 2005), and are formed during coffee roasting from molecules present in the coffee bean (Maillard, 1912). Melanoidins contain proteins, sugars, polyphenols and possibly also lipid molecules, and their size distribution and composition are affected by roasting conditions. There are indications that light-roasted coffee beans contain mainly medium-sized melanoidins as a result of the Maillard reaction between protein fragments and sucrose with additional CGA incorporation, whereas darker roasting produces larger-sized melanoidins with arabinogalactans and galactomannans being the main components (Bekedam *et al.*, 2008a).

Because of the plethora of assays that have been reported to measure antioxidants, we have focused on identifying a small group of validated assays that can reliably measure the TAC in coffee brews, and the antioxidant capacity of various coffee components that have been associated with beneficial health effects. However, interpretation of results is complicated by the complexity of coffee brews. One approach could be chromatographic separation of the components of the brews prior to analysis, but this is time consuming, costly, and not straightforward. Additionally, a fast assessment of the antioxidant capacity is important, because of the possible occurrence of unwanted oxidation processes during the separation, and also because of possible synergistic effects of antioxidant mixtures in real food samples (Blauz *et al.*, 2008). Our validated and recommended methods for fast and reliable antioxidant analyses of coffee brews are presented in this chapter.

PROCEDURE USED FOR COFFEE PREPARATION

Coffee (*Coffea arabica* from Costa Rica) was roasted, ground and brewed with a French press, which enabled reproducible preparation of the coffee beverage. The coffee brew was subsequently diluted so that its components fell within the dynamic ranges of the individual assays. The antioxidant assays recommended in this chapter are performed using flow injection analysis (FIA) or sequential injection analysis (SIA) with a FIAlab-3200 (FIAlab Instruments Inc., USA). The basic setup consists of a main unit with two integrated syringe pumps, an eight-port multiposition valve and a detector flow cell connected to a fiber-optic spectrometer. Additional equipment that adds to its versatility consists of a photomultiplier detector for fluorescence and luminescence assays, an external digital I/O controller unit, two solenoid valves, a six-port injection valve, two additional external syringe pumps, a Cetac ASX-260 autosampler, temperature-controlled heated reaction coil, and numerous tubing and connections.

In FIA, samples are introduced into a constant flow of the carrier (or reagent) through an injection loop on the six-port injection valve and then dispersed in the carrier inside the tubes by using a reaction coil. Commonly, another reagent is continuously added through a confluence point located after the sample injection port and before the mixing coil, where the reaction takes place. Finally, the reaction products reach the detector flow cell where the detection signal is acquired. The main advantages of this flow procedure are high sample throughput, precise sample introduction, controlled dispersion, and reproducible timing. In SIA, sample and reagent(s) are aspirated in a holding coil sequentially by selection from the multiposition valve and the stacked arrangement is then mixed by pumping the carrier in both flow directions to induce controlled dispersion, before the reaction products are pumped to the flow-through detector. One drawback of this FIA instrument is the lack of an online degasser. The use of solvents saturated with air at room temperature results in the formation of bubbles in the heated holding coil, and in the detector cell, where the pressure in the system drops. To avoid the undesired effects of bubbles in the system, the carrier solution should always be degassed under vacuum before use. To purge the system of any possible small bubbles trapped in the detector cell or in the holding coil tubing, freshly degassed ethanol is used, since this readily removes any air trapped in the system and also dissolves any fats that could adsorb on the hydrophobic tubing, syringe, and detector cell materials.

ASSAYS BASED ON ELECTRON TRANSFER REACTIONS

The principle of these assays is the measurement of the reduction of an oxidizing agent by an electron transfer reaction. This most commonly involves a change in color upon reduction, which can be monitored photometrically. The assays employed in this study are the Folin–Ciocalteu (FC) and the ABTS assays, although many more have been described in the literature, e.g., the ferric reducing antioxidant power (FRAP) assay.

Folin–Ciocalteu (FC) Assay

The FC reagent consists of a mixture of phosphomolybdate and phosphotungstate, although the exact chemical nature of the active component is not known (Singleton *et al.*, 1999). In an alkaline medium, phenolic compounds undergo autoxidation and electrons are transferred from them and other reducing agents to the molybdenum to form the characteristic blue complexes that can be detected at wavelengths around 750nm (Singleton *et al.*, 1999). According to that publication, the blue complexes are formed independent of the structure of phenolic compounds, therefore ruling out the possibility of formation of coordination complexes involving the metal and the phenolic compounds.

Although the FC assay is often stated to measure 'total phenolic content', the reagent is not specific for phenolic compounds and it can also be reduced by many non-phenolic compounds [e.g., aromatic amines, sulfur dioxide, ascorbic acid, Cu(I), Fe(II), etc. (Prior *et al.*, 2005), and it is more accurately described as measuring the total reducing capacity of a sample (Huang *et al.*, 2005)]. The high pH (pH$\geq$10) used in this assay favors the reaction of polyphenols, although it is unnaturally high for direct relevance to biological systems. However, the FC assay is operationally simple, reproducible, and easy to use for measuring antioxidant capacities of foods, since the reagent is commercially available. The procedure is standardized, and the absorption of the product at a long wavelength minimizes interferences from the sample matrix.

The FC assay is measured with a multi syringe FIA (MSFIA) setup, where the sample or antioxidant standard is injected (injection loop 100µl) into the flow stream (flow rate 30µl/s) of the FC reagent (0.2M concentration). After mixing with NaOH (0.25M concentration, flow rate 30µl/s) to raise the solution pH for higher polyphenol autoxidation, dispersion in the reaction coil (1m tubing length) mixes the components, and the reaction product (blue-colored metal complexes) is measured photometrically (λ=765nm, slit 10nm) (Figure 26.1A).

ABTS [2,2'-azinobis-(3-ethylbenzothiazoline-6-sulfonate)] Assay

In the ABTS assay, also known as Trolox equivalent antioxidant capacity (TEAC) assay, the green–blue stable radical cationic chromophore, 2,2-azinobis-(3-ethylbenzothiazoline-6-sulfonate) ($ABTS^{\bullet+}$) is produced by oxidation, and has absorption maxima at 414, 645, 734, and 815nm (Prior *et al.*, 2005). In the original assay, metmyoglobin was first reacted with H_2O_2 to generate the ferrylmyoglobin radical, which was then reacted with ABTS to form the $ABTS^{\bullet+}$ (Miller *et al.*, 1993). More recently, different strategies have been used for $ABTS^{\bullet+}$ generation, including reaction with manganese dioxide (Miller *et al.*, 1996), 2,2'-azobis-2-amidinopropane,dihydrochloride (AAPH), or potassium persulfate (van den Berg *et al.*, 1999), enzymatic reaction using horseradish peroxidase (Cano *et al.*, 1998), or electrochemical oxidation (Alonso *et al.*, 2002). There are also considerable variations in the reported assay conditions, e.g., reaction times ranging between 1 and 30min (Magalhaes *et al.*, 2008). Measurements are preferably made at 734nm, because of possible interferences from other components in coffee at the other maxima (Prior *et al.*, 2005); it is also necessary to minimize sample turbidity (Arnao, 2000). In general, the test sample has to be added after the $ABTS^{\bullet+}$ has been generated and the amount of $ABTS^{\bullet+}$ remaining after reaction with antioxidants in the sample is measured after a certain reaction time; this is expressed as Trolox equivalents or gallic acid equivalents (GAE) depending on the standard chosen for the comparison. The assay is technically simple and has been widely used for screening and routine determinations. $ABTS^{\bullet+}$ is soluble in both water and organic solvents, which enables the antioxidant capacity of both hydrophilic and lipophilic compounds to be determined with the same basic methodology. However, ABTS is not found naturally, so there is possible criticism that the assay is not directly relevant to any biological function. Furthermore, thermodynamically any compound with a redox potential lower than that of $ABTS^{\bullet+}$ may react with the radical (Scott *et al.*, 1993).

The ABTS assay is performed using an SIA setup to minimize consumption of the reagent, which is oxidized with potassium persulfate before the analysis, since the oxidation is too slow for $ABTS^{\bullet+}$ to be prepared *in situ*. The sample or antioxidant standard (injection volume 40µl), the $ABTS^{\bullet+}$ solution (injection volume 20µl) and carrier solution (0.1M sodium phosphate, pH7.4) are aspirated sequentially in a heated holding coil, which is thermostated at 35°C. By aspirating and dispensing the components through the holding coil, an even dispersion is achieved, and by stopping the instrument a reaction delay of 30s is applied to extend reaction times. After dispensing the mixture to the detector cell (flow rate 30µl/s), the absorbance of the $ABTS^{\bullet+}$ solution is measured photometrically (λ=734nm) (Figure 26.1B).

ASSAYS BASED ON HYDROGEN ATOM TRANSFER REACTIONS

These assays are based on the measurement of changes in light emission from fluorescent or colored probes that result from reactions with oxidizing agents. The oxidation reaction involves the transfer of a hydrogen atom to the oxidant and a reduction in the fluorescence or ultraviolet (UV) signal. In many assays where hydrogen transfer is involved, the antioxidant and the substrate compete for thermally generated peroxyl radicals after their formation by decomposition of azo compounds (e.g., AAPH); therefore, this is a competitive reaction scheme (Huang *et al.*, 2005). An often-promoted assay of this family to

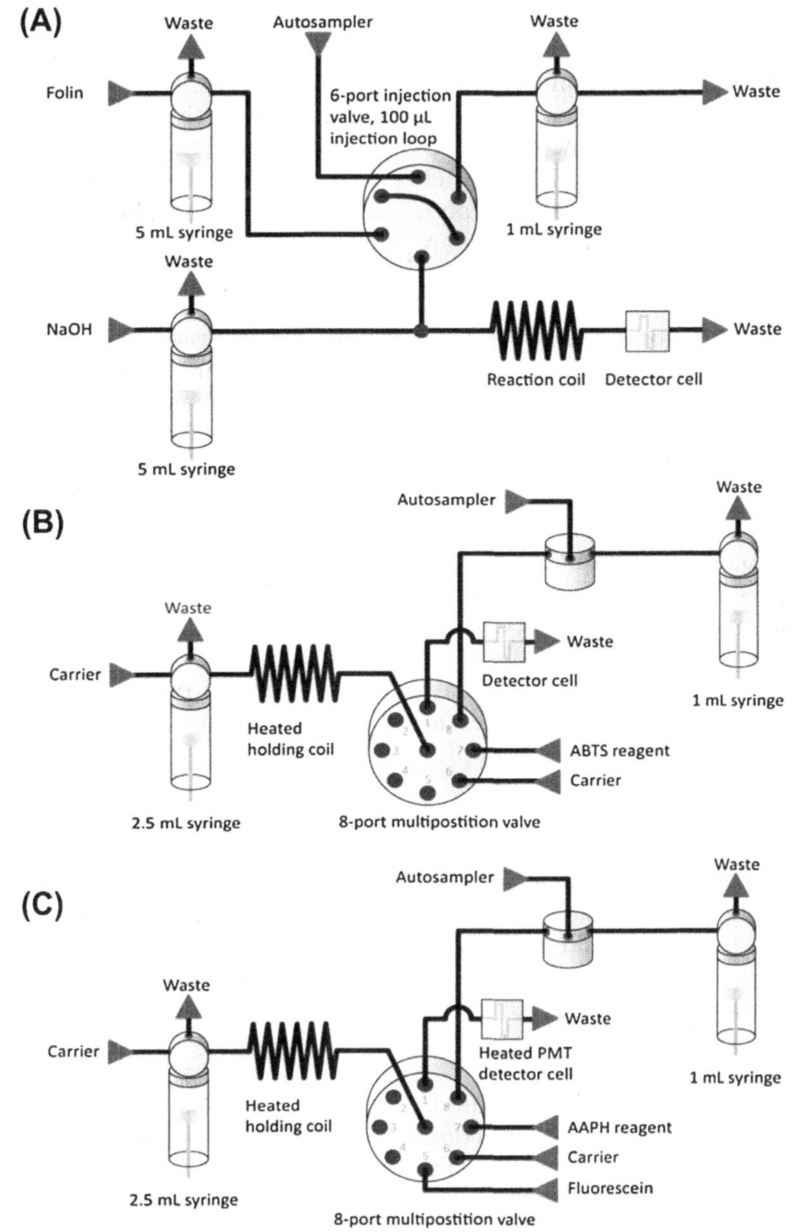

FIGURE 26.1 Schemes of the three assays. A) Folin–Ciocalteu assay; B) ABTS assay; and C) ORAC assay.

assess the capacity to quench ROS is the oxygen radical absorbance assay (ORAC) that can be used for standardized measurements of TAC (Prior *et al.*, 2005).

Oxygen Radical Absorbance Capacity (ORAC) Assay

In the ORAC assay the decrease in fluorescence of a target (probe) is measured under a reproducible and constant flux of peroxyl radicals. When a sample containing chain-breaking antioxidants is present, the decay of fluorescence is inhibited (Glazer, 1990). Initially, β-phycoerythrin was used as the fluorescent target, and this reacts with $ROO^{\bullet}$ to form a non-fluorescent product (Cao *et al.*, 1993). However, because of unwanted interactions of this protein with antioxidants, it was subsequently replaced by fluorescein as the fluorescent target (Ou *et al.*, 2001). Based on this procedure, Huang *et al.* (2002) developed a high-throughput assay using a multichannel liquid-handling system coupled with a microplate fluorescence reader in a 96-well format. Drawbacks of the ORAC assay for application to FIA are: (i) the reaction needs to be monitored for long periods (≥30 min); and (ii) quantification is based on the area under a curve (AUC), the integration of which represents the oxidation of the probe during the period of time of the measurements. Thus, the assay is time-consuming and has technical limitations related to the formation of the peroxyl radicals; for example, radical-forming compounds such as APPH are sensitive to temperature and can undergo spontaneous decay. In air-saturated solutions, it is thought that the AAPH radical reacts rapidly with O_2 to give a more stable peroxyl radical ROO•, and loss of fluorescence is then an indication of the extent of damage resulting from reaction with the radicals. When antioxidants are present, ROO• abstracts a hydrogen atom from an antioxidant to form a hydroperoxide (ROOH) and a stable antioxidant-derived radical (ArO•). The fluorescence decrease is independent of fluorescein concentrations, but has first-order dependence on the AAPH concentration (Ou *et al.*, 2001). The crucial step in repeatability of the assay is the control of the initial formation of radicals from AAPH. Also, oxidative degradation and antioxidant protection of a fluorescent target such as fluorescein does not necessarily mimic a biological substrate (Frankel and Meyer, 2000). However, this assay, together with the total radical-trapping antioxidant parameter (TRAP) assay, meets most of the requirements stated by Prior *et al.* (2005) for antioxidant capacity measurements. By applying AUC-quantification, this assay can be used for antioxidants that exhibit distinct lag phases and for those that have no lag phases. Quantification includes the initial reaction rate as well as the total amount of inhibition, which includes slow-reacting or secondary antioxidant products (Ou *et al.*, 2002). A lag phase could be the consequence of an equilibrium between fluorescein radicals and antioxidant compounds, and the actual action of the antioxidants represents a kind of 'repair' reaction of the fluorescein radicals (Bisby *et al.*, 2008).

With a SIA setup, the sample or antioxidant standard, fluorescein solution, and the AAPH solution (injection volumes of 25, 25, and 50 µl) are aspirated sequentially into the holding coil. Introducing the components and the carrier into the holding coil produces an even dispersion and subsequently the reagents–sample mixture is dispensed to the detector cell. At the peak of the fluorescein signal, the flow is stopped and the degradation of fluorescein is monitored for 1 h (Figure 26.1C). The fluorescence excitation peak wavelength is at $\lambda = 494$ nm and the emission wavelength peak at $\lambda = 514$ nm. We use broadband filters for both excitation and emission, with a blue-colored filter as an exication source and an interference green filter for emission.

VALIDATION OF RESULTS

To obtain a sufficient number of replications for each of these assays, every coffee was roasted in triplicate and each roast prepared three times in order to assess the variance in preparing the coffee brew. Then each prepared coffee brew was injected (once for the ORAC assay and three times for FC and ABTS assays) to estimate the variation in coffee analysis. Measurements were performed to test the reproducibility of results from standards within a day, between days, the limit of detection (LoD), limit of quantification (LoQ), linearity range and recovery.

Folin–Ciocalteu Assay

For the FC assay, the coffee was diluted 1:50, and a calibration curve produced by analysis of gallic acid (GA) standards (gallic acid monohydrate, purity >99%, Sigma–Aldrich, SZBB0130V) at 765 nm. A stock solution was prepared by dissolving 50 mg GA in 100 ml degassed water, and then diluting with degassed water to provide working standard solutions of 10, 20, 30, 40, 50, and 60 ppm. Intra-day reproducibility was assessed by repetitive injections of a single standard sample. Inter-day reproducibility was performed over a period of 3 days, and involved the preparation of a calibration curve of freshly prepared standards and reagents. The LoD and LoQ were investigated using 2–8 ppm GA standard solutions, and the dynamic range for the analyses was determined from the linearity of the calibration curve and the LoD. Assessment of recovery was made by spiking coffee samples (160 ppm freeze-dried, soluble coffee) with GA (30 ppm), dissolved in 100 ml degassed water. The measured outcome of a calibration curve and one dark-roasted coffee sample are presented in Figure 26.2, and the validation results summarized in Table 26.1; the measured reproducibility was very good with a deviation of only 2% relative standard deviation (RSD), and the antioxidant capacity of the presented dark-roasted coffee was calculated to be 2103 GAE. Compared to the other two assays, this assay showed the lowest sensitivity, which is compensated with lower dilution of the sample, but it had the highest dynamic range.

ABTS Assay

For the ABTS assay the coffee was diluted 1:200, and a calibration curve was produced from the $ABTS^{\bullet+}$ absorbance at 734 nm with GA concentrations of 0.5, 1, 2, 3, 4, and 5 ppm (Figure 26.3). Measurements of intra-day repeatability of standards, their reproducibility between days, LoD, LoQ, linearity range, recovery and robustness of the assay are summarized in Table 26.2. Intra-day repeatability was assessed by repetitive injections of a single standard. Inter-day reproducibility was determined from measurements made over 3 days and a calibration curve was produced using freshly prepared standards and reagents. The LoQ and LoD were calculated from the calibration curve and the dynamic range was determined as its linear region. Recovery was assessed by

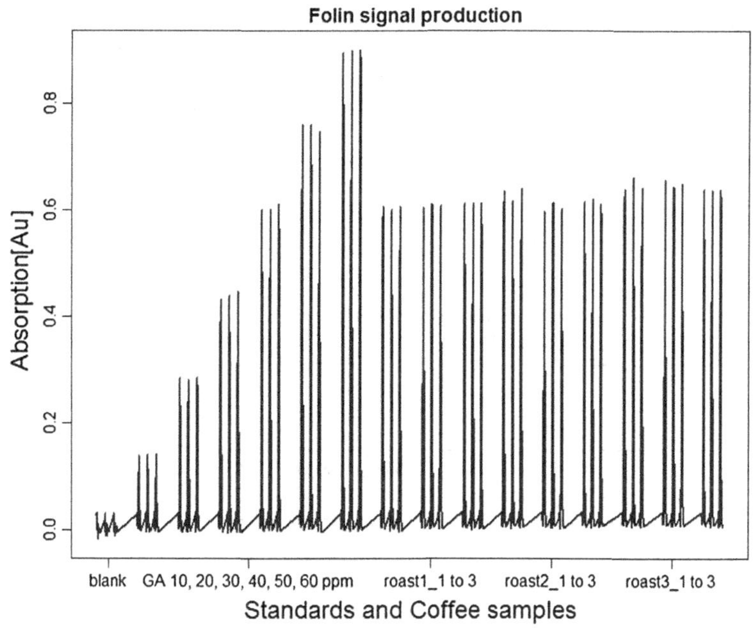

FIGURE 26.2 Results of coffee analyses with Folin–Ciocalteu assay.

TABLE 26.1 Validation Parameters for the FC Assay using gallic acid (GA) as Standard

	Intra-Day RSD (%)	Inter-Day RSD (%)	LoD (ppm)	LoQ (ppm)	Linearity Range (ppm)	R^2	Recovery
Gallic acid	0.9 (at 30 ppm)	2.1 (at 30 ppm)	1.6	3.2	2–60	0.9996	92.6 ± 0.7% (of 2 ppm)

All obtained values correspond to ppm GA concentrations (given in brackets). The relative standard deviation (% RSD) is a measure for variation used to determine the intra-day and inter-day variation. The limit of detection (LoD) and limit of quantification (LoQ) are given as well as the linearity range. The coefficient of determination R^2 is a value that explains how well the linear regression line fits the data. The recovery value is a way to determine the accuracy of a method and 100% would be the maximum accuracy.

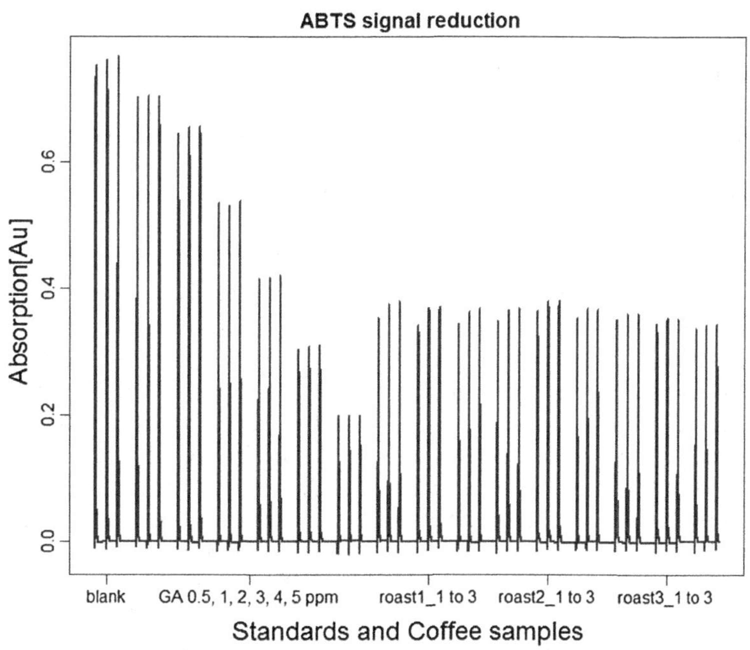

FIGURE 26.3 Results of coffee analyses with ABTS assay.

TABLE 26.2 Validation Parameters for the ABTS Assay using gallic acid (GA) as Standard

(a)	Intra-Day RSD (%)	Inter-Day RSD (%)	LoD (ppm)	LoQ (ppm)	Linearity Range (ppm)	R^2	Recovery
Gallic acid	0.4 (at 3 ppm)	4.4 (at 3 ppm)	0.4 (inhibition %)	0.12	0.5–5	0.9987	89% (Inh. % at 2 ppm)
Gallic acid	0.4 (at 3 ppm)	4.4 (at 3 ppm)	0.4 (absorbance drop)	0.12	0.5–5	0.9987	62% (absorbance drop at 2 ppm)

(b)	Initial ABTS$^{\bullet+}$ Absorbance (AU)	Absorbance Drop Deviation (% RSD)	Inh. % Deviation (% RSD)
Initial ABTS$^{\bullet+}$ concentration	High 0.880	–0.5	–15.8
	Low 0.603	–4.2	+18.2

(c)	Experiment Time (h)	Slope of ABTS$^{\bullet+}$ Degradation (AU/10 h)	R^2	Inh. % (% RSD)	Absorbance Drop (% RSD)
ABTS$^{\bullet+}$ degradation	8 (n = 8)	–0.026	0.82	0.6	1.2
	13 (n = 13)	–0.023	0.71	0.8	1.8

All obtained values correspond to ppm GA concentrations (given in brackets). The relative standard deviation (% RSD) is a measure for variation used to determine the intra-day and inter-day variations. The limit of detection (LoD) and limit of quantification (LoQ) are given as well as the linearity range. The coefficient of determination R^2 is a value that explains how well the linear regression line fits the data. The recovery value is a way to determine the accuracy of a method and 100% would be the maximum accuracy. Here two different ways are used to analyze the data; either the decrease in the ABTS$^{\bullet+}$absorbance is taken as the result, or the inhibition percentage is calculated relative to the blank value. In addition, several robustness parameters were investigated in order to determine their influence on the assay. Part (b) shows the variation in ABTS$^{\bullet+}$concentration and the effect on the deviation of absorbance decrease or inhibition percentage (Inh. %). The maximum allowable deviation of initial ABTS$^{\bullet+}$ absorbance was calculated as ±0.025 absorption units (AU), which is within a 3% error. Part (c) shows the degradation of the ABTS$^{\bullet+}$ reagent, and the deviation and slope of degradation (AU/10 h) including coefficient of determination R^2 were calculated for different experimental times.

spiking a coffee sample solution (freeze-dried, soluble coffee) with GA; measurements were made on 40 ppm of coffee, spiked with 0, 1, 2, and 3 ppm of GA.

Variable concentrations of the ABTS$^{\bullet+}$ are produced from its synthesis reaction. Therefore, the working concentration of the ABTS$^{\bullet+}$ reagent was adjusted to produce a response with absorbance of 0.75 ± 0.025 absorption units (AU) by adding water or ABTS$^{\bullet+}$ stock solution prior to measurements. If this AU value drops out of the specified range during the measurement sequence it has failed the validation criteria and the adjustment must be repeated. Stability of ABTS$^{\bullet+}$ was tested by injecting a blank sample and 3 ppm GA standard every hour during measurements. Because the ABTS$^{\bullet+}$ reagent degrades slowly, it is useful to set the initial blank absorbance above 0.75 AU.

Two different methods were used to quantify the results: (i) the inhibition percentage, which is calculated as a percentage drop of the sample signal compared to the blank ABTS$^{\bullet+}$ signal; and (ii) the absorbance drop, which is an absolute difference in absorbance between a blank and a sample signal. As can be seen in Table 26.2b and c, there are differences in the robustness of the two methods, and we recommend use of inhibition % despite its stricter conditions, since the absorbance drop method failed in the recovery validation (Table 26.2a), and gave inconsistent results for robustness test of the initial ABTS$^{\bullet+}$ concentrations (Table 26.2b and c).

Repeatability was excellent, and the reproducibility of the assay was in acceptable ranges, although better reproducibility values can usually be produced with a FIA system [e.g., 1.3% (Pellegrini *et al.*, 2003) and 1.95% (Ivekovic *et al.*, 2005]. The stability of the ABTS$^{\bullet+}$ was sufficient over the time range of our experiment, which is usually less than 12 h for measuring three batches of three roast degrees with three replicates each (in total 27 coffee samples, Figure 26.3). The antioxidant capacity of the presented dark-roasted coffee was calculated to be 710 GAE.

ORAC Assay

The coffee brew was diluted 1:400 prior to analysis, and a calibration curve was produced for the delay of fluorescein degradation with GA concentrations of 5, 10, 15, 20, and 25 ppm (Figure 26.4). Measurements of the repeatability of standards within a day, the reproducibility of standards between days, the LoD, LoQ, linearity range, recovery and robustness of the AAPH and fluorescein concentrations are summarized in Table 26.3. Intra-day repeatability was assessed by repetitive injections of a single standard. Inter-day reproducibility was performed over 5 days, and freshly prepared standards and reagents were used to produce a calibration curve. This was used to calculate the LoD

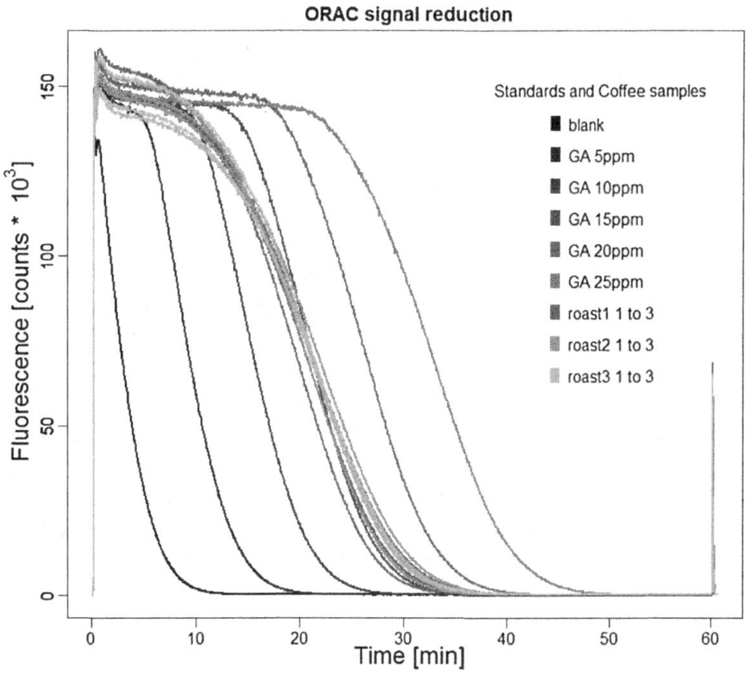

FIGURE 26.4 Results of coffee analyses with ORAC assay.

TABLE 26.3 Validation Parameters for the ORAC Assay using gallic acid (GA) as Standard

(a)	Intra-Day RSD (%)	Inter-Day RSD (%)	LoD (ppm)	LoQ (ppm)	Linearity Range (ppm)	R^2	Recovery
Gallic acid	4.7 (at 10 ppm)	8.2 (at 10 ppm)	0.43	2.1	5–25	0.9993	78% (of 20 ppm)
(b)	**Concentration Relative to Applied Value**		**Response Ratio (10 ppm GA)**	**Response Ratio (15 ppm GA)**		**Response Ratio (20 ppm GA)**	**Average Response Ratio**
Fluorescein	0.9		0.886	0.864		0.908	0.886
	1.1		1.17	1.10		1.06	1.11
AAPH	0.9		1.18	1.01		1.10	1.07
	1.1		0.88	0.81		0.89	0.86
(c)	**Experiment Time (h)**		**RSD (%)**				
AAPH stability measured by repeated injection of 20 ppm GA	30		3				
	40		4.3				

In Part (a), all obtained values correspond to ppm GA concentrations (given in brackets). The relative standard deviation (% RSD) is a measure of variation used to determine the intra-day and inter-day variability. The limit of detection (LoD) and limit of quantification (LoQ) are given as well as the linearity range. The coefficient of determination R^2 is a value that explains how well the linear regression line fits the data. The recovery value is a way to determine the accuracy of a method and 100% would be the maximum accuracy. Several robustness parameters were investigated in order to determine their influence on the assay. Part (b) shows the variation in fluorescein and APPH concentrations (relative to 1) at three GA standard concentrations as well as the average response ratio was calculated. Part (c) shows the AAPH stability and the relative standard deviation (% RSD) of the antioxidant capacity of 20 ppm GA for experimental times of 30 h and 40 h.

and LoQ, and its linear region was taken to represent the dynamic range. Recovery was determined by spiking a coffee sample (freeze-dried, soluble coffee) with GA dissolved in 100 ml of degassed water (spiking 2000 ppm of coffee with 15 ppm of GA). The antioxidant capacity of the presented dark roast was then calculated as 5868 GAE.

Stability of AAPH was tested hourly by injecting a 20-ppm GA standard. For 30 h and 30 injections the relative standard deviation (% RSD) was 3%, for 40 h and 40 injections it was 4.3%. The main drawback of this assay is the amount of time required to measure samples, but the validation procedure gave a time window of ≥40 h during which the assay conditions remained stable. The main factor that affects the performance of the assay is the stability of AAPH. AAPH is used as a radical generator and is temperature sensitive (Prior *et al.*, 2003). In comparison to Ou *et al.* (2001), who introduced the now widely accepted ORAC assay, we preferred to use GA instead of Trolox, because of problems associated with Trolox solubility in our aqueous samples. Also the AAPH preparation prior to analysis was changed and the incubation step at 37°C was omitted. As we needed a valid time period of 40 h, it was important to ensure the same AAPH quality for this time range. Therefore, APPH was stored in ice water and was purged before each run to have fresh AAPH solution for radical generation. Despite the long reaction time of 1 h, we were able to get reproducible results. The reproducibility of our results (intra-day RSD 4.7% and inter-day RSD 8.2%), which were made in series with SIA analyses, was in the range of other studies. The literature reports values for the intermediate precision relative standard deviations (RSD_{int}) of 1.8–9.4% and reproducibility relative standard deviations (RSD_R) of 4.4–13.8% (Watanabe *et al.*, 2012) and 15% RSD (Huang *et al.*, 2005; Ou *et al.*, 2001), which were made in parallel using a fluorescence plate reader.

DISCUSSION AND CONCLUSIONS

The principal aim of this work has been to develop a platform of antioxidant assays that can be used routinely for assessment of coffee brews and quality control. However, it must be appreciated that all assays are based on chemical reactions, and it is a big, and not altogether justifiable, step to equate their results to any specific health consequences (either beneficial or detrimental) from consumption of the beverage. Nevertheless, because of the currently held common view that antioxidants in foods convey health benefits on the consumer, it is important to have a set of simple and reliable assays that can be used for characterization and quality control of the product. The methods presented in this chapter have been chosen to measure different aspects of the hydrophilic antioxidant activity in coffee brews, and thus to produce a more complete picture of their antioxidant activities than can be obtained with a single assay. Furthermore, we considered it important to establish a set of assays, which could be performed with the same apparatus.

As a representative assay involving redox reactions with transition metal species, we chose the FC assay, because of its widespread applications, ease of use, and special sensitivity to polyphenolic compounds, which are considered to be one of the most important groups of antioxidants in foods. This assay proved to have excellent reproducibility, but relatively low sensitivity (10–60 ppm). The FC assay is preferred to the commonly used FRAP assay, which has the disadvantage that it reacts slowly with polyphenols and is therefore not suitable when a single end-point absorption is measured with FIA (Pulido *et al.*, 2000). Furthermore, the ferrous ions produced in the FRAP assay can catalyze the formation of hydroxyl radicals in the presence of hydrogen peroxide, which is a common component of coffee brews (e.g., Hegele *et al.*, 2009).

We chose the ABTS assay as a representative method for scavenging relatively stable free radicals, since we experienced validation problems with the assay based on the DPPH radical (Prior *et al.*, 2005). For assays based on the TEMPO [(2,2,6,6-tetramethylpiperidin-1-yl)-oxyl] radical there are solubility problems with coffee solutions, and in addition it has been reported to react predominantly with the melanoidins (Bekedam *et al.*, 2008b). However, the $ABTS^{\bullet+}$ radical exhibited adequate stability and the reproducibility of the assay was good, although the dynamic range was quite small (0.5–5 ppm).

The ORAC assay exhibits its polyphenolic antioxidant action via a hydrogen atom transfer mechanism, which is more relevant to biological systems than electron transfer mechanisms (Wright *et al.*, 2001). It is also an assay that measures the AUC of the whole antioxidant reaction, which is more reliable than measuring inhibition degree or only at a fixed time point (Prior *et al.*, 2005).

The range of values obtained for the TAC from the various assays are a reflection both of the degree of selectivity of the individual assays and the relevance of gallic acid as a representative molecule for probing the molecules covered by the assays. The fact that the highest number in terms of GAE is obtained from the ORAC assay does not necessarily mean that the assay provides the most complete assessment of antioxidant molecules in the coffee brew, since the conditions may not be the most favorable for gallic acid reaction. There is also the possibility of underestimation of antioxidant capacity in the FC reaction, because of rapid secondary reactions of the autoxidation products of the phenols, especially when the radical is a phenoxyl rather than a semiquinone (as occurs with gallic acid). Finally, the ABTS assay simply determines compounds with a redox potential lower than that of the $ABTS^{\bullet+}$ radical. Thus these three assays determine different aspects of the antioxidant activity of the coffee beverage, and provide a fuller picture of its antioxidant behavior, even though their respective relevance to biological systems is still unclear.

An important aspect of the development of an assay for routine use is the validation procedure for the product under investigation; otherwise there is always doubt as to the relevance of the results. It is also essential to recognize that one antioxidant assay will not cover all of the antioxidants found in complex mixtures of compounds such as a coffee brew. Conversely, because of the dynamic nature of the chemistry of beverages such as coffee, it is essential to perform all assays within a fixed (and preferably short) time range. Therefore, we have selected what we consider to be a reasonable set of complementary antioxidant assays for our platform. If necessary, this could be expanded in the future, although it should also be appreciated that performing too many assays will result in similar and thus redundant information.

Validation of the various selected antioxidant assays is important in order to have confidence in the reliability of the assays, and to have a strict control on the results. Only by performing such strict assessment of the quality of the results can the information generated on the classification and comparison of different coffee brews be considered to be relevant to quality criteria for the beverage. Nevertheless, it is still difficult to form confident links between the results from the various antioxidant assays and the impact of various coffee preparations on the health of consumers. Indeed, even though such associations are becoming increasingly common as marketing tools for various food products, it is only assays for proven detrimental compounds (e.g., known toxins) that can confidently predict a health effect. Differences in the functioning of the digestive systems of different types of mammal limit the value of using animals with short life-spans to predict beneficial effects.

Acknowledgments

We acknowledge valuable assistance by Sergio Petrozzi in establishing the experimental setup for FIA analysis and Babette Klopprogge for preparing coffee brews. We also thank the Scientific Exchange Program between Switzerland and the New Member States of the EU (Sciex-NMS^{ch}) for financial support to the PhD thesis work of Samo Smrke. Finally we acknowledge the Commission for Technology and Innovation (CTI) of Switzerland and Bühler AG for their respective contributions to the CTI-project 'Modulating the Coffee Roasting Process to Optimize the Antioxidant Potential of Coffee'(Project Nr. 13897.1 PFIW-IW).

References

Alonso, A.M., Dominguez, C., Guillen, D.A., Barroso, C.G., 2002. Determination of antioxidant power of red and white wines by a new electrochemical method and its correlation with polyphenolic content. J. Agric. Food Chem. 50, 3112–3115.

Arnao, M.B., 2000. Some methodological problems in the determination of antioxidant activity using chromogen radicals: a practical case. Trends Food Sci. Technol 11, 419–421.

Bekedam, E.K., Loots, M.J., Schols, H.A., van Boekel, M.A.J.S., Smit, G., 2008a. Roasting effects on formation mechanisms of coffee brew melanoidins. J. Agric. Food Chem. 56, 7138–7145.

Bekedam, E.K., Schols, H.A., Cämmerer, B., Kroh, L.W., van Boekel, M.A.J.S., Smit, G., 2008b. Electron spin resonance (ESR) studies on the formation of roasting-induced antioxidative structures in coffee brews at different degrees of roast. J. Agric. Food Chem. 56, 4597–4604.

Bisby, R.H., Brooke, R., Navaratnam, S., 2008. Effect of antioxidant oxidation potential in the oxygen radical absorption capacity (ORAC) assay. Food Chem. 108, 1002–1007.

Blauz, A., Pilaszek, T., Grzelak, A., Dragan, A., Bartosz, G., 2008. Interaction between antioxidants in assays of total antioxidant capacity. Food Chem. Toxicol. 46, 2365–2368.

Borrelli, R.C., Visconti, A., Mennella, C., Anese, M., Fogliano, V., 2002. Chemical characterization and antioxidant properties of coffee melanoidins. J. Agric. Food Chem. 50, 6527–6533.

Cano, A., Hernandez-Ruiz, J., Garcia-Canovas, F., Acosta, M., Arnao, M.B., 1998. An end-point method for estimation of the total antioxidant activity in plant material. Phytochem. Anal. 9, 196–202.

Cao, G.H., Alessio, H.M., Cutler, R.G., 1993. Oxygen-radical absorbency capacity assay for antioxidants. Free Radic. Biol. Med. 14, 303–311.

Clifford, M.N., 1997. The nature of chlorogenic acids. Are they advantageous compounds in coffee? In: Colloque Scientifique International sur le Café (17. Nairobi). ASIC, Paris, pp. 79–90.

Delgado-Andrade, C., Morales, F.J., 2005. Unraveling the contribution of melanoidins to the antioxidant activity of coffee brews. J. Agric. Food Chem. 53, 1403–1407.

Delgado-Andrade, C., Rufian-Henares, J.A., Morales, F.J., 2005. Assessing the antioxidant activity of melanoidins from coffee brews by different antioxidant methods. J. Agric. Food Chem. 53, 7832–7836.

Dorea, J.G., da Costa, T.H., 2005. Is coffee a functional food? Br. J. Nutr. 93, 773–782.

Frankel, E.N., Meyer, A.S., 2000. The problems of using one-dimensional methods to evaluate multifunctional food and biological antioxidants. J. Sci. Food Agric. 80, 1925–1941.

Goodman, B.A. (Ed.), 2012. Coffee consumption and health. Nova Science Publishers, Inc.. Hauppauge, New York.

Halliwell, B., Gutteridge, J.M.C., 1999. Free radicals in biology and medicine. Oxford University Press, Oxford.

Huang, D.J., Ou, B.X., Hampsch-Woodill, M., Flanagan, J.A., Prior, R.L., 2002. High-throughput assay of oxygen radical absorbance capacity (ORAC) using a multichannel liquid handling system coupled with a microplate flourescence reader in 96-well format. J. Agric. Food Chem. 50, 4437–4444.

Huang, D., Ou, B., Prior, R.L., 2005. The chemistry behind antioxidant capacity assays. J. Agric. Food Chem. 53, 1841–1856.

Iso, H., Date, C., Wakai, K., Fukui, M., Tamakoshi, A., 2006. The relationship between green tea and total caffeine intake and risk for self-reported type 2 diabetes among Japanese adults. Ann. Intern. Med 144, 554–562.

Ivekovic, D., Milardovic, S., Roboz, M., Grabaric, B.S., 2005. Evaluation of the antioxidant activity by flow injection analysis method with electrochemically generated ABTS radical cation. Analyst 130, 708–714.

Jaiswal, R., Patras, M.A., Eravuchira, P.J., Kuhnert, N., 2010. Profile and characterization of the chlorogenic acids in green robusta coffee beans by LC-MSn: identification of seven new classes of compounds. J. Agric. Food Chem. 58, 8722–8737.

Magalhaes, L.M., Segundo, M.A., Reis, S., Lima, J.L.F.C., 2008. Methodological aspects about in vitro evaluation of antioxidant properties. Anal. Chim. Acta 613, 1–19.

Maillard, L.C., 1912. Action of amino acids on sugars. Formation of melanoidins in a methodical way. C. R. Hebd. Seances Acad. Sci. 154, 66–68.

Miller, N.J., RiceEvans, C., Davies, M.J., Gopinathan, V., Milner, A., 1993. A novel method for measuring antioxidant capacity and its application to monitoring the antioxidant status in premature neonates. Clin. Sci. 84, 407–412.

Miller, N.J., Sampson, J., Candeias, L.P., Bramley, P.M., RiceEvans, C.A., 1996. Antioxidant activities of carotenes and xanthophylls. FEBS Lett. 384, 240–242.

Nawrot, P., Jordan, S., Eastwood, J., Rotstein, J., Hugenholtz, A., Feeley, M., 2003. Effects of caffeine on human health. Food Addit. Contam., Part A 20, 1–30.

Ou, B., Hampsch-Woodill, M., Prior, R.L., 2001. Development and validation of an improved oxygen radical absorbance capacity assay using fluorescein as the fluorescent probe. J. Agric. Food Chem. 49, 4619–4626.

Ou, B.X., Huang, D.J., Hampsch-Woodill, M., Flanagan, J.A., Deemer, E.K., 2002. Analysis of antioxidant activities of common vegetables employing oxygen radical absorbance capacity (ORAC) and ferric reducing antioxidant power (FRAP) assays: A comparative study. J. Agric. Food Chem. 50, 3122–3128.

Pellegrini, N., Del Rio, D., Colombi, B., Bianchi, M., Brighenti, F., 2003. Application of the 2,2′-azinobis(3-ethylbenzothiazoline-6-sulfonic acid) radical cation assay to a flow injection system for the evaluation of antioxidant activity of some pure compounds and beverages. J. Agric. Food Chem. 51, 260–264.

Prior, R.L., Cao, G.H., 1999. In vivo total antioxidant capacity: Comparison of different analytical methods. Free Radical Biol. Med. 27, 1173–1181.

Prior, R.L., Hoang, H., Gu, L.W., Wu, X.L., Bacchiocca, M., Howard, L., Hampsch-Woodill, M., Huang, D.J., Ou, B.X., Jacob, R., 2003. Assays for hydrophilic and lipophilic antioxidant capacity (oxygen radical absorbance capacity (ORAC(FL))) of plasma and other biological and food samples. J. Agric. Food Chem. 51, 3273–3279.

Prior, R.L., Wu, X.L., Schaich, K., 2005. Standardized methods for the determination of antioxidant capacity and phenolics in foods and dietary supplements. J. Agric. Food Chem. 53, 4290–4302.

Pulido, R., Bravo, L., Saura-Calixto, F., 2000. Antioxidant activity of dietary polyphenols as determined by a modified ferric reducing/antioxidant power assay. J. Agric. Food Chem. 48, 3396–3402.

Ruhl, C.E., Everhart, J.E., 2005. Coffee and tea consumption are associated with a lower incidence of chronic liver disease in the United States. Gastroenterology 129, 1928–1936.

Saaksjarvi, K., Knekt, P., Rissanen, H., Laaksonen, M.A., Reunanen, A., Mannisto, S., 2007. Prospective study of coffee consumption and risk of Parkinson's disease. Eur. J. Clin. Nutr. 62, 908–915.

Scott, S.L., Chen, W.J., Bakac, A., Espenson, J.H., 1993. Spectroscopic parameters, electrode-potentials, acid ionization-constants, and electron-exchange rates of the 2,2′-azinobis(3-Ethylbenzothiazoline-6-Sulfonate) radicals and ions. J. Phys. Chem. 97, 6710–6714.

Singleton, V.L., Orthofer, R., Lamuela-Raventos, R.M., 1999. Analysis of total phenols and other oxidation substrates and antioxidants by means of Folin–Ciocalteu reagent. Methods Enzymol. 299, 152–178.

Tao, K.S., Wang, W., Wang, L., Cao, D.Y., Li, Y.Q., Wu, S.X., Dou, K.F., 2008. The multifaceted mechanisms for coffee's anti-tumorigenic effect on liver. Med. Hypotheses 71, 730–736.

Tverdal, A., Skurtveit, S., 2003. Coffee intake and mortality from liver cirrhosis. Ann. Epidemiol. 13, 419–423.

van Dam, R.M., Hu, F.B., 2005. Coffee consumption and risk of type 2 diabetes: A systematic review. JAMAJ. Am. Med. Assoc. 294, 97–104.

van den Berg, R., Haenen, G.R.M.M., van den Berg, H., Bast, A., 1999. Applicability of an improved Trolox equivalent antioxidant capacity (TEAC) assay for evaluation of antioxidant capacity measurements of mixtures. Food Chem 66, 511–517.

Watanabe, J., Oki, T., Takebayashi, J., Yamasaki, K., Takano-Ishikawa, Y., Hino, A., Yasui, A., 2012. Method validation by interlaboratory studies of improved hydrophilic oxygen radical absorbance capacity methods for the determination of antioxidant capacities of antioxidant solutions and food extracts. Anal. Sci. 28, 159–165.

Wright, J.S., Johnson, E.R., DiLabio, G.A., 2001. Predicting the activity of phenolic antioxidants: Theoretical method, analysis of substituent effects, and application to major families of antioxidants. J. Am. Chem. Soc. 123, 1173–1183.

Wu, J., Ho, S.C., Zhou, C., Ling, W.H., Chen, W.Q., Wang, C.L., Chen, Y.M., 2009. Coffee consumption and risk of coronary heart diseases: A meta-analysis of 21 prospective cohort studies. Int. J. Cardiol. 137, 216–225.

Yamaji, T., Mizoue, T., Tabata, S., Ogawa, S., Yamaguchi, K., Shimizu, E., Mineshita, M., Kono, S., 2004. Coffee consumption and glucose tolerance status in middle-aged Japanese men. Diabetologia 47, 2145–2151.

Yanagimoto, K., Lee, K.G., Ochi, H., Shibamoto, T., 2002. Antioxidative activity of heterocyclic compounds found in coffee volatiles produced by Maillard reaction. J. Agric. Food Chem. 50, 5480–5484.

CHAPTER

27

Off-Line HPLC Integrated to Total Antioxidant Capacity Measurement of Beverages

Esma Tütem, Reşat Apak, Kevser Sözgen Başkan

Istanbul University, Faculty of Engineering, Department of Chemistry, Avcilar, Istanbul, Turkey

CHAPTER POINTS

- The integrated HPLC-TAC (total antioxidant capacity) determination approach involves the calculation of contributions to the TAC of individual phenolics and ascorbic acid by multiplying the HPLC-determined concentrations of these compounds with their Trolox equivalent antioxidant capacity (TEAC) coefficients, followed by summing up the products.
- Both off-line HPLC integrated to CUPRAC (cupric reducing antioxidant capacity) (combined HPLC-CUPRAC) and off-line HPLC integrated to ABTS [2,2′-azino-bis(3-ethylbenzothiazoline-6-sulfonic acid)] assays were successfully applied to seven fresh apple juices and three synthetic mixtures comprising various phenolic compounds.
- Off-line HPLC integrated to ABTS assay was also applied to various wine samples. However, due to the limitations in the chromatographic quantification of polyphenolic compounds, the percentages of experimental TAC as accounted for by HPLC-ABTS estimates were found to be rather low (around 16%).
- Off-line HPLC integrated to LPO (lipid peroxidation) assay was applied to Jonagold apple juices prepared with different production methods.
- Off-line HPLC integrated to spectrophotometric TAC assays are useful both in the determination of individual antioxidant components and TAC values only in a successful chromatographic run.

INTRODUCTION

Although reactive oxygen and nitrogen species (ROS/RNS) which can damage biological molecules such as proteins, lipids or DNA are naturally formed during normal metabolism, and the human body has an antioxidant defensive system ensuring an appropriate balance between pro-oxidants (i.e., free radicals) and antioxidants via certain enzymes and proteins, the defenses can be overwhelmed under certain circumstances to give way to oxidative stress and its consequent harmful effects. Antioxidants are accepted to be health-beneficial compounds reinforcing the defenses against ROS/RNS and so preventing oxidative stress-related diseases such as cancer, cardiovascular and neurodegenerative diseases, and cell aging. Epidemiological data have strongly suggested an inverse correlation between the intake of fruits and vegetables, naturally rich in antioxidants, and the incidence of certain diseases such as cancer, cardiovascular disorders and diabetes (Litescu and Radu, 2000; Blomhoff, 2005). Plant foods, particularly fruits and vegetables and beverages prepared from them, are rich in antioxidants mainly comprised of polyphenols, followed by vitamins and carotenoids. Also, investigation into the antioxidant capacities of common beverages has shown that the hydrophilic fraction normally contributes more to total antioxidant capacity (TAC) than the lipophilic fraction (Pulido *et al.*, 2003). Fruit juices are increasingly being used by people looking for healthy options as part of the World Health Organization (WHO) five-a-day dietary recommendations. Fruit juice consumption in some Western European countries has increased up to 28 l/person per year (Borges *et al.*, 2010). Beverages are the main contributors to the TAC of a whole diet such as the Mediterranean diet (Saura-Calixto and Goñi, 2006).

Processing and Impact on Antioxidants in Beverages
http://dx.doi.org/10.1016/B978-0-12-404738-9.00027-1

Pérez-Jiménez *et al.* (2008) stated that the antioxidant capacity present in beverages may be more bioaccessible than the capacity associated with solid plant foods, where enzymatic action is necessary to release antioxidant compounds from the food matrix. Therefore, both the determination of TAC and of its constituents contributing to TAC are important and common for food and beverages of plant origin. Determining the TAC values and antioxidant composition (especially phenolic composition) of beverages has become an expanding area of research. There have been many studies on the antioxidant capacity of tea, wine, coffee, beer, and fruit juices, individually and/or comparatively (Karaman *et al.*, 2010; Van der Sluis *et al.*, 2000, 2002, 2004; Miller and Rice-Evans, 1997; De Beer *et al.*, 2006; Šeruga *et al.*, 2011; Borges *et al.*, 2010; Porgalı and Büyüktuncel 2012; Di Majo *et al.*, 2008; Tabart *et al.*, 2010).

Many TAC assays have been developed in the last two decades. The ABTS/TEAC [2,2′-azino-bis(3-ethylbenzothiazoline-6-sulfonic acid)/Trolox equivalent antioxidant capacity] (Re *et al.*, 1999), ORAC (oxygen radical absorbance capacity) (Cao *et al.*, 1995), DPPH (2,2-diphenyl-1-picrylhydrazyl) (Brand-Williams *et al.*, 1995), FRAP (ferric reducing antioxidant power) (Benzie and Strain, 1996), and CUPRAC (cupric reducing antioxidant capacity) (Apak *et al.*, 2004) are the most widely used assays for determining *in vitro* antioxidant capacity.

TAC is a very common and important parameter concerning health benefits of beverages, but identification and quantification of antioxidant composition of beverages is necessary from many points of view such as quality and authentication. Also some classes of antioxidant compounds may have cellular effects *in vivo* which are not directly related to their antioxidant properties (Virgili and Marino, 2008). Therefore, chromatographic (especially high-performance liquid chromatographic, HPLC) analysis is applied to plant foods and beverages with plant origins such as fruit juices, tea, and wine. Unfortunately, in a number of studies concerning TAC and polyphenolic contents of beverages, TAC assays and chromatographic analyses are evaluated independently (Šeruga *et al.*, 2011; Porgalı and Büyüktuncel, 2012). Chromatographic analysis results in these studies are mainly used to compare with total phenolics (TP) values determined by Folin–Ciocalteu assay. TAC values are also correlated with TP. However, methods integrated with chromatography and spectrophotometric TAC assays are useful both in the determination of antioxidant components and TAC values only in a successful chromatographic run. These assays are on-line (Shi *et al.*, 2009; Çelik *et al.*, 2010; Borges *et al.*, 2010) or off-line procedures (Karaman *et al.*, 2010; Van der Sluis *et al.*, 2002, 2004; Miller and Rice-Evans, 1997; De Beer *et al.*, 2006). Off-line HPLC integrated spectrophotometric TAC assays do not require additional equipment (i.e., post-column system) and TAC reagents. This chapter will stress the importance of off-line HPLC integrated TAC assays, which do not require the consumption of additional chemicals and time to establish the TAC values when most of the antioxidant components can be identified and quantified.

OFF-LINE HPLC INTEGRATED TO TOTAL ANTIOXIDANT CAPACITY ASSAYS

Off-Line HPLC Integrated to CUPRAC (Combined HPLC-CUPRAC) Assay

An assay for the evaluation of plant foods and beverages regarding both individual antioxidant constituents and total antioxidant capacity (TAC) was named the 'combined HPLC-CUPRAC (cupric ion reducing antioxidant capacity) assay' for the first time by the authors' research group (Yıldız *et al.*, 2008). The assay is based on separation, identification and quantification of individual antioxidants, especially phenolics, in the sample with the aid of a mathematical procedure comprising multiplication of the HPLC-determined concentration of each antioxidant with its TEAC (Trolox equivalent antioxidant capacity, defined as the mM Trolox equivalent concentration of 1 mM solution of the tested antioxidant) coefficient and summation of these products to yield the theoretical TAC value by virtue of the additivity of absorbances of constituents in a mixture. Thus, the theoretical TAC of the investigated material could be estimated using the equation:

$$TAC_{\text{theoretical}} = \sum_{i=1}^{n} c_i(\text{TEAC})_i \qquad \text{(Eq. 27. 1)}$$

where c_i is the concentration of antioxidant constituent i found with the help of HPLC, and $(\text{TEAC})_i$ is the TEAC coefficient of constituent i with respect to CUPRAC.

For the calculation of TEAC coefficients required for theoretical TAC estimation and the comparison of experimental and theoretical TAC values, CUPRAC methods [the normal CUPRAC ($CUPRAC_N$) and the incubated CUPRAC ($CUPRAC_I$)] (Apak *et al.*, 2004) were applied to various antioxidant standards and samples. In the normal CUPRAC ($CUPRAC_N$) procedure, x ml beverage (diluted if necessary) and $(1-x)$ ml distilled water were added to 1 ml copper(II) (1.0×10^{-2} M in distilled water), 1 ml neocuproine (7.5×10^{-3} M in 96% ethanol), and 1 ml ammonium acetate buffer (1.0 M in distilled water, pH 7) mixture solution (total volume 4.0 ml), and then were mixed well by vortexing. This final mixture in a stoppered test tube was left to stand at room temperature for 30 min. At the end of this period, the absorbance at 450 nm was measured against a reagent blank. The incubated CUPRAC method ($CUPRAC_I$) was also applied with 20 or 30 min (according to the sample composition) incubation time at 50°C instead of room temperature as described above. TAC results were expressed as mmol Trolox/l beverage. The unitless TEAC coefficients ($TEAC_{CUPRAC}$) calculated by dividing the CUPRAC molar absorptivity of the antioxidant standard to that of Trolox measured under identical conditions are tabulated in Table 27.1.

TABLE 27.1 TEAC Coefficients of Various Phenolics and Ascorbic Acid with Respect to the CUPRAC Method

Antioxidant	$TEAC_N$	$TEAC_I$
Ascorbic acid	1.00	1.32
HYDROXYBENZOIC ACIDS		
Gallic acid	2.99	3.25
HYDROXYCINNAMIC ACIDS		
Chlorogenic acid	3.15	3.64
Caffeic acid	3.12	3.67
p-Coumaric acid	0.55	1.00
Ferulic acid	1.20	1.23
Rosmarinic acid	5.65	6.02
FLAVONES		
Luteolin	2.90	3.14
Apigenin	0.27	0.82
FLAVONOLS		
Quercetin	4.64	5.88
Myricetin	4.27	4.79
Kaempferol	1.47	1.92
Rutin	3.24	3.68
Isoquercitrin	3.54	4.44
FLAVANOLS		
Catechin	3.46	4.22
Epicatechin	4.04	4.80
FLAVANONES		
Naringenin	—	2.28
Naringin	0.02	0.13
Hesperetin	0.99	1.05
Hesperidin	0.97	1.11
DIHYDROCHALCONES		
Phloridzin	1.47	2.99
PROCYANIDINS		
Procyanidin B2	7.11	8.89

TEAC, Trolox equivalent antioxidant capacity; $TEAC_N$, TEAC values calculated by the normal cupric reducing antioxidant capacity ($CUPRAC_N$) method; $TEAC_I$, TEAC values calculated by the incubated cupric reducing antioxidant capacity ($CUPRAC_I$) method.
References: Yıldız et al. (2008), Karaman et al. (2010.)

As the theoretical TAC values of samples could be estimated by multiplying the HPLC-determined concentration of each antioxidant with its TEAC coefficient and adding up the products according to Eq. 27.1 (Yıldız *et al.*, 2008), separation, identification, and quantification of antioxidant constituents in complex mixtures by HPLC were necessary.

Gradient elution programs were applied with two different solutions, methanol (A) and 0.2% of *o*-H_3PO_4 (v/v) in bi-distilled water (B), allowed to pass through a C18 column. The gradient program was normally variable with respect to the sample (complex matrix) and/or antioxidant constituent. For example, for polyphenolic compounds in antioxidant standard mixtures (Yıldız *et al.*, 2008), *Gradient Elution Program 1* was: in the first 8 min 7% (A), slope (0.0); 8–13 min from 7 to 30% (A), slope (−4.0); 13–48 min from 30 to 66% (A), slope (1.0); 48–55 min from 66 to 75% (A), slope (−4.0). Detection wavelength was 280 nm, and flow rate 1 ml/min.

Ascorbic acid determination was performed by using isocratic elution for 8 min, the mobile phase being composed of 7% methanol (A) and 93% bi-distilled water containing 0.2% of *o*-H_3PO_4 (B). The detection wavelength was 215 nm, and the elution rate was the same as for phenolics.

Another gradient elution program (*Gradient Elution Program 2*) for phenolic constituents in apple juices (Karaman *et al.*, 2010) was: in the first 5 min 30% (A), slope (0.0); 5–40 min from 30 to 66% (A), slope (1.0).

For ascorbic, malic, and fumaric acids, *Gradient Elution Program 3* in apple juices (Karaman *et al.*, 2010) was: in the first 3 min 2% (A), slope (0.0); 3–7 min from 2 to 7% (A), slope (4.0); 7–10 min isocratically 7% (A).

Detection wavelengths were 280 nm for phenolics and 215 nm for ascorbic, malic, and fumaric acids, and flow rates were 1 ml/min for both elutions.

Using the above working modes, the calibration curves and linear equations (Table 27.2) of peak area (y) *versus* concentration (c) were determined for the phenolic antioxidants of interest and for ascorbic, malic, and fumaric acids.

With the aid of these calibration curves, apple juices freshly prepared from seven varieties, King Luscious, Amasya, Ervin Spur, Sky Spur, Arap Kizi, Lutz Golden, and Granny Smith (Karaman *et al.*, 2010), and three synthetic mixtures (Yıldız, *et al.*, 2008) were analyzed.

To prepare the fresh apple juices, the apple flesh obtained by peeling and deseeding the apple fruit was grated with the aid of a glass grater. This puree was bolted through cheesecloth and separated apple juice was centrifuged at 5000 rpm for 15 min; the supernatant juice was decanted, and clarified by passing through a GF/PET (glass fiber/polyethyleneterephtalate) 1.0/0.45-μm microfilter.

TABLE 27.2 HPLC Calibration Equations (as Peak Area *versus* Molar Concentration) of Various Antioxidants

Antioxidant	Calibration Equation	(r)
Ascorbic acid	$y=3.16\times10^9\,c+1.67\times10^4$	0.9999
HYDROXYBENZOIC ACIDS		
Gallic acid	$y=9.60\times10^9\,c-1.10\times10^4$	0.9997
HYDROXYCINNAMIC ACIDS		
Chlorogenic acid	$y=1.00\times10^{10}\,c+2.50\times10^4$	0.9996
Caffeic acid	$y=1.10\times10^{10}\,c-0.20\times10^4$	0.9999
p-Coumaric acid	$y=1.48\times10^{10}\,c+3.06\times10^4$	0.9999
Ferulic acid	$y=1.05\times10^{10}\,c-2.48\times10^4$	0.9998
Rosmarinic acid	$y=9.49\times10^9\,c+6.43\times10^4$	0.9995
FLAVONES		
Luteolin	$y=9.02\times10^9\,c-7.95\times10^4$	0.9990
Apigenin	$y=1.04\times10^{10}\,c-0.17\times10^4$	0.9999
FLAVONOLS		
Quercetin	$y=7.90\times10^9\,c-1.78\times10^4$	0.9996
Myricetin	$y=5.93\times10^9\,c-4.20\times10^4$	0.9996
Kaempferol	$y=7.23\times10^9\,c-3.61\times10^4$	0.9996
Rutin	$y=8.50\times10^9\,c-1.67\times10^4$	0.9993
Isoquercitrin	$y=8.40\times10^9\,c+0.08\times10^4$	0.9998
FLAVANOLS		
Catechin	$y=3.80\times10^9\,c-0.10\times10^4$	0.9997
Epicatechin	$y=3.71\times10^9\,c-0.35\times10^4$	0.9997
FLAVANONES		
Naringenin	$y=1.47\times10^{10}\,c+0.60\times10^4$	0.9999
Naringin	$y=1.69\times10^{10}\,c+0.91\times10^4$	0.9999
Hesperetin	$y=1.43\times10^{10}\,c-8.15\times10^4$	0.9983
Hesperidin	$y=1.65\times10^{10}\,c-4.45\times10^4$	0.9988
DIHYDROCHALCONES		
Phloridzin	$y=1.70\times10^{10}\,c+3.96\times10^4$	0.9996
PROCYANIDINS		
Procyanidin B2	$y=6.70\times10^9\,c+6.31\times10^4$	0.9995

HPLC, high-performance liquid chromatography; r, correlation coefficients of linear calibration equations.
References: Yıldız et al. *(2008), Karaman* et al. *(2010).*

The tested synthetic mixture compositions were as follows:

Synthetic mixture 1: contained in final solution 2.0×10^{-4}, 4.0×10^{-5}, and 1.0×10^{-4} M of *p*-coumaric acid, myricetin, and apigenin, respectively.
Synthetic mixture 2: contained in final solution 2.0×10^{-4}, 4.0×10^{-5}, 3.4×10^{-5}, and 1.0×10^{-4} M of chlorogenic acid, myricetin, luteolin, and apigenin, respectively.
Synthetic mixture 3: contained in final solution equimolar (2.0×10^{-4} M) concentrations of catechin and chlorogenic acid.

The contents of phenolic constituents (mg/l) calculated from chromatograms (only two chromatograms of the King Luscious and Amasya apple juices are given in Figure 27.1) of apple juices and the experimental and theoretical TAC values of synthetic mixtures and apple juices (mmol Trolox/l) can be seen in Table 27.3 and Table 27.4, respectively. Percentages of experimental TAC as accounted for by HPLC-CUPRAC estimates are also shown in Table 27.4.

The ascorbic acid contents of the analyzed apple juices except Arap Kizi, Lutz Golden, and Granny Smith were calculated to lie between 12 ± 0.24 and 88 ± 0.91 mg/l with respect to apple varieties from their chromatograms and linear calibration equation of ascorbic acid. Due to the high malic acid contents of the three varieties mentioned, the peak of the ascorbic acid, the relatively lower component, could not be resolved in their chromatograms. So, the theoretical TAC values of fresh apple juices of the other four varieties in Table 27.4 contained the TAC arising from ascorbic acid which could be separately determined.

For synthetic mixtures, the experimental CUPRAC and theoretical HPLC-CUPRAC results for TAC (Table 27.4) were very close to each other. This finding confirms that if all the antioxidants in a complex mixture were identified and quantified with the help of HPLC techniques, their contribution to the overall capacity could be envisaged, and the experimentally found TAC values (with the use of the spectrophotometric CUPRAC antioxidant assay) could be correctly estimated by theoretically calculating the TAC using the principle of additivity of individual antioxidant capacities by combined HPLC-CUPRAC assay.

Off-Line HPLC Integrated to ABTS

The approach adopted in the combined HPLC-CUPRAC assay was also applied as HPLC-ABTS to the same seven fresh apple juices and the same three synthetic mixtures by the authors' research group (Yıldız *et al.*, 2008; Karaman *et al.*, 2010). But $(TEAC)_i$ in Eq. 27.1 is the TEAC coefficient of constituent i with respect to the ABTS method in this assay.

For the determination of TEAC coefficients required for theoretical TAC estimation and the comparison of experimental and theoretical TAC values, the ABTS/persulfate (Re *et al.*, 1999) method was applied to antioxidant standards and synthetic mixtures (Yıldız *et al.*, 2008), and the slightly modified (in terms of reagent

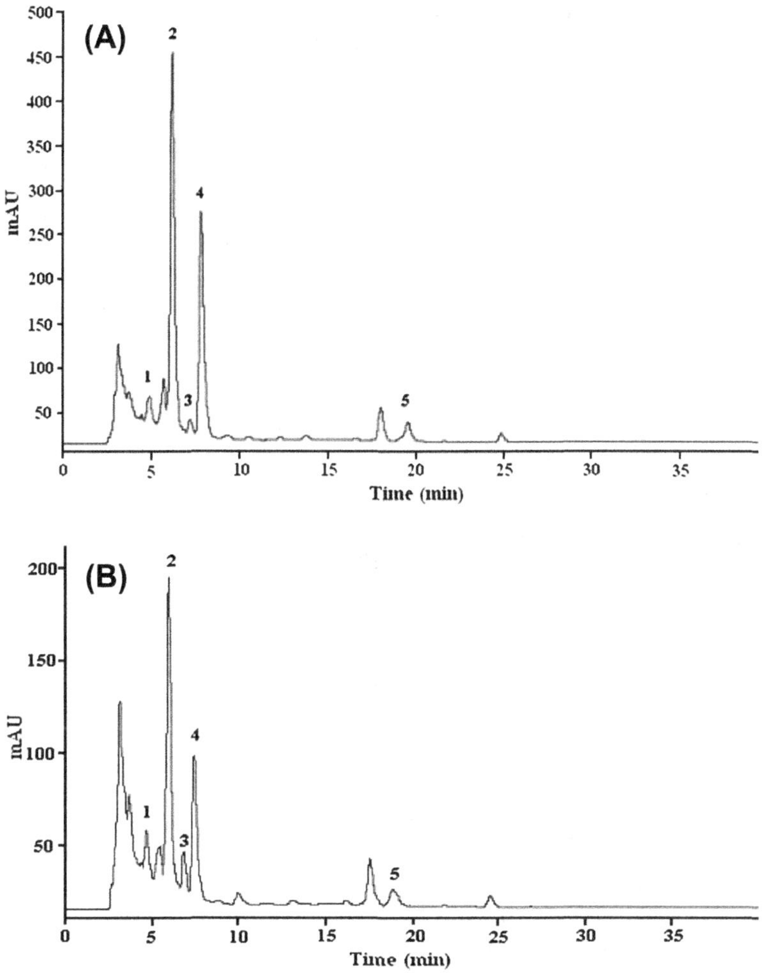

FIGURE 27.1 The chromatograms of King Luscious (A) and Amasya (B) apple juices. 1: catechin + procyanidin B2; 2: chlorogenic acid; 3: epicatechin; 4: caffeic acid; 5: phloridzin; *References: Karaman et al. (2010), Özyürek et al. (2011).*

TABLE 27.3 The Contents of Phenolic Constituents (mg/l) of Fresh Apple Juices

Apple Variety	Catechin	Chlorogenic Acid	Epicatechin	Caffeic Acid	Phloridzin
King Luscious	115.30 ± 1.14	276.36 ± 2.80	16.63 ± 0.33	79.09 ± 1.19	12.43 ± 0.25
Amasya	89.41 ± 0.91	126.84 ± 0.13	43.54 ± 0.65	27.20 ± 0.41	6.61 ± 0.13
Ervin Spur	81.39 ± 0.85	111.25 ± 1.14	66.76 ± 0.70	20.72 ± 0.31	20.13 ± 0.41
Sky Spur	9.09 ± 0.18	72.99 ± 0.81	45.28 ± 0.68	21.80 ± 0.33	11.80 ± 0.24
Arap Kizi	67.52 ± 0.69	242.30 ± 2.50	28.04 ± 0.42	12.14 ± 0.24	0.52 ± 0.01
Lutz Golden	41.62 ± 0.62	71.92 ± 0.74	11.58 ± 0.17	8.22 ± 0.16	3.57 ± 0.07
Granny Smith	76.77 ± 0.82	41.10 ± 0.45	3.38 ± 0.07	3.04 ± 0.08	1.15 ± 0.03

Reference: Karaman, et al. *(2010).*

amount and time) ABTS/HRP (horseradish peroxidase) (Arnao *et al.*, 2001) method was applied to antioxidant standards and apple juices (Karaman *et al.*, 2010) in this off-line HPLC procedure integrated to TAC assay.

In the ABTS/persulfate method (Re *et al.*, 1999), the chromogenic radical reagent ABTS, at 7.0 mM concentration, was prepared by dissolving this compound in water and adding $K_2S_2O_8$ to this solution such that the final persulfate concentration in the mixture was 2.45 mM. The resulting ABTS radical cation solution was left to mature at room temperature in the dark for 12–16 h, and then used for ABTS/TEAC assays. The volumes of $(4-x)$ ml EtOH and x ml sample solution were taken. The reagent blank was prepared with 4 ml EtOH.

TABLE 27.4 The Experimental and Theoretical TAC Values of Synthetic Mixture Solutions of Antioxidants and Apple Juices (mmol Trolox/l) and the Percentages of Experimental TAC as Accounted for by HPLC-CUPRAC Estimates (in Parentheses)

Sample	$CUPRAC_N$	$CUPRAC_I$	HPLC- ($CUPRAC_N$)	HPLC- ($CUPRAC_I$)
Mixture 1	0.32	0.43	0.29 (91)	0.47 (109)
Mixture 2	0.80	0.90	0.79 (99)	0.92 (102)
Mixture 3	1.29	1.41	1.19 (92)	1.34 (95)
King Luscious[a]	8.26	11.4	5.83 (71)	7.16 (63)
Amasya[a]	5.43	7.67	4.19 (77)	5.45 (71
Ervin Spur[a]	9.74	13.70	6.76 (69)	9.03 (66)
Sky Spur[a]	5.85	9.06	4.21 (72)	7.62 (84)
Arap Kizi	4.70	7.77	3.50 (74)	4.15 (53)
Lutz Golden	3.39	5.78	1.42 (42)	1.69 (29)
Granny Smith	4.37	5.83	1.75 (40)	2.09 (36)

TAC, total antioxidant capacity; $CUPRAC_N$, experimental TAC values obtained by the normal cupric reducing antioxidant capacity method; $CUPRAC_I$, experimental TAC values obtained by the incubated cupric reducing antioxidant capacity method; HPLC-$CUPRAC_N$, theoretical (calculated) TAC values of complex samples obtained by multiplying the concentration of each constituent (determined chromatographically) with its TEAC coefficient (determined by $CUPRAC_N$ method) and summing up the products; HPLC-$CUPRAC_I$, theoretical (calculated) TAC values of complex samples obtained by multiplying the concentration of each constituent (determined chromatographically) with its TEAC coefficient (determined by $CUPRAC_I$ method) and summing up the products.

[a]*The TAC arising from ascorbic acid was added to the theoretically calculated total antioxidant capacities.*

References: *Yıldız* et al. *(2008) for synthetic mixture solutions of antioxidants, Karaman* et al. *(2010) for apple juices.*

One milliliter of 1:10 diluted (with ethanol) ABTS radical cation solution was added to the mixture, and mixed well (total volume 5.0 ml). The absorbance of solution was recorded at 734 nm against ethanol at the end of sixth minute. The absorbance of the reagent blank (A_0) diminished in the presence of antioxidants, the absorbance decrease (ΔA) being proportional to antioxidant concentration.

In the ABTS/HRP procedure, the formation of ABTS radical cation in the mixture containing 0.5 ml each of ABTS (4.0×10^{-3} M in distilled water) and H_2O_2 (1.2×10^{-4} M in 96% EtOH) was started by the addition of 0.5 ml of HRP (2.4×10^{-5} M in a mixture (1:1, v/v) of EtOH and H_2O). Stabilization of absorbance due to the achievement of a constant concentration of ABTS radical cation ($ABTS^{\bullet+}$) was observed in 5 min, then x ml sample and $(0.5-x)$ ml EtOH were added to the mixture. After 5 min, the absorbance at 730 nm was recorded against EtOH. The absorbance difference (ΔA) was found by subtracting the absorbance of the antioxidant sample from that of the reagent blank. The decrease in absorbance corresponded to the disappearance of $ABTS^{\bullet+}$ that reacted with antioxidants, equivalent to antioxidant content of the sample. TAC results were expressed as mmol Trolox/l beverage.

The reason for our preference for the ABTS/HRP method for apple juice studies was that the ABTS capacities for various antioxidants are quite dependent on the methodology of the colored ABTS radical cation production (Re *et al.*, 1999) and that lower TEAC values were obtained with ABTS/persulfate for the major antioxidant components of apple juices, especially phloridzin (Karaman *et al.*, 2010). The TEAC coefficients ($TEAC_{ABTS}$) calculated by dividing the ABTS molar absorptivity of the antioxidant standard to that of Trolox measured under identical conditions are tabulated in Table 27.5.

Theoretical TAC values of synthetic mixtures and apple juices (mmol Trolox/l) calculated via Eq. 27.1 by using $TEAC_{ABTS}$ coefficients (Table 27.5), chromatographically determined contents of phenolics (Table 27.3), and ascorbic acid, compared to experimental TAC_{ABTS} values are tabulated in Table 27.6. Percentages of experimental TAC as accounted for by HPLC-ABTS estimates are also shown in Table 27.6.

For synthetic mixtures, theoretical HPLC-ABTS calculations slightly overestimated the experimental ABTS/persulfate results (Table 27.6). However, it can be seen from the same table that the HPLC-ABTS/HRP procedure was successful in predicting the experimentally observed capacities of apple juices by the spectrophotometric method (with the exception of Lutz Golden apples, which have probably more than one unidentified constituent through HPLC), and percentages obtained with this assay and HPLC-CUPRAC assay were rather similar (except Granny Smith). Naturally, there can not be 100% estimation due to the limited availability of all concerned flavonoid glycoside standards. It can be stated that the experimentally found TAC values (with the use of the spectrophotometric ABTS/HRP antioxidant assay) could be correctly estimated by theoretically calculating the TAC using the principle of additivity of individual antioxidant capacities by combined HPLC-ABTS/HRP assay.

Miller and Rice-Evans (1997) studied long-life orange and apple juices and blackcurrant drink in order to establish the relative contributions of ascorbic acid and phenolic antioxidants to total antioxidant activity (TAA), and followed a mathematical procedure by multiplying the ABTS-TEAC coefficients of individual constituents with their concentrations, and summing up these products to approximate the TAA of these juices with respect to the ABTS method.

Ferrylmyoglobin/ABTS spectrophotometric assay generating the chromogenic $ABTS^{\bullet+}$ radical cation from the interaction between ABTS (150 μM), metmyoglobin (2.5 μM) and hydrogen peroxide (375 μM) (Miller *et al.*, 1995) was used to measure TAA of beverages and Trolox

TABLE 27.5 TEAC Coefficients of Various Phenolics and Ascorbic Acid with Respect to the ABTS Method

Antioxidant	$TEAC_{ABTS/Persulfate}$	$TEAC_{ABTS/HRP}$
Ascorbic acid	1.02	1.06
HYDROXYBENZOIC ACIDS		
Gallic acid	2.52	3.64
HYDROXYCINNAMIC ACIDS		
Chlorogenic acid	0.98	1.38
Caffeic acid	1.05	1.11
p-Coumaric acid	1.63	—
Ferulic acid	2.16	—
Rosmarinic acid	2.30	—
FLAVONES		
Luteolin	1.01	—
Apigenin	0.65	—
FLAVONOLS		
Quercetin	2.89	4.01
Myricetin	3.43	—
Kaempferol	0.86	—
Rutin	0.93	3.19
Isoquercitrin	1.01	3.21
FLAVANOLS		
Catechin	1.44	4.56
Epicatechin	1.01	4.58
FLAVANONES		
Naringenin	0.64	—
Naringin	0.62	—
Hesperetin	1.11	—
Hesperidin	1.40	—
DIHYDROCHALCONES		
Phloridzin	0.12	1.83
PROCYANIDINS		
Procyanidin B2	—	6.61

TEAC, Trolox equivalent antioxidant capacity; $TEAC_{ABTS/Persulfate}$, TEAC values obtained by decolorization of ABTS [2,2′-azino-bis(3-ethylbenzothiazoline-6-sulfonic acid)] radical cation prepared by oxidizing ABTS with persulfate to $ABTS^{\bullet +}$; $TEAC_{ABTS/HRP}$, TEAC values obtained by decolorization of ABTS [2,2′-azino-bis(3-ethylbenzothiazoline-6-sulfonic acid)] radical cation prepared by oxidizing ABTS with H_2O_2 in the presence of HRP (horseradish peroxidase) as catalyst to $ABTS^{\bullet +}$.
References: Yıldız et al. *(2008), Karaman* et al., *(2010).*

equivalent antioxidant capacity (TEAC) of pure compounds. Absorbances were read at 734 nm.

Reversed-phase HPLC analyses of beverages and phenolic standards present (or potentially present) in these beverages were carried out by the different elution programs as in our studies. Anthocyanins and free gallic acid in blackcurrant drink were analyzed by aqueous acetonitrile gradient at pH 1.5 via monitoring by diode array detector at 280 and 520 nm. For soluble polyphenols (hesperidin, narirutin) in orange juice, a gradient between 0.5% acetic acid in water to 100% acetonitrile was used. Composition of apple juice was identified and quantified by water–acetonitrile gradient acidified to pH 2.5 (Lea, 1988).

TABLE 27.6 The Experimental and Theoretical TAC Values of Synthetic Mixture Solutions of Antioxidants and Apple Juices (mmol Trolox/l) and the Percentages of Experimental TAC as Accounted for by HPLC-ABTS Estimates (in Parentheses)

Sample	ABTS	HPLC-ABTS
Mixture 1	0.33	0.54 (>>100)
Mixture 2	0.40	0.47 (118)
Mixture 3	0.76	0.90 (118)
King Luscious[a]	5.32	3.37 (63)
Amasya[a]	3.91	3.44 (88)
Ervin Spur[a]	7.59	6.34 (84)
Sky Spur[a]	4.55	3.76 (83)
Arap Kizi	3.08	2.40 (78)
Lutz Golden	2.84	1.12 (39)
Granny Smith	3.06	1.84 (60)

TAC, total antioxidant capacity; ABTS, experimental TAC values obtained by ABTS [2,2′-azino-bis(3-ethylbenzothiazoline-6-sulfonic acid)] method; HPLC-ABTS, theoretical (calculated) TAC values of complex samples obtained by multiplying the concentration of each constituent (determined chromatographically) with its TEAC coefficient (determined by ABTS method) and summing up the products.
[a]*The TAC arising from ascorbic acid was added to the theoretically calculated total antioxidant capacities.*
References: Yıldız et al. (2008) for synthetic mixture solutions of antioxidants, Karaman et al. (2010) for apple juices.

Ascorbic acid was measured spectrophotometrically using the Fe(III)-ferrozine assay (Butts and Mulvihill, 1975), but specificity is a problem, as other reducing agents also respond to this reagent.

Contributions to the TAC (referred to as TAA in this source) of individual phenolics and ascorbic acid were calculated by multiplying the concentrations of these compounds with their TEAC values. Calculated TAA was given as μM by summing up individual contributions. All results (TEAC coefficients obtained by ferrylmyoglobin/ABTS, chromatographically determined concentrations of phenolics, concentration of ascorbic acid, measured and calculated TAA values) are summarized in Table 27.7. Percentages of experimental TAA as accounted for by HPLC-ABTS estimates are also shown in Table 27.7.

The authors noted the large difference between the calculated and measured TAA of blackcurrant drink, leading to the speculation that there might be a major unidentified antioxidant.

TABLE 27.7 TEAC Coefficients and Concentrations (μmol/l) of Ascorbic Acid and some Polyphenols, Calculated and Measured Total Antioxidant Activities (TAA, μmol Trolox/l) of some Fruit Beverages, and the Percentages of Measured TAA as Accounted for by Calculated TAA with the aid of HPLC-ABTS Estimates (in Parentheses)

Sample	Constituent	TEAC	Concentration Mean (μmol/l)	Activity (conc. × TEAC) μM
Longlife apple juice	Chlorogenic acid	1.24	274	340
	p-Coumaroyl quinic acid	2.22[a]	74	164
	Phloridzin	2.38	23	55
	Phloretin xyloglucoside	2.38[b]	35	83
	Epicatechin	2.50	<14	35
	Ascorbic acid	1.00	51	51
	Calculated TAA			728 (87)
	Measured TAA			840
Longlife orange juice	Hesperidin[c]	1.08	141	152
	Narirutin[c]	0.76	62	47
	Ascorbic acid	1.00	2270	2270
	Calculated TAA			2469 (95)
	Measured TAA			2610
Blackcurrant drink	Delphinidin-3-glucoside	2.47[d]	5.9	150
	Delphinidin-3-rutinoside	3.25[e]	19.9	65
	Cyanidin-3-glucoside	2.47	2.4	6
	Cyanidin-3-rutinoside	3.25	15.0	49
	Gallic acid	3.01	Undetectable	
	Ascorbic acid	1.00	3726	3726
	Calculated TAA			3861 (76)
	Measured TAA			5070

TEAC, Trolox equivalent antioxidant capacity; TAA, total antioxidant activity.
[a]*Pure substance not available for testing: activity taken as that of* p-*coumaric acid.*
[b]*Pure substance not available for testing: activity taken as that of phloridzin.*
[c]*Soluble fraction only.*
[d]*Pure substance not available for testing: since TEAC values for the anthocyanidins delphinidin (4.44) and cyanidin (4.42) were nearly identical, the activity was taken as that of cyanidin-3-glucoside.*
[e]*Pure substance not available for testing: activity taken as that of cyanidin-3-rutinoside.*
Reference: Miller and Rice-Evans (1997).

TABLE 27.8 TEAC Coefficients, Average Content (mg/l) and TAC Contribution (mM TE) of Individual Phenolic Compounds in Pinotage Wines (n = 139)

Compound	TEAC	Content (average)	TAC Contribution (average)
Gallic acid	2.79	11.9	0.20
Procyanidin B1	2.30	21.6	0.09
(+)-Catechin	0.96	15.1	0.05
Caffeoyltartaric acid	0.90	178.2	0.51
Caffeic acid	0.98	3.4	0.02
p-Coumaroyltartaric acid[a]	NA	17.2	NA
p-Coumaric acid	<0.01	1.7	[b]
Quercetin-3-gal	0.96	2.9	0.01
Quercetin-3-glc	0.92	14.2	0.03
Quercetin-3-rham	0.91	8.8	0.02
Myricetin	2.67	2.6	0.02
Quercetin	1.75	3.9	0.02
Kaempferol	0.88	0.8	0.00
Isorhamnetin	0.95	0.6	0.00
Delphinidin-3-glc	2.40	15.0	0.07
Petunidin-3-glc	2.06	22.6	0.09
Peonidin-3-glc	1.49	7.5	0.02
Malvidin-3-glc	1.46	221.3	0.61
Delphinidin-3-glc-ac[c]	NA	5.3	0.03
Vitisin A[c]	NA	5.7	0.02
Petunidin-3-glc-ac[c]	NA	5.5	0.02
Peonidin-3-glc-ac[c]	NA	5.1	0.02
Malvidin-3-glc-ac[c]	NA	59.4	0.16
Malvidin-3-glc-coum[c]	NA	21.2	0.06
Polymers[d]	NA	150.4	NA
Total		820.1	2.04
TAC measured			13.15
%TAC accounted[e]			16.10

NA, not available; TEAC, Trolox equivalent antioxidant capacity; TAC, total antioxidant capacity; TE, Trolox equivalent.
[a]*mg* p-*coumaric acid equivalents/l.*
[b]*Negligible.*
[c]*mg corresponding anthocyanin-3-glc equivalents/l.*
[d]*mg (+)-catechin equivalents/l*
[e]*%TAC accounted for = TAC calculated × 100/TAC measured.*
Reference: De Beer et al. *(2006).*

Another study on the contributions of individually quantified phenolic compounds to the TAC values calculated using their HPLC-quantified concentrations and TEAC coefficients is the investigation for TAC and phenolic composition of 139 Pinotage wines (2002 and 2003 vintages) by De Beer and co-workers (De Beer *et al.*, 2006).

The TAC of wines and the TEAC of phenolic standards were determined by the $ABTS^{\bullet+}$ scavenging assay (Re *et al.*, 1999). For wine samples, 1 ml $ABTS^{\bullet+}$ solution was added to a 50-µl sample (50-times diluted with 10% ethanol) and the absorbance was read after exactly 4 min incubation at 37°C. $ABTS^{\bullet+}$ concentration in the control and samples was calculated using the absorbances and the molar extinction coefficient of $ABTS^{\bullet+}$ ($\varepsilon = 16{,}000$) (Re *et al.*, 1999). A graph of remaining $ABTS^{\bullet+}$ concentration against Trolox concentration in the standard samples was used to determine the TAC of the wines as mM Trolox equivalent (TE).

Chromatographic identification and quantification was carried out using a reversed-phase HPLC method adapted from Peng *et al.* (2002). The column oven was adjusted to 30°C during the run, and the flow rate was 0.9 ml/min. Data were obtained in the wavelength range of 250–600 nm from the photodiode array detector. The gradient elution program with two mobile phases [A: 1.5% (v/v) aqueous phosphoric acid and B: 1.5% (v/v) phosphoric acid in acetonitrile/water (80/20)] was as follows: from 94 to 69% solvent A in the first 73 min, reduced from 69 to 38% from 73 to 78 min, held isocratic at 38% from 78 to 86 min, increased from 38 to 94% from 86 to 90 min, and equilibration at 94% for 20 min.

The TAC contribution (mM TE) of individual phenolic compounds to the wine TAC was calculated from their content (mg/l) and TEAC values. TEAC coefficients, average content and TAC contribution of individual phenolic compounds in Pinotage wines can be seen in Table 27.8.

In this study, between 11 and 24% (average 16%) of the measured TAC of Pinotage wines was explained by the sum of the calculated contributions of their HPLC-quantified phenolic compounds comprising monomeric phenolic compounds and procyanidin B1. Ultrafiltration was carried out to attempt separation of monomeric and polymeric phenolic compounds.

Analysis of ultrafiltration permeates and retentates enabled estimation of the TAC contribution of large molecular weight (MW) unknown compounds (46%) (>50 kDa), including oligomeric and polymeric phenolic compounds and small MW unknown compounds (34%) (<50 kDa).

Using the major monomeric polyphenol contents of Šeruga and co-workers (Šeruga *et al.*, 2011) found in red wines produced in Croatia and $TEAC_{ABTS/persulfate}$ coefficients determined previously by our research group (Yıldız *et al.*, 2008), theoretical TAC values were calculated and compared with experimental TAC values determined by ABTS method in the study.

In the study of Šeruga and co-workers, ABTS assay was applied as given by Villano and co-workers (Villano *et al.*, 2004). A stock solution of the $ABTS^{\bullet+}$ radical cation was generated by mixing 0.2 ml of ammonium peroxodisulfate solution (65 mmol/l) with 50 ml of ABTS solution (1 mmol/l, prepared in 0.1 mol/l phosphate buffer, pH 7.4). The mixture was left overnight, then 0.5 ml of $ABTS^{\bullet+}$ stock solution was mixed with 2 ml of phosphate buffer (pH 7.4) in a cuvette, and finally the absorbance at 734 nm (A_{ABTS}) was read. Subsequently, 0.1 ml of diluted wine was added to the cuvette, the solution was mixed quickly, and the absorbance at 734 nm (A_{wine}) was measured after 60 s. The decrease in absorbance ($\Delta A = A_{ABTS} - A_{wine}$) after 60 s was calculated for each diluted wine sample. Antioxidant activity of each diluted wine sample was calculated utilizing the calibration curve obtained by Trolox standard and expressed in mmol/l of Trolox equivalents (mmol TE/l).

The contents of individual wine polyphenols (flavonols, flavan-3-ols, and phenolic acids) were determined by the RP-HPLC method. The flavonols, flavan-3-ols, and phenolic acids present in wines were separated using phosphoric acid (0.1%) as solvent A and 100% methanol as solvent B forming the total eluent mixture. The elution conditions were: 0–30 min from 5% B to 80% B; 30–33 min 80% B; 33–35 min from 80% B to 5% B; flow rate = 0.8 ml/min. Column temperature was maintained at 20°C and injection volume was 10 µl. UV-Vis spectra were recorded in the wavelength range 190–600 nm with photodiode array (PDA) detector. The detection wavelength was 280 nm for gallic acid, (+)-catechin and (−)-epicatechin, 320 nm for caffeic acid and *p*-coumaric acid and 360 nm for quercetin. The researchers determined the total polyphenols (TP) content of wine from the total area of the RP-HPLC chromatograms at 280 nm and expressed it as mg gallic acid equivalents GAE/l of wine (TP_{HPLC}).

Table 27.9 shows experimental and calculated antioxidant activities using polyphenol contents of Šeruga and co-workers (Šeruga *et al.*, 2011) and $TEAC_{ABTS/persulfate}$ coefficients reported in Table 27.5, and also the percentages of experimental TAC as accounted for by HPLC-ABTS estimates (in parentheses).

The percentages of experimental TAC as accounted for by HPLC-ABTS estimates in Table 27.9 were consistent with the results of De Beer and co-workers (average 16%) obtained from Pinotage wines (De Beer *et al.*, 2006). These low values are due to limited quantification of only major monomeric polyphenols although there are many antioxidant constituents, proanthocyanidins, oligomeric, and polymeric phenolic compounds in wine.

Off-Line HPLC Integrated to Inhibition of Lipid Peroxidation Activity

Van der Sluis *et al.* tried to establish a method of antioxidant activity estimation by finding the inhibition of lipid peroxidation (LPO) activity of individual

TABLE 27.9 TEAC Coefficients and Quantities (mg/l) of some Polyphenols, Measured and Calculated TAA (Total Antioxidant Activity) (mmol TE/l) of Some Croatian Wines, and the Percentages of Measured TAA as Accounted for by HPLC-ABTS Estimates (in Parentheses)

Wine	Gallic Acid[a] (2.52)[b]	Caffeic Acid[a] (1.05)[b]	*p*-Coumaric Acid[a] (1.63)[b]	(+)-Catechin[a] (1.44)[b]	(−)-Epicatechin[a] (1.01)[b]	Quercetin[a] (2.89)[b]	AA_{ABTS}-TEAC	HPLC-ABTS
Ivan Dolac	179.4±7.6	18.1±0.6	2.9±0.7	138.0±7.7	29.3±3.6	6.8±0.3	24.2±0.8	3.64 (15)
Dingač	143.8±3.9	18.6±0.8	2.5±0.4	130.3±6.0	37.4±2.3	7.0±0.6	22.9±0.6	3.11 (14)
Plavac Hvar	128.8±3.1	12.4±0.4	2.5±0.5	135.2±7.5	22.4±1.3	4.6±0.2	22.2±0.5	2.80 (13)
Zweigelt	100.1±3.4	4.3±0.2	2.2±0.2	118.7±9.9	37.7±1.0	1.6±0.1	16.1±0.6	2.26 (14)
Pinot Noir	97.6±2.8	5.1±0.5	1.8±0.2	117.0±9.5	34.2±1.1	1.3±0.1	12.8±0.3	2.21 (17)
Babić	94.8±2.2	4.2±0.3	3.3±0.3	107.7±4.1	10.6±0.7	4.3±0.4	12.5±0.7	2.07 (17)
Kaštelet	92.5±1.7	3.2±0.2	2.3±0.3	79.8±8.9	7.9±1.2	1.5±0.2	11.7±0.4	1.85 (16)
Teran	86.2±1.3	5.6±0.4	1.8±0.2	62.2±4.2	12.8±1.3	2.8±0.3	11.0±0.6	1.71 (15)
Klikun noir	77.2±1.5	8.2±0.6	2.9±0.4	56.6±7.0	15.3±1.5	1.2±0.1	10.0±0.2	1.56 (16)
Merlot	55.5±1.4	5.0±0.2	4.2±0.9	46.2±3.4	7.8±0.4	4.8±0.5	9.2±0.1	1.19 (13)
Frankovka	51.2±2.2	5.2±0.4	4.4±0.7	31.0±2.6	9.8±0.5	1.5±0.1	7.9±0.4	1.04 (13)

TEAC, Trolox equivalent antioxidant capacity; AA, antioxidant activity; TE, Trolox equivalent; AA_{ABTS}-TEAC, experimental antioxidant activity determined by ABTS [2,2′-azino-bis(3-ethylbenzothiazoline-6-sulfonic acid)] method; HPLC-ABTS, theoretical (calculated) TAC values of complex samples obtained by multiplying the concentration of each constituent (determined chromatographically) with its TEAC coefficient (determined by ABTS method) and summing up the products.

[a]*Results represent mean values ± standart deviation of three independent measurements.*

[b]*TEAC values from Table 27.5.*

References: Yıldız et al. *(2008), Šeruga* et al. *(2011).*

constituents in the apple juice and extending their results to total activity assay (Van der Sluis *et al.*, 2002, 2004).

The calculated (or predicted) antioxidant activity of a mixture of known antioxidants could be calculated by Eq. 27.2. The ratio in Eq. 27.2 equals the calculated dilution factor of the mixture of components to give 50% inhibition of lipid peroxidation ($DF_{50,\ mixture}$).

$$\frac{\sum_{i=1}^{n} C_i}{IC_{50,mixture}} = \sum_{i=1}^{n} \frac{C_i}{IC_{50,i}} = DF_{50,mixture} \qquad \text{(Eq. 27. 2)}$$

where C_i is the concentration of component i [mg/kg of fresh weight (fw)] and $IC_{50,i}$ is the IC_{50} value of component i (mg/l).

Antioxidant activity was determined by an *in vitro* method based on the inhibition of LPO (Van der Sluis *et al.*, 2000). LPO was initiated by Fe^{2+} (10 μM) and ascorbic acid (200 μM) in rat liver microsomes. Reaction products are measured spectrophotometrically by the thiobarbituric assay. Absorbances were read at 540 nm (for colored product) and at 620 nm (for turbidity correction). The mean absorbance ($A_{540-620}$) of the blank must be around 1.0. The antioxidant concentration at which 50% inhibition of LPO occurs (IC_{50}) was calculated using the calibration curve of six different antioxidant concentrations ranging from no (0%) to full (100%) inhibition of LPO.

The experimental antioxidant activity of an apple juice is expressed as the dilution factor that gives 50% inhibition ($DF_{50,\ sample}$) according to Eq. 27.3:

$$DF_{50,\ sample} = \frac{1000}{IC_{50,\ sample}} \qquad \text{(Eq. 27. 3)}$$

Chromatographic analysis of polyphenolic constituents were carried out by a Merck Lichrosorb RP18 column and a linear 20-min solvent gradient from 0 to 21% acetonitrile, with a 10-min hold at the final concentration, whereas 10% acetic acid in water was the other elution solution (Van der Sluis *et al.*, 2000, 2001). Column temperature was maintained at 35°C and the flow rate was 1.0 ml/min. Quercetin, its glycosides, and chlorogenic acid were monitored at 350 nm, phloridzin and the catechins at 280 nm, and cyanidin galactoside at 525 nm.

Concentrations of flavonoids and chlorogenic acid [mg/kg fresh weight (fw)], and measured and calculated antioxidant activities of raw apple juices prepared with different production methods from Jonagold apple are tabulated in Table 27.10.

LPO method has some disadvantages, such as very low activity for phloridzin (an important contributor to TAA in apple juices) and the inadequacy of the extrapolation that it was not possible to compose a concentration range from 0% to 100% inhibition to calculate the antioxidant activity.

TABLE 27.10 Concentrations (mg/kg fw) and IC_{50} Values (μM) of Flavonoids and Chlorogenic Acid, Measured ($1000/IC_{50}$, l/mg fw) and Calculated [$\Sigma(C/IC_{50})$] Antioxidant Activities of Raw and Enriched Apple Juices Prepared with Different Production Methods from Jonagold Apple and the Percentages of Measured Activity as Accounted for by Calculated Activity (in Parentheses)

Constituent	IC_{50} Mean	Straight Pressing		Pulp Enzyming	
		Raw Juice (n = 3) Mean ± SD	Enriched Juice (n = 2) Mean ± SD	Raw Juice (n = 1)	Enriched Juice (n = 1)
Cyanidin galactoside	25.5	3.0 ± 0.0	7.0 ± 1.0	2.0	6.0
Phloridzin	1925.5	4.0 ± 1.0	35.0 ± 3.0	2.0	25.0
Chlorogenic acid	124.5	133.0 ± 15.0	180.0 ± 1.0	63.0	99.0
Total quercetin glycosides	9.7[a]	13.0 ± 1.0	117.0 ± 2.0	11.0	96.0
Total catechins	15.6[b]	16.0 ± 1.0	67.0 ± 7.0	6.0	40.0
Measured activity		21.2 ± 18.4	104.3 ± 3.0	6.2	49.1
Calculated activity		9.5 ± 1.2 (45)	39.5 ± 1.7 (38)	4.7 (76)	27.0 (55)

IC_{50} values, the antioxidant concentration at which 50% inhibition of lipid peroxidation (LPO) occurs.
[a] *This value is of quercetin (Van der Sluis* et al., *2000).*
[b] *This value is of catechin (Van der Sluis* et al., *2000).*
References: Van der Sluis et al., *2000, 2002, 2004.*

CONCLUSION

Spectrophotometric TAC methods are very useful for the beverage industry, especially in the control of processing, production, and storage cycles. TAC determination methods in particular, such as CUPRAC, are very convenient in the sense that they can respond to a wide variety of antioxidants within a relatively short time. In contrast, more comprehensive analyses of beverages gain importance in detecting and monitoring their nutritional values. HPLC is the most convenient analytical method in this regard, making its use essential for beverages under current consumption or at development stage. The current practice usually comprises the identification and quantification of constituents by HPLC and determination of their TAC levels using at least two different methods. This study proposes an economic and efficient alternative to TAC determination by performing a thorough HPLC analysis, and approximately calculating the possible contributions of these accurately quantified constituents with the use of their TEAC coefficients supplied from relevant literature on CUPRAC (Yıldız *et al.*, 2008; Karaman *et al.*, 2010) and ABTS (Yıldız *et al.*, 2008; Karaman *et al.*, 2010) methods of integrated HPLC-TAC assays (having proven additivity and reproducibility of responses to various antioxidants).

It becomes feasible to roughly estimate the experimental TAC of complex antioxidant mixtures such as beverages without requiring either additional equipment (such as a post-column reactor for on-line HPLC methods), reagents (for TAC assays) and time on the condition that all or most of the antioxidants contained therein be properly identified and quantified by HPLC and their TEAC coefficients known. It becomes possible to evaluate the antioxidant values of beverages with regard to both individual antioxidant constituents and TAC values in a single chromatographic run.

References

Apak, R., Güçlü, K., Özyürek, M., Karademir, S.E., 2004. Novel total antioxidant capacity index for dietary polyphenols and vitamins C and E, using their cupric ion reducing capability in the presence of neocuproine: CUPRAC method. J. Agric. Food Chem. 52, 7970–7981.

Arnao, M.B., Cano, A., Acosta, M., 2001. The hydrophilic and lipophilic contribution to total antioxidant activity. Food Chem. 73, 239–244.

Benzie, I.F.F., Strain, J.J., 1996. The ferric reducing ability of plasma (FRAP) as a measure of "antioxidant power": The FRAP assay. Anal. Biochem. 239, 70–76.

Blomhoff, R., 2005. Dietary antioxidants and cardiovascular disease. Curr. Opin. Lipidol. 16, 47–54.

Borges, G., Mullen, W., Crozier, A., 2010. Comparison of the polyphenolic composition and antioxidant activity of European commercial fruit juices. Food Funct. 1, 73–83.

Brand-Williams, W., Cuvelier, M.E., Berset, C., 1995. Use of a free-radical method to evaluate antioxidant activity. Food Sci. Tech — Lebensm. Wiss. u. Technol. 28, 25–30.

Butts, W.C., Mulvihill, H.J., 1975. Centrifugal analyzer determination of ascorbate in serum or urine with Fe^{3+}/ferrozine. Clin. Chem. 21, 1493–1497.

Cao, G., Verdon, C.P., Wu, A.H.B., Wang, H., Prior, R.L., 1995. Automated assay of oxygen radical absorbance capacity with the COBAS FARA II. Clin. Chem. 41, 1738–1744.

Çelik, S.E., Özyürek, M., Güçlü, K., Apak, R., 2010. Determination of antioxidants by a novel on-line HPLC-cupric reducing antioxidant capacity (CUPRAC) assay with post-column detection. Anal. Chim. Acta 674, 79–88.

De Beer, D., Joubert, E., Marais, J., Manley, M., 2006. Unravelling the total antioxidant capacity of Pinotage wines: Contribution of phenolic compounds. J. Agric. Food Chem. 54, 2897–2905.

Di Majo, D., La Guardia, M., Giammanco, S., La Neve, L., Giammanco, M., 2008. The antioxidant capacity of red wine in relationship with its polyphenolic constituents. Food Chem. 111, 45–49.

Karaman, Ş., Tütem, E., Sözgen Başkan, K., Apak, R., 2010. Comparison of total antioxidant capacity and phenolic composition of some apple juices with combined HPLC-CUPRAC assay. Food Chem. 120, 1201–1209.

Lea, A.G.H., 1988. HPLC of natural pigments in foodstuffs. In: Macrae, R. (Ed.), HPLC in Food Analysis. Academic Press, New York, pp. 277–333.

Litescu, S., Radu, G., 2000. Estimation of the antioxidative properties of tocopherols – an electrochemical approach. Eur. Food Res. Technol. 211, 218–221.

Miller, N.J., Diplock, A.T., Rice-Evans, C.A., 1995. Evaluation of the total antioxidant activity as a marker of the deterioration of apple juice on storage. J. Agric. Food Chem. 43, 1794–1801.

Miller, N.J., Rice-Evans, C.A., 1997. The relative contributions of ascorbic acid and phenolic antioxidants to the total antioxidant activity of orange and apple fruit juices and blackcurrant drink. Food Chem. 60, 331–337.

Özyürek, M., Güçlü, K., Tütem, E., Sözgen Başkan, K., Erçağ, E., Çelik, S.E., Baki, S., Yıldız, L., Karaman, Ş., Apak, R., 2011. A comprehensive review of CUPRAC methodology. Anal. Methods 3, 2439–2453.

Peng, Z., Iland, P., Oberholster, A., Sefton, M.A., Waters, E.J., 2002. Analysis of pigmented polymers in red wine by reversed phase HPLC. Aust. J. Grape Wine Res. 8, 70–75.

Pérez-Jiménez, J., Arranz, S., Tabernero, M.M., Díaz-Rubio, E., Serrano, J., Goñi, I., Saura-Calixto, F., 2008. Updated methodology to determine antioxidant capacity in plant foods, oils and beverages: Extraction, measurement and expression of results. Food Res. Int. 41, 274–285.

Porgalı, E., Büyüktuncel, E., 2012. Determination of phenolic composition and antioxidant capacity of native red wines by high performance liquid chromatography and spectrophotometric methods. Food Res. Int. 45, 145–154.

Pulido, R., Hernández-García, M., Saura-Calixto, F., 2003. Contribution of beverages to the intake of lipophilic and hydrophilic antioxidants in the Spanish diet. Eur. J. Clin. Nutr. 57, 1275–1282.

Re, R., Pellegrini, N., Proteggente, A., Pannala, A., Yang, M., Rice-Evans, C.A., 1999. Antioxidant activity applying an improved ABTS radical cation decolorization assay. Free Radic. Biol. Med. 26, 1231–1237.

Saura-Calixto, F., Goñi, I., 2006. Antioxidant capacity of the Spanish Mediterranean diet. Food Chem. 94, 442–447.

Šeruga, M., Novak, I., Jakobek, L., 2011. Determination of polyphenols content and antioxidant activity of some red wines by differential pulse voltammetry, HPLC and spectrophotometric methods. Food Chem. 124, 1208–1216.

Shi, S.-Y., Zhou, H.-H., Zhang, Y.-P., Jiang, X.-Y., Chen, X.-Q., Huang, K.-L., 2009. Coupling HPLC to on-line, post-column (bio)chemical assays for high-resolution screening of bioactive compounds from complex mixtures. TrAC Trends Anal. Chem. 28, 865–877.

Tabart, J., Kevers, C., Pincemail, J., Defraigne, J.-O., Dommes, J., 2010. Evaluation of spectrophotometric methods for antioxidant compound measurement in relation to total antioxidant capacity in beverages. Food Chem. 120, 607–614.

Van der Sluis, A.A., Dekker, M., Verkerk, R., Jongen, W.M.F., 2000. An improved, rapid *in vitro* method to measure antioxidant activity. Application on selected flavonoids and apple juice. J. Agric. Food Chem. 48, 4116–4122.

Van der Sluis, A.A., Dekker, M., de Jager, A., Jongen, W.M.F., 2001. Activity and concentration of polyphenolic antioxidants in apple: Effect of cultivar, harvest year and storage conditions. J. Agric. Food Chem. 49, 3606–3613.

Van der Sluis, A.A., Dekker, M., Skrede, G., Jongen, W.M.F., 2002. Activity and concentration of polyphenolic antioxidants in apple juice. 1. Effect of existing production methods. J. Agric. Food Chem. 50, 7211–7219.

Van der Sluis, A.A., Dekker, M., Skrede, G., Jongen, W.M.F., 2004. Activity and concentration of polyphenolic antioxidants in apple juice. 2. Effect of novel production methods. J. Agric. Food Chem. 52, 2840–2848.

Villano, D., Fernandez-Pachon, M.S., Troncoso, A.M., Garcia-Parrilla, M.C., 2004. The antioxidant activity of wines determined by the $ABTS^{\bullet+}$ method: Influence of sample dilution and time. Talanta 64, 501–509.

Virgili, F., Marino, M., 2008. Regulation of cellular signals from nutritional molecules: A specific role for phytochemicals, beyond antioxidant activity. Free Radic. Biol. Med. 45, 1205–1216.

Yıldız, L., Sözgen Başkan, K., Tütem, E., Apak, R., 2008. Combined HPLC-CUPRAC (cupric ion reducing antioxidant capacity) assay of parsley, celery leaves, and nettle. Talanta 77, 304–313.

CHAPTER

28

Antioxidant Screening of Beverages using the Online HPLC–DPPH• Assay Incorporating Active Flow Technology Chromatography Columns

*M. Camenzuli**, *G.R. Dennis**, *H.J. Ritchie*†, *R.A. Shalliker**

*Australian Centre for Research on Separation Science (ACROSS), School of Science and Health, University of Western Sydney (Parramatta), Sydney, NSW, Australia, †Thermo Fisher Scientific, Runcorn, UK

CHAPTER POINTS

- High-efficiency HPLC separations with DPPH antioxidant detection were evaluated.
- High sensitivity in the analysis of antioxidants with HPLC and DPPH reagent was evaluated.
- A new column technology referred to as Active Flow Technology was discussed.
- The analysis of coffee with multiplexed detection using active flow technology columns was evaluated.
- Reaction flow chromatography was applied to the analysis of coffee using on-line DPPH reagent for the detection of antioxidants.
- No mixing coil was required in the post-column decolorization reaction when using reaction flow columns.

INTRODUCTION

The popularity of antioxidant-rich products from food sources has triggered an increase in the screening of natural products for antioxidants (Niederländer *et al.*, 2008), which commonly involves testing extracts with a number of antioxidant assays (Antolovich *et al.*, 2002; Niederländer *et al.*, 2008); mostly these are conducted on a 'whole of sample' basis. One popular assay reagent used to assess antioxidants in foods and natural products is diphenylpicrylhydrazyl radical (DPPH•), which predominately reacts with phenolic antioxidants. The reaction between DPPH• and phenolic antioxidants results in bleaching of the DPPH• reagent, which can be detected spectrophotometrically. The kinetics of this reaction varies depending on the strength of the antioxidant (Bondet *et al.*, 1997). In order to obtain information on the strength and number of individual antioxidants within a sample, and thus potentially evaluate the nature of that antioxidant, it is necessary to isolate each antioxidant within the extract prior to using the DPPH• assay.

Isolation of antioxidants from natural sample extracts is usually achieved using high-performance liquid chromatography (HPLC), where different compounds can be collected and then tested using the DPPH• assay. However, this process is labor-intensive and the isolation of the antioxidant from the extract can result in sample degradation.

A preferred option is to couple the HPLC separation with the DPPH• assay (Koleva *et al.*, 2000; Dapkevicius *et al.*, 2001; Bandonienė *et al.*, 2002; Bandonienė and Murkovic, 2002a, b; Pukalskas *et al.*, 2002; Kosar *et al.*, 2003; Nuengchamnong *et al.*, 2005; Pukalskas *et al.*, 2005; Exarchou *et al.*, 2006; Oki *et al.*, 2006; Wu *et al.*, 2007; Bartasiute *et al.*, 2007; Niederländer *et al.*, 2008; Zhang *et al.*, 2009; Mnatsakanyan *et al.*, 2010; McDermott *et al.*, 2011). Compounds as they elute from the HPLC column flow into a mixing coil, react with the DPPH• reagent and the response recorded, usually at around 520 nm. This technique is known as the online HPLC–DPPH• assay and was introduced 12 years ago by (Koleva *et al.*, 2000). In their initial experimental

Processing and Impact on Antioxidants in Beverages
http://dx.doi.org/10.1016/B978-0-12-404738-9.00028-3

setup the HPLC separation with ultraviolet (UV) detection was followed by a reaction with DPPH• reagent delivered through a *T*-piece and then reaction in a mixing coil. To ensure sufficient reaction time a 15-m-long coil (736 µl) was used. This design has been adopted by numerous researchers with minor variation. Some have used shorter reaction coils to reduce band broadening (Bandonienė and Murkovic, 2002b; Kosar *et al.*, 2003; Nuengchamnong *et al.*, 2005; Pukalskas *et al.*, 2005; Exarchou *et al.*, 2006; Bartasiute *et al.*, 2007; Zhang *et al.*, 2009; Mnatsakanyan *et al.*, 2010; McDermott *et al.*, 2011), but these have still all been greater than 100 µl (Mnatsakanyan *et al.*, 2010), most nearer to 500 µl. The use of large reaction coils is no longer compatible with modern HPLC columns, which tend to be short and packed with sub-3-µm particles having around 180,000–250,000 plates/meter. Large mixing coils decay almost entirely the benefit from these columns.

In this chapter we reflect on the processes employed in the DPPH• hyphenated method of detection and introduce a new HPLC column technology referred to as active flow technology (AFT). In AFT there are various modes of operation, which include parallel segmented flow (PSF) (Camenzuli *et al.*, 2012a, b, c; Shalliker *et al.*, 2012; Camenzuli *et al.*, 2013) curtain flow (CF) (Camenzuli *et al.*, 2011, 2012d), and reaction flow (RF) chromatographies.

Parallel Segmented Flow Chromatography

Parallel segmented flow chromatography columns separate the radial central flow region in the column from the wall or peripheral flow region. This new column technology overcomes the loss of separation performance associated with the radial variations in column bed packing density (Schwartz and Smith, 1953; Horne *et al.*, 1966; Knox *et al.*, 1976; Eon, 1978; Baur *et al.*, 1988; Baur and Wightman, 1989; Farkas *et al.*, 1994; Tallarek *et al.*, 1996; Yun and Guiochon, 1997; Farkas *et al.*, 1997; Farkas and Guiochon, 1997; Miyabe and Guiochon, 1999; Shalliker *et al.*, 2000, 2002, 2003; Yew *et al.*, 2003; Abia *et al.*, 2009), the wall effect (Shalliker *et al.*, 2000), and the distortion of bands as a consequence of thermal gradients generated by frictional heating (Lin and Horváth, 1981; Gritti and Guiochon, 2006; de Villiers *et al.*, 2006; Gritti and Guiochon, 2007a,b). The latter is especially important when very fine particles are employed at high pressures, becoming more significant as the particle size decreases and pressure increases (Gritti and Guiochon, 2007b).

Radial heterogeneity in flow velocity causes the solute band to migrate along the column in a parabolic-like profile; bands resemble partially filled bowls rather than solid, thin discs. Generally the flow along the radial central region of the column is faster than near the wall. Hence at elution, the solute band is diluted by the diffuse tail and the separation power of the column is not fully realised.

The recent developments afforded by AFT are important for 'bioassay' type detection processes. Effectively bioassay or reaction-type detectors, such as the DPPH• antioxidant assay, require mixing chambers to be incorporated into the system design so that chemicals can be reacted with solute molecules as they emerge from the column. The addition of a reaction coil greatly reduces the effectiveness of the HPLC column because of the additional system dead volume, which negates the benefits of the high plate numbers achievable on modern HPLC columns containing fine particles. Therefore, in order to take advantage of the separation performance afforded by the modern column, assays used to discover compound activity should not be at the detriment of the separation, otherwise larger-particle-size columns that provide half as many plates may just as effectively be employed. A redesign in on-line assay techniques must therefore be made to keep speed with HPLC developments.

The design of AFT columns has been described in detail previously (Camenzuli *et al.*, 2012a; Shalliker *et al.*, 2012). For clarity, key aspects are discussed here. In AFT columns the flow between the wall and radial central regions are separated at the column exit using a new design in column end fittings and frits. The new frit design is annular, having a central porous region that is separated from an outer porous region by a solid impermeable ring that prevents cross flow between the wall and radial central regions. The frit is housed in an outlet fitting that has multiple exit ports: A central port captures flow eluting from the central, inner frit region, and three exit ports capture flow eluting from the outer annulus or wall region. Either flow portion can be processed further, i.e., passed through a detector or fraction collector. Multiplexed detection processes are therefore enabled. The ratio of central flow to wall flow (segmentation ratio) can be easily adjusted by regulating the pressure in any of the respective outlet ports. This on-column flow splitting process is distinctly different from post-column splitting processes because the AFT process selectively extracts from the column more concentrated sample zones, as the sample is not diluted by tailing sections of the band (Camenzuli *et al.*, 2012b). Furthermore, the on-column splitting process has a zero dead volume contribution, requiring no *T*-piece section or additional tubing to split the flow.

Multiplexed Detection using PSF Chromatography

Since PSF columns can be utilized for multiplexed detection without the need for post-column flow stream splitting, or setting up detectors in a linear series with the last detector being the 'destructive' detector, on-line bioassay-type detectors can be employed simultaneously with other detection modes, even if all detectors are destructive, i.e., the DPPH• reagent for antioxidants. Using standard column technology, the antioxidant detector is ordinarily the last detector in a series. As a consequence, the sample must migrate through several detectors and the additional mixing coil substantially broadens the band. Parallel segmented flow columns maximize the detection capabilities with minimal extra-column contributions. Furthermore, the multiplexed simultaneous detection afforded by PSF columns enables the exact matching of retention times between detectors. Figure 28.1 illustrates a comparison in the design of a system incorporating standard column technology (a and b) to that of a system employing AFT enabling multiplexed detection (c). In Figure 28.1a (standard system) analysis is undertaken using a single mode of detection—fluorescence. In Figure 28.1b, the antioxidant detection response using the DPPH• reagent is assessed (standard system). The DPPH• reagent is added to the flow stream after the column through a mixing '*T*-piece', which is followed by a reaction coil for efficient mixing. An ultraviolet–visible (UV/Vis) detector measures the decolorization response. In both Figures 28.1a and 28.1b, only a single mode of detection is illustrated since adding the DPPH• detection process post the fluorescence detector may result in the relatively fragile fluorescence flow cell cracking under the added pressure of the DPPH• reagent flow stream. Hence if dual detection is required here, a splitting '*T*-piece' should be added prior to the introduction of the DPPH• solution (not shown). Figure 28.1c illustrates the system design when an AFT column is employed. The eluent flow stream is directed simultaneously to the fluorescence detector and the antioxidant detector.

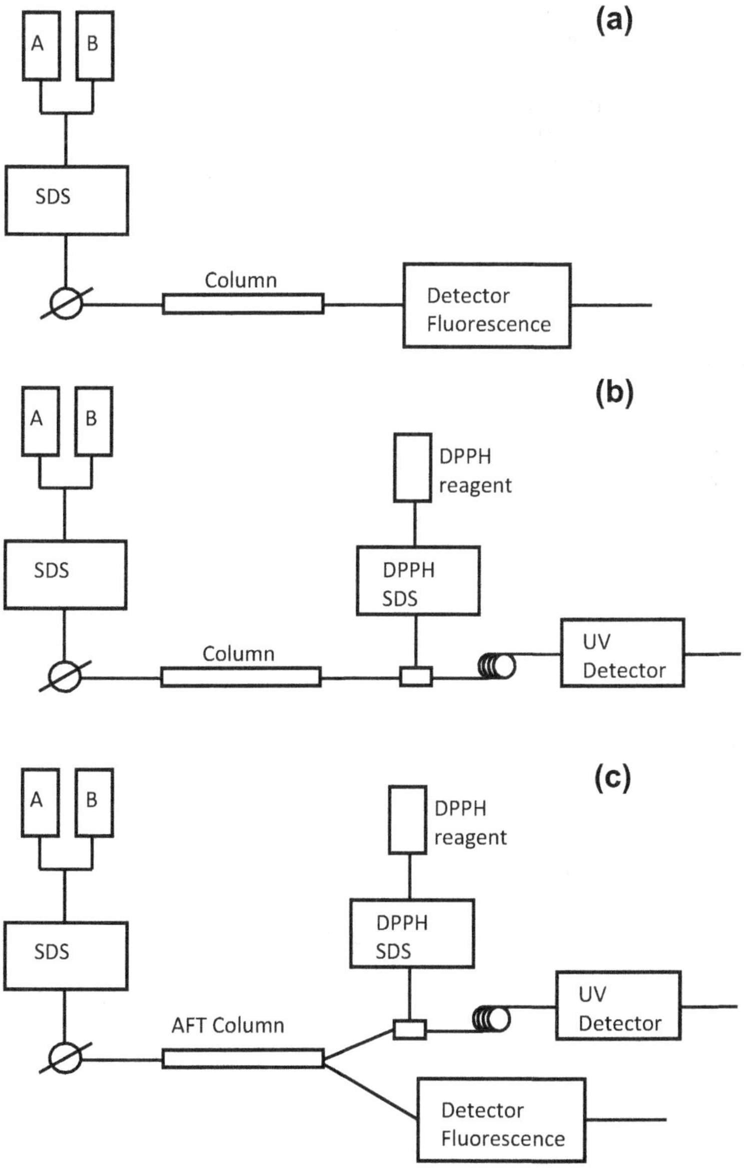

FIGURE 28.1 System configurations for DPPH• detection processes. (a) Standard system with fluorescence detection, (b) standard system with antioxidant detection, and (c) PSF with multiplexed detection incorporating fluorescence and antioxidant detections.

Reaction Flow Chromatography with DPPH• Solution

The multi-exit port design of the PSF column also enables new concepts for reaction-type chromatography involving post-column derivatizations. Potentially one or several of the exit ports could be employed as a feed for solvent flow into the outlet fitting of the column; the outlet fitting and frit subsequently act as an in-built zero dead-volume mixer, negating the need for a mixing coil. We have referred to this mode of operation as 'reaction flow (RF)' chromatography. The design of the AFT chromatography column in RF mode is illustrated in Figure 28.2; the DPPH• reagent is pumped into a peripheral port of the outlet end fitting, and the decolorized reagent exits from a different exit port. In this design multiplexed detection is also feasible with a detector measuring response from the central exit port unadulterated by the DPPH• reagent (as illustrated).

THE ANALYSIS OF COFFEE USING LIQUID CHROMATOGRAPHIC SEPARATION AND DPPH• ANTIOXIDANT DETECTION: A COMPARISON BETWEEN STANDARD AND PARALLEL SEGMENTED FLOW CHROMATOGRAPHIES

We start the discussion on the antioxidant detection process illustrating separations that are obtained using systems fitted with standard columns. We also introduce the new PSF chromatography column showing gains in sensitivity and reduced extra column volume.

Throughout, we have used Pentafluorophenyl (PFP) Hypersil GOLD columns (100×4.6mm) from Thermo-Fisher Scientific (Runcorn, Cheshire, United Kingdom). Some of these columns were fitted out as the new PSF and RF columns. DPPH• reagent solutions were prepared fresh daily in 100% HPLC-grade methanol (unbuffered), ultrasonicated prior to use and protected from light.

The analysis of espresso coffee is illustrated throughout; however, any food or beverage sample could be tested in the same manner. The coffee sample was prepared using *Ristretto* and *Decaffeinato Intenso* espresso coffees (Nespresso, Sydney, NSW, Australia) where one *Ristretto* or *Decaffeinato* capsule was used to prepare a 30-ml shot using a Delonghi Nespresso coffee machine. The coffee, including the crema, was diluted four-fold with water containing 0.1% formic acid (v/v) and filtered through a 0.22-μm syringe filter prior to analysis. The chromatographic system was an Agilent 1290 with diode array (DAD) and fluorescence (Fl) detectors (Figure 28.1). All analyses were conducted under generic gradient conditions; the initial mobile phase composition was 85/15 (A/B), where A was 0.1% formic acid in water and B was 0.1% formic acid in methanol. The initial composition was held for 1min followed by a 10-min gradient to 100% B, which was held for 3min, then returned to initial conditions. The mobile phase flow rate was 1.0ml/min. When using either standard HPLC or PSF columns the DPPH• reagent solution was mixed with the column post-column eluent via a *T*-piece then into a reaction coil prior to UV detection (520nm). Reaction coils ranged from 5 to 500μl as specified within the text. For consistency, the volume ratio of mobile phase to DPPH• reagent was kept constant for the comparison between standard and active flow technologies. In the case of the standard HPLC the flow rate of the DPPH• solution was 1.5ml/min (mobile phase flow rate=1.0ml/min) (Figure 28.1b). This mobile phase flow rate: DPPH• solution flow rate ratio yielded the optimal decolorization response. When the PSF columns were used the flow rate of the DPPH• reagent was adjusted to match the segmentation ratio employed (i.e., using a segmentation ratio of 25% from the column center, the DPPH• reagent flow rate was 0.37ml/min, i.e., 25% of 1.5ml/min) (Figure 28.1c). A Shimadzu LC-10ATvp pump fitted with an on-line degassing unit was used to supply DPPH• reagent.

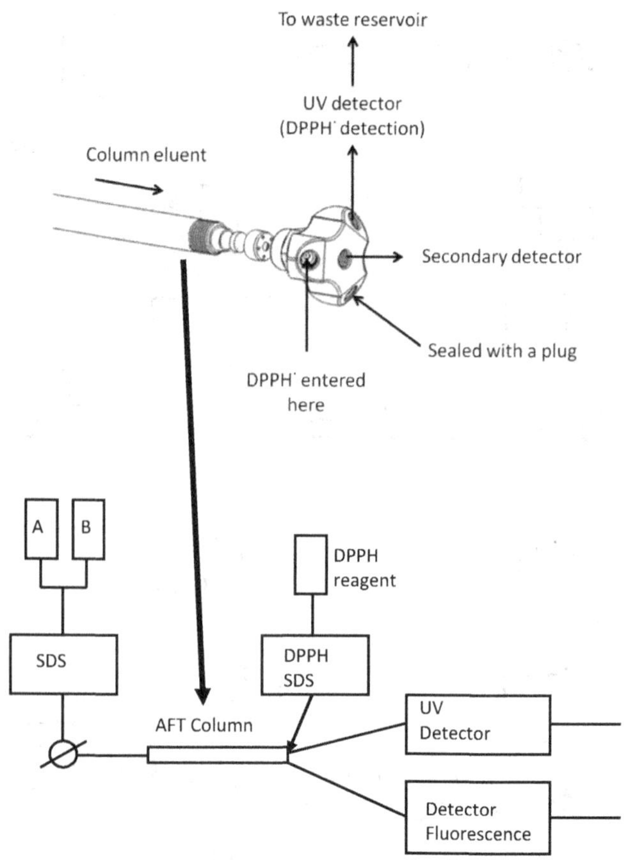

FIGURE 28.2 A system design for the active flow technology column operating in a reaction flow mode using multiplexed detection.

Fluorescence chromatograms were acquired using excitation and emission wavelengths of 278 and 366 nm, respectively. Rather than having the UV/Vis detector for the DPPH• antioxidant detection in series with the fluorescence detector, each analysis was repeated in order to gauge separately the fluorescence and antioxidant responses. That way there was no risk of damaging the fluorescence flow cell due to the back pressure associated with running the DPPH• reagent in series with the fluorescence detector. The antioxidant chromatograms were acquired at 517 nm.

Standard Chromatography Incorporating DPPH• Detection

'Typical' literature-reported conditions employed for DPPH• antioxidant detection were used as the starting point to illustrate the on-line antioxidant detection process, the initial focus being response and resolution using reaction coil volumes that are commonly employed. The analysis of the *Ristretto* coffee sample with fluorescence and DPPH• detections is shown in Figure 28.3. The reaction coil tested initially was 500 µl (0.75 × 1130 mm), which was midrange with respect to many of the recent literature-reported applications. The chromatograms in Figure 28.3 are 5-min portions of the separation using both Fl detection (positive response) and DPPH• antioxidant detection (negative responses). This region is the most interesting portion of the chromatogram, discussion around which will be developed later in this chapter. An important component is the Fl chromatogram peak (i) (caffeine). Using the standard mode of DPPH• antioxidant detection employing a 500-µl loop the large negative response labeled as (a) suggests that caffeine yields a strong antioxidant response. Yet previous studies have shown that caffeine does not decolorize the DPPH• reagent (Mnatsakanyan *et al.*, 2010). This false response is a result of the large post-column dead volume delaying the DPPH• antioxidant detection response relative to the Fl detector. While the delay between detectors can be corrected, judgement may be biased due to what may normally be considered as an 'appropriate' offset response that relates components (i) and (a).

The use of smaller reaction coils minimizes the delay and the loss in resolution, and improves the assignment of antioxidant peaks, without sacrificing DPPH• detection response. This is shown in Figure 28.3 for the chromatogram obtained using a 50-µl reaction coil (10% of the original reaction coil), far smaller than any coil currently reported. The DPPH• detection response was improved substantially; both sensitivity and peak-to-peak resolution increased, evidenced by the numerous, sharply defined bands. Because time off-set delay was effectively eliminated it was apparent that the peak labeled as (a) in the DPPH• detection mode does not

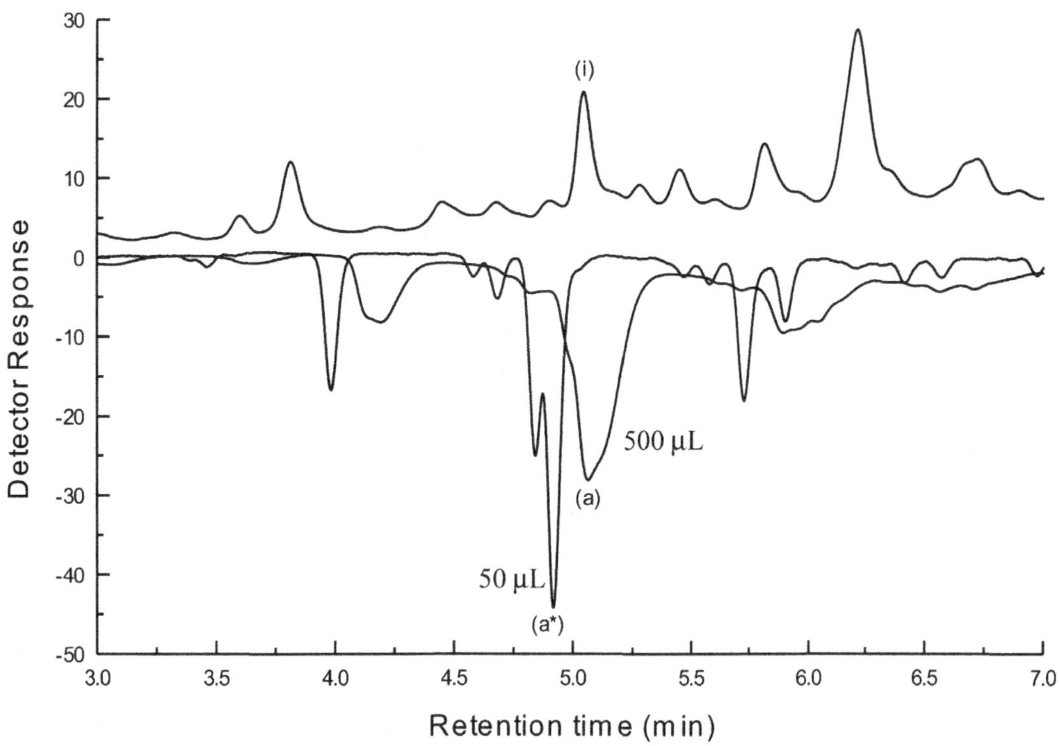

FIGURE 28.3 Fluorescence (positive) and DPPH• (negative) chromatograms of a *Ristretto* coffee. The post-column reaction coils for the DPPH• reagent were 500 and 50 µl.

correspond to the peak labeled as (i) when using fluorescence detection, rather, it relates to components that are not visible in Fl detection; under these conditions. The corresponding band (a*) elutes prior to component (i), not after. This is a very interesting aspect of selectivity in detection and is a point that will be developed further as the implementation of the various modes of parallel segmented flow columns for the analysis of coffee are discussed.

Parallel Segmented Flow Chromatography Incorporating DPPH• Detection

In this study the segmentation ratio on the PSF column was set to 25% with one-quarter of the flow eluting from each of the four exit ports. This results in a 75% reduction in the peak volume and the flow rate from any particular port (Camenzuli *et al.*, 2012a). An advantage, therefore of the PSF mode is that less DPPH• reagent per sample injection is required, since the volumetric delivery of the DPPH• reagent is reduced to maintain constant proportion with the volumetric flow through the reaction coil following elution from the HPLC column, a feature that is important given that the reagent lifetime is relatively short. Further, the reduction in the volumetric flow rate effectively increases the residence time of a sample within the reaction coil (inversely proportional to the segmentation ratio), giving the opportunity for greater derivatization without sacrificing the analysis throughput. This allowed for further optimization in the detection conditions, such that the reaction coil for the PSF column could be reduced to just 20 μl. The chromatograms obtained using the on-line DPPH• antioxidant detection using the standard HPLC column and PSF column are illustrated in Figure 28.4. The DPPH• responses using the standard column (solid) and PSF column (dotted) were almost identical despite the reaction coil on the PSF system being only 20 μl, and that only 25% of the sample eluting from the PSF column was subjected to the DPPH• reagent. Clearly the use of large-volume reaction coils is unwarranted when on-line DPPH• assays are employed. The reason for this is that the mobile phase that elutes from the HPLC column contains a significant proportion of water. In aqueous systems the DPPH• decolorization rapidly reaches a steady state and is thermodynamically driven, rather than kinetically. Hence, residence time is less important and consequently the volume or length scale dimensions of the reaction coil are less important (McDermott *et al.*, 2010). This is illustrated in the chromatographic separations of the coffee samples: by decreasing the size of the reaction coil, the band broadening was reduced and peak (a) (Figure 28.3) was separated into two distinct components. In addition, sensitivity increased.

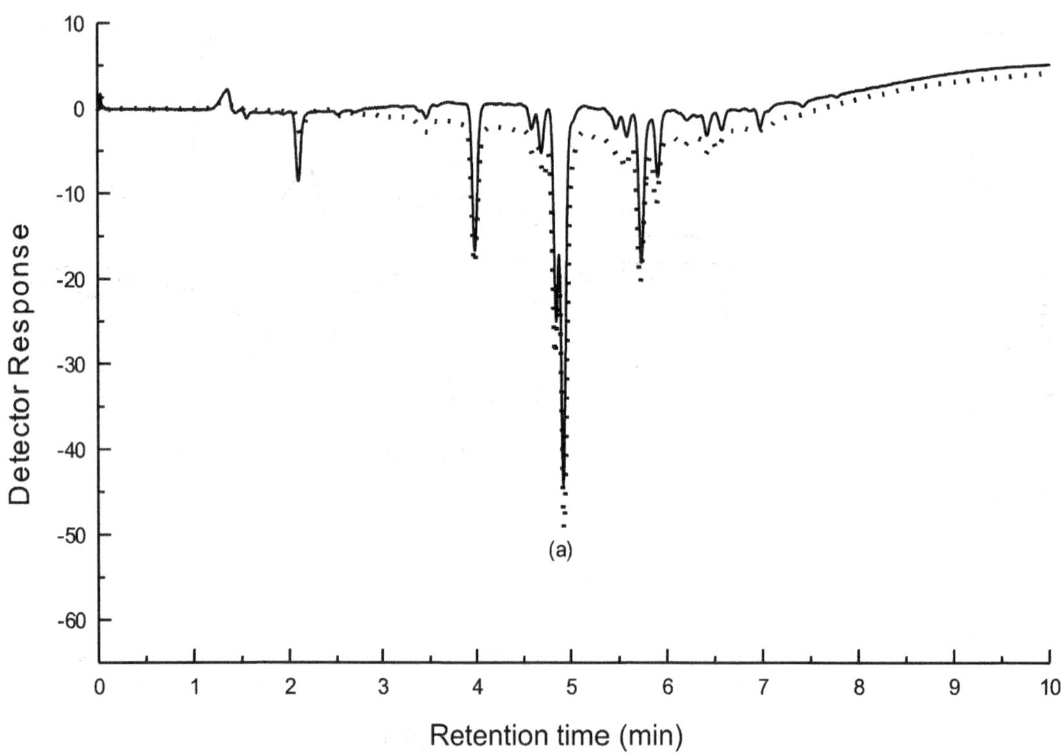

FIGURE 28.4 Antioxidant detection responses for coffee samples analyzed using a standard column (solid trace) and a PSF column (dotted trace).

Three important outcomes can be noted:

1. Slower volumetric flow emanating from the PSF column conserves the derivatizing reagent.
2. The reduced sample load to the DPPH• reaction coil and detection process did not result in a decrease in the signal response.
3. PSF chromatography enabled a reduction in the size of the mixing coils, increasing sensitivity and preserving chromatographic performance.

THE ANALYSIS OF COFFEE USING PARALLEL SEGMENTED FLOW CHROMATOGRAPHY INCORPORATING MULTIPLEXED FLUORESCENCE AND DPPH• ANTIOXIDANT DETECTION

In this section the analysis of the coffee sample using PSF columns with multiplexed fluorescence and DPPH• antioxidant detections is assessed. The PSF column was operated such that the central flow stream was directed to the fluorescence detector and the peripheral flow was mixed with the DPPH• reagent solution through a *T*-piece and then through a 20-μl mixing coil prior to the UV/Vis detector. The segmentation ratio was adjusted so that 25% of flow left each outlet port. The DPPH• reagent flow rate was 0.37 ml/min. A diagram of the system is shown in Figure 28.1c

In PSF chromatography with multiplexed detection the flow from both the central and the peripheral streams are utilized for detection. Chromatograms obtained using fluorescence detection for the two coffee samples are shown in Figure 28.5 (positive response). The bold positive trace is the decaffeinated coffee chromatogram. Caffeine in the *Ristretto* sample is labeled as peak (i). The decaffeinated coffee has no detectable quantity of caffeine. Another difference factor between the *Ristretto* and decaffeinated coffees is that there is generally a higher concentration of the components in the *Ristretto* coffee.

The antioxidant profiles of the two coffee samples obtained using the multiplexed PSF column are also shown in Figure 28.5—(negative response). Aside from the difference in relative intensity, these antioxidant profiles were almost identical. The component labeled as (a) is the same component that appeared to present antioxidant response for caffeine when the conventional mode of detection with a 500-μl reaction coil was employed. However, when the reaction coil volume was decreased to 50 μl, band broadening was decreased and it became apparent that the caffeine band was not responsible for the antioxidant response. Here in Figure 28.5, it is shown beyond doubt that component (a) is not a response observed from caffeine; rather, it is derived from a component that has negligible response to fluorescence detection. Furthermore, the simultaneous 'multiplexed' detection allows

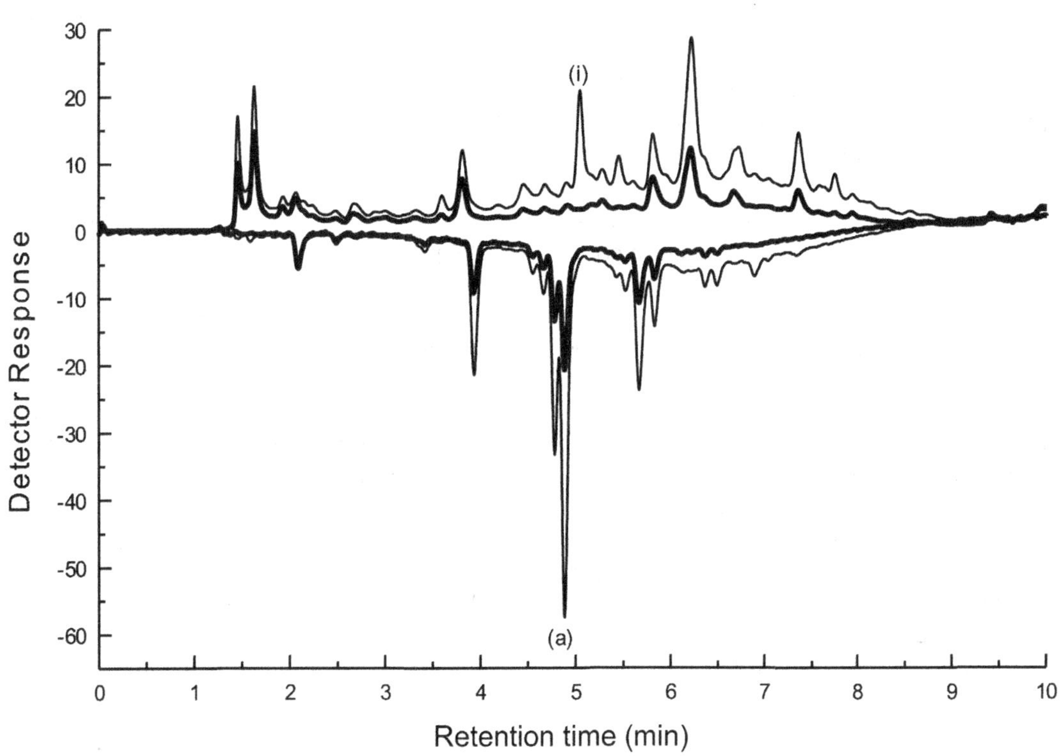

FIGURE 28.5 Fluorescence and antioxidant detection responses acquired using multiplexed PSF columns for *Decaffeinato Intenso* (bold traces) and *Ristretto* espresso coffees.

an absolute assignment of components with respect to retention time through each mode of detection. No adjustments need be made to offset delays associated with sequential 'in-line' detection trains. Under the conditions employed with multiplexed detection the assignment of the appropriate chemical and/or physical or biological function of the eluting molecules can be easily undertaken, minimizing uncertainty in component isolation and drug discovery.

Another important feature of the multiplexed sample analysis was that it yielded essentially identical chromatograms to the non-multiplexed detection (compare Figures 28.5 and 28.4).

Four important outcomes can be derived from these studies:

1. Lower volumetric flow from the PSF column conserves derivatizing reagent, in-line with the outcomes from Section 2.
2. The PSF column operated in a multiplexed detection mode provided rapid information regarding selectivity in detection.
3. The multiplexed mode of detection provided almost exactly the same magnitude in the detection response compared to the non-multiplexed PSF mode of detection.
4. Multiplexed PSF could be used with modern columns without unduly sacrificing their performance because low dead-volume mixing coils could be used yielding high sensitivity.

THE ANALYSIS OF COFFEE USING REACTION FLOW CHROMATOGRAPHY WITH MULTIPLEXED DETECTION

This section illustrates the new concept of reaction flow chromatography, which enables the reaction coil to be removed from the system design entirely. The analysis of the coffee is explored in more detail.

In the current design configuration, RF chromatography can be multiplexed using two different detection modes. The post-column reagent is delivered into one peripheral outlet port against the flow of mobile phase. Solute traversing the chromatographic bed in the wall region mixes with the post-column derivatizing reagent in the outer porous region of the frit. The mixture exits the outlet of the column through a second peripheral outlet port and passes through the antioxidant detector. The Fl detector analyzes the solute that leaves the column from the central port. Since the outlet frit is annular in construction with the central and peripheral regions separated by an impermeable barrier the solute that exits here is free from contamination by the derivatizing reagent. Hence, simultaneously, the native and the derivatized solutes can be analyzed.

In this study experiments were conducted in the same manner as described previously, but the defining difference between multiplexed detection using PSF column technology and RF chromatography was that in the RF mode the mixing coil was removed since the frit in the outlet fitting provides efficient mixing, greatly reducing system dead volume. The column could be connected directly to the UV/Vis detector using the absolute minimum length of tubing, in this case (250 × 0.13 mm, 3.32 μl). The design of this system is illustrated in Figure 28.2.

The optimal proportions of the DPPH• solution (volumetric flow and concentration) differed from the multiplexed PSF system. Hence optimization experiments were undertaken to maximize the response. The optimal conditions employed a DPPH• reagent with a concentration of 0.1 mg/ml using a delivery flow rate of 0.5 ml/min. Using these conditions, the antioxidant response (Figure 28.6) of the coffee samples were tested in RF mode. In both cases the signal intensity was essentially the same as that observed in Figure 28.5 where a multiplexed PSF system was employed, but here there was no reaction coil. More importantly, the band broadening and signal intensity in the RF system were greatly improved (see Figure 28.3 for comparison). Also shown in Figure 28.6 are the fluorescence detection responses. They show close similarity with the fluorescence responses obtained using the multiplexed PSF system (Figure 28.5), confirming that the central flow stream and the peripheral flow streams in this mode of operation are independent.

Four important outcomes can be derived from these studies using the reaction flow mode of operation:

1. Mixing between the post-column derivatizing reagent and the solute inside the peripheral region of the outlet frit is very efficient. Subsequently the proportion of DPPH• reagent: mobile phase is decreased, conserving DPPH• reagent.
2. The direct feed of DPPH• reagent into the outlet fitting of the RF column eliminates the need for mixing *T*-pieces, reducing the post-column dead volume to no more than required in chromatography using standard modes of detection.
3. The RF mode of operation with dual multiplexed detection allows for an absolute assignment of biological function with component identity, removing the need to compensate for detection delay that is observed in systems that employ sequential detection trains.
4. The RF multiplexed process could be used with modern columns without sacrificing their performance because post-column dead volume is minimized to the point consistent with standard modes of detection in HPLC and ultra-high-performance liquid chromatography (UHPLC).

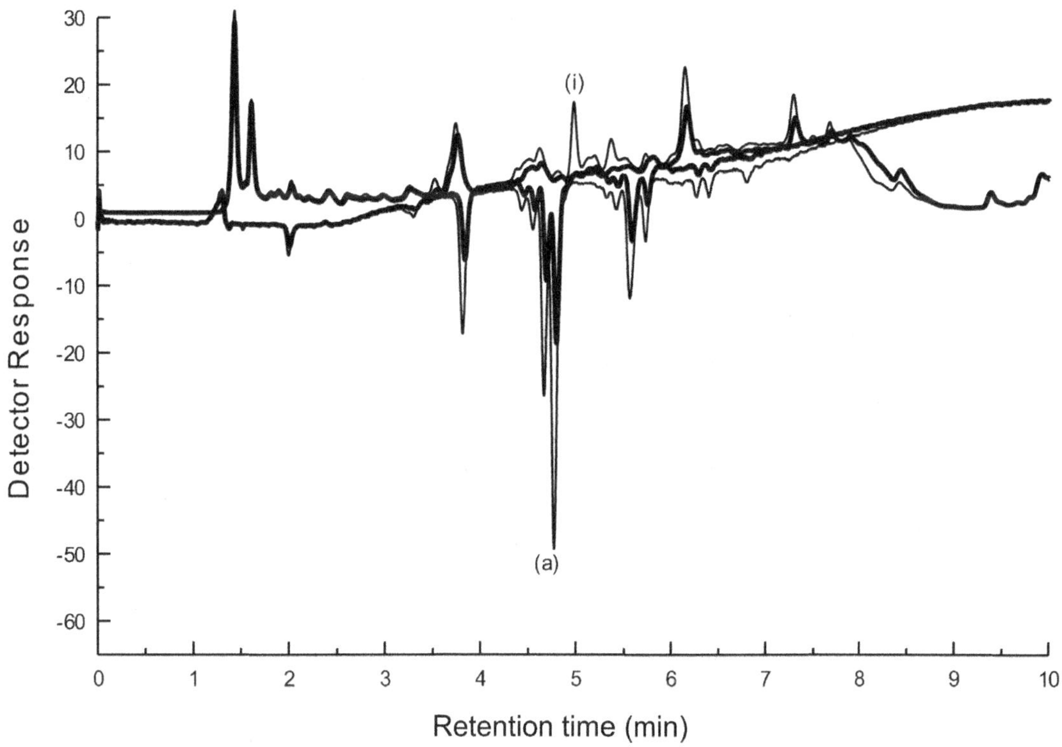

FIGURE 28.6 Fluorescence and DPPH• detection responses acquired using reaction flow chromatography with multiplexed detection for *Decaffeinato Intenso* (bold traces) and *Ristretto* espresso coffees.

GENERAL CONCLUSION

The analysis of beverages, foods, drugs, and natural products is often complex because they contain multitudes of components with widely varying structures and properties. This places a high demand on separation performance, which often requires an assessment of biological function. To elucidate biological function or activity, tests must be undertaken that probe these specific responses. For the case discussed here, the property of interest is the antioxidant capabilities of the molecules that constitute the sample. Hence, chromatography is necessary for separation of the various molecular species and some form of analysis is required to test antioxidant function, i.e., decolorization of a DPPH• reagent. This test can be coupled directly with chromatographic separation for speedy analysis. However, the process of undertaking the antioxidant analysis requires the mixing of two flow streams and subsequent time for reaction in order to observe decolorization. This process has, to date, been undertaken using large-volume mixing coils, which negates the separation power of modern chromatography columns. However, by using a new column technology known as active flow technology, reaction mixing coils are no longer required, and separation performance is not sacrificed. Furthermore, detection can be multiplexed providing detection selectivity.

Acknowledgments

MC acknowledges the support of a UWS ACROSS Postgraduate Research Award.

References

Abia, J.A., Mriziq, K.S., Guiochon, G.A., 2009. Radial heterogeneity of some analytical columns used in high-performance liquid chromatography. J. Chromatogr. A 1216, 3185–3191.

Antolovich, M., Prenzler, P.D., Patsalides, E., McDonald, S., Robards, K., 2002. Methods for testing antioxidant activity. Analyst 127, 183–198.

Bandonienė, D., Murkovic, M., Pfannhauser, W., Venskutonis, P., Gruzdienė, D., 2002a. Detection and activity evaluation of radical scavenging compounds by using DPPH free radical and on-line HPLC-DPPH methods. Eur. Food Res. Technol. 214, 143–147.

Bandonienė, D., Murkovic, M., 2002. The detection of radical scavenging compounds in crude extract of borage (Borago officinalis L.) by using an on-line HPLC-DPPH method. J. Biochem. Biophys. Methods 53, 45–49.

Bandonienė, D., Murkovic, M., 2002b. On-Line HPLC-DPPH Screening Method for Evaluation of Radical Scavenging Phenols Extracted from Apples (Malus domestica L.), J. J. Agric. Food Chem. 50, 2482–2487.

Bartasiute, A., Westerink, B.H.C., Verpoorte, E., Niederländer, H.A.G., 2007. Improving the in vivo predictability of an on-line HPLC stable free radical decoloration assay for antioxidant activity in methanol–buffer medium. Free Radic. Biol. Med. 42, 413–423.

Baur, J.E., Kristensen, E.W., Wightman, R.M., 1988. Radial dispersion from commercial high-performance liquid chromatography columns investigated with microvoltammetric electrodes. Anal. Chem. 60, 2334–2338.

Baur, J.E., Wightman, R.M., 1989. Microcylinder electrodes as sensitive detectors for high-efficiency, high-speed liquid chromatography. J. Chromatogr. A 482, 65–73.

Bondet, V., Brand-Williams, W., Berset, C., 1997. Kinetics and Mechanisms of Antioxidant Activity using the DPPH Free Radical Method. LWT Food Sci. Technol. 30, 609–615.

Camenzuli, M., Ritchie, H.J., Ladine, J.R., Shalliker, R.A., 2011. The design of a new concept chromatography column. Analyst 136, 5127–5130.

Camenzuli, M., Ritchie, H.J., Ladine, J., Shalliker, R.A., 2012a. Enhanced separation performance using a new column technology: Parallel segmented outlet flow. J. Chromatogr. A 1232, 47–51.

Camenzuli, M., Goodie, T.A., Bassanese, D.N., Francis, P.S., Barnett, N.W., Ritchie, H., LaDine, J., Shalliker, R.A., Conlan, X.A., 2012b. The use of parallel segmented outlet flow columns for enhanced mass spectral sensitivity at high chromatographic flow rates. Rapid Commun. Mass Spectrom. 26, 943–949.

Camenzuli, M., Ritchie, H.J., Shalliker, R., 2012c. Gradient elution chromatography with segmented parallel flow column technology: A study on 4.6 mm analytical scale columns. J. Chromatogr. A 1270, 204–211.

Camenzuli, M., Ritchie, H.J., Ladine, J.R., Shalliker, R.A., 2012d. Active flow management in preparative chromatographic separations: A preliminary investigation into enhanced separation using a curtain flow inlet fitting and segmented flow outlet fitting. J. Sep. Sci. 35, 410–415.

Camenzuli, M., Ritchie, H.J., Shalliker, R.A., 2013. Improving HPLC separation performance using parallel segmented flow chromatography. Microchem. J. 111, 3–7.

Dapkevicius, A., van Beek, T.A., Niederländer, H.A.G., 2001. Evaluation and comparison of two improved techniques for the on-line detection of antioxidants in liquid chromatography eluates. J. Chromatogr. A 912, 73–82.

Eon, C.H., 1978. Comparison of broadening patterns in regular and radially compressed large-diameter columns. J. Chromatogr. A 149, 29–42.

Exarchou, V., Fiamegos, Y.C., van Beek, T.A., Nanos, C., Vervoort, J., 2006. Hyphenated chromatographic techniques for the rapid screening and identification of antioxidants in methanolic extracts of pharmaceutically used plants. J. Chromatogr. A 1112, 293–302.

Farkas, T., Chambers, J.Q., Guiochon, G., 1994. Column efficiency and radial homogeneity in liquid chromatography. J. Chromatogr. A 679, 231–245.

Farkas, T., Sepaniak, M.J., Chambers, J.Q., Guiochon, G., 1997. Radial distribution of the flow velocity, efficiency and concentration in a wide HPLC column. AIChE J. 43, 1964–1974.

Farkas, T., Guiochon, G., 1997. Contribution of the radial distribution of the flow velocity to band broadening in HPLC columns. Anal. Chem. 69, 4592–4600.

Gritti, F., Guiochon, G., 2006. Effects of the thermal heterogeneity of the column on chromatographic results. J. Chromatogr. A 1131, 151–165.

Gritti, F., Guiochon, G., 2007a. Measurement of the axial and radial temperature profiles of a chromatographic column. Influence of thermal insulation on column efficiency. J. Chromatogr. A 1138, 141–157.

Gritti, F., Guiochon, G., 2007b. Consequences of the radial heterogeneity of the column temperature at high mobile phase velocity. J. Chromatogr. A 1166, 47–60.

Horne, D.S., Knox, J.H., McLaren, L., 1966. A comparison of mobile-phase peak dispersion in gas and liquid chromatography. Sep. Sci. Technol. 1, 531–554.

Knox, J.H., Laird, G.R., Raven, P.A., 1976. Interaction of radial and axial dispersion in liquid chromatography in relation to the "infinite diameter effect". J. Chromatogr. A 122, 129–145.

Koleva, I.I., Niederländer, H.A.G., van Beek, T.A., 2000. An On-Line HPLC Method for Detection of Radical Scavenging Compounds in Complex Mixtures. Anal. Chem. 72, 2323–2328.

Kosar, M., Dorman, H.J.D., Bachmayer, O., Baser, K.H.C., Hiltunen, R., 2003. An Improved On-Line HPLC-DPPH Method for the Screening of Free Radical Scavenging Compounds in Water Extracts of Lamiaceae Plants. Chem. Nat. Comp. 39, 161–166.

Lin, H.-J., Horváth, S., 1981. Viscous dissipation in packed beds. Chem. Eng. Sci. 36, 47–55.

McDermott, G.P., Noonan, L.K., Mnatsakanyan, M., Shalliker, R.A., Conlan, X.A., Barnett, N.W., 2010. High performance liquid chromatography with post-column 2,2′-diphenyl-1-picrylhydrazyl radical scavenging assay: methodological considerations and application to complex samples'. Anal. Chem. Acta 675 (1), 76–82.

McDermott, G.P., Conlan, X.A., Noonan, L.K., Costin, J.W., Mnatsakanyan, M., Shalliker, R.A., Barnett, N.W., Francis, P.S., 2011. Screening for antioxidants in complex matrices using high performance liquid chromatography with acidic potassium permanganate chemiluminescence detection. Anal. Chim. Acta 684, 134–141.

Miyabe, K., Guiochon, G., 1999. Estimation of the column radial heterogeneity from an analysis of the characteristics of tailing peaks in linear chromatography. J. Chromatogr. A 830, 29–39.

Mnatsakanyan, M., Goodie, T.A., Conlan, X.A., Francis, P.S., McDermott, G.P., Barnett, N.W., Shock, D., Gritti, F., Guiochon, G.,R.A., Shalliker, R.A., 2010. High performance liquid chromatography with two simultaneous on-line antioxidant assays: Evaluation and comparison of espresso coffees. Talanta 81, 837–842.

Niederländer, H.A.G., van Beek, T.A., Bartasiute, A., Koleva, I.I., 2008. Antioxidant activity assays on-line with liquid chromatography. J. Chromatogr. A 1210, 121–134.

Nuengchamnong, N., de Jong, C.F., Bruyneel, B., Niessen, W.M.A., Irth, H., Ingkaninan, K., 2005. HPLC coupled on-line to ESI-MS and a DPPH-based assay for the rapid identification of anti-oxidants in Butea superba. Phytochem. Anal. 16, 422–428.

Oki, T., Kobayashi, M., Nakamura, T., Okuyama, A., Masuda, M., Shiratsuchi, H., Suda, I., 2006. Changes in Radical-scavenging Activity and Components of Mulberry Fruit During Maturation. J. Food Sci. 71, C18–C22.

Pukalskas, A., van Beek, T.A., Venskutonis, R.P., Linssen, J.P.H., van Veldhuizen, A., de Groot, Æ., 2002. Identification of Radical Scavengers in Sweet Grass (Hierochloe odorata). J. Agric. Food Chem. 50, 2914–2919.

Pukalskas, A., van Beek, T.A., de Waard, P., 2005. Development of a triple hyphenated HPLC–radical scavenging detection–DAD–SPE–NMR system for the rapid identification of antioxidants in complex plant extracts. J. Chromatogr. A 1074, 81–88.

Schwartz, C.E., Smith, J.M., 1953. Flow Distribution in Packed Beds. Ind. Eng. Chem. 45, 1209–1218.

Shalliker, R.A., Broyles, B.S., Guiochon, G., 2000. Physical evidence of two wall effects in liquid chromatography. J. Chromatogr. A 888, 1–12.

Shalliker, R.A., Wong, V., Broyles, B.S., Guiochon, G., 2002. Visualization of bed compression in an axial compression liquid chromatography column. J. Chromatogr. A 977, 13–223.

Shalliker, R.A., Broyles, B.S., Guiochon, G., 2003. Axial and radial diffusion coefficients in a liquid chromatography column and bed heterogeneity. J. Chromatogr. A 994, 1–12.

Shalliker, R.A., Camenzuli, M., Pereira, L., Ritchie, H.J., 2012. Parallel segmented flow chromatography columns: Conventional analytical scale column formats presenting as a 'virtual' narrow bore column. J. Chromatogr. A 1262, 64–69.

Tallarek, U., Albert, K., Bayer, E., Guiochon, G., 1996. Measurement of transverse and axial apparent dispersion coefficients in packed beds. AIChE J 42, 3041–3054.

de Villiers, A., Lauer, H., Szucs, R., Goodall, S., Sandra, P., 2006. Influence of frictional heating on temperature gradients in ultra-high-pressure liquid chromatography on 2.1 mm I.D. columns. J. Chromatogr. A 1113, 84–91.

Wu, J.-H., Huang, C.-Y., Tung, Y.-T., Chang, S.-T., 2007. Online RP-HPLC-DPPH Screening Method for Detection of Radical-Scavenging Phytochemicals from Flowers of Acacia confusa. J. Agric. Food Chem. 56, 328–332.

Yew, B.G., Drumm, E.C., Guiochon, G., 2003. Mechanics of column beds: I. Acquisition of the relevant parameters. AIChE J 49, 626–641.

Yun, T., Guiochon, G., 1997. Visualization of the heterogeneity of column beds. J. Chromatogr. A 760, 17–24.

Zhang, Q., van der Klift, E.J.C., Janssen, H.-G., van Beek, T.A., 2009. An on-line normal-phase high performance liquid chromatography method for the rapid detection of radical scavengers in non-polar food matrixes. J. Chromatogr. A 1216, 7268–7274.

CHAPTER

29

Analytical Methods for Determination of Polyphenols in Beer

M.L. Fernández de Córdova, A. Ruiz Medina

Department of Physical and Analytical Chemistry, University of Jaén, Jaén, Spain

CHAPTER POINTS

- Among beer antioxidants, polyphenols are of particular interest to brewers due to their essential role in the brewing process by delaying or preventing oxidation, and influencing the flavor and physical stability of beer.
- Assays for polyphenols in beer can be classified as those determining the total content of polyphenols or a specific class of polyphenols (global assays), and those allowing the determination of the phenolic profile.
- Most analytical methods for quantifying the total polyphenol content in beer are based on the reaction of phenolic compounds with a colorimetric reagent and measurement of the absorption of the colored product in the visible region.
- Reported methods for analysis of the total flavanoid content are usually based on colorimetric derivatization. Assays using polymeric precipitation have also been proposed for the particular case of analysis of proanthocyanidins.
- The most commonly used technique for separation and determination of individual phenolic compounds in beer is high-performance liquid chromatography with photodiode array or electrochemical detection.

INTRODUCTION

Beer is an alcoholic beverage made from malt, yeast, water, and hops. This beverage itself is known to have endogenous antioxidant activity due to the presence of certain compounds with strong antiradical or antioxidant properties. These substances come mainly from the raw materials used in its production, being already present in those or obtained by transformation and modification of natural constituents. According to reported studies on antioxidant activity of beer components, the constituents potentially active are mainly polyphenolic compounds, sulfite, and Maillard reaction products (Vanderhaegen *et al.*, 2006). The major endogenous source of sulfite is the conversion of sulfate by yeast, and it has been highlighted as one of the most effective antioxidants in beer (Andersen and Skibsted, 2001). Melanoidins and reductones constitute the main antioxidants generated from Maillard reactions that take place in the wort boiling and malt roasting, their levels being low in light malts, but significant in dark malts.

Among beer antioxidants, polyphenols are of particular interest to brewers as they play an essential role in the brewing process by delaying or preventing oxidation processes through mechanisms involving both free radical scavenging and metal chelation, and influencing the flavor and physical stability of beer. Phenolic compounds in beer come either from barley and other raw or malted grains used as adjuncts or from the hops. Once dissolved in the wort, polyphenols may undergo alterations during the fermentation process and the subsequent stages of clarification of beer. Therefore, the final phenolic content in beer will depend on both the raw materials and the brewing process. The common characteristic of polyphenols is the presence in their structure of one or more hydroxylated aromatic rings, which, in fact, seems to be responsible for their properties as radical scavengers. The beneficial impacts of these antioxidant compounds on the stability of beer flavor have been demonstrated by several authors, whereas there is currently controversy concerning the effectiveness of some phenolic

Processing and Impact on Antioxidants in Beverages
http://dx.doi.org/10.1016/B978-0-12-404738-9.00029-5

compounds as potential antioxidants in beer and wort (Oñate-Jaén *et al.*, 2006). In particular, lower-molecular-weight polyphenols seem to be excellent antioxidants, whereas a decrease in the reducing power is obtained with increasing molecular weight (Vanderhaegen *et al.*, 2006). During the oxidation process, polyphenols can polymerize forming tannins, which are able to react with beer proteins and originate insoluble compounds, the presence of which is known as haze.

Beer contains a complex mixture of polyphenols derived from barley (70–80%) and hops (20–30%) (Gerhäuser, 2005). The differences in the polyphenol levels in beer relate to raw materials (malt and hop varieties), and malting and brewing processes. Polyphenols in beer include several classes such as:

phenolic acids, derived from hydroxycinnamic acid (*p*-coumaric, caffeic, sinapic, ferulic, chlorogenic acids, etc.) and hydroxybenzoic acid (syringic, gallic, protocatechuic, gentisic, *p*-hydroxybenzoic, vanillic acids, etc.);
flavanols or *flavanoids*, including *flavan-3-ols* or *catechins* [(+)-catechin, (−)-epicatechin, (−)-galloepicatechin, (−)-gallocatechin, etc.], which are monomeric compounds; and flavan-3-ols oligomers or *proanthocyanidins*, which are dimers (prodelphinidins B, procyanidins B, propelargonidin B, etc.) and trimers (procyanidins C and prodelphinidins C);
flavonols (kaempferol, quercetin, rutin, etc., and their glycosides);
flavones (apigenin);
isoflavones (genistein, daidzein, etc.);
prenylchalcones (xanthohumol, desmethylxanthohumol, etc.) and derived *flavanones* (isoxanthohumol, 8-prenylnaringenin, etc.);
alpha-acids (*n*-humulone, cohumulone and adhumulone) and *iso-alpha-acids* (*iso-n*-humulone, *iso*-cohumulone and *iso*-adhumulone); and
stilbenes (*cis*-piceid, *trans*-piceid, *cis*-resveratrol and *trans*-resveratrol).

Assays for phenolic compounds in beer can be classified as those determining the total phenolic content or a specific class or group of phenolic compounds content (global assays) and those allowing the determination of the phenolic profile. Global assays proposed in the literature for analysis of polyphenol content are usually based on the formation of a colored compound between the phenolic fraction of the sample and a chromogenic reagent. These assays show very different specificity and the contribution of other groups of phenolic compounds or different species of beer has to be often taken into account. Conversely, the most commonly used technique for the separation and determination of individual phenolic compounds of beer is high-performance liquid chromatography (HPLC), with ultraviolet (UV), photodiode array or electrochemical detection, although in the last few years other techniques such as electrophoresis and chemiluminescence have also been proposed. Some of these reported methods have been proposed for the determination of both the total and individual polyphenol content. The aim of this chapter is to review and discuss these analytical methods.

GLOBAL ASSAYS

Determination of Total Phenolic Content

There are several analytical procedures for the quantification of the total polyphenol content in beers, most of them being based on the reaction of phenolic compounds with a colorimetric reagent and measurement of the absorption of the colored product in the visible region of the spectra. These assays are not specific for particular groups of phenolic compounds and are widely used mainly due to their simplicity and high sensitivity. A common difficulty related to these different colorimetric methods of analysis is the selection of a suitable standard, which is a consequence of the heterogeneity of the phenolic fraction. Table 29.1 shows the total polyphenolic content found in several commercial beers by using different assays.

Folin–Ciocalteu (FC) Assay

This is the most frequently applied method (Singleton and Rossi, 1965) for determining the total phenolic content, being an improved version of the Folin–Denis method for measuring tyrosine in proteins. This determination is also called *FC polyphenol index*. The method is based on the reduction of the FC reagent, consisting of a mixture of phosphomolybdate and phosphotungstate (6+ valence state of the metal), by the polyphenols and the measuring of the absorbance at 725–760 nm. It is believed that the blue color developed is due to tungsten and molybdene reduction products (5+ valence state of the metal). The reduction reaction is enhanced under alkaline conditions, provided by sodium carbonate solution. Total phenol content is usually expressed as gallic acid equivalents (GAE), although caffeic acid and (+)-catechin have been also used for this purpose. By using this methodology, the total mass of phenolic compounds in the sample is quantified. Therefore, the polymerization of these compounds does not affect the total phenolic content. Conversely, it is necessary to take into account that the FC assay does not only allow measuring total phenols but FC reagent also reacts with other reducing substances (Singleton *et al.*, 1999). Possible contributors mentioned in literature include aldehydes, ketones, amines, amino acids, nucleotides,

TABLE 29.1 Total Polyphenolic Content Found In Several Commercial Beers

Beer	Total Phenolic Content (GAE mg/l)	Reference
Brown ale	280–526[b]	Granato *et al.* (2011)[d]
Pilsen	250–300[b]	Siqueira *et al.* (2011)[e]
Lager and dark	376–473[b]	Lugasi (2003)[f]
Non-alcoholic	(366 ± 73)[a,b]	Piazzon *et al.* (2010)
Non-alcoholic	80[c]	Dvořáková *et al.* (2007)
Lager	(452 ± 86)[a,b]	Piazzon *et al.* (2010)
Lager	157[c]	Dvořáková *et al.* (2007)
Lager	152–339 [b]	Zhao *et al.* (2010)
Lager	120–200 [b]	Granato *et al.* (2011)[d]
Lager	(388 ± 165)[a,b]	Floridi *et al.* (2003)
Wheat	(504 ± 44)[a,b]	Piazzon *et al.* (2010)
Abbey	(622 ± 77)[a,b]	Piazzon *et al.* (2010)
Draught	125[c]	Dvořáková *et al.* (2007)

[a]*Average value ± standard deviation (SD).*
[b]*FC assay (Singleton and Rossi, 1965)*
[c]*Bishop assay (Bishop, 1972; Jerumanis, 1973)*
[d]*Granato, D., Favalli Branco, G., Fonseca Faria, J., and Gomes Cruz, A. (2011). Characterization of Brazilian lager and Brown ale beers based on color, phenolic compounds, and antioxidant activity using chemometrics.* J. Sci. Food Agric. *91, 563–571*
[e]*Siqueira, P.B., Bolini, H.M.A., and Macedo, G.A. (2011). Polyphenols and antioxidant properties in forced and naturally aged Brazilian beer.* J. Brew. Distilling *2, 45–50*
[f]*Lugasi, A. (2003). Polyphenol content and antioxidant properties of beer.* Acta Aliment. *32, 181–192.*

proteins, carbohydrates, thiols, and vitamins. Therefore, the assay should not be viewed as a measure of total phenolic content, but rather a measure of overall antioxidant capacity. Thus, recent studies have shown that the phenolic content determined by FC method can be correlated to antioxidant activity determined by several methods (Roginsky and Lissi, 2005).

The automation of the classical FC assay, making use of flow-injection analysis (FIA), has also been proposed (Peris *et al.*, 1991). The application of FIA methodology has made it possible to improve both reproducibility [relative standard deviation (RSD), 1.58%] and sampling frequency (42/h). Some modifications in sodium carbonate and FC reagent concentrations had to be introduced in order to avoid the appearance of a precipitate that hindered the spectrophotometric measurements and obstructed the tubes of the manifold. It was found that the use of gallic acid as standard provided results differing considerably from those obtained with the batch method. Consequently, coumaric acid was selected for quantification purposes. FC batch and FIA procedures provided similar results. Multi-syringe FIA (MSFIA) has also been reported for the automation of FC assay (Magalhães *et al.*, 2006). The flow procedure was used for the determination of reducing capacity using gallic acid as reference standard and it provided results similar to those obtained by the conventional batch FC assay. The best strategy for mixing the sample and FC reagent was the merging zones approach, which allowed low reagent consumption and high throughput (12/h). The strict control of the reaction conditions and the reduction in the intervention of the operator achieved reliable results and good repeatability (RSD < 1.3%).

Prussian Blue Assay

This method (Price and Butler, 1977) is probably one of the most popular assays for quantifying total phenolics in plant materials because it is rapid, simple, and economical, and it has low interference by non-phenolic compounds. Polyphenols react with ferricyanide ion $[Fe(CN)_6]^{3-}$, this latter being reduced to ferrocyanide ion $[Fe(CN)_6]^{4-}$. Then, ferrocyanide ion reacts with an excess of iron (III) to form ferric ferrocyanide $Fe_4[Fe(CN)_6]_3$, commonly known as Prussian blue, whose absorbance at 700 nm is measured. Several working conditions affect the results of the assay, such as temperature, pH and order in which reagents are added. The method presents two important drawbacks, namely, the formation of a precipitate after short incubation periods, and the increase in color density with time. The addition of orthophosphoric acid has been proposed to maintain color stability by reducing the pH value and complexing the excess of iron (III). In addition, gum acacia, a protective colloid, is used for preventing precipitation by impeding the coalescence of the colloidal particles formed (Graham, 1992).

Bishop Assay

This is another spectrophotometric method for the determination of the total polyphenol content (Bishop, 1972; Jerumanis, 1973), which was adopted by the European Brewery Convention (EBC) as an official method (EBC method 9.11). The assay is based on the chelation of polyphenols with iron (III) ion, in alkaline solution (pH 10 with ammonia), to form red complexes. The sample is mixed with carboxymethyl cellulose/ethylenediaminetetraacetic acid reagent (CMC/EDTA) and ferric reagent and the absorbance at 600 nm is measured. CMC is added to the reaction medium to improve transparency, whereas EDTA is used as an antiseptic (Callemien and Collin, 2010). No color development is observed with phenolic acids such as syringic, vanillic, and ferulic acids, as they do not include interaction sites. Reductones, ascorbic acid, cysteine, colorants, and melanoidins interfere in the determination. The calibration is carried out by using polyphenols isolated from malt and hops.

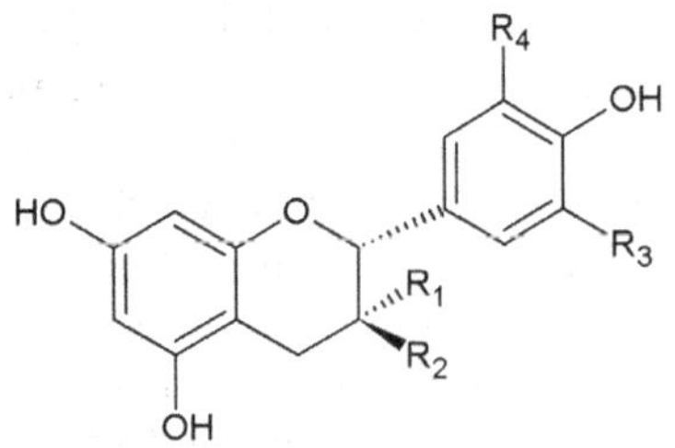

Flavan-3-ol monomers	R_1	R_2	R_3	R_4
(+)-Catechin	- H	-OH	-H	-OH
(-)-Epicatechin	-OH	-H	-H	-OH
(-)-Gallocatechin	- H	-OH	-OH	-OH
(-)-Catechin gallate	-H	Gallate	-H	-OH

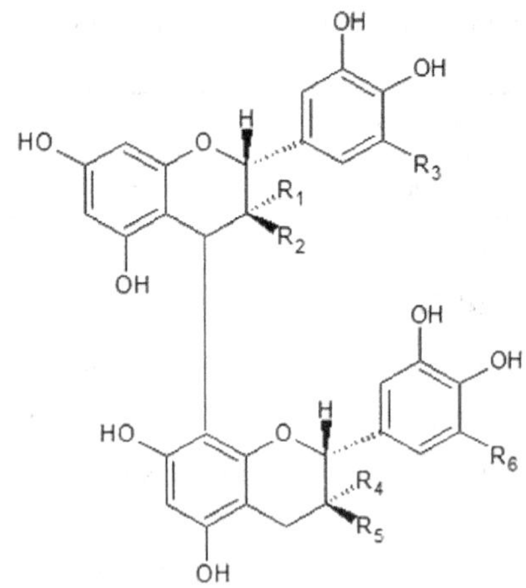

Flavan-3-ol dimers	R_1	R_2	R_3	R_4	R_5	R_6
Procyanidin B_1	-OH	-H	-H	-H	-OH	-H
Prodelphinidin B_3	-H	-OH	-OH	-H	-OH	-H

FIGURE 29.1 Flavanoids structure: flavan-3-ol monomers and dimers.

When compared to FC assay, Bishop assay results tend to be more selective and, therefore, provide lower polyphenol concentrations.

The Bishop assay has been automated by using a FIA device (Peris *et al.*, 1991). The results obtained were generally lower than those provided with the batch procedure. Coumaric acid was used for constructing calibration curves, as the kinetics of this compound was the most similar to those of the polyphenols analyzed in beer.

Determination of Groups of Phenolic Compounds

Some assays have been proposed for the determination of flavanoid content. This group of polyphenols includes flavan-3-ol monomers and proanthocyanidins. These latter compounds, also called condensed tannins, are polymers composed of flavan-3-ol monomer subunits, with variable degree of polymerization, and they have the characteristic ability of reacting with proteins and forming high-molecular-weight species and haze, being responsible for colloidal instability of beer during storage. Therefore, they are of additional concern to brewers. However, very little haze is produced with flavan-3-ol monomers. Some of flavanoids identified in beer are shown in Figure 29.1.

Reported methods for analysis of flavanoids are usually based on colorimetric derivatization although, in the particular case of proanthocyanidins, some assays using polymeric precipitation have also been proposed. In general terms, colorimetric methods are cheap and fast and work well with purified samples, yielding method-specific results. Nevertheless, they lack specificity for individual compounds and show poor reproducibility between different kinds of samples.

Colorimetric Assays for Flavan-3-ol and Proanthocyanidins

BUTANOL-HCL ASSAY

This is one of the most common colorimetric assays for determination of proanthocyanidin content, being based on the oxidative depolymerization of proanthocyanidins in a hot acid-alcohol medium (Porter *et al.*, 1986). First, condensed tannins are cleaved into the flavan-3-ol unit (terminal unit) and the carbocation unit (extension unit). Then, the carbocations are rapidly converted into red anthocyanidins by autoxidation (λ_{max}: 550 nm) (Figure 29.2a). The yield of anthocyanidin is critically affected by the presence of trace impurities of transition metal ions. This can be avoided by adding Fe^{3+} ion, which favors the autoxidation step. The tannin structure influences the yield and the kinetics of the reaction and the results of quantification strongly depend on the standard used for calibration. The yield of anthocyanidin is also affected by small amounts of water in the reaction mixture. The assay is performed by mixing the sample with butanol-HCl and $NH_4Fe\ (SO_4)_2.12\ H_2O$ solutions, maintaining the mixture at 95°C for 40 min and measuring the absorbance at 550 nm.

VANILLIN ASSAY

This method is based on the reaction of vanillin with flavan-3-ols and proanthocyanidins to yield red complexes (Burns, 1971). Sulfuric acid and (+)-catechin are usually used as the catalyst and reference standard, respectively. The results are influenced by the type of solvent used, the nature and concentration of the acid,

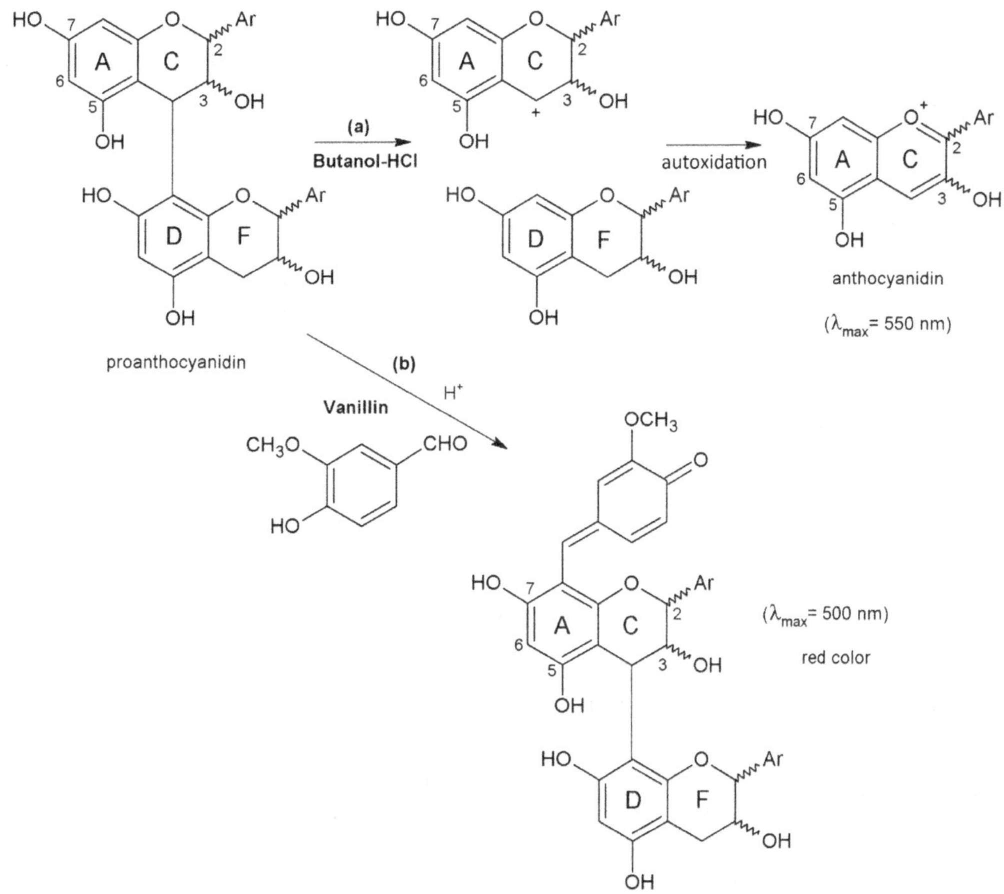

FIGURE 29.2 (a) Butanol-HCl assay for proanthocyanidins; (b) Vanillin assay for flavan-3-ols and proanthocyanidins. Chemical reactions involved in both assays are shown.

the reaction time, temperature, vanillin concentration, and type of reference standards used. According to the studies carried out, the vanillin reagent only reacts with flavanoids containing free *meta*-oriented hydroxy groups in the A-ring and a single bond at the 2,3-position of the C-ring (Figure 29.2b). Taking into account these structural requirements, the vanillin reagent is not completely specific for flavanoids and a significant color development can be expected by dihydrochalcones. In fact, it was found that the color intensity developed by the dihydrochalcone phloretin was of the same order as that for catechin (Sarkar and Howarth, 1976). The assay is carried out by adding to the sample the 1% vanillin reagent and measuring the absorbance at 500 nm (Broadhurst and Jones, 1978).

p-DIMETHYLAMINOCINNAMALDEHYDE ASSAY

This method consists of the use of the chromogen *p*-dimethylaminocinnamaldehyde (DMACA), which specifically reacts with proanthocyanidin polymers, small oligomers, and the immediate proanthocyanidin precursors, flavan-3,4-diols and flavan-3-ols (Delcour and Janssens de Varebeke, 1985). The reaction of these flavanoids with DMACA originates the development of a green color (λ_{max}: 640 nm). The structural requirements for reaction with vanillin seem to be also applicable for reactivity with DMACA. Thus, it seems that DMACA reagent, like vanillin, only reacts with flavanoids containing *meta*-oriented di- or trihydroxy substituted A-rings and a single bond at the 2,3 position of the C-ring (Figure 29.3). (+)-Catechin is used as reference for quantification purposes.

ANTHOCYANOGEN ASSAY

In a similar way to the butanol-HCl assay, proanthocyanidins yield red anthocyanidin when they are heated in the presence of hydrochloric acid (McFarlane, 1961). The procedure consists of the complete adsorption of polyphenols on polyvinylpyrrolidone (PVP) and the subsequent desorption with N-methyl-2-pyrrolidone containing hydrochloric acid and iron. This latter enhances the color development, which is different for cyanidins and delphinidins (545/557 nm) (Callemien and Collin, 2010). Leucocyanidin is used as standard.

Precipitation Assays for Proanthocyanidins

Proanthocyanidins are, by definition, protein-binding and precipitating agents and assays based on polymer

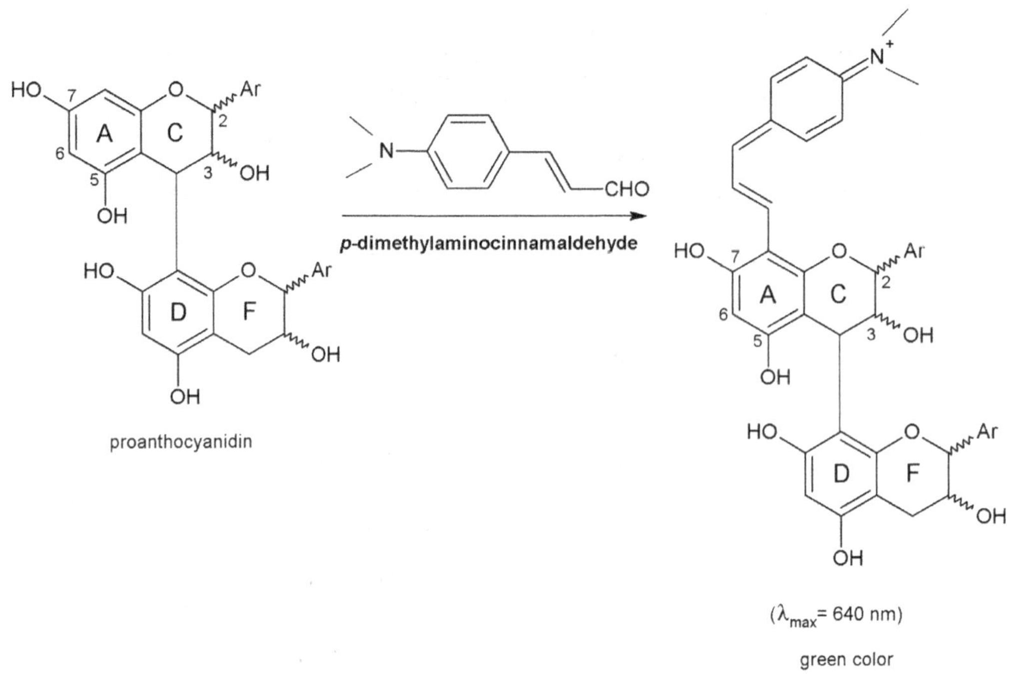

FIGURE 29.3 *p*-Dimethylaminocinnamaldehyde assay. Chemical reaction involved in the assay is shown.

precipitation represent an interesting and popular option for routine applications. In the application of this kind of method, it is necessary to take into account the fact that protein precipitation of condensed tannins is dependent on variables such as isoelectric point, ionic strength, pH, protein and tannin structure, and temperature (De Freitas and Mateus, 2001).

One of the most widely cited protein precipitation assays uses bovine serum albumin (BSA) as a precipitant agent (Hagerman and Butler, 1978). Condensed tannins and BSA are mixed under defined conditions of pH value and ionic strength to form an insoluble protein–tannin complex. The precipitate is dissolved in an alkaline sodium dodecyl sulfate (SDS) buffer solution and the absorbance is measured at 510 nm, with previous addition of iron (III) ion. Several modifications of this assay have been reported in literature (Makkar, 1989), but all of them are time-consuming and are potentially compromised by their inability to directly measure the removed condensed tannins.

Proanthocyanidins present an absorption maximum at approximately 280 nm and, therefore, the measurement of the absorbance at this wavelength before and after tannin precipitation could allow their selective determination. Nevertheless, the spectral measurement suffers interference from the added protein precipitant agent, which limits the practical application. In order to circumvent this drawback and obtain a direct spectral measurement of tannins, several non-proteinaceous polymers have been assayed for precipitations of tannins, such as PVP and polyvinylpolypyrrolidone (PVPP) (Sarneckis *et al.*, 2006). Nevertheless, these precipitant agents also absorb at 280 nm, and the reaction is not specific for tannins. The use of methyl cellulose to precipitate tannins has also been reported (Montedoro and Fantozzi, 1974; Sarneckis *et al.*, 2006), the assay being both robust and specific for plant tannins as there is no interference of other phenolics absorbing at 280 nm.

Condensed tannins have also been determined by non-tannins subtracted from the total phenols (Fumi *et al.*, 2011). Non-tannins were evaluated by the Folin–Ciocalteu method in the remaining supernatant after the selective precipitation of tannins with methyl cellulose (Montedoro and Fantozzi, 1974).

DETERMINATION OF INDIVIDUAL PHENOLIC COMPOUNDS

Because of high sensitivity and selectivity, HPLC is the most commonly used technique for the separation and individual determination of phenolic compounds in beer. Reported HPLC methods include normal-phase, reversed-phase, and size-exclusion chromatography columns coupled with photodiode array (DAD) (Montanari *et al.*, 1999; Szwajgier, 2009; Dvořáková *et al.*, 2007; Zhao *et al.*, 2010), fluorimetric (Madigan and McMurrough, 1994), electrochemical (ECD) (Floridi *et al.*, 2003; Vanbeneden *et al.*, 2006; Piazzon *et al.*, 2010), mass spectrometry (MS) (Česlová *et al.*, 2009), and tandem mass spectrometry (MS/MS) (Stevens *et al.*, 1999). Due

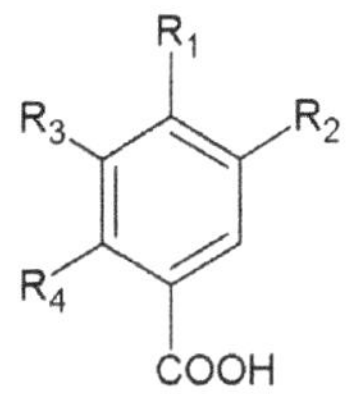

Hydroxybenzoic acids	R_1	R_2	R_3	R_4
p-Hydroxybenzoic acid	- OH	-H	-H	-H
Vanillic acid	- OH	-H	$-OCH_3$	-H
Gallic acid	- OH	-OH	-OH	-H
Syringic acid	- OH	$-OCH_3$	$-OCH_3$	-H

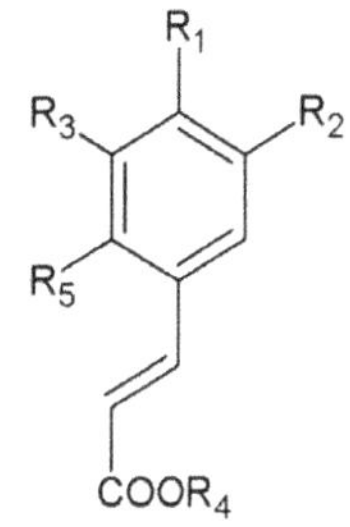

Hydroxycinnamic acids	R_1	R_2	R_3	R_4	R_5
p-Coumaric acid	- OH	-H	-H	-H	-H
Caffeic acid	- OH	-OH	-H	-H	-H
Sinapic acid	- OH	$-OCH_3$	$-OCH_3$	-H	-H
Ferulic acid	- OH	$-OCH_3$	-H	-H	-H

FIGURE 29.4 Benzoic and cinnamic acid structures.

to the structural similarity of most polyphenols and the complex nature of beer matrix, gradient elution is often required to obtain an appropriate separation by HPLC. In the case of charged polyphenols, capillary electrophoresis (CE) represents an interesting alternative to HPLC (Moane *et al.*, 1998).

Gas chromatography (GC) is the technique of preference for separation of volatile polyphenols like hydroxycinnamic-acid-derived flavors (Thurston and Tubb, 1981; Callemien *et al.*, 2006). In the case of phenolic compounds showing lower volatility, a previous derivatization is required before GC analysis (Tekel *et al.*, 1999). This latter can be carried out with trifluoroacetic acid, bis-trimethyl-silylacetamide or N-O-bis(trimethylsilyl) trifluoroacetamide. Luminescent techniques, namely, fluorescence and chemiluminescence, have also been proposed for the determination of individual phenolic compounds (García-Sánchez *et al.*, 1988, 1995).

Extraction of polyphenols in beer is usually carried out by liquid–liquid extraction, with organic solvents such as ethyl acetate, *n*-hexane, isooctane and acetone/water, or solid-phase extraction (SPE) (Alonso García *et al.*, 2004)

Methods reported for the separation and/or individual determination of several groups of polyphenols are described below.

Phenolic Acids

Phenolic acids are a major class of phenolics that are found in numerous drinks. They are derived from hydroxycinnamic acid (caffeic acid, ferulic acid, *p*-coumaric acid, etc.) and hydroxybenzoic acid (syringic acid, vanillic acid, gentisic acid, etc.), having one or more hydroxyl groups and a carboxylic acid function at the benzene ring. Their structures are shown in Figure 29.4. Recently, a study reported that ferulic, caffeic, sinapic, and syringic acids are the major contributors to the antioxidant activity of beer (Piazzon *et al.*, 2010). Phenolic acids are extracted mostly from malt, being also present in small amounts in hops. The most abundant phenolic acid in beer is ferulic acid, followed by sinapic, vanillic, caffeic, and *p*-coumaric. These phenolics occur in food mainly in the esterified form as glycosides (Herrmann, 1989), and the studies performed about their content of bound forms are scarce. Therefore, the methods reported in literature for analysis of phenolic acids usually refer to the determination of the free forms. When the analysis of the total (free plus conjugated) and/or conjugated amount of phenolic acids is required, most of the procedures take advantage of alkaline hydrolysis to release bound forms. This treatment leads to a significant loss of some phenolic acids derivatives, particularly dihydroxy-derivatives (Maillard and Berset, 1995). The loss is even more dramatic when phenolic acids are released by acidic hydrolysis.

The most commonly used technique for determination of phenolic acids is HPLC with DAD (Montanari *et al.*, 1999; Szwagjier, 2009; Bartolomé *et al.*, 2000) or ECD (Montanari *et al.*, 1999; Floridi *et al.*, 2003) detection. In this latter case, both amperometric and coulometric detection have been used for reversed-phase HPLC, usually with an octadecyl silica column and an aqueous-organic mobile phase containing an acid or acidic buffer in order to suppress the dissociation of weakly acidic phenolic compounds.

Other non-chromatographic techniques have also been proposed for determination of phenolic acids. Thus, the determination of *p*-coumaric acid has been carried out in beer by measurement of its native fluorescence ($\lambda_{ex}/\lambda_{em}$, 330/435 nm) (García-Sánchez *et al.*, 1988). The samples, adjusted to pH 2.0, were mixed with hexane,

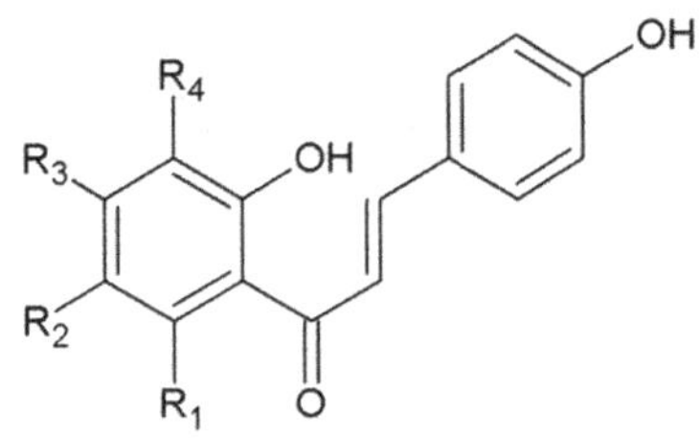

Prenylchalcones	R_1	R_2	R_3	R_4
Xanthohumol	$-OCH_3$	-H	-OH	Isoprenyl
Desmethylxanthohumol	-OH	-H	-OH	Isoprenyl

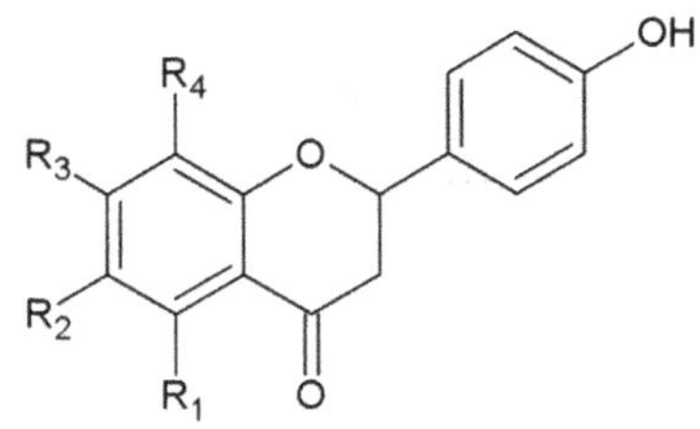

Derived flavanones	R_1	R_2	R_3	R_4
Isoxanthohumol	$-OCH_3$	-H	-OH	Isoprenyl
Hopein	-OH	-H	-OH	Isoprenyl

FIGURE 29.5 Prenylchalcone and prenylflavanone structures.

and then the phenolic acid was extracted with ethyl acetate. Before measuring the fluorescence of the extracts, the pH had to be adjusted to 10.7 as this latter influenced both the shape of the spectra and the fluorescence intensity. The slight matrix effect produced by other phenolic acids and other compounds could be circumvented by applying the standard addition method. Chemiluminescence has also been explored for the analysis of phenolic acids. Numerous compounds, among them *p*-coumaric acid, increase the chemiluminescence of the luminol-H_2O_2-horseradish peroxidase system. Studies carried out with several phenolic acids showed that only *p*-coumaric acid has a significant enhancing effect on this system and this has been used for its determination in beers in the range 0–12.5 nM using a stopped-flow technique (García Sánchez *et al.*, 1995).

Flavanoids

To date, beer flavanoids, especially proanthocyanidins, have been considered mostly for their involvement, with proteins, in haze development. Therefore, improvement of beer shelf-life requires a deeper knowledge of the amounts of these polyphenols and all the chemicals involved.

One of the techniques of choice for the analysis of these polyphenols is reversed-phase HPLC with DAD detection. Nevertheless, when using this detection system the interference of other UV-absorbing phenolic components of the sample is usual. In addition, if the sample has a complex proanthocyanidin composition, no good simultaneous separation of prodelphinidins and procyanidins is usually achieved. An alternative detection system, which allows a sensitive and highly selective determination of flavanoids in beer and other beverages, consists of a post-column derivatization with DMACA (De Pascual-Teresa *et al.*, 1998). The colored adduct formed shows maximum absorption between 632 and 640 nm, thus preventing the interference of other colored compounds such as anthocyanins.

Normal-phase HPLC–electrospray ionization (ESI)–MS/MS has also been proposed for the analysis of flavanoids in extracts of lager beers, revealing the presence of (+)-catechin, (−)-epicatechin, procyanidin dimers, and prodelphinidin trimers. Later, a reversed-phase HPLC–ESI–MS/MS method was used to study the detailed structures of those dimers and trimers, revealing the existence of four dimers, three procyanidins (B1, B3, and B4), and one prodelphinidin (B3) (Callemien and Collin, 2008). In this study, three trimers previously detected in hops were identified in beer for the first time. The isolation of dimer and trimer fractions was accomplished by thioacidolysis, and the results slowed that, as expected, most beer proanthocyanidins carried catechin as the terminal unit.

The potential of biosensors for determination of catechins and their dimers and trimers is based on the incorporation of several plant tissues containing polyphenol oxidase (banana, apple, and potato) in carbon paste electrodes (Eggins *et al.*, 1997; Cummings *et al.*, 2001). Nevertheless, the fact that these devices are not able to differentiate between the different flavanoids has also been explored by some authors.

Prenylchalcones and Flavanones

Prenylchalcones, xanthohumol and desmethylxanthohumol, predominate in hops, whereas prenylflavanones, isoxanthohumol and prenylnaringenins are more abundant in beer (Figure 29.5). This can be explained by taking into account the fact that isomerization in the brew kettle converts most xanthohumol into isoxanthohumol and generally all desmethylxanthohumol into its isomeric prenylflavanones, 6- and 8-prenylnaringenin.

HPLC–MS/MS has been successfully applied to the quantitative analysis of prenylflavonoids in beer without sample clean-up (Stevens *et al.*, 1999). The method only allows the accurate determination of isoxanthohumol (0.04–3.44 mg/l), 6-prenylnaringenin (0.001–0.56 mg/l) and 8-prenylnaringenin (0.001–0.24 mg/l), the major prenylflavanones in beer. An underestimation in the concentrations of xanthohumol, desmethylxanthohumol and the apolar flavanone 6-geranylnaringenin was

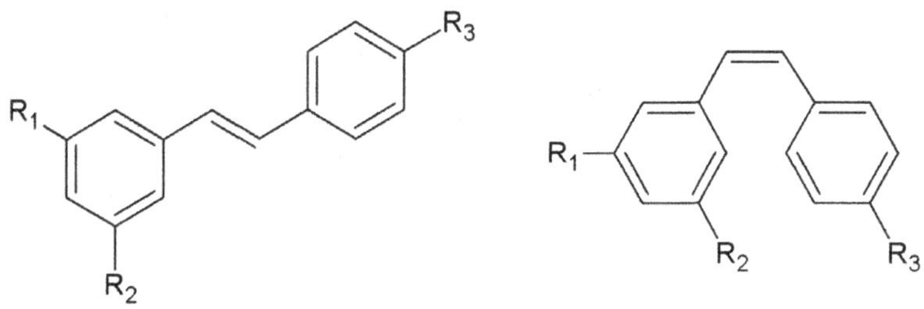

Stilbenes	R_1	R_2	R_3
trans-Resveratrol	-OH	-OH	-OH
trans-Piceid	Glucose	-OH	-OH
cis-Resveratrol	-OH	-OH	-OH
cis-Piceid	Glucose	-OH	-OH

FIGURE 29.6 Stilbenoid structures.

obtained. These results have been attributed to the complexation of the most apolar prenylflavonoids with macromolecules of beer, such as proteins or carbohydrates.

A non-aqueous reverse polarity CE method has been proposed for the determination of xanthohumol in hop pellets from different crops and cultivars, after extraction of the prenylchalcone with methanol (Kac *et al.*, 2007). The xanthohumol amounts found varied from 0.14 to 0.42%. The method was also applied to the analysis of beer samples but xanthohumol could only be detected in two of the analyzed samples, because its concentration was below the quantification limit.

Stilbenoids

Beside major polyphenols, usually, lager beers also contain traces of stilbenoids. These phenolic compounds are phytoalexins usually synthesized by plants in response to injury or fungal attack (Figure 29.6). Stilbenoids, even in low concentrations, exhibit very interesting biological effects such as estrogenic, antiplatelet, anti-inflammatory, cardioprotective, and antitumor properties. To date, the presence of this kind of polyphenol in beer has been scarcely studied. *trans*-Resveratrol and its glucosides, *trans*-piceid and *cis*-piceid, were identified recently in hops for the first time; before then *trans*-resveratrol was the only known source of stilbenoids in beer (Callemien *et al.*, 2003); moreover, *cis*-resveratrol was undetectable. Later, *trans*-resveratrol and *trans*-piceid were identified in red sorghum grains, which are used in African countries for local production (Bröhan *et al.*, 2011). Stilbenoids, specifically *trans*-resveratrol, *cis*-resveratrol, *trans*-piceid, and *cis*-piceid, have also been found in beer (Jerkovic *et al.*, 2008; Chiva-Blanch *et al.*, 2011).

Normally, the extraction of phenolic compounds from beer has to include a previous delipidation stage, that is, the removal of hop lipids and resins in order to obtain an appropriate extraction recovery. Diethyl ether is the solvent often used for this previous clean-up step. However, ether is not appropriate for resveratrol extraction because this compound is much more hydrophobic than other hop polyphenols and, therefore, it presents a significant solubility in this solvent. After the pre-cleaning of beer samples, the extraction of stilbenoids can be performed by using liquid/liquid extraction or SPE.

To date, only a few chromatographic methods have been proposed for determination of resveratrol and piceid in beer. Two procedures based on solid-phase microextraction (SPME) combined with GC–MS and HPLC–MS/MS, with atmospheric pressure chemical ionization (APCI), were compared for analysis of both stilbenoids (Jerkovic *et al.*, 2008). The results showed that SPME–GC–MS allows the direct determination of *trans*-resveratrol (detection limit (LOD): 5 μg/l), without previous pre-cleaning and long extraction steps, although a derivatization by silylation has to be performed due to the low volatility of this stilbenoid. A limitation of this technique which should be taken into account was the scarce reproducibility between successive fibers. With respect to the procedure using HPLC–MS/MS technique, it allowed the identification of *trans*-resveratrol (LOD: 5 μg/l) and/or *trans*-piceid (LOD: 15 μg/l) after a previous delipidation of the sample with toluene and cyclohexane followed by SPE. *cis*-Forms of resveratrol and piceid were undetectable with the application of both procedures.

An HPLC–ESI–MS/MS method, which allows the determination of *trans*-resveratrol, *trans*-piceid, and their *cis*-forms, has also been proposed (Chiva-Blanch *et al.*, 2011). The pre-cleaning of the samples was carried out according to the procedure previously described by Jerkovic *et al.*, but the SPE procedure was improved and optimized for piceid recovery, using 1 M acetic acid in methanol and 1 M acetic acid in ethyl acetate for elution instead of ethanol. Resveratrol and piceid concentrations in the beers analyzed were in the range 1.34–77.0 μg/l and 1.80–27 μg/l, in 79% and 33% of them, respectively.

To date, only one spectroscopic method has been proposed for analysis of stilbenoids in beer. Thus, resveratrol has been determined in beer by means of an automatic fluorimetric optosensor (Molina-García *et al.*, 2011). The method is based on the measurement of the fluorescence of the photoproduct generated from the stilbenoid under UV irradiation. In this case, the total amount of resveratrol (*cis*- plus *trans*-forms) is determined, as *trans*- and *cis*- isomers originate the same photoproduct. Both the generation of the photoproduct and the measurement of its fluorescence were carried out on-line in a multicomuted-flow injection assembly. The analytical signal was measured in the solid-phase, after the retention of the monitored species on an appropriate solid-support (Sephadex QAE A-25) filling the flow-cell. Delipidation of the samples was carried out by following a previously reported method (Jerkovic *et al.*, 2008) and, next, resveratrol was extracted by a SPE procedure, using C_{18} cartridges and methanol as eluent. The interference of piceid was circumvented by eluting this from the cartridge with 40% methanol before eluting resveratrol. Resveratrol concentrations found in the beer samples analyzed ranged from 4.1 to 14.1 μg/l.

References

Alonso García, A., Cancho Grande, B., Simal Gándara, J., 2004. Development of a rapid method based on solid-phase extraction and liquid chromatography with ultraviolet absorbance detection for the determination of polyphenols in alcohol-free beers. J. Chromatogr. A. 1054, 175–180.

Andersen, M.L., Skibsted, L.H., 2001. Modification of the levels of polyphenols in wort and beer by addition of hexamethylenetetramine or sulfite during mashing. J. Agric. Food Chem. 49, 5232–5237.

Bartolomé, B., Peña-Neira, A., Gómez-Cordovés, C., 2000. Phenolics and related substances in alcohol-free beers. Eur. Food Res. Technol. 210, 419–423.

Bishop, L.R., 1972. Analysis Committee of the European Brewery Convention. Measurement of total polyphenols in wort and beers. J. Inst. Brew. 78, 37–38.

Broadhurst, R.B., Jones, W.T., 1978. Analysis of condensed tannins using acidified vanillin. J. Sci. Food Agric. 29, 783–794.

Bröhan, M., Jerkovic, V., Collin, S., 2011. Potentiality of red sorghum for producing stilbenoid-enriched beers with antioxidant activity. J. Agric. Food Chem. 59, 4088–4094.

Burns, R.E., 1971. Methods for estimation of tannin in grain sorghum. Agron. J. 63, 511–512.

Callemien, D., Cawet, Q., Counet, C., Collin, S., 2003. Hop has a determinant nutrition key for health? Proceedings of the 29th European Brewery Convention (Dublin). Fachverlag Hans Carl, Nürnberg, Germany.

Callemien, D., Dasnoy, S., Collin, S., 2006. Identification of a stale-beer-like odorant in extracts of naturally aged beer. J. Agric. Food Chem. 54, 1409–1413.

Callemien, D., Collin, S., 2008. Use of RP-HPLC-ESI(−)-MS/MS to differentiate various proanthocyanidin isomers in lager beer extracts. J. Am. Soc. Brew. Chem. 66, 109–115.

Callemien, D., Collin, S., 2010. Structure, organoleptic properties, quantification methods, and stability of phenolic compounds in beer—A review. Food Rev. Int. 26, 1–84.

Česlová, L., Holčapek, M., Fidler, M., Drštičková, J., Lísa, M., 2009. Characterization of prenylflavonoids and hop bitter acids in various classes of Czech beers and hop extracts using high-performance liquid chromatography–mass spectrometry. J. Chromatogr. A 1216, 7249–7257.

Chiva-Blanch, G., Urpi-Sarda, M., Rotchés-Ribalta, M., Zamora-Ros, R., Llorach, R., Lamuela-Raventós, R.M., Estruch, R., Andrés-Lacueva, C., 2011. Determination of resveratrol and piceid in beer matrices by solid-phase extraction and liquid chromatography–tandem mass spectrometry. J. Chromatogr. A 1218, 698–705.

Cummings, E.A., Linquette-Mailley, S., Mailley, P., Cosnier, S., Eggins, B.R., McAdams, E.T., 2001. A comparison of amperometric screen-printed, carbon electrodes and their application to the analysis of phenolic compounds present in beers. Talanta 55, 1015–1027.

De Freitas, V., Mateus, N., 2001. Structural features of procyanidin interactions with salivary proteins. J. Agric. Food Chem. 49, 940–945.

Delcour, J.A., Janssens de Varebeke, D., 1985. A new colorimetric assay for flavanoids in pilsner beers,. J. Inst. Brew. 91, 37–40.

De Pascual-Teresa, S., Treutter, D., Rivas-Gonzalo, J.C., Santos-Buelga, C., 1998. Analysis of flavanols in beverages by high-performance liquid chromatography with chemical reaction detection. J. Agric. Food Chem. 46, 4209–4213.

Dvořáková, M., Hulín, P., Karabín, M., Dostálek, P., 2007. Determination of polyphenols in beer by an effective method based on solid-phase extraction and high performance liquid chromatography with diode array detection. Czech J. Food Sci. 25, 182–188.

Eggins, B.R., Hickey, C., Toft, S.A., Zhou, D.M., 1997. Determination of flavanols in beer with tissue biosensors. Anal. Chim. Acta. 347, 281–288.

Floridi, S., Montanari, L., Marconi, O., Fantozzi, P., 2003. Determination of free phenolic acids in wort and beer by coulometric array detection. J. Agric. Food Chem. 51, 1548–1554.

Fumi, M.D., Galli, R., Lambri, M., Donadini, G., De Faveri, D.M., 2011. Effect of full-scale brewing process on polyphenols in Italian all-malt and maize adjunct lager beers. J. Food Compos. Anal. 24, 568–573.

García-Sánchez, F., Navas Díaz, A., González García, J.A., 1995. Study of the enhanced chemiluminescence from the luminol–horseradish peroxidase–hydrogen peroxide–*p*-coumaric acid system at very short times: stopped flow selective determination of *p*-coumaric acid in beers. Anal. Chim. Acta 310, 399–406.

García-Sánchez, F., Carnero, C., Heredia, A., 1988. Fluorimetric determination of *p*-coumaric acid in beer. J. Agric. Food Chem. 36, 80–82.

Gerhäuser, C., 2005. Beer constituents as potential cancer chemopreventive agents. Eur. J. Cancer 41, 1941–1954.

Graham, H.D., 1992. Stabilization of the Prussian Blue color in the determination of polyphenols. J. Agric. Food Chem. 40, 801–805.

Hagerman, A.E., Butler, L.G., 1978. Protein precipitation method for the quantitative determination of tannins. J. Agric. Food Chem. 26, 809–812.

Herrmann, K., 1989. Occurrence and content of hydroxycinnamic and hydroxybenzoic acid compounds in foods. Crit. Rev. Food Sci. Nutr. 28, 315–347.

Jerkovic, V., Nguyen, F., Timmermans, A., Collin, S., 2008. Comparison of procedures for resveratrol analysis in beer: assessment of stilbenoids stability through wort fermentation and beer aging. J. Inst. Brew. 114, 143–149.

Jerumanis, J., 1973. Methods of determination of polyphenols in the brewery. Bull. de l'Assoc. Royale des Anciens Etudiants en Brasserie de l'Univ. de Louvain 69, 1–14.

Kac, J., Zakrajšek, J., Mlinarič, A., Kreft, S., Filipič, M., 2007. Determination of xanthohumol in hops (Humulus lupulus L.) by non-aqueous CE. Electrophoresis 28, 965–969.

Madigan, D., McMurrough, I., 1994. Rapid determination of 4-vinyl guaiacol and ferulic acid in beers and worts by high-performance liquid chromatography. J. Am. Soc. Brew. Chem. 52, 152–155.

Magalhães, L.M., Segundo, M.A., Reis, S., Lima, J.L.F.C., Rangel, A.O.S.S., 2006. Automatic method for the determination of Folin–Ciocalteu reducing capacity in food products. J. Agric. Food Chem. 54, 5241–5246.

Maillard, M.N., Berset, C., 1995. Evolution of antioxidant activity during kilning: role of insoluble bound phenolic acids of barley and malt. J. Agric. Food Chem. 43, 1789–1793.

Makkar, H.P.S., 1989. Protein precipitation methods for quantitation of tannins: a review. J. Agric. Food Chem. 37, 1197–1202.

McFarlane, W.D., 1961. Determination of anthocyanogens. J. Inst. Brew. 67, 502–506.

Moane, S., Park, S., Lunte, C.E., Smyth, M.R., 1998. Detection of phenolic acids in beverages by capillary electrophoresis with electrochemical detection. Analyst 123, 1931–1936.

Molina-García, L., Ruiz-Medina, A., Fernández-de Córdova, M.L., 2011. A novel muticommuted fluorimetric optosensor for determination of resveratrol in beer. Talanta 83, 850–856.

Montanari, L., Perretti, G., Natella, F., Guidi, A., Fantozzi, P., 1999. Organic and phenolic acids in beer. Lebensm. Wiss. Technol. 32, 535–539.

Montedoro, G., Fantozzi, P., 1974. Dosage des tannins dans le mouts et les vins a l'aide de la methylcellulose et evaluation d'autres fractions phenoliques. Lebensm. Wiss. Technol. 7, 155–161.

Oñate-Jaén, A., Bellido-Milla, D., Hernández-Artiga, M.P., 2006. Spectrophotometric methods to differentiate beers and evaluate beer aging. Food Chem. 97, 361–369.

Peris, M., Müller, D., Maquieira, A., 1991. Determination of total polyphenols in beers by flow injection analysis. Food Chem. 40, 1–8.

Piazzon, A., Forte, M., Nardini, M., 2010. Characterization of phenolics content and antioxidant activity of different beer types. J. Agric. Food Chem. 58, 10677–10683.

Porter, L.J., Hrstich, L.N., Chan, B.G., 1986. The conversion of procyanidins and prodelphinidins to cyanidin and delphinidin. Phytochemistry 25, 223–230.

Price, M.L., Butler, L.G., 1977. Rapid visual estimation and spectrophotometric determination of tannin content of sorghum grain. J. Agric. Food Chem. 25, 1268–1273.

Roginsky, V., Lissi, A.E., 2005. Review of methods to determine chain-breaking antioxidant activity in food. Food Chem. 92, 235–254.

Sarkar, S.K., Howarth, R.E., 1976. Specificity of the vanillin test for flavanols. J. Agric. Food Chem. 24, 317–320.

Sarneckis, C.J., Dambergs, R.G., Jones, P., Mercurio, M., Herderich, M.J., Smith, P.A., 2006. Quantification of condensed tannins by precipitation with methyl cellulose: development and validation of an optimised tool for grape and wine analysis. Aust. J. Grape Wine Res. 12, 39–49.

Singleton, V.L., Rossi, J.A., 1965. Colorimetry and total phenolics with phosphomolybdic-phosphotungstic acid reagents. Am. J. Enol. Vitic. 16, 144–158.

Singleton, V.L., Orthofer, R., Lamuela-Raventós, R.M., 1999. Analysis of total phenols and other oxidation substrates and antioxidants by means Folin–Ciocalteu reagent. Methods Enzymol. 299, 152–178.

Stevens, J.F., Taylor, A.W., Deinzer, M.L., 1999. Quantitative analysis of xanthohumol and related prenylflavonoids in hops and beer by liquid chromatography–tandem mass spectrometry. J. Chromatogr. A 832, 97–107.

Szwajgier, D., 2009. Content of individual phenolic acids in worts and beers and their possible contribution to the antiradical activity of beer. J. Inst. Brew. 115, 243–252.

Tekel, J., De Keukeleire, D., Rong, H., Daeseleire, E., Van Peteghem, C., 1999. Determination of the hop-derived phytoestrogen, 8-prenylnaringenin, in beer by gas chromatography/mass spectrometry. J. Agric. Food Chem. 47, 5059–5063.

Thurston, P.A., Tubb, R.S., 1981. Screening yeast strains for their ability to produce phenolic off-flavors: a simple method for determining phenols in wort and beer. J. Inst. Brew. 87, 177–179.

Vanbeneden, N., Delvaux, F., Delvaux, F.R., 2006. Determination of hydroxycinnamic acids and volatile phenols in wort and beer by isocratic high-performance liquid chromatography using electrochemical detection. J. Chromatogr. A 1136, 237–242.

Vanderhaegen, B., Neven, H., Verachtert, H., Derdelinckx, G., 2006. The chemistry of beer aging—A critical review. Food Chem. 95, 357–381.

Zhao, H., Chen, W., Lu, J., Zhao, M., 2010. Phenolic profiles and antioxidant activities of commercial beers. Food Chem. 119, 1150–1158.

CHAPTER

30

Deriving a Global Antioxidant Score for Commercial Juices by Multivariate Graphical and Scoring Techniques: Applications to Blackcurrant Juice

Jessica Tabart, Claire Kevers*, Nadia Dardenne†, Valérie Schini-Kerth‡, Adelin Albert†, Jacques Dommes*, Jean-Olivier Defraigne§, Joël Pincemail§*

***Plant Molecular Biology and Biotechnology B22, University of Liège, Plant Biology Institute, Sart Tilman, Liège, Belgium, †Department of Medical Informatics and Biostatistics B23, University of Liège, CHU Sart Tilman, Liège, Belgium, ‡Faculty of Pharmacy, University of Strasbourg, Illkirch, France, §CREDEC, University of Liège, CHU Sart Tilman, Liège, Belgium**

CHAPTER POINTS

- Blackcurrant juices are known to have a high antioxidant capacity.
- Ten juices are compared with respect to their antioxidant capacity (five assays) and content of phenolics and ascorbic acid.
- No single test is able to properly assess the actual antioxidant capacity of blackcurrant juices.
- A large battery of antioxidant assays allows the establishment of a global antioxidant score (GAS).
- The GAS is correlated with graphical representations like Chernoff faces or stars.
- The use of the GAS makes it possible to improve the discrimination of a food matrix.
- Industry can use the GAS to benchmark their antioxidant products with those of the market.

BLACKCURRANTS AND ANTIOXIDANT COMPOUNDS

Small berries constitute one of the important sources of potential health-promoting phytochemicals (Karjalainen *et al.*, 2009; Torronen *et al.*, 2012) due to their antioxidant properties (Hung *et al.*, 2004). These fruits are rich sources of phenolic compounds (Szajdek and Borowska, 2008) such as flavonoids (Hakkinen *et al.*, 1999) and other polyphenols (Zadernowski *et al.*, 2005). Blackcurrants have been shown to be one of the richest species in phenolic compounds (Bermùdez-Soto and Tomàs-Barberàn, 2004; Kucharska *et al.*, 2011). Among phenolics, flavonoids, a broad class of low molecular weight, are present in plant tissues, where they are synthesized and perform several important functions: (i) protecting against ultraviolet (UV) radiation, pathogens, and herbivores; (ii) providing flavor and plant color in flowers, fruits, and leaves; and (iii) contributing to many aspects of plant physiology. The flavonoid group comprises anthocyanins, flavonols, flavones, catechins, flavanones, and chalcones. Anthocyanins are responsible for the red, violet, and blue colors of most berries and fruits (Szajdek and Borowska, 2008; Kahkonen *et al.*, 2003). Especially high levels of anthocyanins and high antioxidant capacity were found in bilberries and blackcurrants (Moyer *et al.*, 2002; Benvenuti *et al.*, 2004). Myricetin, quercetin, and kaempferol are the main flavonols in blackcurrant, another group of flavonoids. Their amounts change greatly during blackcurrant fruit ripening (Raffo *et al.*, 2004). Phenolic acids in blackcurrant

Processing and Impact on Antioxidants in Beverages
http://dx.doi.org/10.1016/B978-0-12-404738-9.00030-1

berries are represented by cinnamic acid and benzoic acid derivatives (Zadernowski *et al.*, 2005). They occur mainly in bound forms as esters or glycosides (Hakkinen *et al.*, 1999).

Besides the phenolic compounds, blackcurrant fruits are the richest source of vitamin C among all berry fruit species (Hagg *et al.*, 1995; Nes *et al.*, 2012). The presence of phenolic antioxidants has been shown to protect vitamin C against oxidative decomposition, mainly in blackcurrant (Miller and Rice-Evans, 1997).

The content of phenolics in berries is affected by the degree of maturity at harvest (Viskelis *et al.*, 2010), genetic differences (Tabart *et al.*, 2006), pre-harvest environmental conditions, post-harvest storage conditions and processing (Shahidi and Naczk, 2004), and growing season (Tabart *et al.*, 2006). The potential health benefits of these fruits and their products may be greatly increased if berries from cultivars with high contents of antioxidant compounds are used as raw materials (Mikkonen *et al.*, 2001).

CONSUMPTION OF BLACKCURRANTS

Domestic berries, both wild and cultivated species, are consumed in abundance. However, the consumption of blackcurrants as fresh product is restricted due to their short seasonal availability and relatively high cost. Therefore, only small portions of berries are consumed fresh, with most of their intake coming from processed foods, such as frozen products, juices (Woodward *et al.*, 2011), jams, and jellies (Caswell, 2009). Blackcurrant juice concentrates are generally quite rich in phenolic compounds (±15 g/l), mainly anthocyanins. Moreover they have a strong antioxidant capacity, meaning that blackcurrant juice may have huge potential as an ingredient in functional juices (Bermùdez-Soto and Tomàs-Barberàn, 2004).

The process used to obtain juice requires a succession of steps that may modify juice composition. Woodward *et al.* (2011) observed that anthocyanins remain stable during commercial blackcurrant juice processing. No significant loss in anthocyanins was observed throughout the commercial processing of blackcurrants, from whole berry through milling, to pectin hydrolysis and sodium bisulphate addition. Likewise, no significant loss in anthocyanins was observed following the subsequent processing of pressed juice, through pasteurization, decantation, filtration, and concentration. Storage of the juice is another step where composition modification may occur. However, total phenolics (Casati *et al.*, 2012) and antioxidant capacity values were found to be rather stable during storage at 4°C and 20°C. By contrast (Würth *et al.*, 2009), a decrease was observed for the monomeric anthocyanins, although minor at 4°C but becoming larger with increasing temperature. The half-life of monomeric anthocyanins was between 17 and 21 days at 37°C.

EVALUATION OF TOTAL ANTIOXIDANT ACTIVITY

In their noteworthy paper "How to standardize the multiplicity of methods to evaluate natural antioxidants?" Frankel and Finley (2008) described all the complexity to evaluate the total antioxidant activity in foods. Indeed, a large variety of techniques of inducing and catalyzing oxidation and measuring the endpoint of oxidation have been used. Globally, the methods to determine the total antioxidant capacity in foods can be divided into two major groups: (1) those based on single electron transfer reaction, namely the Trolox equivalent antioxidant capacity (TEAC) assay, the ferric reducing ability of plasma (FRAP) assay, the copper reduction (CUPRAC) assay and the 2,2-diphenyl-1-picrylhydrazyl radical scavenging capacity (DPPH) assay; and (2) those based on hydrogen atom transfer reaction, namely the crocin bleaching assay, the total peroxyl radical-trapping antioxidant parameter (TRAP) assay and the popular oxygen radical absorbance capacity (ORAC) assay. Additional methods have also been developed like the total oxidant scavenging capacity (TOSC) assay, the chemiluminescence assays and electrochemiluminescence assays. From one test to another, there are changes in the conditions of oxidation, the kinetics and the composition of the system, the partitioning properties of the antioxidants between different phases and their interface and the physical state of the substrate. The selection of valid reference compounds for relative *in vitro* activity is also a controversial problem. In particular, the common use of Trolox as a reference to evaluate the relative antioxidant activity of compounds such as flavonoids is a matter of debate. Ultimately, this gives rise to a disparity of results, further emphasized by the absence of standardization among assays. As an example, studies by the authors' group have shown that a given food can exert a high antioxidant activity with the ORAC test but not with the DPPH assay (Kevers *et al.*, 2007; Tabart *et al.*, 2009) and *vice versa*. Thus, we believe that a better approach would consist of developing a methodology that integrates results from a wide range of assays into a global antioxidant score (GAS).

In this chapter, we shall demonstrate how multivariate graphical techniques (e.g., principal component analysis, Chernoff faces and stars) can help develop such a GAS for 10 commercial blackcurrant juices found on the German and Belgian markets by means of seven parameters measuring antioxidant capacity and antioxidant compounds.

TABLE 30.1 Antioxidant Capacity, Ascorbic Acid (AA), and Total Phenolic (TP) Contents of 10 Commercial Blackcurrant Juices

Juices	TEAC µmol ET/l	ORAC µmol ET/l	DPPH µmol ET/l	FRAP µmol ET/l	TP mgEGA/l	AA mg/l	ESR USOD/ml
A	155	11939	4232	6868	1207	230	50570
M	126	10862	5138	10357	1569	585	44710
Na	117	5057	3737	5327	1228	386	54140
R	160	6536	4801	8070	1193	467	33280
G	133	6753	3882	7138	1071	331	39710
J	132	8092	4597	8133	1414	549	37020
Jb	133	7665	5793	12033	2142	620	41280
S	140	12004	3892	6932	1118	293	39570
V	132	13853	3466	8351	1511	465	46430
Z	131	13053	4266	7260	2026	290	44140
Mean	136	9581	4380	8047	1448	422	43085
SD	13	3109	712	1909	374	136	6264
CV(%)	9.7	32.5	16.3	23.7	25.8	32.3	14.5

Antioxidant activity measured by various assays (TEAC, ORAC, DPPH, FRAP, and ESR) and ascorbic acid (AA) and total phenolic (TP) contents in 10 blackcurrant juices. Mean, standard deviation (SD) and coefficient of variation (CV) calculated for each variable. ET = Trolox equivalent; EGA = gallic acid equivalent; ESR = electron spin resonance; USOD = unity of superoxide dismutase.

ANTIOXIDANT ACTIVITIES OF TEN BLACKCURRANT JUICES AND CORRELATIONS

The study material is given in Table 30.1. For each blackcurrant juice, the group assayed *in vitro* TEAC, DPPH, ORAC, electron spin resonance (ESR) (Tabart *et al.*, 2012), and FRAP (Benzie and Strain, 1996), as well as total phenolic (TP) content (Tabart *et al.*, 2012) and ascorbic acid (AA) level (Pincemail *et al.*, 2012). Results of the variables were summarized as mean (M), standard deviation (SD) and coefficient of variation (CV).

The correlation matrix of the seven assays is displayed in Table 30.2. The correlation coefficients were used to measure the association between all pairs of variables. Good correlations were found between two assays of antioxidant capacity, DPPH and FRAP, between total phenolics and the antioxidant capacity estimated by FRAP assay, and between ascorbic acid content and the measurement of the antioxidant capacity by DPPH or FRAP.

STATISTICAL ANALYSIS

To establish the individual antioxidant profile of the different juices, each variable (X) was transformed into a T_{score} by means of the formula $T_{score} = (X - min)/(max - min)$, where min and max, respectively, represent the smallest and largest values of variable X among the 10 juices. Thus, the T_{score} takes values from 0 (when X = min) to 1 (when X = max). The original values of the seven variables were then replaced by the seven corresponding T_{score}, all between 0 and 1. This transform is useful to associate with each juice a T_{score} profile so that juices can be more easily compared. For each juice, the seven T_{score} were summed up, yielding a global antioxidant score (GAS) ranging from 0 to 7. This allowed us to establish a ranking of the blackcurrant juices, the highest GAS corresponding to the 'best' juice (Table 30.3). In particular, juices Jb and M exhibited the best values while juices G and Na had the lowest global scores.

The second transform used in this work was $Z_{score} = (X - M)/SD$ which standardizes each value of a given variable X to a score with zero mean and unit standard deviation. The transform is utilized for principal component analysis (PCA) on the correlation matrix.

TABLE 30.2 Correlation Matrix between the Various Variables

	TEAC	ORAC	DPPH	FRAP	TP	AA	ESR
TEAC	1	0.11	0.12	−0.023	−0.31	−0.28	−0.43
ORAC		1	−0.24	0.058	0.25	−0.29	0.18
DPPH			1	**0.83**	0.57	**0.67**	−0.37
FRAP				1	**0.65**	**0.79**	−0.33
TP					1	0.43	0.032
AA						1	−0.31
ESR							1

Correlations between the various essays of antioxidant activity and the content in ascorbic acid (AA) or total phenolics (TP).

TABLE 30.3 T_{score}, GAS and Ranking of the 10 Different Blackcurrant Juices

	TEAC	ORAC	DPPH	FRAP	TP	AA	ESR	GAS	Ranking
A	0.89	0.78	0.33	0.23	0.13	0.00	0.83	3.19	5
M	0.21	0.66	0.72	0.75	0.46	0.91	0.55	4.26	2
Na	0.00	0.00	0.12	0.00	0.15	0.40	1.00	1.67	9
R	1.00	0.17	0.57	0.41	0.11	0.61	0.00	2.87	7
G	0.38	0.19	0.18	0.27	0.00	0.26	0.31	1.59	10
J	0.35	0.35	0.49	0.42	0.32	0.82	0.18	2.93	6
Jb	0.37	0.30	1.00	1.00	1.00	1.00	0.38	5.05	1
S	0.78	0.79	0.18	0.31	0.05	0.19	0.43	2.73	8
V	0.35	1.00	0.00	0.45	0.41	0.60	0.63	3.44	3
Z	0.32	0.91	0.34	0.29	0.89	0.15	0.52	3.42	4

T_{score} [(X − min)/(max − min)] of the different juices for each assay, total T_{score} for the seven assays as GAS (global antioxidant score) and ranking.

PCA is a graphical method for displaying multivariate observations on a two-dimensional space (plane). It consists in determining two orthogonal weighted sums of the Z_{score} of the original variables. The first one, called the first principal component PC_1 (X-axis), has the largest dispersion among all possible weighted sums, whereas the other one, called the second principal component PC_2 (Y-axis), has the second largest dispersion. The quality of the two-dimensional representation of the data can be measured by the sum of the first two largest latent roots also called eigenvalues of the correlation matrix divided by the number of variables (here seven). The two principal components can be interpreted in some way by looking at the weights given to the variables. As seen in Figure 30.1, the 10 juices appeared quite dispersed on the plane but some clusters could be clearly discerned: (C1) Jb and M, the best juices obtained by the ranking, (C2) J and R, (C3) V and Z, (C4) A, S, and G, and (C5) Na, the juice with the worst rank.

GRAPHICAL REPRESENTATIONS

The 'Chernoff faces' method is a graphical representation of a multivariate observation by a human face. Each variable is associated with a characteristic of the face (smile, shape, mouth, eyes, nose, ears, and so on) making it possible to compare juices and possibly rank them according to their antioxidant property. In this work (Figure 30.2), the associations between T_{score} and face features were performed for each juice as follows: TEAC = nose length, ORAC = surface of the face, DPPH = shape of the face, FRAP = eye location, TP = smile shape, AA = mouth length, ESR = mouth location. This allows the comparison of juices in a more amusing way and identifies salient differences between the various faces, here juices. For instance, juice Jb has the largest mouth while juice A has the smallest, because ascorbic acid content, the variable that determines the mouth length is maximal for Jb and minimal for A.

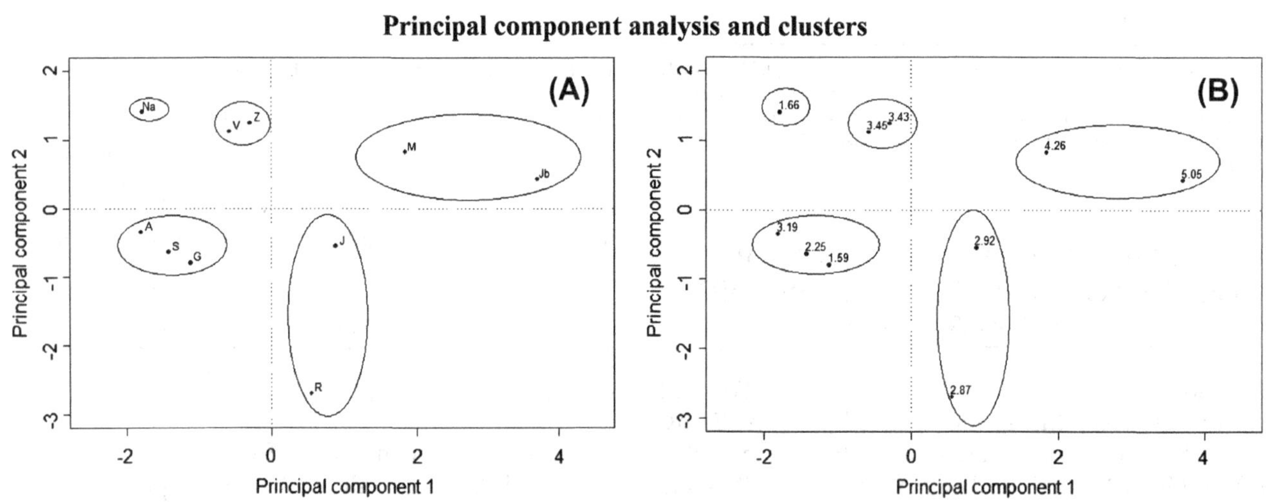

FIGURE 30.1 Representation of the first two main components and formation of clusters. A, juice code; B.: GAS). $CP_1 = -0.054\, Z_{TEAC} - 0.087\, Z_{ORAC} + 0.51\, Z_{DPPH} + 0.53\, Z_{FRAP} + 0.39\, Z_{TP} + 0.49\, Z_{AA} - 0.23\, Z_{ESR}$; $CP_2 = -0.63\, Z_{TEAC} + 0.27\, Z_{ORAC} - 0.13\, Z_{DPPH} + 0.031\, Z_{FRAP} + 0.40\, Z_{TP} + 0.041\, Z_{AA} + 0.59\, Z_{ESR}$, where Z_{TEAC} (*resp.* Z_{ORAC}, Z_{DPPH}, Z_{FRAP}, Z_{TP}, Z_{AA} *et* Z_{ESR}) is the standardized transform of TEAC (resp. ORAC, DPPH, FRAP, TP, AA and ESR).

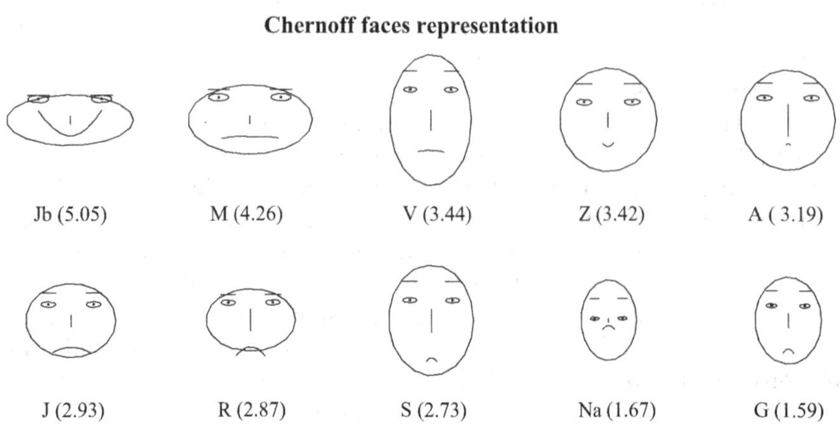

FIGURE 30.2 Chernoff faces representation of the 10 commercial blackcurrant juices by decreasing order of global antioxidant score (GAS) as calculated by T_{score}. TEAC, nose length; ORAC, surface of the face; DPPH, shape of the face; FRAP, eye location; TP, smile shape; AA, mouth length; ESR, mouth location.

The 'stars' method is an alternative graphical representation which associates with each variable a radius in a unit circle, each radius being separated from its neighbors by an angle of 360° divided by the number of variables (here seven). Thus there are as many radiuses as there are variables. For a given blackcurrant juice, the values of the seven variables are represented by their T_{score} (Table 30.3) on each radius and the corresponding apexes are joined together to form a polygonal figure that looks like a 'star'. This allows comparison of the juices by comparing the corresponding stars (Figure 30.3). Thus, each star is a blackcurrant juice. The apex located at the right corresponds to the variable ORAC. Then, turning anticlockwise, the variables are DPPH, TEAC, ESR, TP, AA, and FRAP, respectively. The simultaneous comparison of the stars of the 10 juices allows differentiation and some ranking. For instance, the ESR T_{score} for juice Na lies on the circle (value equal to 1), whereas for juice R its T_{score} is equal to 0 and there is no corresponding radius.

GLOBAL ANTIOXIDANT SCORE (GAS) AND RANKING

As shown in Table 30.1, raw data arising from the seven investigated parameters did not adequately allow discrimination of one blackcurrant juice from another. For instance, juice A exhibits one of the best antioxidant activities for TEAC while its concentration in vitamin C is lowest among all juices. Likewise, juice V has the highest ORAC value but its antioxidant capacity as assessed by the DPPH test is extremely low. From a general point of view, Table 30.2 clearly indicates large differences in the correlation coefficients between parameters. Only the well-correlated DPPH and FRAP assays exhibited a relationship with total phenolics and ascorbic acid. Surprisingly, the ORAC assay, still considered as a reference for evaluating the antioxidant capacity of a food matrix, correlated poorly with all the other tests. Therefore, evaluating an antioxidant capacity in a food matrix on the basis of only one or two tests can lead to misinterpretation.

By transforming the juice values to T_{score}, we were able to establish a GAS allowing us to rank the juices and to potentially discriminate them. As shown in Table 30.3, Jb turned out to be the best juice with a GAS of 5.05. This can be explained by the fact that juice Jb exhibited the best antioxidant activity in four of the seven investigated tests (DPPH, FRAP, TP, and AA). For the popular ORAC assay, juice Jb only ranked fourth among all juices, thus underestimating its actual antioxidant capacity. By contrast, juice V, which yielded the highest ORAC value, only ranked third when assessed by the GAS (3.44). This might be explained by the fact that its antioxidant activity was lowest when using the DPPH assay. While ranking in position seven with respect to TEAC, juice G finally got the lowest GAS (1.59) when combining all assays. In sum, the GAS appears to be a discriminating tool for comparing the relative antioxidant capacity of similar food matrices. Table 30.3 shows that by decreasing the order of magnitude of the GAS, the juices may be ranked as follows: Jb > M > V = Z > A > J > R > S > Na > G.

CLUSTERING

To obtain a two-dimensional representation of the 10 juices that accounts for the associations between the variables, we applied principal component analysis (PCA) to the data, more specifically to the correlation matrix (see Table 30.2). The first two eigenvalues of the correlation matrix amounted to 68% of the total, yielding a quite satisfactory graphical plot of the 10 juices studied (see Figure 30.1). The first principal component (*X*-axis) is written as:

$$CP_1 = -0.054\ Z_{TEAC} - 0.087\ Z_{ORAC} + 0.51\ Z_{DPPH} + 0.53\ Z_{FRAP} + 0.39\ Z_{TP} + 0.49\ Z_{AA} - 0.23\ Z_{ESR}$$

where $Z = (X - \text{mean})/SD$ represents the standardized variable. It is seen that the weights of DPPH, FRAP, TP, and AA are markedly positive whereas the coefficient of ESR is negative. The coefficients of TEAC and ORAC are close to zero. It follows that on the *X*-axis the 10 juices

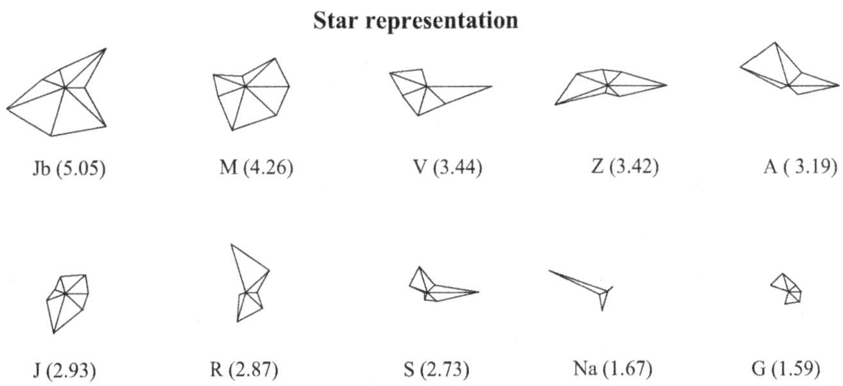

FIGURE 30.3 Star representation of the 10 commercial blackcurrant juices by decreasing order of global antioxidant score (GAS) as calculated by T_{score}. The apex located at the right corresponds to the variable ORAC. Then, turning anti-clockwise, the variables are DPPH, TEAC, ESR, TP, AA and FRAP, respectively.

ranging from left to right have increasing values of TP, AA, DPPH, and FRAP but decreasing values of ESR.

The second principal component (*Y*-axis) is written as:

$$CP_2 = -0.63\ Z_{TEAC} + 0.27\ Z_{ORAC} - 0.13\ Z_{DPPH} + 0.031\ Z_{FRAP} + 0.40\ Z_{TP} + 0.041\ Z_{AA} + 0.59\ Z_{ESR}$$

In this equation, a contrast is seen between variables ESR, TP, AA, and ORAC (all positive weights) on the one hand and variables TEAC and DPPH (negative weights) on the other hand, variable FRAP being negligible (weight close to 0). Thus on the *Y*-axis the 10 juices ranging from bottom to top have increasing values of ESR, TP, AA, and ORAC but decreasing values of TEAC and DPPH.

A close examination of Figure 30.1 leads to the definition of five distinct clusters of juices in which the GAS values are quite similar. The first cluster incorporates juices Jb (GAS 5.05) and M (4.26), the second cluster contains juices V (3.45) and Z (3.43), the third cluster includes juices J (2.92) and R (2.87), the fourth cluster encloses juices A (3.19), S (2.25), and G (1.59), and the final cluster entails juice Na (1.66). Thus, there appears to be a nice ordering between the cluster and the GAS of the juices.

VISUALIZATION OF THE GAS (CHERNOFF FACES AND STARS)

As shown in Figure 30.2, the use of Chernoff faces condensing high dimensional data into simple human faces provides an interesting way for visualizing the GAS of each blackcurrant juice. Indeed, humans are quite sensitive to facial expressions; quoting Pierre de Marivaux (*Marianne, Part 8*) "As soon as we see people whose faces appeal to us, we always tend to be more obliging towards them than to others." Thus, Chernoff faces associate facial characteristics with a set of specific variables. In our model, we ranked the variables by order of importance. From a biological point of view, ESR is first because it directly highlights the scavenging effect of a food matrix on a physiological free radical that is superoxide anion. In the other assays, free radicals are synthetic or generated by pure chemical reactions. Next, to further evidence the content of antioxidants, we placed in second position total phenolics and ascorbic acid, these assays providing practical and precise information about the nutritional quality of a food. A third group consisted of ORAC and DPPH, since they are the two most popular assays in the literature for evaluating the global antioxidant capacity of a food. Finally, we considered both FRAP and TEAC assays as the less interesting variables. Indeed, FRAP only measures the ferric ability of plasma and not an antioxidant capacity as such. Moreover, in TEAC assay, the formation of $ABTS^{\bullet+}$ radical from ABTS requires the use of enzymes or drastic chemical conditions. According to their importance, the seven parameters were dedicated to specific facial features. As the 'smile' usually expresses pleasure and, thus, plays an important social role, it was logical to attribute the location, shape and width of the smile to the three most important variables, namely ESR, total phenolics, and ascorbic acid. Further, as a 'round face' generally features the bonhomie of a person, the surface and the width of the head were attributed to the classical ORAC and DPPH assays. The less interesting variables TEAC and FRAP were, respectively, allocated to the length of the nose and the location of the eyes, generally face characteristics with a lower impact. As a consequence, a round face characterized by a large smile should correspond to a blackcurrant juice with the best antioxidant activity. Figure 30.2 shows that juices J, R, S, Na, and G have less pleasing faces than juices Jb and Z, which definitely exhibit large smiling and highly jovial appearances. The third group consisting of juices A, M, and V demonstrated more neutral faces without any significant expression, more particularly at the level of the smile. With the exception of juices V and Z, it was interesting to note that the facial appearance was well correlated to the GAS. From the considerations above, there is no doubt that juice Jb demonstrated the best antioxidant capacity. By contrast, the antioxidant capacities of J, R, and S were definitely lower, while those of juices Na and G were quite low. Interestingly, all these juices are sold as having similar percentages of blackcurrant (around 25%).

The 'star' approach yielded a similar picture to the Chernoff faces regarding the antioxidant capacity of the 10 juices. Indeed, the largest star was clearly the one associated with juice Jb and the two following ones with juices M and Z. The stars got lower and lower for juices V, A, J, R, S, Na, and G. Except for juice J, there was an excellent association between the GAS score and the size of the stars.

CONCLUSIONS

In conclusion, no single test, even the most popular ORAC assay, is able to compare the antioxidant capacity of similar food matrices such as blackcurrant juices. The only antioxidant variable that appears to be correlated with the content of antioxidant is the DPPH assay. By combining elements from a large battery of tests assessing antioxidant capacity, it is possible to improve discrimination of a food matrix by establishing a global antioxidant score (GAS) that correlates well with graphical representations like Chernoff faces or stars. The latter approach may help food industry managers and authorities to compare their antioxidant products with other similar products on the market.

References

Benvenuti, S., Pellati, F., Melegari, M., Bertelli, D., 2004. Polyphenols, anthocyanins, ascorbic acid, and radical scavenging activity of Rubus, Ribes, and Aronia. J. Food Sci. 69, FCT164–FCT169.

Benzie, I.F.F., Strain, J.J., 1996. The ferric reducing ability of plasma (FRAP) as a measure of 'antioxidant power': The FRAP assay. Anal. Biochem. 239, 70–76.

Bermúdez-Soto, M.J., Tomás-Barberán, F.A., 2004. Evaluation of commercial red fruit juice concentrates as ingredients for antioxidant functional juices. Eur. Food Res.Techn. 219, 133–141.

Casati, C.B., Sánchez, V., Baeza, R., Magnani, N., Evelson, P., Zamora, M.C., 2012. Relationships between colour parameters, phenolic content and sensory changes of processed blueberry, elderberry and blackcurrant commercial juices. Int. J. Food Sci. Technol. 47, 1728–1736.

Caswell, H., 2009. The role of fruit juice in the diet: An overview. Nutrition Bull. 34, 273–288.

Frankel, E.N., Finley, J.W., 2008. How to standardize the multiplicity of methods to evaluate natural antioxidants. J. Agric. Food Chem. 56, 4901–4908.

Hägg, M., Ylikoski, S., Kumpulainen, J., 1995. Vitamin C Content in Fruits and Berries Consumed in Finland. J. Food Comp. Anal. 8, 12–20.

Häkkinen, S., Heinonen, M., Kärenlampi, S., Mykkänen, H., Ruuskanen, J., Törrönen, R., 1999. Screening of selected flavonoids and phenolic acids in 19 berries. Food Res. Int. 32, 345–353.

Hung, H.C., Joshipura, K.J., Jiang, R., Hu, F.B., Hunter, D., Smith-Warner, S.A., 2004. Fruit and vegetable intake and risk of major chronic disease. J. Nat. Cancer Inst. 96, 1577–1584.

Kähkönen, M.P., Heinämäki, J., Ollilainen, V., Heinonen, M., 2003. Berry anthocyanins: Isolation, identification and antioxidant activities. J. Sci. Food Agric. 83, 1403–1411.

Karjalainen, R., Anttonen, M., Saviranta, N., Hilz, H., Törrönen, R., Stewart, D., 2009. A review on bioactive compounds in black currants (Ribes nigrum L.) and their potential health-promoting properties. Acta Hort. 839, 301–307.

Kevers, C., Falkowski, M., Tabart, J., Defraigne, J.O., Dommes, J., Pincemail, J., 2007. Evolution of antioxidant capacity during storage of selected fruits and vegetables. J. Agric. Food Chem. 55, 8596–8603.

Kucharska, A.Z., Sokół-Łętowska, A., Gabrielska, J., Bąkowska-Barczak, A., Włoch, A., Dudra, A., 2011. Antioxidant properties of polyphenolic extracts from chokeberry, blackcurrant, blackberry, and raspberry fruits. Annales Universitatis Mariae Curie-Sklodowska, Sectio DDD: Pharmacia. 24, 183–188.

Mikkonen, T.P., Määttä, K.R., Hukkanen, A.T., Kokko, H.I., Törrönen, A.R., Kärenlampi, S.O., 2001. Flavonol content varies among black currant cultivars. J. Agric. Food Chem. 49, 3274–3277.

Miller, N.J., Rice-Evans, C.A., 1997. The relative contributions of ascorbic acid and phenolic antioxidants to the total antioxidant activity of orange and apple fruit juices and blackcurrant drink. Food Chem. 60, 331–337.

Moyer, R.A., Hummer, K.E., Finn, C.E., Frei, B., Wrolstad, R.E., 2002. Anthocyanins, phenolics, and antioxidant capacity in diverse small fruits: Vaccinium, Rubus, and Ribes. J. Agric. Food Chem. 50, 519–525.

Nes, A., Espelien, H.G., Wold, A.B., Remberg, S.F., 2012. Cropping and chemical composition of black currant (Ribes nigrum L.) cultivars in Norway. Acta Hort. 946, 119–122.

Pincemail, J., Kevers, C., Tabart, J., Defraigne, J.O., Dommes, J., 2012. Cultivars, culture conditions, and harvest time influence phenolic and ascorbic acid contents and antioxidant capacity of strawberry (Fragaria x ananassa). J. Food Sci. 77, C205–C210.

Raffo, A., Paoletti, F., Antonelli, M., 2004. Changes in sugar, organic acid, flavonol and carotenoid composition during ripening of berries of three seabuckthorn (Hippophae rhamnoides L.) cultivars. Eur. Food Res. Techn. 219, 360–368.

Shahidi, F., Naczk, M., 2004. Phenolic compounds in fruits and vegetables. In: Phenolics in food and nutraceutical. LLC: CRC, pp. 131–156.

Szajdek, A., Borowska, E.J., 2008. Bioactive compounds and health-promoting properties of Berry fruits: A review. Plant Food Hum. Nutr. 63, 147–153.

Tabart, J., Franck, T., Kevers, C., Pincemail, J., Serteyn, D., Defraigne, J.O., Dommes, J., 2012. Antioxidant and anti-inflammatory activities of Ribes nigrum extracts. Food Chem. 131, 1116–1122.

Tabart, J., Kevers, C., Pincemail, J., Defraigne, J.O., Dommes, J., 2006. Antioxidant capacity of black currant varies with organ, season, and cultivar. J. Agric. Food Chem. 54, 6271–6276.

Tabart, J., Kevers, C., Pincemail, J., Defraigne, J.O., Dommes, J., 2009. Comparative antioxidant capacities of phenolic compounds measured by various tests. Food Chem. 113, 1226–1233.

Törrönen, R., McDougall, G.J., Dobson, G., Stewart, D., Hellström, J., Mattila, P., 2012. Fortification of blackcurrant juice with crowberry: Impact on polyphenol composition, urinary phenolic metabolites, and postprandial glycemic response in healthy subjects. J. Funct. Foods 4, 746–756.

Viskelis, P., Anisimoviene, N., Rubinskiene, M., Jankovska, E., Sasnauskas, A., 2010. Physical properties, anthocyanins and antioxidant activity of blackcurrant berries of different maturities. J. Food Agric. Env. 8, 159–162.

Woodward, G.M., McCarthy, D., Pham-Thanh, D., Kay, C.D., 2011. Anthocyanins remain stable during commercial blackcurrant juice processing. J. Food Sci. 76, S408–S414.

Würth, K., Bonerz, D., Will, F., Patz, C.D., Quast, P., Hillebrand, S., 2009. Anthocyanin deterioration in blackcurrant juice and concentrate — Part 1: Kinetics of anthocyanin loss during storage. Deutsche Lebensmittel Rundschau 105, 176–182.

Zadernowski, R., Naczk, M., Nesterowicz, J., 2005. Phenolic acid profiles in some small berries. J. Agric. Food Chem. 53, 2118–2124.

Index

Note: Page numbers followed by "f" denote figures; "t" tables.

P

Q

R

S

Edwards Brothers Malloy
Ann Arbor MI. USA
April 17, 2014